Dating Torrential Processes on Fans and Cones

ADVANCES IN GLOBAL CHANGE RESEARCH

VOLUME 47

For further volumes:
http://www.springer.com/series/5588

Michelle Schneuwly-Bollschweiler
Markus Stoffel • Florian Rudolf-Miklau
Editors

Dating Torrential Processes on Fans and Cones

Methods and Their Application for Hazard and Risk Assessment

Springer

Editors
Michelle Schneuwly-Bollschweiler
Laboratory of Dendrogeomorphology
Institute of Geological Sciences
University of Berne
Baltzerstrasse 1+3
Berne
Switzerland
and
Chair for Climate and Climatic Impacts
Environmental Sciences
University of Geneva
7, route de Drize Carouge-Geneva
Switzerland

Markus Stoffel
Laboratory of Dendrogeomorphology
Institute of Geological Sciences
University of Berne
Baltzerstrasse 1+3
Berne
Switzerland
and
Chair for Climate and Climatic Impacts
Environmental Sciences
University of Geneva
7, route de Drize Carouge-Geneva
Switzerland

Florian Rudolf-Miklau
Federal Ministry for Agriculture
Forestry Environment and Water
Management
Department IV/5 – Torrent
and Avalanche Control
Marxergasse 5
Vienna
Austria

ISSN 1574-0919
ISBN 978-94-007-4335-9 ISBN 978-94-007-4336-6 (eBook)
DOI 10.1007/978-94-007-4336-6
Springer Dordrecht Heidelberg New York London

Library of Congress Control Number: 2012941287

Cover image title: Debris-flow event in the Holzerbach torrent on 29 June 1959

Copyright line/text: Department of Hydraulic Engineering, Autonomous Province of Bolzano, Italy

Cover image source: Municipality of Predoi – Prettau

Printed on acid-free paper

Springer is part of Springer Science+Business Media (www.springer.com)

Preface

Water-related processes (floods, debris floods, flash floods, and debris flows) form alluvial fans and cones and represent major geomorphic hazards in most mountain areas of the world. The gentle gradients and workable materials of fans and cones render this environment a desirable site for residential developments and transportation corridors. Over the past decades, human pressure has increased in these regions, leading to an aggravation of conflicts between natural hazards and infrastructure needs. However, detailed knowledge on frequency and magnitude of past flood or debris-flow events on alluvial fans and cones still remains scarce, although it is widely accepted that such knowledge is of crucial importance for the assessment of hazards, mitigation of risks, and land-use planning. Modern methods of historical dating of past debris-flow and flood events such as dendrochronology, radiocarbon dating, and lichenometry can provide valuable insights into past process activity and therefore complement historical records. In the field of natural hazard management, these methods are mostly applied by scientists, but only rarely for planning or engineering purposes.

It is the aim of this book to fill this gap by providing a detailed overview on methods for the dating of historical events and by fostering the discussion on the impact of past and potential future climatic changes on torrential processes. The book summarizes state-of-the-art knowledge on dating methods with a clear focus on the applicability of these methods in practice. The survey of methods is complemented with potential fields of application and case studies and takes account of the specific limitations of dating methods in case of excessive natural and human interventions on fans and cones.

This book was realized within "AdaptAlp", a project funded by the Alpine Space Program of the European Commission. The AdaptAlp project (www.adaptalp.org) contributes to a growing body of scientific research on the effect of climate change within the Alpine region and on how our societies can adapt to the increasing risk of natural disasters. An important issue tackled by AdaptAlp is the provision of reliable data and design events for the Alpine space under current and possible future conditions. There is a great need for innovative methods enabling better and especially faster modelling, forecast, and thus prevention of impacts in a changing world. The book goes far beyond the scope of this project and aims at the comprehensive compilation of all dating methods applicable on alluvial fans in the quality of a "state-of-the-art" report.

Furthermore, this book was also partly realized within the ACQWA project (www.acqwa.ch) and benefitted from support of the Federal Ministry for Agriculture, Forestry, Environment and Water Management of Austria and the Austrian Service for Torrent and Avalanche Control. We gratefully acknowledge DI Maria Patek, M.B.A. for paving the way for this fruitful cooperation. We would also like to thank Prof. Dr. Martin Beniston, editor-in-chief of *Advances in Global Change Research*, for giving us the opportunity to publish this book in his Springer series. DI Andreas Pichler is warmly acknowledged for his helpful comments on the book content. To enumerate all persons who contributed to the elaboration of this book would exceed the scope of this preface and we therefore greatly acknowledge them collectively.

The book is structured in four parts. After a short introduction (*chapter "Dating Past Events on Fans and Cones – An Introduction"*), the fan and cone formation is described (*part "Material Transport and Fan or Cone Formation"*). The central part of the book (*part "Dating Past Events"*) then presents dating methods that can be applied to track past hydrogeomorphic events on fans and cones. The basic principles of all methods are described and their application is given with examples, also illustrating advantages and limitations of the methods. In *part "Documentation and Monitoring"*, the documentation of current-day events and the monitoring of active channels is presented as current-day events provide a perfect occasion to learn for any future events. In *part "Application of Event Dating in Practice"*, we illustrate the use and application of data on past events for the assessment of hazard and risk. The chapter also provides insights on how to cope with risk induced by events on fans and cones. Possible impacts of climate change on hydrogeomorphic activity are addressed in *chapter "Climate Research and Adaptation Strategies – Examples from the European Alps"*. Finally, *chapter "Dating Methods Overview: Checklist for Practitioners"* provides an overview on the dating methods with a checklist for practitioners summarizing the main characteristics, advantages, limitations, and costs of all methods. The book closes then with a short summary and an outlook (*chapter "Summary and Outlook"*).

Berne and Vienna

Michelle Schneuwly-Bollschweiler
Markus Stoffel
Florian Rudolf-Miklau

Contents

List of Figures

List of Tables

Dating Past Events on Fans and Cones – An Introduction

Markus Stoffel, Michelle Schneuwly-Bollschweiler, and Florian Rudolf-Miklau

1 Alluvial Fan Environment and Processes

Alluvial fans and debris cones are fan- or cone-shaped sedimentary structures that are formed where a stream or torrent flattens, slows, and spreads (AGI 1962). Moreover cones are also formed by scree deposits, rock avalanches, slope-type debris flows, mudflows, lahars, landslides, sheet floods and other forms of rapid mass movements. Alluvial fans and debris cones are typically located in mountain fronts and develop at the base of headwater basins. The apex of the depositional environment is usually located within a canyon or ravine mouth that serves as the outlet for a mountain drainage system (Bull 1977) or at the juncture of steeper hillslope tributaries with the main stream in a flatter plain (e.g. valley floor).

In this book, the terms 'alluvial fan' and 'debris cone' define different depositional environments with the first being used to define flatter accumulations and the latter to describe steep fan- or cone-shaped accumulation of debris-flow material. As pointed out by Jakob (2005) there is no unique threshold to differentiate the two environments. Fans with a mean slope of less than 4° are unlikely to be formed by debris flows, but this can vary depending on the geologic setting and material involved. Muddy debris flows and flows of volcanic origin (lahars) can produce

M. Stoffel (✉) • M. Schneuwly-Bollschweiler
Laboratory of Dendrogeomorphology (dendrolab.ch), Institute of Geological Sciences, University of Berne, CH-3012 Berne, Switzerland

Chair for Climate and Climatic Impacts, Environmental Sciences, University of Geneva, CH-1227 Carouge-Geneva, Switzerland
e-mail: markus.stoffel@dendrolab.ch; michelle.schneuwly@dendrolab.ch

F. Rudolf-Miklau
Federal Ministry for Agriculture, Forestry, Environment and Water Management, Vienna, Austria
e-mail: florian.rudolfmiklau@lebensministerium.at

M. Schneuwly-Bollschweiler et al. (eds.), *Dating Torrential Processes on Fans and Cones*, Advances in Global Change Research 47, DOI 10.1007/978-94-007-4336-6_1,

fan gradients as low as 1–2°. The formation of alluvial fans and debris cones is therefore ultimately linked to the presence of breaks in slope between the steep canyon or ravine and a flatter depositional environment. Such breaks in slope will result in a reduction or complete cessation of sediment transport by the tributary and sediment will be released in the form of unconsolidated, poorly-sorted material with predominantly coarse grains (alluvium; Blair and McPherson 2009; Harvey et al. 2005). As this deposition of material also reduces the capacity of the channel, changes in channel direction and shifts in the channel course are likely to occur over time, therefore gradually building up a conical shape. Catchment characteristics control the processes occurring on fans and cones. As a consequence, catchment characteristics also control the regime of fan and cone processes and the overall fan properties (Suwa and Okuda 1983; Beaty 1990; Whipple and Dunne 1992). Debris-flow dominated cones are normally associated with small steep catchments (Harvey 1984; Kostaschuk et al. 1986; Rickenmann and Zimmermann 1993), and fluvially-dominated fans with larger or flatter catchments (see Harvey 2012, *this volume* and Rickenmann and Scheidl 2012, *this volume* for more details).

A variety of processes supplies sediment to tributary channels and ravines and ultimately to fans and cones. These 'preparatory' processes include rockfall, snow avalanche, debris-flow, and fluvial processes such as sheet flood and channelized stream flow (Blair and McPherson 1994; Harvey 2012, *this volume*). The supply of sediment to fans and cones is not continuous but almost always episodic and generally related to major rain- and hailstorms. Other triggers of sediment transport by hydrogeomorphic processes include the breach of moraine or temporary landslide dams, the sudden release of water from pockets in glacierized headwaters or through an anthropogenic input of water into the torrential system (e.g. through defects in water conduits, tubes). Catchment size is a critical parameter for the generation of hydrogeomorphic processes and influences flood magnitude. Within larger catchments, a greater likelihood exists for the storage of coarser sediment, resulting in water–sediment ratios being usually higher and sediment size smaller in the case of larger catchments. Catchment steepness will favor increased erosion rates and facilitate slope failures and therefore contribute to increased sediment supply (Harvey 2001). In addition, geology may have an effect on sediment supply as well, as some settings tend to favor slope failure while others are more prone to erosion. Land cover and land use (changes) also affect runoff and sediment yield (also see Harvey 2012, *this volume*). Fans and cones can therefore be seen as the expression of the "history" of their watersheds because they have been created by and represent a summary of hydrologic, geomorphic, climatic, biologic, and anthropogenic processes (e.g. agriculture, forestry) in the mountains and hillslopes upstream of fans and cones (Shroder et al. 2012). Inherited and present-day processes in the watershed actively influence water and sediment regimes and can thus lead to significant changes in the timing, frequency, and magnitude of hydrogeomorphic events.

2 Human Activities and Risks on Alluvial Fans and Cones

The gentle gradients and workable materials of fans and cones render this environment a desirable site for residential developments and transportation corridors (Bloetzer et al. 1998; Wilford et al. 2009). In many parts of the Alps, fans and cones have been preferred over hillslopes and plains as their distal parts were apparently less affected by smaller but more frequent natural disasters such as snow avalanches, rockfalls or floods.

At the same time, a broad range of folk memory and archival records report the recurrence of less frequent but not less devastating hydrogeomorphic processes on fans and cones and the resulting damage and casualties (Lanz-Stauffer and Rommel 1936). Torrents like the Bretterwandbach (Tyrol, Austria), which has devastated the municipality of Matrei 27 times since 1347 (INTERPRAEVENT 2010), have had a major impact on the development and structure of settlements on fans and cones. The village of Randa (Valais, Switzerland), illustrated in Zimmermann (2012, *this volume*), is yet another fine example on how mountain populations tried to avoid the most frequent processes by building their houses in the distal part of cones where the risk of snow avalanches, rockfalls, floods and debris flows tends to be lower than elsewhere on the cone or in the river plain.

Fans and cones have been demonstrated to be of practical and economic importance to society, as they may be the principal if not only groundwater source for irrigation farming and the sustenance of life, dryer subsoils and elevated locations. In some instances, entire villages, such as Zell am See (Salzburg, Austria) or even cities, such as Los Angeles (USA), have been built on alluvial fans.

Recent intensification of human activities on and the construction of infrastructure over wide parts of fans and cones have in many instances increased risks and vulnerability of the population and their goods to hydrogeomorphic disasters. With this development the strategy of hazard-adapted land use on alluvial fans based on folk experience and tradition was lost. As a consequence, the current increase in frequency and magnitude of hydrogeomorphic events, as observed in many catchments of the European Alps, is not only the result of ongoing climate change but mainly reflective of the scarcity of disasters between the 1920s through 1970s (e.g. Pfister 1999; Stoffel and Beniston 2006). Interestingly, widespread colonization of fans and cones and the construction of residential homes and industrial plants in the vicinity of canyon mouths or next to (temporarily) abandoned channels primarily occurred during this period of the twentieth century and changes in hazard potential and damage can at least partly be attributed to human activities rather than to consequences of climate change. Changing climatic conditions and recent hydrogeomorphic disasters (such as the floods in the Austrian, German, and Swiss Alps in August 2005) have, however, demonstrated clearly that altered natural conditions combined with intensified land use may indeed exacerbate conflicts between natural processes and human assets and lead to increased hazards and risks with a potential for considerable loss of life and high financial costs in case that hydrogeomorphic events affect inhabited segments of the fan and cone surface (Sidle et al. 1985; Jakob and Hungr 2005).

3 Principles of Hazard and Risk Assessment

3.1 The Intensity-Frequency Concept

Hazards are defined as potentially damaging events or phenomena that may harm people or cause damage to the environment or property (UNDHA 1992). Consequences of such events are expressed with "risk" which combines hazard (expressed as frequency and process intensity) and related consequences (expressed as the potential for e.g. loss of life, economic damage, damage to the reputation of a company whose operations may be affected by debris flows and environmental losses). In mathematical terms risk is defined as the product of hazard and vulnerability, which corresponds to the degree to which a person, object, infrastructure or socio-economic system is either susceptible or resilient to the impact of hazards. Vulnerability may be expressed as a percentage of the value of the affected facility or structure. Vulnerable objects and living space need protection, which comprises a set of measures or natural conditions reducing an existing risk (Loat and Meier 2003; Rudolf-Miklau 2009). Consequently traditional protection concepts aim at a reduction of the probability and magnitude of damaging events. The "integral risk management" approach has been developed in the U.S. and Europe and incorporates all measures contributing to a reduction of damage caused by natural hazards. These include particularly emergency management during disasters, the maintenance of protective structures, repair work, the maintenance of protective forests or structural mitigation measures. This system is based on the principles of the "risk concept" (Plate and Merz 2001; Ammann 2006; Bründl et al. 2009; Bründl 2012, *this volume*).

Comprehensive data on the frequency of processes in the area under investigation is crucial for a realistic estimation of scenarios in hazard and risk assessments and for the choice and dimensioning of mitigation measures. For engineering purposes the principles of hazard assessment may be reduced to the following basic questions (for details see Zimmermann 2012, *this volume*): What type of process does occur? How big? How far? How fast? How often? To solve these questions accurate and precise data on statistical and physical properties for processes causing extreme events are needed but are frequently missing due to the episodic occurrence of torrential disasters. Fragmentary information results in large uncertainties which lie behind each of the relevant parameters. A comprehensive overview on methods for frequency analysis of floods and other hydrogeomorphic hazards is for instance given by Kidson and Richards (2005) or Khaliq et al. (2006).

Hazard assessments are usually based on the intensity-frequency concept developed by Wolman and Miller (1960). Intensity refers to the power or magnitude of a process or an event. Intensity of torrential events can be expressed by physical criteria like peak discharge, volume, runout distance, lateral spreading, flow velocity, flow depth, flow energy or pressure. In engineering practice these physical criteria are normally generated by numerical process models that simulate virtual events corresponding to a determined frequency of occurrence (Rudolf-Miklau and

Suda 2012, *this volume*). Frequency is defined as the number of times a natural event of a given magnitude is likely to occur over a given period of time (ONR 24800:2008). In hydrology, frequency can be expressed as return period which is equal to the reciprocal of exceedance probability of extreme precipitation or discharge. Here, the frequency of an event corresponds to an expected value for a rainfall, flood discharge or debris mass with a defined return period (including a band width or confidence interval) and can be derived from hydraulic extreme value statistics (Rudolf-Miklau and Sereinig 2009).

3.2 The Significance of Uncertainties

Statistical analyses resulting in probabilities are 'uncertain' by definition. Nevertheless, working with stochastic methods is inevitable in natural hazard assessment as deterministic approaches have considerable limitations due to missing data on physical processes driving hydrogeomorphic events. When applying extreme value statistics for safety planning, expression and description of uncertainties is essential. In addition, it appears noteworthy that such statistics are based on the assumption that the observation period is representative for the long-term regime of the watershed or catchment area. However, extreme events are often qualified as 'statistical outliers' that are not represented by the measured data collective as a result of limited observation periods (Merz 2007). Uncertainties in extreme value statistics result either from the variability of natural processes (aleatoric uncertainties) or from the gaps in knowledge about the examined process (epistemic uncertainty), with the most important sources of uncertainty being:

- variability: Due to limited observation period and sample size (data) it is not possible to identify and describe every variability in scarce natural processes (events);
- incidental or systematic measurement errors;
- model uncertainty and approximation: Models describe natural processes with rough simplification. Uncertainties increase if an extrapolation is performed with limited statistical series.

To explain the impact of uncertainties of particular parameters on results, an integration of uncertainties into calculations appears crucial and confidence intervals should be used to mark a spectrum of possible results. The 'true' value with a defined probability (usually 95% or 98%) will be located within this spectrum.

With time, stochastic models of natural hazards will need to be replaced by deterministic approaches, especially in small and/or remote torrential environments where data and measurements are often unavailable. The consideration of uncertainties in deterministic models can be obtained by varying the values of specific or all parameters for 'best-case', 'realistic-case' and 'worst-case' scenarios, and through a description of the reasons for the selection of a certain scenario as the 'worst case' (Mudersbach 2009).

As a result, epistemic uncertainty can only be reduced if process understanding can be improved with additional data and research findings (Ang and Tang 2007). Consequently, supporting additional historic information with temporal, spatial or causal details is not only essential but a necessary prerequisite for improved natural hazard analysis (Blöschl and Merz 2008). Especially the dating of historic torrential events from chronicles or traces in nature can provide precious additional information on return periods or process intensity (Ballesteros et al. 2011a, b). A return periods or process intensitycausal supplement of information is gained if observed events are analyzed with respect to their emergence regarding weather conditions, the role of precipitation and the disposition of the catchment area (Merz 2007; Stoffel et al. 2011). The same holds true for tracks and marks of historic events on alluvial fans and debris cones (Bollschweiler and Stoffel 2010), which are the core theme of this book.

4 Data on Historic Torrential Events: Importance and Gaps

The scarcity of event occurrence and the widespread absence of hydrogeomorphic disasters over several decades in the twentieth century represent a major handicap for an appropriate assessment of the frequency and magnitude of past and potential future events. The incompleteness of historical archives does not help the appraisal of hazards and risks either. As a result, the history of fans and cones is only badly known in general (e.g. limited data availability, quality, length or completeness of existing records) and major gaps persist for realistic assessments of event frequency and magnitude, spread, or reach for most of the European Alps and even more for other mountainous regions of the World (Bollschweiler and Stoffel 2010; Stoffel and Wilford 2012). In addition, archival data are often descriptive and do not contain quantitative information on deposited volumes, runout distance or lateral spread, neither do they indicate locations where avulsion took place. In a similar way, records are often limited to catastrophic past events whereas smaller events were not necessarily noted in archival databases. However, through the intensification of land use and the construction of infrastructure in more exposed segments of fans and cones, smaller but more frequent events can become relevant as well.

An urgent need therefore exists for a better documentation and a more detailed appraisal of past and potential future hydrogeomorphic events through the analysis of silent witnesses present in the field or the reconstruction of past events at the local and/or regional scales. In the Alpine countries joint efforts have resulted in the establishment of general standards (PLANALP 2006; Hübl et al. 2006) and institutional systems (databases) for the documentation of recent events. In addition, comprehensive documentations and analysis reports have been regularly published after major disasters in Austria (BMLFUW 2006; INTERPRAEVENT 2010) and Switzerland (e.g. BAFU 2007). These initiatives guarantee an as complete as possible documentation of current events, but historic event data is still fragmentary. In the Valais Alps, tree-ring studies have shown that archival data remains fragmentary even for the more recent period with one in four events being noted for the Ritigraben torrent (Valais, Switzerland) for the past 80 years (Stoffel et al. 2008) or one in three outbreak events entering the archives at Bruchi (Naters, Switzerland) since the 1960s.

Records of past hydrogeomorphic activity are stored in many different types of archives. Besides the classical anthropogenic archives, nature provides many sources of information that can be extracted using the appropriate methods. A broad spectrum of methods allow the improvement of data series on past hydrogeomorphic events on fans and cones and are described in this book (Schneuwly-Bollschweiler et al. 2012).

A first possible method is the use of 'silent witnesses' (Kaitna & Huebl, 2012, *this volume*) that include all types of geomorphic and biologic evidence of past mass-movement activity from very small (e.g. grain-size distribution) to very large (e.g. catchment form and properties) scales. Historical sources are described in D'Agostiono 2012, *this volume* and generally form the basis of each study. Digital images (Staley & Wasklewicz 2012, *this volume*) can be used to represent past events in space or to describe the evolution of fans and cones. These approaches provide invaluable data about the recent past; however, they normally fail to provide longer records. Here, tree-ring records (dendrogeomorphology; Schneuwly-Bollschweiler & Stoffel 2012, *this volume*; Stoffel 2012, *this volume*) may help to close this gap as past hydrogeomorphic events are recorded in the growth-ring series of trees and shrubs standing along channels or on fans and cones and therefore enabling reconstruction of event frequency or the lateral spread and reach of past activity. The distribution of vegetation (Baroni et al. 2012, *this volume*; Pierson 2012, *this volume*) on a fan or cone may complement information on currently active or inactive sectors. Lichens (Jomelli 2012, *this volume*) on blocks or lake deposits (Gilli et al. 2012, *this volume*) further assist the reconstruction of past hydrogeomorphic activity. The above-mentioned methods normally cover a few decades up to some centuries. If data on older events (several centuries to millennia) is sought, cosmogenic nuclide (Ivy-Ochs et al. 2012, *this volume*), radiocarbon (Chiverell & Jakob 2012, *this volume*) or luminescence dating (Lang 2012, *this volume*; Schlunegger et al. 2012, *this volume*) approaches can be applied. All those dating methods aim at helping the understanding of past activity of a torrent and therefore also provide indirect data on future hazards and risks.

There is a wide range of tasks in the management of natural hazards and disasters which are based and rely on data of historic events. Process models, planning procedures, engineering measures, forecasting and warning systems or emergency management organizations all highly depend on an appropriate and as detailed as possible data about frequency and magnitude of extreme hydrogeomorphic events (see part "Application of Event Dating in Practice"). However, data and model parameters usually used for these purposes suffer from a disproportional inaccuracy stemming from fragmentary knowledge of past torrential events. While technology for measurements, event documentation and now-casting of present events has reached a high level of development, comparably basic methods (e.g. silent witnesses, historic chronicles) are still being applied to track and date older events. The rich and dense information contained in 'natural archives' of alluvial fan and debris cones was only recently used for engineering purposes (e.g. Jakob and Friele 2010; Bollschweiler and Stoffel 2010), but dating methods remain widely unknown to practitioners although used extensively by the scientific community.

This book aims to fill this gap between science and practice by addressing the above topics and by providing an overview on how more data on past events can be achieved, how the different dating methods work and what results they can yield. The different chapters are illustrated with case-study examples and followed by a checklist for practitioners at the end in one of the final chapters so that the value, temporal reach and efforts (time, costs) of the different approaches can be estimated. Besides its value for undergraduate and graduate students, this book will be useful for practitioners involved in the assessment of hazards and risks on fans and cones.

References

AGI – American Geological Institute (1962) Dictionary of geological terms. Dolphin Books, New York

Ammann WJ (2006) The risk concept. In: Dannenmann S, Ammann WJ, Vulliet L (eds) Risk 21 – coping with risks due to natural hazards in the 21st century. Taylor & Francis, London, pp 3–23

Ang AH-S, Tang WH (2007) Probability concepts in engineering – emphasis on applications to civil and environmental engineering. Wiley, New York

BAFU (2007) Ereignisanalys Hochwasser 2005. Bundesamt für Umwelt, Bern

Ballesteros JA, Bodoque JM, Díez A, Sánchez M, Stoffel M (2011a) Calibration of floodplain roughness and estimation of palaeoflood discharge based on tree-ring evidence and hydraulic modeling. J Hydrol 403:103–115

Ballesteros JA, Eguibar M, Bodoque JM, Díez A, Stoffel M, Gutiérrez I (2011b) Estimating flash flood discharge in an ungauged mountain catchment with 2D hydraulic models and dendrogeomorphic paleostage indicators. Hydrol Process 25:970–979

Baroni G, Gentili R, Armiraglio S (2012) Vegetation analysis on composite debris cones. In: Schneuwly-Bollschweiler M, Stoffel M, Rudolf-Miklau M (eds) Dating torrential processes on fans and cones – methods and their application for hazard and risk assessment. Advances in Global Change Research. Springer, Dordrecht/Heidelberg/London/New York

Beaty CB (1990) Anatomy of a White Mountains debris flow – the making of an alluvial fan. In: Rachocki AH, Church M (eds) Alluvial fans: a field approach. Wiley, Chichester, pp 69–89

Blair TC, McPherson JG (1994) Alluvial fan processes and forms. In: Abrahams AD, Parsons AJ (eds) Geomorphology of desert environments. Chapman & Hall, London, pp 354–402

Blair TC, McPherson JG (2009) Processes and forms of alluvial fans. In: Parsons AJ, Abrahams AD (eds) Geomorphology of desert environments, 2nd edn. Springer, New York, pp 413–467

Bloetzer W, Egli T, Petrascheck A, Sauter J, Stoffel M (1998) Klimaänderungen und Naturgefahren in der Raumplanung – Methodische Ansätze und Fallbeispiele. Vdf Hochschulverlag AG, Zürich

Blöschl G, Merz R (2008) Bestimmung von Bemessungshochwässern gegebener Jährlichkeit – Aspekte einer zeitgemäßen Strategie. Wasserwirtschaft 11:2–11

BMLFUW (2006) Ereignisdokumentation Hochwasser 2005. Bundesministerium für Land- und Forstwirtschaft, Umwelt und Wasserwirtschaft, Vienna

Bollschweiler M, Stoffel M (2010) Tree rings and debris flows – trends and challenges. Prog Phys Geogr 34:625–645

Bründl M (2012) Dealing with natural hazard risks in Switzerland – the influence of hazard mapping on risk-based decision making. In: Schneuwly-Bollschweiler M, Stoffel M, Rudolf-Miklau M (eds) Dating torrential processes on fans and cones – methods and their application for hazard and risk assessment, Advances in global change research. Springer, Dordrecht/Heidelberg/London/New York

Bründl M, Romang H, Bischof N, Rheinberger CM (2009) The risk concept and its application in natural hazard risk management in Switzerland. Nat Hazard Earth Syst Sci 9(3):801–813

Bull WB (1977) The alluvial-fan environment. Prog Phys Geogr 1:222–270

Chiverell R, Jakob M (2012) Radiocarbon dating : alluvial fan / debris cone evolution and hazards. In: Schneuwly-Bollschweiler M, Stoffel M, Rudolf-Miklau M (eds) Dating torrential processes on fans and cones – methods and their application for hazard and risk assessment. Advances in Global Change Research. Springer, Dordrecht/Heidelberg/London/New York

D'Agostiono V (2012) Assessment of past torrential events through historical sources. In: Schneuwly-Bollschweiler M, Stoffel M, Rudolf-Miklau M (eds) Dating torrential processes on fans and cones – methods and their application for hazard and risk assessment. Advances in Global Change Research. Springer, Dordrecht/Heidelberg/London/New York

Gilli A, Anselmetti FS, Glur L, Wirth SB (2012) Lake sediments as archives of recurrence rates and intensities of past flood events. In: Schneuwly-Bollschweiler M, Stoffel M, Rudolf-Miklau M (eds) Dating torrential processes on fans and cones – methods and their application for hazard and risk assessment. Advances in Global Change Research. Springer, Dordrecht/Heidelberg/London/New York

Harvey AM (1984) Debris flows and fluvial deposits in Spanish Quaternary alluvial fans: implications for fan morphology. In: Koster EH, Steel R (eds) Sedimentoloqy of gravels and conglomerates, Memoir 10. Canadian Society of Petroleum Geologists, Calgary, pp 23–132

Harvey AM (2001) Coupling between hillslopes and channels in upland fluvial systems: implications for landscape sensitivity, illustrated from the Howgill Fells, northwest England. Catena 42:225–250

Harvey AM (2012) Processes of sediment supply to alluvial fans and debris cones. In: Schneuwly-Bollschweiler M, Stoffel M, Rudolf-Miklau M (eds) Dating torrential processes on fans and cones – methods and their application for hazard and risk assessment, Advances in global change research. Springer, Dordrecht/Heidelberg/London/New York

Harvey AM, Mather AE, Stokes M (2005) Alluvial fans: geomorphology, sedimentology, dynamics – introduction. A review of alluvial-fan research. In: Alluvial fans: geomorphology, sedimentology, dynamics, Geological Society special publication 251. Geological Society, London, pp 1–7

Hübl J, Kienholz H, Loipersberger A (2006) DOMODIS: Dokumentation alpiner Naturereignisse [Documentation of mountain disasters]. Internationale Forschungsgesellschaft INTERPRAEVENT, Klagenfurt

INTERPRAEVENT (2010) Alpine Naturkatastrophen. Verlag Leopold Stocker, Graz

Ivy-Ochs S, Dühnforth M, Densmore AL, Alfimov V (2012) Dating fan deposits with cosmogenic nuclides. In: Schneuwly-Bollschweiler M, Stoffel M, Rudolf-Miklau M (eds) Dating torrential processes on fans and cones – methods and their application for hazard and risk assessment. Advances in Global Change Research. Springer, Dordrecht/Heidelberg/London/New York

Jakob M (2005) Debris-flow hazard analysis. In: Jakob M, Hung O (eds) Debris-flow hazards and related phenomena. Springer, Berlin, pp 411–444

Jakob M, Friele P (2010) Frequency and magnitude of debris flows on Cheekye River, British Columbia. Geomorphology 114:382–395

Jakob M, Hungr O (2005) Introduction. In: Jakob M, Hung O (eds) Debris-flow hazards and related phenomena. Springer, Berlin

Jomelli V (2012) Lichenometric dating of debris avalanche deposits with an example from the French Alps. In: Schneuwly-Bollschweiler M, Stoffel M, Rudolf-Miklau M (eds) Dating torrential processes on fans and cones – methods and their application for hazard and risk assessment. Advances in Global Change Research. Springer, Dordrecht/Heidelberg/London/New York

Kaitna R, Huebl J (2012) Silent witnesses for torrential processes. In: Schneuwly-Bollschweiler M, Stoffel M, Rudolf-Miklau M (eds) Dating torrential processes on fans and cones – methods and their application for hazard and risk assessment. Advances in Global Change Research. Springer, Dordrecht/Heidelberg/London/New York

Khaliq MN, Ouarda TMBJ, Ondo J-C, Gachon P, Bobée B (2006) Frequency analysis of a sequence of dependent and/or non-stationary hydro-meteorological observations: a review. J Hydrol 329:534–552

Kidson R, Richards KS (2005) Flood frequency analysis: assumptions and alternatives. Prog Phys Geogr 3:392–410

Kostaschuk RA, Macdonald GM, Putnam PE (1986) Depositional processes and alluvial fan – drainage basin morphometric relationships near Banff, Alberta, Canada. Earth Surf Process Landf 11:471–484

Lang A (2012) Luminescence dating of alluvial fans and cones. In: Schneuwly-Bollschweiler M, Stoffel M, Rudolf-Miklau M (eds) Dating torrential processes on fans and cones – methods and their application for hazard and risk assessment. Advances in Global Change Research. Springer, Dordrecht/Heidelberg/London/New York

Lanz-Stauffer H, Rommel C (1936) Elementarschäden und Versicherung. Studie des Rückversicherungs-Verbandes kantonal-schweizerischer Feuerversicherungen zu Förderung der Elementarschaden-Versicherung. Rückversicherungsverband, Bern

Loat R, Meier E (2003) Dictionary of flood protection. Bundesamt für Wasser und Geologie, Biel

Merz R (2007) Hochwasserstatistik – Das Ausreisserproblem. Wiener Mitteilungen: Wasser, Abwasser, Gewässer. Institut für Wasserbau und Ingenieurhydrologie, Technische Universität Wien, Vienna, pp 181–194

Mudersbach C (2009) Untersuchungen zur Ermittlung von hydrologischen Bemessungsgrößen mit Verfahren der instationären Extremwertstatistik. Dissertationsschrift, Universität Siegen, Fachbereich Bauingenieurwesen

ONR (24800:2008) Schutzbauwerke der Wildbachverbauung – Begriffe und ihre Definition sowie Klassifizierung. Austrian Standards Institute, Vienna

Pfister C (1999) Wetternachhersage. Paul Haupt, Bern

Pierson T (2012) Using Age of colonizing Douglas fir for the dating of young geomorphic surfaces – a case study. In: Schneuwly-Bollschweiler M, Stoffel M, Rudolf-Miklau M (eds), Dating torrential processes on fans and cones – methods and their application for hazard and risk assessment. Advances in Global Change Research. Springer, Dordrecht/Heidelberg/London/New York

Plate EJ, Merz B (2001) Naturkatastrophen: Ursachen, Auswirkungen, Vorsorge. Verlag Schweizerbart, Stuttgart

Plattform Naturgefahren der Alpenkonvention (PLANALP) (2006) Dokumentation von Naturereignissen – Feldanleitung. Ständiges Sekretariat der Alpenkonvention, Innsbruck

Rickenmann D, Scheidl C (2012) Debris-flow runout and deposition on the fan. In: Schneuwly-Bollschweiler M, Stoffel M, Rudolf-Miklau M (eds) Dating torrential processes on fans and cones – methods and their application for hazard and risk assessment, Advances in global change research. Springer, Dordrecht/Heidelberg/London/New York

Rickenmann D, Zimmermann M (1993) The 1987 debris flows in Switzerland: documentation and analysis. Geomorphology 8:175–189

Rudolf-Miklau F (2009) Naturgefahren-Management in Österreich. Verlag LexisNexis Orac, Wien

Rudolf-Miklau F, Sereinig N (2009) Festlegung des Bemessungshochwassers: Prozessorientierte Harmonisierung für Flüsse und Wildbäche. ÖWAW (7–8):29–32

Rudolf-Miklau F, Suda J (2012) Design criteria for torrential barriers. In: Schneuwly-Bollschweiler M, Stoffel M, Rudolf-Miklau M (eds) Dating torrential processes on fans and cones – methods and their application for hazard and risk assessment, Advances in global change research. Springer, Dordrecht/Heidelberg/London/New York

Schlunegger F, Norton K, Steffen D (2012) Timing the interface between mass wasting and fluvial processes with OSL. In: Schneuwly-Bollschweiler M, Stoffel M, Rudolf-Miklau M (eds) Dating torrential processes on fans and cones – methods and their application for hazard and risk assessment. Advances in Global Change Research. Springer, Dordrecht/Heidelberg/London/New York

Schneuwly-Bollschweiler M, Stoffel M (2012) Dendrogeomorphology – tracking past events with tree rings. In: Schneuwly-Bollschweiler M, Stoffel M, Rudolf-Miklau M (eds) Dating torrential processes on fans and cones – methods and their application for hazard and risk assessment. Advances in Global Change Research. Springer, Dordrecht/Heidelberg/London/New York

Schneuwly-Bollschweiler M, Stoffel M, Rudolf-Miklau M (2012) Dating torrential processes on fans and cones – methods and their application for hazard and risk assessment, Advances in global change research. Springer, Dordrecht/Heidelberg/London/New York

Shroder J Jr, Marston RA, Stoffel M (2012) Mountain and hillslope geomorphology. Treatise on geomorphology. Academic, San Diego

Sidle RC, Pearce AJ, O'Loughlin CL (1985) Hillslope stability and land use, Water Resources Monograph 11. American Geophysical Union, Washington, DC, pp 1–140

Staley DM, Wasklwicz TA (2012) The use of airborne laser swath mapping on fans and cones: an example from the Colorado Front Range. In: Schneuwly-Bollschweiler M, Stoffel M, Rudolf-Miklau M (eds) Dating torrential processes on fans and cones – methods and their application for hazard and risk assessment. Advances in Global Change Research. Springer, Dordrecht/Heidelberg/London/New York

Stoffel M (2012) Tree-ring based record of debris-flow dynamics and triggering rainstorms at Ritigraben (Swiss Alps). In: Schneuwly-Bollschweiler M, Stoffel M, Rudolf-Miklau M (eds) Dating torrential processes on fans and cones – methods and their application for hazard and risk assessment. Advances in Global Change Research. Springer, Dordrecht/Heidelberg/London/New York

Stoffel M, Beniston M (2006) On the incidence of debris flows from the early Little Ice Age to a future greenhouse climate: a case study from the Swiss Alps. Geophys Res Lett 33:L16404

Stoffel M, Wilford DJ (2012) Hydrogeomorphic processes and vegetation: disturbance, process histories, dependencies and interactions. Earth Surf Process Landf 37(1):9–22

Stoffel M, Conus D, Grichting MA, Lièvre I, Maitre G (2008) Unraveling the patterns of late Holocene debris-flow activity on a cone in the Swiss Alps: Chronology, environment and implications for the future. Glob Planet Chang 60(3–4):222–234

Stoffel M, Bollschweiler M, Beniston M (2011) Rainfall characteristics for periglacial debris flows in the Swiss Alps: past incidences – potential future evolutions. Clim Chang 105:263–280

Suwa H, Okuda S (1983) Deposition of debris flows on a fan surface, Mt. Yakedake, Japan. Zeitschrift fuer Geomorphologie, Supplementband 46:79–101

UNDHA/IDNDR (1992) Glossary of disaster management. UNDHA/IDNDR, 34p

Whipple KX, Dunne T (1992) The influence of debris-flow rheology on fan morphology, Owens Valley, California. Geol Soc Am Bull 104(7):887–900

Wilford DJ, Sakals ME, Grainger WW, Millard TH, Giles TR (2009) Managing forested watersheds for hydrogeomorphic risks on fans, Land management handbook 61. B.C. Ministry of Forest and Range, Victoria, pp 1–62

Wolman MG, Miller JP (1960) Magnitude and frequency of forces on geomorphic processes. J Geol 68(1):54–74

Zimmermann M (2012) Hazard assessment. In: Schneuwly-Bollschweiler M, Stoffel M, Rudolf-Miklau M (eds) Dating torrential processes on fans and cones – methods and their application for hazard and risk assessment, Advances in global change research. Springer, Dordrecht/Heidelberg/London/New York

Part I
Material Transport and Fan or Cone Formation

Processes of Sediment Supply to Alluvial Fans and Debris Cones

Adrian Harvey

1 Introduction

Debris cones and alluvial fans involve a range of landform sizes from individual debris-flow lobes, through debris cones and "classic" alluvial fans, to enormous fluvial "megafans" (Harvey 2011). Within the context of this book the focus is on the intermediate scale, debris cones to "classic alluvial fans". Such landforms are found in three main settings (Harvey 2010): mountain-front, intra-montane, and tributary-junction settings. They occur in all climatic environments, but again, within the context of this book the focus is on dry-region, and temperate upland and mountain environments.

A spectrum of processes supplies sediment to fans and cones, including "pure mass-movement" processes, e.g. rockfall, snow avalanche, mass-flow processes, i.e. debris-flow, and fluvial processes, including sheetflood and channelised stream-flow, which may involve reworking of fan sediment by usually braided or wandering single-thread gravel-bed channels (Blair and McPherson 1994). Fans and cones may be debris-flow dominant, mixed, or fluvially-dominant.

Almost always, sediment supply is episodic, generally related to major rainstorm events. In alpine environments rapid snowmelt may be significant, and in some cases sediment supply may more seasonal and less episodic. For example on fluvially-dominant tributary-junction fans in glaciated mountain environments such as the Canadian Rockies, sediment supply may relate to stream floods following snow and glacial melt. Otherwise, the generally storm-induced, episodic nature of sediment supply to fans and cones renders such sites hazardous, prone to flash floods and

A. Harvey (✉)
School of Environmental Sciences (Geography), University of Liverpool, Roxby Building, Liverpool L69 7ZT, UK
e-mail: amharvey@liverpool.ac.uk

M. Schneuwly-Bollschweiler et al. (eds.), *Dating Torrential Processes on Fans and Cones*, Advances in Global Change Research 47, DOI 10.1007/978-94-007-4336-6_2,

debris flows. It is significant that many of the most severe flood disasters in Europe have occurred on alluvial fans fed by steep mountain catchments (e.g. Gutierrez et al. 1998).

2 Process Regimes

Sediment transport processes and depositional characteristics on fans and cones depend on the water:sediment ratio and particle-size characteristics of the sediment mix supplied to the fan from the feeder catchment. Low water:sediment ratios result in debris-flow processes, high water:sediment ratios in fluvial processes.

2.1 Debris-Flow Processes and Sediment Properties

Hillslope, gully-floor or slope-failure debris may include an unsorted range of sediment sizes together with a variable moisture content (Johnson and Rodine 1984). Under loading, dependant on the gradient, and above a yield limit that is dependant on the strength of the material, the material will fail and deform. When the moisture content of the water:sediment mix is lower than its liquid limit, internal deformation takes place by plastic flowage. When moisture content is low, near the plastic limit, and when the material has a relatively high clay content, and therefore relatively high internal strength, shearing also may take place. On the other hand, when moisture content is high, near the liquid limit and therefore near the point where the material would drain under its own weight, the flow becomes much more fluid, and is known as wet debris flow or hyperconcentrated flow.

Debris flows range between cohesive debris flows, favoured by low water content especially when the clay content is high, and fluid or wet debris flows, favoured by high water content. Behaviour and depositional properties differ between these two types.

2.1.1 Cohesive Debris Flows

Cohesive debris flows have greater internal strength and move more slowly than wet debris flows. The movement is maintained by loading from behind with the flow moving forward as a lobe. On entering an unconfined fan environment the lobe may spread and thin, losing momentum. The movement is retarded by friction. On the flow margins this leads to the deposition of levees, parallel with the flow direction. A local reduction of gradient or an increase in basal friction may cause the forward movement of the lobe to cease temporarily. This leads to compression towards the front of the lobe, which may cause lobe breaching, and perhaps the diversion of the flow direction. With progressive loss of momentum, increasing friction or reduction

Fig. 1 Cohesive debris flows: (**a**) lobe and levee topography, 1982 debris-flow deposits on Thrush Gill, Howgill Fells, sheep gives scale. (**b**) Push fabric at the front of a bouldery debris flow (*upper right* of photo), late Pleistocene debris-flow deposits, Zzyzx fans, California

of gradient the flow stops moving. The total runout distance (D'Agostino 2010) onto the fan depends primarily on the mass of the flow, its fluidity, and the fan gradient.

On deposition, cohesive debris flows create characteristic forms (Fig. 1a), known as lobe and levee topography (Blair and McPherson 1998; Blair 1999), and have characteristic internal sedimentary features. Because cohesive debris flows have sufficient internal strength to support clasts within the matrix, the deposits tend to have a matrix-supported fabric. During transport, clasts may ride to the upper part or the surface of the flow by dispersive sorting, therefore on deposition an individual flow unit may show a crude inverse grading of the larger clasts. During transport the clasts may interact with one another to produce a distinctive compressional fabric, where towards the front of the lobe clasts are orientated across the axis of the lobe with their major planes dipping steeply against the flow direction (Fig. 1b). This gives a quasi- imbricate fabric (Wells and Harvey 1987), but with the major particle planes dipping much more steeply than in a conventional fluvial imbricate fabric. At the base of the flow major clast planes may be sub parallel with flow base, but rise towards the lobe front especially if basal shearing has occurred during the flow. Clasts on the levees tend to have their long-axes aligned with the flow direction.

2.1.2 Fluid (Wet) Debris Flows

With their greater fluidity, wet debris flows move at higher velocities than cohesive debris flows. During transport, clasts are more likely to be concentrated near the base and show clast alignment with the flow direction. Depositional forms may show less relief than those of cohesive debris flows with flatter, lower lobes, and any levees may be of clasts only. The greater fluidity may result in greater runout distances. During the flow the rheology of hyperconcentrated flows may switch backwards and forwards between quasi-plastic and liquid conditions, so that on deposition pockets of the sediment may locally show grading and stratification. After deposition the fluid matrix may drain down, either into the base of the flow or out altogether, resulting, especially in the upper part of the flow in a collapsed fabric, one of an openwork of clasts with no preferred orientation, simply the appearance of a collapsed mass (Wells and Harvey 1987).

2.2 Fluvial Processes and Sediment Properties on Fans

When the water content is high, well above the liquid limit of the water-sediment mix, sediment transport and deposition are by fluvial processes with fine sediment in suspension and coarse sediment moved by traction along the bed. On fans there are two styles: unconfined sheetflows, shallow flows in a wide "stream" over the fan surface; and channelized fluvial flows, of much greater depth within defined channels on the fan surface. The channels themselves may be single thread channels or braided channel networks. On non-trenched fans shallow unconfined sheetflows may occur from the apex spreading and thinning downfan. On trenched fans the same may occur downfan from the intersection point (Bull 1977), the point at which a channel within the fanhead trench emerges onto the surface of the fan. On such proximally trenched fans channelized flows are common within the fanhead trench but also may persist downfan from the intersection point or give way to sheetflows there (Harvey 1984a). On the distal fan surface the distinction between sheetflows and very wide braided channel flows is rather arbitrary, given the tendency for flows and channels to form a distributary pattern on the fan surface.

Fig. 2 Intersection point bar deposition fluvial lobate deposition, southeast Spain. Flow towards the camera. Note coarser clasts towards the front of the lobe; general clast alignment across the flow

Blair and McPherson (1994) argued that channelized flows merely rework fan sediment rather than deliver fresh sediment downfan. This may be the case on some of the smaller debris-flow and sheetflood-dominated fans in Death Valley, California, but on many larger fluvially-dominant fans braided streams may be the primary sedimentation mechanism. In some cases, especially on very large, low gradient, fluvially-dominant fans even meandering (Gabris and Nagy 2005) or anastomosing channels (Gohain and Parkash 1990) may be the primary sedimentation mechanism.

Fluvial flows deposit sediment on fan surfaces. Sheetflows are shallow with low unit power and transport and deposit gravel and sand sheets especially on distal fan surfaces. In section, such sheets may show fining-up couplets (Blair and McPherson 1994). Channelized flows deposit within-channel bars, but may also be subject to avulsion causing channel switching over the fan surface. Immediately downfan from the intersection point the flows may be less constrained, the channel may widen and lose unit power, causing deposition of bars that radiate downfan from the intersection point (Wasson 1979). These bars are unlike central bars on larger, lower gradient rivers, that characteristically show a distal fining of bar sediments. Bars on steep fan channels may show a distal zone of imbricated cobbles or boulders which trap finer gravels upstream (Fig. 2). Such features have been described as "sieve lobes" (Hooke 1967), though whether sieving of successively finer sediment, as envisaged by Hooke, actually takes place is open to question. Perhaps such features are deposited by sediment laden flows, in the transition zone between hyperconcentrated flows and true fluvial traction. They are clearly primarily fluvial as opposed to debris-flow deposits as they show well developed sorting, some stratification of the finer upstream trapped sediment and a characteristic low angle clast imbrication at the distal end of the bar.

Under flood conditions of sufficient unit power, fluvial flows on fans can carry out erosion. Locally within the fanhead trench, any constriction, flow convergence, or local steepening, perhaps caused by cementation of the fan sediments into which the channel is cut, may cause scour. At the intersection point there may well be an increase in gradient as the channel emerges onto the fan surface. On fans where the fan surface deposits have been cemented by calcrete, as is common in semi-arid areas (Lattman 1973; Harvey 1984a), the usual channel widening may not occur (Van Arsdale 1982), resulting in an increase in unit power. This may lead to mid-fan dissection (Harvey 1987). On fans where aggradation takes place downfan from the intersection point the axial fan slope may have been reduced by aggradation, such that the flanking slopes have higher gradients. Channel switching at this point to the fan flanks may result in an increase in unit power sufficient to trigger erosion.

2.3 *Interactions Between Processes*

Fan sedimentation tends to be episodic, storm-event based, and on any one fan may vary during a storm event in response to short term variations in water and sediment input, and of course may vary between storms. Both trends lead to an inherent variability of fan sediments. In 1982 an extreme storm event occurred over the Howgill Fells in northwest England (Harvey 1986), producing sedimentation on 13 small tributary-junction fans and debris cones. By studying the sedimentology of the fan deposits and their short-term stratigraphic relations we were able to identify the sequence of events on all 13 of the fans and cones affected (Wells and Harvey 1987). We identified four fan styles: cohesive debris-flow dominance; wet debris-flow or transitional-flow dominance, fluvial bar-dominance and fluvial-sheet dominance. For each type (each represented on Fig. 3 by a type fan) we identified a within-storm sequence involving initially erosive streamflows, followed by deposition resulting from progressively lower water:sediment ratios as the major pulses of sediment were delivered to the fans from the feeder catchments, then finally by dilution and back to an erosive regime. All of this occurred in only a few hours.

Within-storm variations are also expressed in the spatial distribution of processes within fans. During the same Howgill storm event, described above (Wells and Harvey 1987), we identified a tendency for downfan flushing on some of the fans, with debris flows giving way downfan to fluvial bar deposits, and cobble and boulder bar deposits giving way downfan to sheets of finer gravels (Fig. 4). Other studies have also demonstrated similar spatial variations in processes on fans during flood events (e.g. Gutierrez et al. 1998, on the Biescas fan, Spanish Pyrennees; Mather and Hartley 2005, in Chile). Downfan flushing particularly was observed in response to a major storm and flood on the Honda fan, southeast Spain in 1980 (Harvey 1984b). A sand-bed braided channel with gravel bars within the wide fanhead trench gave

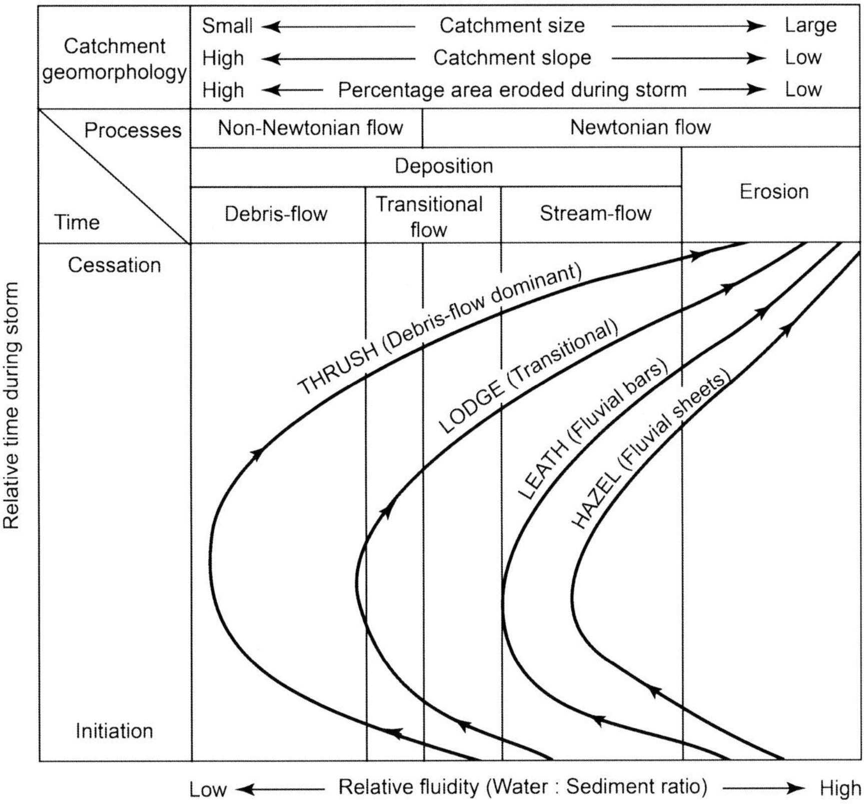

Fig. 3 Interpretive model of sedimentation during the 1982 Howgill flood for the four type cones and fans (modified from Wells and Harvey 1987)

way downfan to thin sand sheets deposited on midfan and distal fan surfaces. Similarly on the Zzyzx fans, California, in response to a major storm and flood in 1988 (Harvey and Wells 1994), although cobble bars were extensively reworked within the fan channels, only fine gravel and sand sheets were flushed out to the fan toes.

The overall control of the range of processes that occur during individual flood events is ultimately the sediment mix fed from the catchment. Smaller steeper catchments tend to supply more sediment whereas larger catchments tend to supply greater flood volumes. Again, this trend was evident for the four groups of Howgill fans (Fig. 4) during the 1982 flood (Wells and Harvey 1987).

3 Short-Term Catchment Controls

Despite the inherent variability of processes on fans during individual storm events and between events, the overall fan regime (i.e. the gross similarity of response to major storm events) responds to catchment controls, provided that these controls

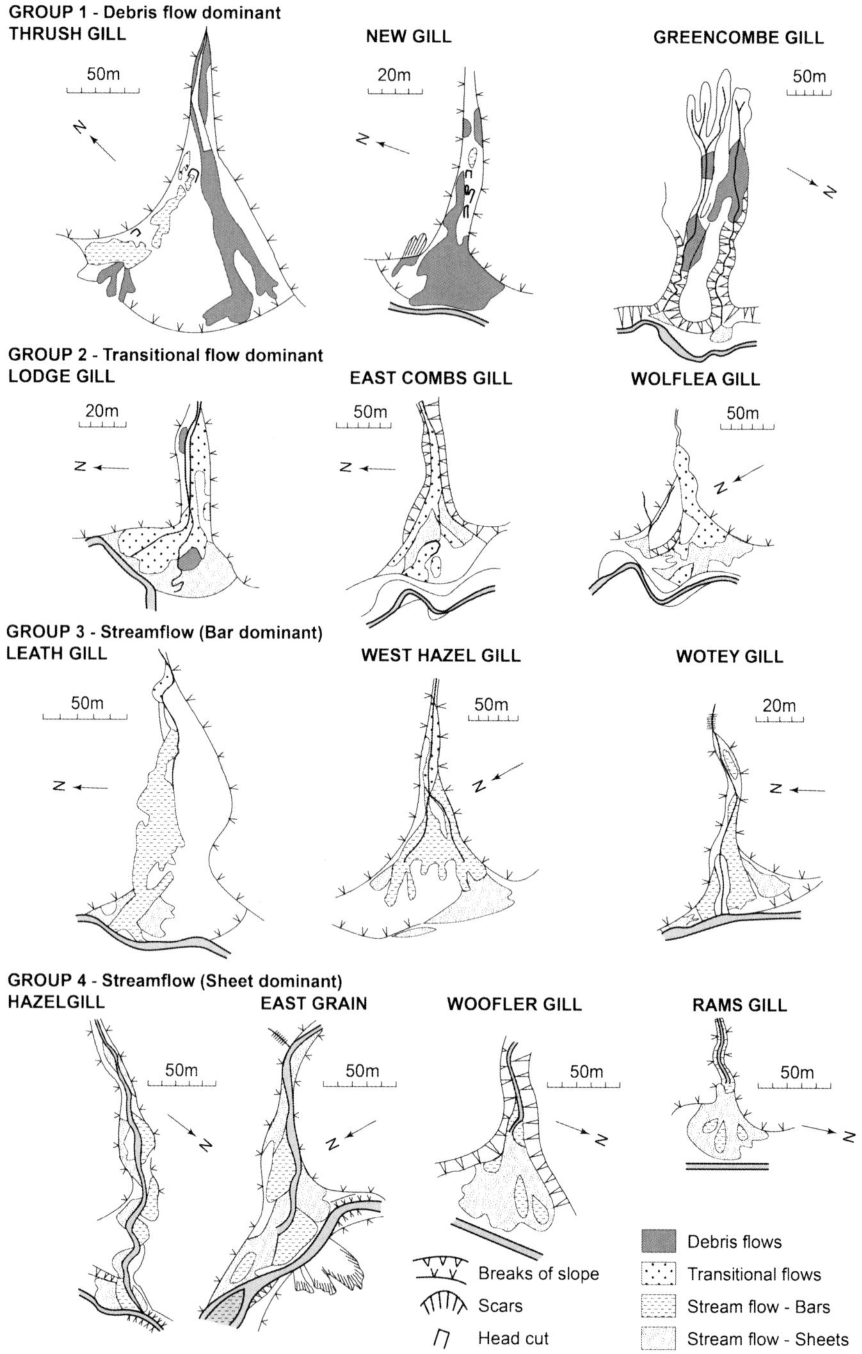

Fig. 4 Spatial distribution of facies types deposited during the 1982 Howgill flood on the 13 fans and cones

remain stable over time (see below, Sect. 4). This is evident both in the surface morphology of fans and in their constituent sedimentary sequences.

The catchment characteristics control the generation of flood discharges and sediment supply, therefore influence the water:sediment ratios generated during storm events. Catchment size influences flood magnitude, and although total sediment would also increase with catchment size, within larger catchments there is a greater likelihood of within-catchment storage particularly of the coarser sediment, therefore water:sediment ratios are likely to be higher for larger catchments, and sediment size may be smaller. Catchment steepness will increase erosion rates, the likelihood of slope failures and the strength of coupling (Harvey 2001) between hillslopes and channels, all factors likely to increase sediment supply. Catchment geology may also have an effect, some geological settings being conducive to slope failure, some lithologies being more prone to erosion, especially in the context of the supply of coarse sediment to the fan environment. Land cover, including the effects of human-induced disturbance, affects runoff rates and sediment yield.

Hence, catchment characteristics control the regime of fan processes, not only the detailed depositional morphology (Suwa and Okuda 1983; Beaty 1990; Whipple and Dunne 1992), but also the overall fan properties. Debris-flow dominated fans are associated with small steep catchments (Harvey 1984c; Kostaschuk et al. 1986; Rickenmann and Zimmerman 1993), and fluvially-dominant fans with larger or less steep catchments.

4 Longer-Term Fan Dynamics

As catchment characteristics control the short-term supply of water and sediment to the fan environment so do they influence the longer-term development of fan morphology. A particularly important morphological property is fan gradient as it both reflects the depositional process and influences the style of deposition. Debris flows stop moving at higher gradients than the threshold depositional slope for tractional fluvial processes. For these processes the threshold depositional gradient is greater for coarser material (Blissenbach 1952) and greater for sheetflows than for channelized flows. Hence the fan gradient is adjusted to the ambient process regime. Since drainage area influences the process regime, many studies of fan morphometry demonstrate an inverse relationship between drainage area and fan gradient (Harvey 1987, 2011; Calvache et al. 1997).

Should the process regime change through time, either relatively abruptly in response to an environmental change affecting flood power and sediment supply, or progressively, reflecting the long-term evolution of the fan setting, then the fan morphology will adjust accordingly. This is perhaps best illustrated by a consideration of fan style (Fig. 5a; Harvey 2002a, 2003). Fan styles range from aggrading fans, through prograding fans, those with a fanhead trench but distal aggradation, to dissecting fans on which dissection might be triggered at the fan apex, in midfan or distally. With the exception of distally-induced dissection in

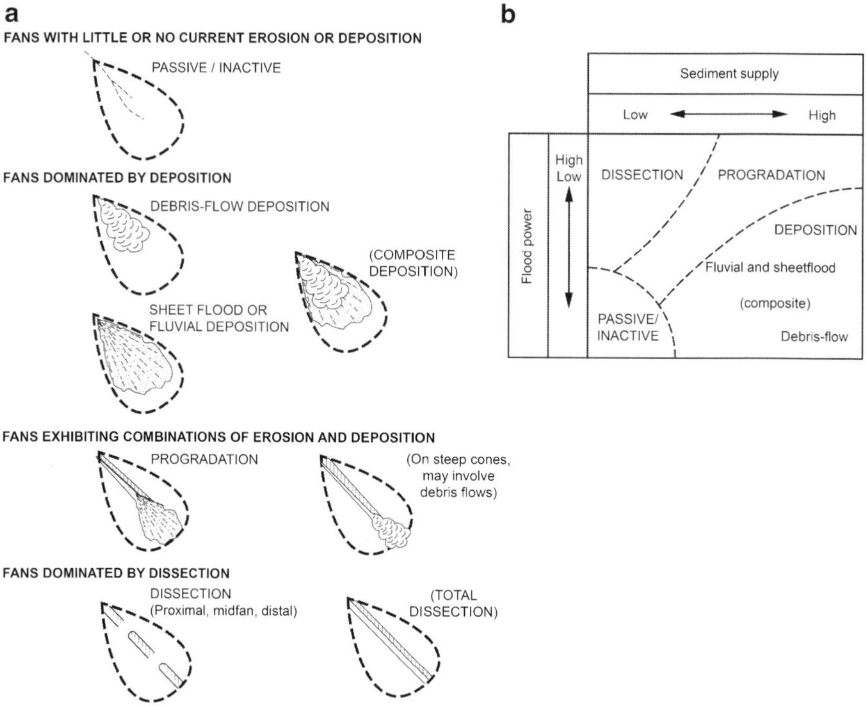

Fig. 5 Proximal controls of fan style: (**a**) Fan aggradational/dissectional styles – (**b**) Conceptual model of the influence of flood power and sediment supply on fan style (modified from Harvey 2002a, 2003)

response to either "toe cutting" (Leeder and Mack 2001) or base-level change (Harvey 2002b), these styles are controlled by the water and sediment input to the fan from the feeder catchment. The relationship between water and sediment supply can be considered as an extension of the threshold of critical stream power (Bull 1979), defined by the relationship between actual (unit) stream power (controlled by gradient and flood magnitude) and the critical stream power, that required to transport the sediment supplied (controlled by sediment volume and calibre). Under conditions of excess sediment supply aggradation takes place, by debris-flow or fluvial processes, depending on the water:sediment mix. Under conditions of excess power dissection takes place (Fig. 5b; Harvey 2002a, b).

If an environmental change occurs that modifies the relationship between flood regime and sediment supply, such as (i) a change in frequency or severity of storm events, (ii) a change in rates or mechanisms of sediment supply or calibre or in the nature of the sediment supplied, such as might be the result of climatic change or land-cover change, then the fan regime may no longer be adjusted to the ambient flood and sediment supply regimes. This can be seen as a shift in plotting position on Fig. 5b, and the establishment of a new fan regime. Such a shift may be expressed by

(i) renewed aggradation on a fan, burying older fan surfaces, filling or partially filling fanhead trenches, or simply a switch from one depositional mechanism to another, or by (ii) trench development on the fan surface, either at the fanhead or in midfan. Responses to base-level change are rather different and initially affect only the distal fan zones.

Past sequences of such "regime changes" may be evident in the fan morphology and particularly are preserved in the sedimentary sequences within a fan. There are numerous published studies of how fan sediment sequences reflect either fan evolution or the response of alluvial fan systems to environmental change. A number of studies have identified that over the long term, often in response to regional tectonics, fans progressively evolve from debris-flow dominance through fluvial and sheetflood aggradation to progradation, by developing a fanhead trench, then ultimately to dissection. This "ageing" process was first described by Eckis (1928), but has since been identified in a number of more recent studies (e.g. Harvey 1984a, in southeast Spain; see also Fig. 6a). The final phases of fan ageing represent a common step in the evolution of uplifted sedimentary basins, whereby basin-marginal fans are dissected and give way to an integrated fluvial system (Maher 2006; Silva 2008; Stokes 2008).

Other than long-term ageing, changes in fan regime reflect environmental changes (e.g. ongoing tectonics, climatic change, human activity/land-cover change, and the special case of base-level change). In southeast Spain fan aggradation phases coincided with Pleistocene global glacials, presumably as a result of increased weathering rates, releasing large quantities of sediment from the feeder catchments (Harvey 1990; Harvey et al. 2003). During such aggradation phases there was a trend from debris-flow or channelized fluvial sedimentation towards sheetflood sedimentation. This trend may be represented in proximal fan sections (Fig. 6b) by a general fining-up of the sedimentary sequence as progressive burial of the fan takes place and the sediments backfill into the mouth of the feeder catchment (Harvey 2011). During interglacials, on the other hand, Spanish fans underwent fan trenching and progradation (Harvey 1990), with soil formation taking place on the abandoned fan surfaces. A similar sequence has been identified on the fans in the Central Valley of California (Weissmann et al. 2005). Under such conditions coarsening upwards sedimentary sequences appear to characterize distal fan environments (Harvey 2011).

In areas outside the Pleistocene glacial limits, alluvial fan morphological and sediment sequences often preserve a record of the geomorphic response to Quaternary climatic changes over timescales of the global glacial/interglacial sequence. This is especially true, as it is in southeast Spain (see above), in dry regions. The Zzyzx fans on the mountain front of the Soda Mountains in the Mojave Desert, California (Fig. 7a) show evidence of late Quaternary adjustment to climatic change. We have no absolute dates on the fan sediments or fan surfaces, but using a multiparameter approach (McFadden et al. 1989) to the correlation and relative dating of fan surface segments, and estimating their ages by considering the relationships with two dated shorelines of late Pleistocene pluvial Lake Mojave (Enzel et al. 2003) we identified changes in fan regime (Fig. 7b), coincident with late Quaternary climatic

Fig. 6 Vertical sediment sequences indicative of fan ageing. (**a**) Ceporro fan, Tabernas, southeast Spain. Proximal fan sediment sequence exposed in the fanhead trench. Note basal debris-flow deposits, locally capped by a paleosol truncated by an erosional horizon, Upper sediments are sheet and channel gravels capped by a well developed soil with calcrete development. (**b**) Proximal fan sediment sequence exposed in the fanhead trench of Mezquita fan, Tabernas, southeast Spain. Note basal fluvial channel gravels, overlain by sheet gravels

changes (Harvey and Wells 1994, 2003; Harvey et al. 1999). A major period of fan aggradation occurred before or during the last glacial maximum, by debris-flow processes on the smaller fans (Fig. 1b) and fluvial processes on the larger fans. This was followed by trenching and progradation during the climatic transition at the end of the Pleistocene (McDonald et al. 2003), then by pulsed phases of progradation following climatic changes during the Holocene.

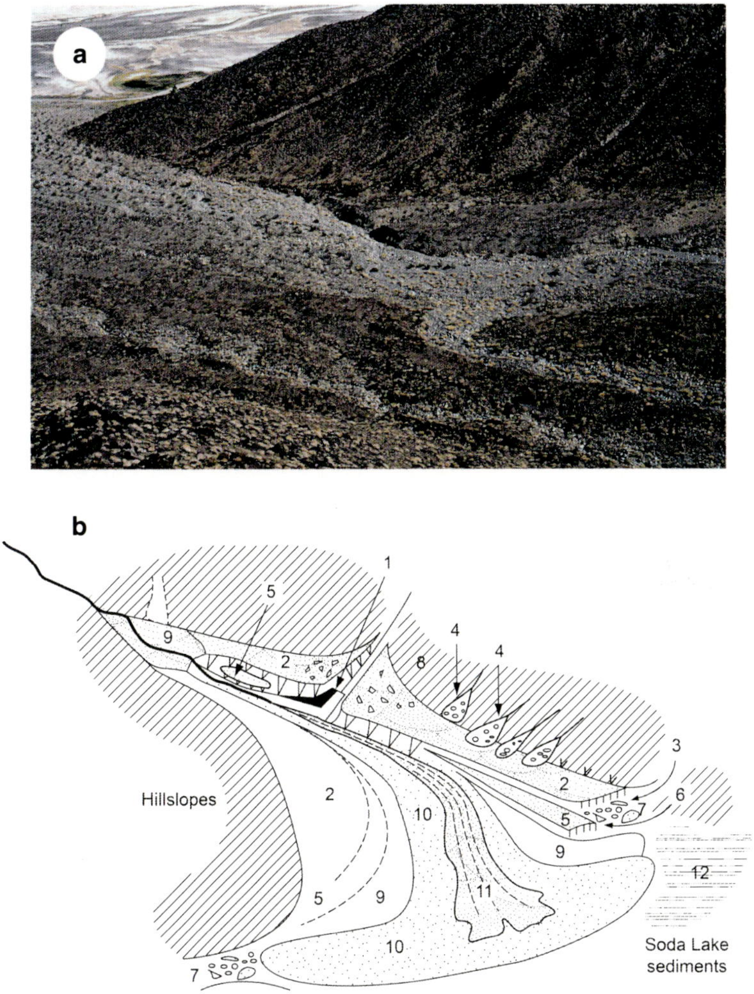

Fig. 7 Zzyzx fans, Soda Mountains, Mojave Desert, California. (**a**) General view of one of the larger fluvially dominant fans, backfilled into the mountain catchment. Note the fanhead trench cut into the older fan surfaces. The active gravel depositional lobe extends downfan. Note also the hillslope debris-flows feeding sediment to the fan margins. (**b**) Interpretive model of the late Quaternary fan aggradation and dissection sequence: (1) Buried, strongly cemented older fan deposits (perhaps mid Pleistocene), (2) Main late Pleistocene aggradation phase (pre-dating the higher lake shoreline (3)). (4) Lat(est) Pleistocene hillslope debris flows. (5) Latest Pleistocene fan surface, follows trenching into surface (2), postdates upper shoreline (3), erosionally cut by youngest (early Holocene) lake shoreline (6) and overlain by early Holocene lake sediments (7). (8) Hillslopes stabilised during the early Holocene. (9) Mid Holocene fan progradation. (10) Late Holocene fan progradation. (11) Modern active depositional lobe

Fig. 8 Late Holocene alluvial fan at Blakethwaite, Howgill Fells. (**a**) General view showing gullied catchment feeding the fan. Note incised channel on *left hand* margin of the fan, depositing gravels on the distal fan surface – *far left*. (**b**) Section through the fan deposits exposed by the incised channel. *Dark layers* are buried soils. Numerous sites like this have provided the basis for the radiocarbon dating of fan deposition in northern Britain

In areas within the limits of the Last Glacial Maximum (LGM) modern alluvial fans date only from deglaciation. In mountain areas where glacial ice persisted until late in the Pleistocene, there was usually a phase of paraglacial (sensu Ryder 1971) fan sedimentation, related to the abundant sediment supply immediately on deglaciation. For example in the Scottish Highlands within the limits of the Younger Dryas (Loch Lomond stage) glacial readvance, a paraglacial debris-flow dominated phase of alluvial-fan and debris-cone sedimentation can be recognized, that persisted until sediment exhaustion at some time during the mid Holocene (Brazier et al. 1988; Brazier and Ballantyne 1989).

Elsewhere in the uplands of northern Britain, within the limits of the LGM but not affected by ice during the Younger Dryas, periglacial hillslope processes dominated the latest Pleistocene, involving solifluction but not alluvial fan nor debris cone formation. In such areas as the Howgill Fells, northwest England (Harvey et al. 1981; Harvey 1996), fan and cone deposition (Fig. 8) was entirely of late Holocene age, with several phases identified through radiocarbon dating of soils interstratified with or buried by fan deposits. A major phase occurred following the tenth century AD and another less precisely dated phase occurred during the last 300 years. Further work in the Howgills and in other upland areas in NW England and SW Scotland (Chiverrell et al. 2007, 2008; Chiverrell and Jakob 2012, *this volume*), based on numerous radiocarbon dates, confirms this general picture. The major fan forming phases appear to relate primarily to gully erosion following human-induced vegetation change, rather than to climatic signals, with the possible exception of the most recent phase, about 300 years ago, that can also be recognised in the Scottish Highlands, that is coincident with the climatic deterioration of the Little Ice Age. In other mountain areas of Europe, there is a similar interaction between climate-led and human-induced changes in flood power and sediment supply to alluvial fans and debris cones, with climate being the most important control during the late Pleistocene and early Holocene (e.g. Salcher et al. 2010) but with human controls becoming increasingly more important during the late Holocene (e.g. Gomez-Villar and Garcia-Ruiz 2000; Columbera and Bersezio 2011).

5 Conclusions

In summary, alluvial fans and debris cones are not only hazardous environments for human occupation and land use, but also in their morphology and sediments preserve a sensitive record of short-term flood behaviour and longer-term environmental change. When it comes to dating of past events it is important to recognise the timescale involved, whether the concern is primarily with modern processes and hazards on fans, or with the response of fans to short-term interactions between climate and human-induced environmental change, or with the use of alluvial fan sediments for interpreting late Quaternary environmental change. Not only do the appropriate dating methods differ according to the timescale involved, but the nature of the evidence itself shifts from the detailed form and sedimentology of individual debris-flow and flood deposits to sedimentary sequences.

Acknowledgements In this chapter, I have relied on my own published work, some of which could not have been done without the cooperation of my co-authors, particularly Richard Chiverrell, Gez Foster, Elizabeth Maher/Whitfield, Anne Mather, Pablo Silva, Martin Stokes, Steve Wells, Peter Wigand. I also thank Sandra Mather of the cartographics section of the School of Environmental Sciences/Geography of the University of Liverpool for assistance in preparing the illustrations.

References

Beaty CB (1990) Anatomy of a White Mountains debris flow – the making of an alluvial fan. In: Rachocki AH, Church M (eds) Alluvial fans: a field approach. Wiley, Chichester, pp 69–89

Blair TC (1999) Sedimentology of the debris-flow-dominated Warm Spring Canyon alluvial fan, Death Valley, California. Sedimentology 46:941–965

Blair TC, McPherson JG (1994) Alluvial fan processes and forms. In: Abrahams AD, Parsons AJ (eds) Geomorphology of desert environments. Chapman & Hall, London, pp 354–402

Blair TC, McPherson JG (1998) Recent debris-flow processes and resultant forms and facies of the dolomite alluvial fan, Owens Valley, California. J Sediment Res 68:800–818

Blissenbach E (1952) Relation of surface angle distribution to particle size distribution on alluvial fans. J Sediment Petrol 22:25–28

Brazier V, Ballantyne CK (1989) Late Holocene debris cone evolution in Glen Feshie, western Cairngorm Mountains, Scotland. Trans R Soc Edinb Earth Sci 80:17–24

Brazier V, Whittington G, Ballantyne CK (1988) Holocene debris cone evolution in Glen Etive, Western Grampian Highlands, Scotland. Earth Process Landf 13:525–531

Bull WB (1977) The alluvial fan environment. Prog Phys Geogr 1:222–270

Bull WB (1979) The threshold of critical power in streams. Geol Soc Am Bull 90:453–464

Calvache ML, Viseras C, Fernandez J (1997) Controls on fan development – evidence from fan morphometry and sedimentology: Sierra Nevada, Spain. Geomorphology 21:69–84

Chiverell RC, Jakob M (2012) Radiocarbon dating: alluvial fan/debris cone evolution and hazards. In: Schneuwly-Bollschweiler M, Stoffel M, Rudolf-Miklau F (eds) Dating torrential processes on fans and cones – methods and their application for hazard and risk assessment, Advances in global change research. Springer, Dordrecht/Heidelberg/London/New York

Chiverrell RC, Harvey AM, Foster GC (2007) Hillslope gullying in the Solway Firth-Morecambe Bay region, Britain: responses to human impact and/or climatic deterioration? Geomorphology 84:317–343

Chiverrell RC, Harvey AM, Hunter (nee Miller) SY, Millington J, Richardson N (2008) Late Holocene environmental change in the Howgill Fells, Northwest England. Geomorphology 100:41–69

Columbera L, Bersezio R (2011) Evolution of an alpine alluvial fan during the last two centuries: impact of debris-flow magnitude and frequency and response to natural and anthropogenic controls. Earth Surf Process Landf 36(12):1632–1646. doi:10.1002/esp.2178

D'Agostino V (2010) Field and laboratory investigations on the runout distances of debris flows in the Dolomites (Eastern Italian Alps). Geomorphology 115:294–304

Eckis R (1928) Alluvial fans of the Cucamunga district, southern California. J Geol 36:225–247

Enzel Y, Wells SG, Lancaster N (2003) Late Pleistocen lakes along the Mojave River, southeast California. In: Enzel Y, Wells SG, Lancaster N (eds) Palaeoenviroments and palaeohydrology of the Mojave and southern Great Basin Deserts, Geological Society of America, special paper 368. Geological Society of America, Boulder, pp 61–77

Gabris G, Nagy B (2005) Climate and tectonically controlled river style changes on the Sajo-Hernad alluvial fan (Hungary). In: Harvey AM, Mather AE, Stokes M (eds) Alluvial fans: geomorphology, sedimentology, dynamics, Geological Society special publication 251. Geological Society, London, pp 61–67

Gohain K, Parkash B (1990) Morphology of the Kosi megafan. In: Rachocki AH, Church M (eds) Alluvial fans: a field approach. Wiley, Chichester, pp 151–178

Gomez-Villar A, Garcia-Ruiz JM (2000) Surface sediment characteristics and present dynamics in alluvial fans of the central Spanish Pyrenees. Geomorphology 34:127–144

Gutierrez F, Gutierrez M, Sancho C (1998) Geomorphological and sedimentological analysis of a catastrophic flash flood in the Aras drainage basin (Central Pyrenees, Spain). Geomorphology 22:277–293

Harvey AM (1984a) Aggradation and dissection sequences on Spanish alluvial fans: influence on morphological development. Catena 11:289–304

Harvey AM (1984b) Geomorphological response to an extreme flood: a case study from southeast Spain. Earth Surf Process Landf 9:267–279

Harvey AM (1984c) Debris flows and fluvial deposits in Spanish Quaternary alluvial fans: implications for fan morphology. In: Koster EH, Steel R (eds) Sedimentoloqy of gravels and conglomerates, Memoir 10. Canadian Society of Petroleum Geologists, Calgary, pp 23–132

Harvey AM (1986) Geomorphic effects of a 100-year storm in the Howgill Fells, northwest England. Zeitschrift fur Geomorphologie 30:71–91

Harvey AM (1987) Alluvial fan dissection: relationships between morphology and sedimentation. In: Frostick L, Reid I (eds) Desert sediments: ancient and modern, Geological Society special publication 35. Geological Society, London, pp 87–103

Harvey AM (1990) Factors influencing Quaternary alluvial fan development in southeast Spain. In: Rachocki AH, Church M (eds) Alluvial fans: a field approach. Wiley, Chichester, pp 247–269

Harvey AM (1996) Holocene hillslope gully systems in the Howgill Fells, Cumbria. In: Brooks SM, Anderson MG (eds) Advances in hillslope processes, vol 2. Wiley, Chichester, pp 731–752

Harvey AM (2001) Coupling between hillslopes and channels in upland fluvial systems: implications for landscape sensitivity, illustrated from the Howgill Fells, northwest England. Catena 42:225–250

Harvey AM (2002a) Factors influencing the geomorphology of alluvial fans: a review. In: Perez-Gonzalez A, Vegas J, Machado MJ (eds) Apertaciones a la Geomorfologia de Espana en el Inicio de Tercer Mileno. Instituto Geologico y Minero de Espana, Madrid, pp 59–75

Harvey AM (2002b) The role of base-level change in the dissection of alluvial fans: case studies from southeast Spain and Nevada. Geomorphology 45:67–87

Harvey AM (2003) The response of dry region alluvial fans to Quaternary climatic change. In: Alsharhan AS, Wood WW, Goudie AS, Fowler A, Ebdelatif EM (eds) Desertification in the third millennium. Swete & Zeitlinger, Lisse, pp 75–90

Harvey AM (2010) Local buffers to the sediment cascade: debris cones and alluvial fans. In: Burt TP, Allison RJ (eds) Sediment cascades: an integrated approach. Wiley-Blackwell, Chichester, pp 153–180

Harvey AM (2011) Dryland alluvial fans. In: Thomas DSG (ed) Arid-zone geomorphology, 3rd edn. Wiley-Blackwell, Chichester, pp 333–371

Harvey AM, Wells SG (1994) Late Pleistocene and Holocene changes in hillslope sediment supply to alluvial fan systems: Zzyzx, California. In: Millington AC, Pye K (eds) Environmental change in drylands: biogeographical and geomorphological perspectives. Wiley, Chichester, pp 66–84

Harvey AM, Wells SG (2003) Late Quaternary variations in alluvial fan sedimentologic and geomorphic processes, Soda Lake basin, eastern Mojave Desert, California. In: Enzel Y, Wells SG, Lancaster N (eds) Palaeoenviroments and palaeohydrology of the Mojave and southern Great Basin Deserts, Geological Society of America, special paper 368. Geological Society of America, Boulder, pp 207–230

Harvey AM, Oldfield F, Baron AF, Pearson G (1981) Dating of post-glacial landforms in the central Howgills. Earth Surf Process Landf 6:401–412

Harvey AM, Wigand PE, Wells SG (1999) Response of alluvial fan systems to the late Pleistocene to Holocene climatic transition: contrasts between the margins of pluvial Lakes Lahontan ad Mojave, Nevada and California, USA. Catena 36:255–281

Harvey AM, Foster G, Hannam J, Mather AE (2003) The Tabernas alluvial fan and lake system, southeast Spain: applications of mineral magnetic and pedogenic iron analyses towards clarifying the quaternary sediment sequences. Geomorphology 50:151–171

Hooke RL (1967) Processes on arid-region alluvial fans. J Geol 75:438–460

Johnson AM, Rodine JR (1984) Debris flows. In: Brunsden D, Prior DB (eds) Slope instability. Wiley, New York, pp 257–361

Kostaschuk RA, Macdonald GM, Putnam PE (1986) Depositional processes and alluvial fan – drainage basin morphometric relationships near Banff, Alberta, Canada. Earth Surf Process Landf 11:471–484

Lattman LH (1973) Calcium carbonate cementation of alluvial fans in southern Nevada. Geol Soc Am Bull 84:3013–3028

Leeder MR, Mack GH (2001) Lateral erosion ("toe cutting") of alluvial fans by axial rivers: implications for basin analysis and architecture. J Geol Soc Lond 158:885–893

Maher E (2006) The Quaternary evolution of the Rio Alias, southeast Spain. PhD thesis, Univerity of Liverpool

Mather AE, Hartley AJ (2005) Flow events on a hyper-arid alluvial fan: Quebrada Tambores, Salar de Atacama, northern Chile. In: Harvey AM, Mather AE, Stokes M (eds) Alluvial fans: geomorphology, sedimentology, dynamics, Geological Society special publication 251. Geological Society, London, pp 9–24

McDonald EV, McFadden LD, Wells SG (2003) Regional response of alluvial fans to the Pleistocene-Holocene climatic transition, Mojave Desert, California. In: Enzel Y, Wells SG, Lancaster N (eds) Palaeoenviroments and palaeohydrology of the Mojave and southern Great Basin Deserts, Geological Society of America, special paper 368. Geological Society of America, Boulder, pp 189–206

McFadden LD, Ritter JB, Wells SG (1989) Use of multiparameter relative-age methods for estimation and correlation of alluvial fan surfaces on a desert piedmont, eastern Mojave Desert, California. Quat Res 32:276–290

Rickenmann D, Zimmerman M (1993) The 1987 debris flows in Switzerland: documentation and analysis. Geomorphology 8:175–189

Ryder JN (1971) The stratigraphy and morphology of paraglacial fans in south central British Columbia. Can J Earth Sci 8:279–298

Salcher BC, Faber R, Wagreich M (2010) Climate as the main factor controlling the sequence development of two alluvial fans in the Vienna Basin (eastern Austria) – a numerical modelling approach. Geomorphology 115:215–227

Silva PG (2008) Transition from alluvial to fluvial systems in the Guadalentin Epression (SE Spain) during the Holocene: Lorca fan versus Guadalentin River. Geomorphology 100:140–153

Stokes M (2008) Plio-Pleistocee drainage development in an inverted sedimentary basin: Vera basin, Betic cordillera, SE Spain. Geomorphology 50:193–211

Suwa H, Okuda S (1983) Deposition of debris flows on a fan surface, Mt. Yakedake, Japan. Z. Geomorph Supplementband 46:79–101

Van Arsdale R (1982) Influence of calcrete on the geometry of arroyos near Buckeye, Arizona. Geol Soc Am Bull 93:20–26

Wasson RJ (1979) Intersection point deposition on alluvial fans: an Australian example. Geogr Ann 56A:83–92

Weissmann GS, Bennett GL, Lansdale AL (2005) Factors controlling sequence development on Quaternary fluvial fans, San Joachim Basin, California, USA. In: Harvey AM, Mather AE, Stokes M (eds) Alluvial fans: geomorphology, sedimentology, dynamics, Geological Society special publication 251. Geological Society, London, pp 169–186

Wells SG, Harvey AM (1987) Sedimentologic and geomorphic variations in storm generated alluvial fans, Howgill Fells, northwest England. Geol Soc Am Bull 98:182–198

Whipple KX, Dunne T (1992) The influence of debris-flow rheology on fan morphology, Owens Valley, California. Geol Soc Am Bull 104:887–900

Susceptibility and Triggers for Debris Flows: Emergence, Loading, Release and Entrainment

Roland Kaitna, Michelle Schneuwly-Bollschweiler, Thomas Sausgruber, Markus Moser, Markus Stoffel, and Florian Rudolf-Miklau

1 Debris-Flow Susceptibility and Initiation: An Overview

Debris flows generally form when unconsolidated material becomes saturated and unstable, either on a hillslope or in a stream channel. The process is defined as a moving mass of loose mud, sand, soil, rock, water and air that travels down a slope under the influence of gravity. Flows can carry material ranging in size from clay to boulders, and may contain a large amount of woody debris. Volumes of

R. Kaitna (✉)
Department of Civil Engineering and Natural Hazards, Institute of Mountain Risk Engineering, University of Natural Resources and Life Sciences, A-1180 Vienna, Austria
e-mail: roland.kaitna@boku.ac.at

M. Schneuwly-Bollschweiler • M. Stoffel
Laboratory of Dendrogeomorphology (dendrolab.ch), Institute of Geological Sciences, University of Berne, CH-3012 Berne, Switzerland

Chair for Climate and Climatic Impacts, Environmental Sciences, University of Geneva, CH-1227 Carouge-Geneva, Switzerland
e-mail: michelle.schneuwly@dendrolab.ch; markus.stoffel@dendrolab.ch

T. Sausgruber
Austrian Service for Torrent and Avalanche Control, Geological Staff Office, Liebeneggstraße 11, A-5020 Innsbruck, Austria
e-mail: thomas.sausgruber@die-wildbach.at

M. Moser
Austrian Service for Torrent and Avalanche Control, Regional Office Lungau, A-5580 Tamsweg, Austria
e-mail: markus.moser@die-wildbach.at

F. Rudolf-Miklau
Department IV/5-Torrent and Avalanche Control, Federal Ministry for Agriculture, Forestry, Environment and Water Management, A-1190 Vienna, Vienna
e-mail: florian.rudolfmiklau@lebensministerium.at

M. Schneuwly-Bollschweiler et al. (eds.), *Dating Torrential Processes on Fans and Cones*, Advances in Global Change Research 47, DOI 10.1007/978-94-007-4336-6_3,

material delivered by single events vary from less than 100 to more than 100,000 m^3. Generally three factors are necessary for a debris flow to develop: water, sufficient inclination, and abundant sediment supply.

From a systemic perspective the initiation of a debris flow is governed by disposition elements (*susceptibility*) in the release area and actual debris-flow triggers. *Susceptibility* is defined as the condition of an area that renders it prone to the occurrence of hazardous mass-movement processes (Loat and Meier 2003) independent of the probability of recurrence.

Triggering events for debris flows in Alpine catchment areas are mainly meteorological events, either short duration-high intensity or long lasting rainfalls. Rapid snow and ice melt often occurs in combination with rainfall and increases the effects of precipitation. A sudden flow of water can be channelled over a steep valley filled with debris that is loose enough to be mobilized. The water soaks down into the debris, adds weight, and triggers a flow (Julien 1995; Iverson 1997). Other, but rather rare factors, like glacier lake outbursts or dam breaks emerge independently from meteorological events. Debris-flows susceptibility is enhanced after wildfires on steep slopes or after logging operations have removed vegetation. Debris flows from volcaniclastic (i.e. lahars) are absent in the Alps.

The initiation of debris flows depends on the stability of the slope, the river bed or natural dams (Costa 1984). Zimmermann (1990) listed four types of typical initiation scenarios for the Alpine region:

- Slope failure or landslide
- Release in the contact area bed rock/loose material
- Liquefaction of river bed (impulsive loading)
- Dam break and entrainment

All these processes show a sudden mobilization of large masses of loose material when the threshold of stability is exceeded and can also interact with each other, reflecting the complex nature of debris flow. Section 2 of this chapter describes the common concept for understanding the susceptibility of a watershed to produce debris flows, Sect. 3 gives an overview of meteorological trigger events, and Sects. 4 and 5 describe relevant processes for debris-flow initiation on the hillslope and in the channel, respectively.

2 Susceptibility of Release Areas

The disposition concept of a catchment to produce debris flows comprises three fundamental elements (Kienholz 1995). The total susceptibility depends on (i) the basic susceptibility, (ii) the variable susceptibility and (iii) an external forcing. A debris flow will only be triggered if the sum of basic and variable susceptibility exceeds a certain threshold and if a sufficiently strong external forcing occurs at that moment (Fig. 1).

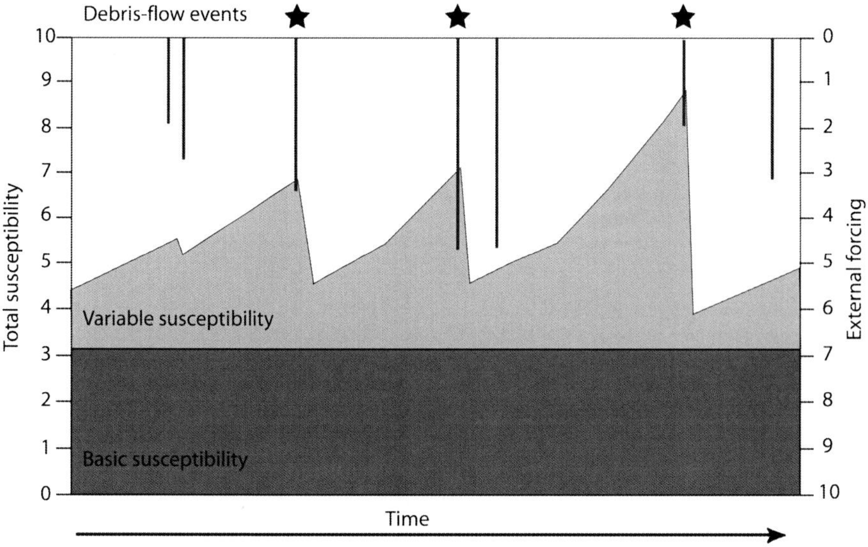

Fig. 1 Sketch of the disposition concept for the occurrence of debris flows: The basic susceptibility remains constant over long time periods. The variable susceptibility changes in short time intervals. If the external forcing exceeds the current threshold value of the total susceptibility, an event will be triggered (adapted from Zimmermann et al. 1997)

2.1 Basic Susceptibility

The basic susceptibility primarily influences the magnitude (*how big?*) and spatial occurrence (*where?*) of debris flows (Zimmermann et al. 1997). The basic susceptibility may vary over long time scales (decades to centuries) and depends mainly on different geologic and hydrologic properties as well as on topography (Moser 1980; Costa 1984), more specifically on the following parameters:

- **Relief:** At least minimal relief energy is needed in a catchment to produce debris flows. Debris flows in a channel and on a slope are generally only triggered if a minimum slope angle is reached. The lower threshold strongly depends on material composition and topography of the catchment (Rickenmann 1999). Altitude and aspect also have an influence on weathering intensity and therefore on material production.
- **Availability of loose material:** Availability of loose material is essential for the occurrence of debris flows. In transport-limited catchments an almost unlimited amount of sediment is available to feed debris flows (so-called "Altschuttwildbach" in the classification scheme of Stiny 1931). Material availability is strongly influenced by past glacial processes (e.g. till, glacial valley-fillings, deep-seated

landslides, scree slopes) and delivers large amounts of sediments to the torrent. In sediment-limited watersheds ("Jungschuttbach", Stiny 1931), material for debris-flow entrainment has to be produced by weathering (Bovis and Jakob 1999; Jakob et al. 2005). After an event, the reservoir is empty and no debris flow will occur until enough material has accumulated in the channel or initiation zone. Time spans between two events are therefore normally longer than in transport-limited basins.

- **Channel bed:** The type of channel bed and channel geometry also influence debris-flow magnitudes considerably. If the channel runs on bedrock, no additional material will be eroded as opposed to channel beds in loose deposits where erosion contributes considerably to debris-flow magnitude. The channel geometry defines the maximum transport capacity of a debris flow.
- **Geology:** geotechnical characteristics of material composition (e.g. porosity, friction angle, cohesion, grain size distribution) not only govern the stability of the deposit but also the flow behaviour of the debris flow.

2.2 *Variable Susceptibility*

The variable susceptibility changes within days to decades and primarily influences the temporal occurrence of debris flows (*when? how often?*). To a lesser extent, also magnitudes and frequency-magnitude relationships are determined through variable susceptibility (Zimmermann et al. 1997), more specifically on the following parameters:

- **Season:** The season influences the occurrence of debris flows greatly. In the Alps, for instance, triggering of debris flows in winter is not common due to frozen grounds and materials. In addition, precipitation will mainly fall as snow and water is therefore stored. In late summer, more loose material is available through a thicker active layer in permafrost sites (Akerman and Johansson 2008).
- **Material availability:** A previous debris flow event may have emptied the material reservoir and a new debris flow cannot be triggered. Sediment supply and channel recharge rate are important factors controlling debris-flow activity (Bovis and Jakob 1999).
- **Land use/land cover:** In a catchment area or channel overgrown with vegetation, much more or more intense precipitation is needed to trigger a debris flow as compared to bare surfaces. Changes in land use or land cover therefore affect the susceptibility of a catchment to produce debris flows. Cannon et al. (2008) report, for instance, that debris flows are triggered at a much lower rainfall threshold after the slope has been burnt by wildfires.
- **Pre-event conditions:** The meteorological pre-event conditions (e.g. antecedent rainfall) greatly influence the occurrence of events. After periods of prolonged precipitation, the soil is saturated and increased pore-water pressure is leading to much lower thresholds of additional rainfall for the release of an event.

2.3 External Forcing

To release a debris flow, an external forcing is needed that exceeds the threshold value defined by the basic and variable susceptibility. Such triggering events can be meteorological factors, tectonics or slope movements. Meteorological events are the most common triggers in the Alpine environment. A rough estimate of the probability of occurrence of future debris-flow events is often made via the probability of rainfall events. It is obvious that the main limitations to applying statistical approaches are the lack of data and the complexity of the initiation process.

3 Meteorological Factors Influencing the Release of Debris Flows

On slopes and in channels where abundant amounts of loose material are available and where the site is susceptible for the triggering of debris flows, meteorological factors often control the triggering of events. Such meteorological triggers include short, intense precipitation events (rain- or hailstorms), prolonged periods of precipitation or intense snow melt. Surface runoff after such meteorological events supplies water to debris masses which have accumulated in channels, thus increasing pore pressure within the debris mass which may initiate a debris flow (Sukamoto et al. 1982).

Intense rainstorms: High-intensity short-duration rainfalls (e.g. summer thunderstorms) are primary trigger of debris flows in Alpine environments. The amount of rainfall necessary to cause instability is highly dependent on site characteristics such as the thickness, porosity and permeability of the local regolith, catchment size, slope angle or material composition. Minimum total rainfall thresholds have been defined for different sites and regions and values vary considerable. For instance, debris flows in tropical areas are normally only triggered after a total of several hundreds of mm in 24 h (e.g. Brand 1995; Jibson 1989). In contrast, several authors report on periglacial debris flows that were triggered with as little as below 20 mm of precipitation in 24 h (e.g. Rapp 1960; Larsson 1982; Stoffel et al. 2011).

Periods of prolonged rainfall: Low-intensity long-duration rainfall may also bring high rainfall totals to catchments and therefore are able to trigger debris flows. The relationship of intensity and duration in the triggering of debris flows has been studied in many regions of the world. Caine (1980) and Guzzetti et al. (2008) assembled worldwide rainfall data to produce intensity-duration relationships. More regional approaches have been chosen in many regions all over the world (e.g. Cannon et al. 2008; Glade 1998).

Rapid snowmelt: Snowmelt enhanced by rainfall or sudden temperature increase can also lead to increased water infiltration (Wieczorek and Glade 2005). Snowmelt not only supplies water over longer time periods than infiltration by rainfall but may also recharge shallow fractured bedrock and raise pore-water pressures beneath shallow soils, thus triggering debris flows (Mathewson et al. 1990). Therefore, rapid snowmelt or snowmelt coupled with rainfall is an important factor in debris-flow triggering (Decaulne et al. 2005).

4 The Role of Slope Processes for Debris-Flow Initiation

Slope processes play a major role for debris flows both in delivering material and triggering the process itself. They may be divided into fluvial slope processes (erosion due to water runoff) and landslides (transport of material down slope in response to gravitational stress).

Surface runoff on soft rocks, especially when vegetation is sparse or absent, results in the development of rills and subsequent of gullies producing a drainage pattern over the time. Within or at the mouth of such gullies often a considerable amount of material is deposited, which may act as a source for debris flows. Nevertheless sudden slope failures are most important to deliver notable volumes of material to the channel feeding debris-flow activity. This material can originate from a single landslide or more usually from several smaller landslides along the channel path, which are spatially and temporally entrained in a complex manner.

The kind of failure strongly depends on the type of rocks or soils involved, the tectonic overprint, adversely orientated discontinuities, the intensity of weathering etc. Many slopes in more or less homogenous soft rocks (e.g. glacial till) show either single or multiple rotational slides. In case weak layers dip out of the slope, translational failures are common. Translational slides often occur along the boundary of strongly weathered rock (regolith) to less altered bedrock. More or less horizontally layered sediment sequences, which are constrained by a weak layer in depth, may show a combination of rotational and translational slides, i.e. a compound slide (Varnes 1978; Hutchinson 1988; Cruden and Varnes 1996). Seldom one can observe debris topple of slightly cohesive material due to undercutting of the creek's banks (Varnes 1978).

Slope failures take place when the strength of a material is exceeded. The shear strength τ of a frictional material is given by the Mohr-Coulomb-failure criterion in the form

$$\tau = c + \sigma \cdot \tan \varphi \tag{1}$$

where c is the cohesion and φ is the angle of internal friction for the soil or rock mass and σ represents the normal stress.

Undercutting of stream banks by erosion can produce steep flanks which tend to fail in sudden modes due to a lack of support of the toe. Slopes, whose toes are repeatedly eroded by the creek, often are characterized by an uphill retrogression of the landslide activity (Fig. 2).

However, water plays a major role in contributing to failure. Following Terzaghi's (1925) effective stress principle an increase of pore-fluid pressure results

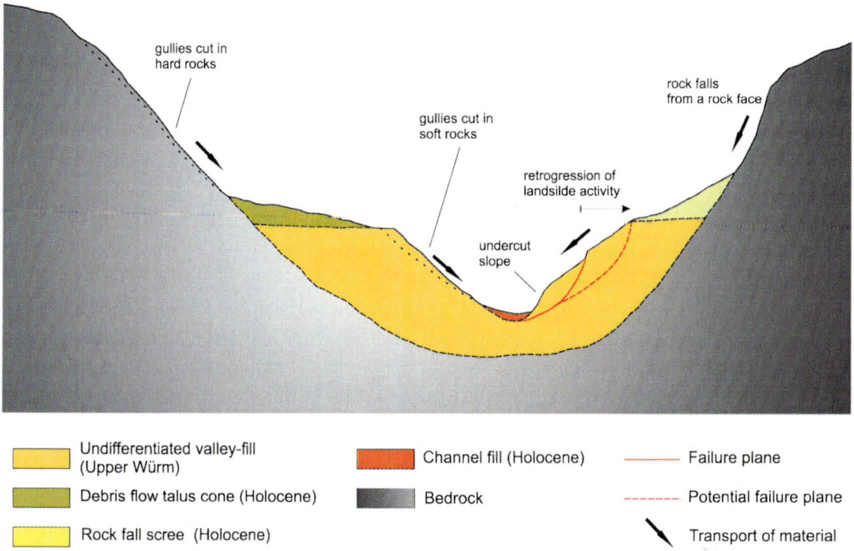

Fig. 2 Schematic cross section through an epigenetic valley in soft rocks of Upper Würm and Holocene age showing different slope processes

in a decrease of the shear strength on potential failure planes and plays therefore a key role for slope stability. The Mohr-Coulomb failure criterion turns to

$$\tau = c + (\sigma - p) \cdot \tan \varphi \tag{2}$$

Herein the total normal stress is reduced by the quantity of the pore-fluid pressure p, reducing the shear resistance on a slip surface in consideration. This mechanism is especially found when loose material is underlain by a less permeable layer like till or bedrock, so the pores of the deposits get quickly filled up by the infiltration of water. Hence fluid pressure increases rapidly and promote failure. This process also can lead to liquefaction of the material due to zero effective stress in the soil mass (Youd 1973; Casagrande 1976; Iverson et al. 1997).

There are many possible triggers for slope failure (WP/WLI 1994 or Cruden and Varnes 1996); the most important ones concerning debris-flow activity in the Alpine environment are undercutting of channel banks and excessive infiltration by prolonged or convective rain fall events. However, the combination of water infiltration from thawing snow along with rain water has turned out to be highly critical causing slopes to fail (e.g. Wieczorek 1996).

In many Alpine catchment areas of the eastern Alps substantial remnants of valley-fills are found, which were mainly deposited in the build-up and decay of glacier ice during the Upper Würm at about 25,000–15,000 years BP, ranging from fluviatile to lacustrine sediments and glacial till (e.g. Penck and Brückner 1909; De Graaff 1996; Pfiffner et al. 1997; van Husen 2000; Reitner 2007). These deposits provide a vast source of sediment of several ten thousand to some million cubic metres for hillslope processes. After deglaciation and lowering of the erosion level in the main valleys tributaries incised into the soft rocks forming epigenetic valleys. The flanks of these valleys are susceptible to a variety of slope processes including erosion due to surface runoff and different kinds of landslides (Fig. 2).

An example of extreme mobilization of hillslope material and delivery to the channel occurred at the Starkenbach creek, western Tyrol, Austria, in August 1999. Here three million cubic metres of scree material became partly liquefied and failed as a debris-flow slide (after the classification of Hungr et al. 2001). The trigger was excessive infiltration of water during several days of intensive precipitation along with an unusual high groundwater table prior to the rainfall event (Huber and Hübl 2000; Huber 2003). The mobilized material run out the valley for some 700 m. Consequently the valley was filled up to a height of 65 m above the old channel bed and dammed the creek (Fig. 3). Within subsequent years convective rainfall events triggered large debris flows by entrainment of the landslide's material.

Channels cutting through toe areas of deep-seated gravitational slope deformations have shown to regularly produce debris flows. Although the rate of movement of such slopes is commonly only within some cm/year, (e.g. at Gradenbach creek (Carinthia, Austria) an average rate of movement of 11 cm/year was recorded, Weidner 2000), slope creep typically results in over-steepened front zones, which become repeatedly unstable and show various failure mechanisms from slides to topples. In the case of the Bretterwandbach-canyon (Tyrol, Austria) many check dams were built since the beginning of the eighteenth century to stabilize the creek's bed to prevent the initiation of debris flows of one of the most active torrents in the eastern Alps. Recent investigations including numerical modelling of the slope with FLAC^{3D} have impressively shown that without the check dams the creek would successively cut into the disintegrated weak rock mass of the creeping slope resulting in rapid slope failures of the canyons flanks. This would provide a total of some 44 million cubic metre of material for debris flows, which could be mobilized over time (Hausmaninger 2004).

5 Channel Processes and Entrainment

Debris flows and torrential floods tend to entrain material in the headwater area and along the steeper reaches of the flow path (transit zone). Massive erosion of the channel bed and the adjacent side-slopes may occur, involving down-cutting of several metres. Trigger mechanisms comprise intensive sediment erosion at steep slopes due to high hydraulic loading, sudden liquefaction of bed material, breakdown of armoured channel beds (coarse grained surface layer or step pool structures) or blockage and subsequent release of saturated material at critical cross-sections. For example, deposition of sediments from contributing channel or gullies can lead to blockage of the main channel (dam). Bursting out of the material, followed by debris flow is possible. Similarly the failure of a natural or artificial dam

Fig. 3 The debris-flow slide of Starkenbach, Austria: Initial debris slide failure (**a**; picture: Rudolf-Miklau), and subsequent debris flow (**b**; picture: die.wildbach)

Fig. 4 Erosion down to the underlying bedrock at the fan apex at Stummerbach (AT) after the passage of a debris flow on July 17th 2010 (picture: Roland Kaitna)

may also transform into a debris flow. In glacial environments the sudden release of water from a moraine-impounded lake (glacial lake outburst flood – GLOF) can turn into a debris flow by intensive *entrainment* of sediment along its way. Debris flows induced by GLOFs are often characterized by high event magnitudes and travel distances (Skermer and VanDine 2005). A special case of debris-flow (mudflow) initiation was reported by Haeberli et al. (2004), where a mudflow of a few million m^3 was initiated subsequent to an ice-rock avalanche event in North Ossetia, Russian Caucasus in September 2002.

Event magnitude, i.e. the total volume of the deposits of a debris-flow surge, is a first order control of the runout distance of debris flows on Alpine fans (e.g. Rickenmann 1999) and can be regarded as key-parameter for runout prediction. Therefore, besides a good assessment of the starting volume of loose debris, also an estimate of possible volume increase by entrainment along its path is an important factor for hazard assessment. Field observations after the passage of debris-flow events often show substantial erosion and even complete stripping of the loose material down to the bedrock underneath, as displayed in Fig. 4. On geologic time scales, the erosive power of debris flows is considered to control channel incision in channels steeper than 3–10% (Stock and Dietrich 2003, 2006). Entrained volumes of loose material by debris flows were assessed by several authors and some values are summarized in Table 1. It can be seen that the increase of volume can be substantial with up to 50 times of the initial volume.

Table 1 Examples of reported initiation and deposition volumes of debris-flow events

Initiation volume [m³]	Deposition volume [m³]	Increase factor	Reference
355	19,000	53 times	King (1996)
3,000	50,000	17 times	Jakob et al. (1997)
266	4,084	16 times	Chen et al. (2006)
600	6,000	10 times	Berti et al. (1999)
25,000	240,000	10 times	Breien et al. (2008)
450[a]	4,000[a]	9 times	Benda (1990)
170,000	520,000	3 times	Scheuner et al. (2009)
30,000	80,000	3 times	Scheidl et al. (2008)

[a]Mean values of several events

Entrainment is the result of mobilization of loose bed material and material originating from the collapse of sidewalls undercut by bed erosion (Hungr et al. 2005). As outlined by Berger et al. (2010), entrainment is expected to be governed by high basal shear stresses due to the hydraulic load during the passage of a debris flow (Takahashi 1991; Armanini and Gregoretti 2000) or by high grain impact (collisional) stresses (Stock and Dietrich 2006; Hsu et al. 2008).

In analogy to the concept of critical dimensionless discharge for bedload transport, a critical discharge for debris-flow initiation as a function of bed slope and characteristic grain diameter has been suggested by different authors (e.g. Tognacca et al. 2000; Takahashi 1987). However, due to the lack of sufficient field data the applicability for practical purposes is restricted.

The mechanism of the erosion process itself may be due to single particle erosion of loose surface material, scour and knick point erosion-migration, or drained or undrained shearing of the bed (Bowman et al. 2010). First time in-situ measurement of debris-flow erosion on an Alpine fan indicates that most of the erosion takes place at the frontal part of a debris flow, where mean and fluctuation stresses are largest (Berger et al. 2010). Subsequent deposition at the later stage of the debris flow led to significant reworking of the channel bed on the fan.

Recent advances in understanding the erosion mechanism of debris flows and its implication for debris-flow runout have been made by Iverson et al. (2011): based on the results of large scale experiments they show that bed material entrainment is positively correlated to bed sediment water content. Simultaneous stress measurements at the bed reveal that the passage of the experimental flow is associated with an increase of fluid pressure due to shear contraction of the bed material, reducing basal friction and enhancing depth scour. Contrary to the expectations, it was observed that entrainment of wet bed material does not lead to a reduction of flow velocity – which would be expected due to the energy consumption for accelerating the entrained material. In fact, the experimental flows gained momentum due to the reduction of basal friction. This important observation reflects the often surprising mobility of debris flows and stresses the importance of material entrainment for runout. Mangeney (2011) argues that, on steep slopes, the fluid pressure effect might be accompanied by granular dynamics between the meta-stable bed material and the overriding highly agitated flows.

Though a realistic estimation of event magnitude and hence potential erosion volumes is needed for debris-flow hazard assessment, only a few guidelines can be found in literature. Hungr et al. (2005) point out that the event volume V can be estimated with

$$V = V_{\mathrm{initial}} + \sum V_{\mathrm{point}} + \sum_{i=1}^{n} Y_i L_i \tag{3}$$

where $V_{initial}$ is the starting volume, V_{point} represents point sources like secondary landslides initiated by the debris flow, and Y_i and L_i are the yield rate and length of n homogeneous channel reaches.

The yield rate may be calculated with $Yi = B_i \times D_i$, where B_i = erosion depth, and D_i = path width. It has to be stated that the sum of secondary landslide contributions as well as the yield rate are hard to assess for future events. Average yield rates of 1–10 m^3/m are frequently found in literature (Hungr et al. 2005), with extreme values exceeding 100 m^3/m (e.g. Rickenmann et al. 2003; Breien et al. 2008). Based on field observations in the Canadian Rockies, Hungr et al. (1984) provide guidelines for deriving values for the yield rate in typical channel reaches based on slope and sediment availability (Table 2). For the Alpine setting, some recommendations have been given by Spreafico et al. (1996). Abanco and Hürlimann (2011) suggest a refined concept for deriving a realistic estimate of the yield rate in homogeneous channel section. The governing factors in this approach are sediment availability, mean slope of the channel reach, cross-section shape in this reach, normalized upstream contributing area, and catchment area. This approach has been tested with some field observations in the Pyrenees and Alps and gives promising results. Another important issue is to define the downstream limits of erosion by debris flows. Hungr et al. (2005) provide an overview of different suggestions given in literature and often a slope around 10–15° is suggested.

Empirical relationships for estimating event volumes show a large scatter of available data (Rickenmann 1999) and should be restricted to indicative estimations in to regions where they have been developed (Marchi and D'Agostino 2004). Therefore, for a realistic estimation of the material volume delivered to the fan, geologic-geomorphic field work is essential. This includes assessment of entrainment along the channel guided by approaches mentioned earlier. Information of past events (if available) and analysis of aerial photos should complete this assessment.

Dynamic debris-flow simulation-tools including initiation, entrainment and deposition are not yet available for engineering purposes. Various flow laws have been proposed in literature to describe the flow and the runout distance of debris flows (e.g. Julien and O'Brien 1997; Hungr 1995; Takahashi 1991; Iverson 1997; Berzi and Jenkins 2008), but only few erosion laws are available (e.g. Egashira et al. 2001; Hungr 1995; Tognacca et al. 2000). In practice, the FLO-2D model has been widely used for debris-flow simulations (Garcia and Lopez 2005;

Table 2 Debris-flow erosion rates in relation to indicative channel parameters modified after Hungr et al. (1984)

Channel type	Gradient [deg]	Bed material	Side slopes	Stability condition	Channel debris yield [m^3/m]	Erodibility coefficient [$m^3/m{\cdot}km$]
A	20–35	Bedrock	Non-erodible	Stable, practically bare of soil cover	0–5	0–5
B	10–20	Thin debris or loose soil over bedrock	Non-erodible (bedrock)	Stable	5–10	5–10
C	10–20	Deep talus or moraine	Less than 5 m high	Stable	10–15	10–15
D	10–20	Deep talus or moraine	Talus, over 5 m high	Side slopes at repose	15–20	15–30
E	10–20	Deep talus or moraine	Talus, over 20 m high	Side slopes potentially unstable (landslide area)	Up to 200 (consider as point source)	Not applicable

Lin et al. 2005; Cetina et al. 2006). Here, a simple rheologic model for predicting viscous and yield stresses as function of sediment concentration is employed and sediment volumes are tracked through the channel system. As sediment concentration changes dilution effects, flow cessation and the remobilization of deposits are simulated. However, for engineering hazard assessment using numerical simulation tools an estimation of the possible volume increase due to entrainment is desirable (Hungr and McDougall 2009).

References

Abanco C, Hürlimann M (2011) Simple geomorphologic approach to estimate debris-flow entrainment. Applications to the Pyrenees and the Alps. In: Genevois R, Hamilton DL, Prestininzi A (eds) 5th international conference on debris-flow hazards mitigation: mechanics, prediction, and assessment, Padua, 14–17 June 2011. Italian Journal of Engineering Geology and Environment-Book, Casa Editrice Universita La Sapienza, Rome, pp 183–197

Akerman HJ, Johansson M (2008) Thawing permafrost and thicker active layers in sub-arctic Sweden. Permafr Periglac Process 19(3):279–292

Armanini A, Gregoretti C (2000) Triggering of derbis-flow by overland flow: a comparison between theoretical and experimental results. In: Wieczorek G, Naeser AA (eds) Proceedings of the 2nd international conference on debris flow hazard mitigation: mechanics, prediction and assessment, Taipei, Taiwan, 16–18 Aug 2000. Balkema, Rotterdam, pp 117–124

Benda L (1990) The influence of debris flows on channels and valley floors in the Oregon Coast Range, USA. Earth Surf Process Landf 15:457–466

Berger C, McArdell BW, Fritschi B, Schlunegger F (2010) A novel method for measuring the timing of bed erosion during debris flows and floods. J Geophys Res 116:F01002. doi:10.1029/2010JF001722

Berti M, Genevois R, Simoni A, Tecca PR (1999) Field observations of a debris flow event in the Dolomites. Geomorphology 29:265–274

Berzi D, Jenkins JT (2008) A theoretical analysis of free-surface flows of saturated granular-liquid mixtures. J Fluid Mech 608:393–410. doi:10.1017/S0022112008002401

Bovis MJ, Jakob M (1999) The role of debris supply conditions in predicting debris flow activity. Earth Surf Process Landf 24(11):1039–1054

Bowman ET, Laue J, Imre B, Springman S (2010) Experimental modelling of debis flow behaviour using a geotechnical centrifuge. Can Geotech J 47:742–762. doi:10.1139/T09-141

Brand EW (1995) Slope instability in tropical areas. In: Bell DH (ed) Proceedings of the 6th international symposium on landslides. Balkema, Rotterdam, pp 2031–2051

Breien H, De Blasio FV, Elverhoi A, Hoeg K (2008) Erosion and morphology of a debris flow caused by a glacial lake outburst flood, Western Norway. Landslides 5:271–280

Caine N (1980) The rainfall intensity – duration control of shallow landslides and debris flows. Geogr Ann Ser A Phys Geogr 62(1–2):23–27

Cannon SH, Gartner JE, Wilson RC, Bowers JC, Laber JL (2008) Storm rainfall conditions for floods and debris flows from recently burned areas in southwestern Colorado and southern California. Geomorphology 96(3–4):250–269

Casagrande A (1976) Liquefaction and cyclic deformation of sands – a critical review, Harvard soil mechanics series 88. Harvard University, Cambridge, 51 pp

Cetina M, Rajar R, Hojnik T, Zakrajesk M, Krzyk M, Mikos M (2006) Case study: numerical simulations of debris flow below Stoze, Slovenia. J Hydraul Eng 132(2):121–130

Chen H, Crosta GB, Lee CF (2006) Erosional effects on runout of fast landslides, debris flows and avalanches: a numerical investigation. Geotechnique 56(5):305–322

Costa JE (1984) Physical geomorphology of debris flows. Springer, Berlin, pp 268–317
Cruden DM, Varnes DJ (1996) Landslide types and processes. In: Turner AK, Schuster RL (eds) Landslides: investigation and mitigation, Special report 247. National Academic Press, Washington, DC, pp 36–45
Decaulne A, Saemundsson P, Petursson O (2005) Debris flow triggered by rapid snowmelt: a case study in the Gleidarhjalli area, northwestern Iceland. Geogr Ann Ser A Phys Geogr 87A(4):487–500
De Graaff LWS (1996) The fluvial factor in the evolution of alpine valleys and of ice-marginal topography in Vorarlberg (W-Austria) during the Upper Pleistocene and Holo-cene. Zeitschrift für Geomorphologie NF Supplement 104:129–159
Egashira S, Honda N, Itoh T (2001) Experimental study on the entrainment of bed material into debris flow. Phys Chem Earth (C) 26(9):645–650
Garcia R, Lopez JL (2005) Debris flows on December 1999 in Venezuela. In: Jakob M, Hungr O (eds) Debris-flow hazards and related phenomena. Springer Verlag Praxis, Berlin
Glade T (1998) Establishing the frequency and magnitude of landslide-triggering rainstorm events in New Zealand. Environ Geol 35(2–3):160–174
Guzzetti F, Peruccacci S, Rossi M, Stark CP (2008) The rainfall intensity-duration control of shallow landslides and debris flows: an update. Landslides 5(1):3–17
Haeberli W, Huggel C, Kääb A, Zgraggen-Oswald S, Polkvoj A, Galushkin I, Zotikov I, Osokin N (2004) The Kolka-Karmadon rock/ice slide of 20 September 2002: an extraordinary event of historical dimensions in North Ossetia, Russian Caucasus. J Glaciol 5(171):533–546
Hausmaninger J (2004) Standsicherheitsuntersuchungen der Einhänge des Bretterwandbaches mittels FLAC3D. Diploma thesis at the Technical University Vienna, Austria
Hsu L, Dietrich WE, Sklar LS (2008) Experimental study of bedrock erosion by granular flows. J Geophys Res 113: F02001. doi:10.1029/2007JF000778
Huber T (2003) Die Großrutschung im Starkenbach am 29.8.1999 in der Gemeinde Schönwies/Tirol. Wildbach- und Lawinenverbau 150:86–95
Huber T, Hübl J (2000) Risk mitigation of a landslide dammed alpine torrent. Geophysical Research Abstracts 2:128 (CD-Rom)
Hungr O (1995) A model for the runout analysis of rapid flow slides, debris flows, and avalanches. Can Geotech J 32:610–623
Hungr O, McDougall S (2009) Two numerical models for landslide dynamic analysis. Comput Geosci 35:978–992
Hungr O, Morgan G, Kellerhals R (1984) Quantitative analysis of debris torrent hazards for design of remedial measures. Can Geotech J 21:663–677
Hungr O, Evans SG, Bovis MJ, Hutchinson JN (2001) A review of the classification of landslides of the flow type. Environ Eng Geosci VII(3):221–238
Hungr O, McDougall S, Bovis M (2005) Entrainment of material by debris flows. In: Jakob M, Hungr O (eds) Debris-flow hazards and related phenomena. Springer Verlag Praxis, Berlin, pp 135–158
Hutchsinson JN (1988) General report: morphological and geotechnical parameters of landslides in the relation to geology and hydrogeology. In: Bonnard C (ed) Proceedings of the 5th international symposium on landslides, vol 1. Balkema, Rotterdam, pp 3–35
Iverson RM (1997) The physics of debris flows. Rev Geophys 35(3):245–296
Iverson RM, Reid ME, LaHusen RG (1997) Debris-flow mobilization from landslides. Annu Rev Earth Planet Sci 25:85–138
Iverson RM, Reid ME, Logan M, LaHusen RG, Godt JW, Griswold JP (2011) Positive feedback and momentum growth during debris-flow entrainment of wet bed sediment. Nat Geosci 4:116–121
Jakob M, Hungr O, Thomson B (1997) Two debris flow with anomalously high magnitude. In: Chen CI (ed) Proceedings of the 1st international conference on debris-flow hazard mitigation, San Francisco, USA, 7–9 Aug 1997. American Society of Civil Engineers (ASCE), New York, pp 382–394

Jakob M, Bovis M, Oden M (2005) The significance of channel recharge rates for estimating debris-flow magnitude and frequency. Earth Surf Process Landf 30(6):755–766

Jibson RW (1989) Debris flows in southern Puerto Rico. In: Schultz AP, Jibson RW (eds) Landslide processes of the eastern United States and Puerto Rico, Geological Society of America special paper 236. Geological Society of America, Boulder, pp 29–55

Julien PY (1995) Erosion and sedimentation. Cambridge University Press, Cambridge

Julien PY, O'Brien JS (1997) Selected notes on debris flow dynamics, recent developments on debris flows, Lecture note in earth sciences. Springer, Berlin, pp 144–162

Kienholz H (1995) Gefahrenbeurteilung und –bewertung: Auf dem Weg zu einem Gesamtkonzept. Schweizer Zeitschrift für das Forstwesen 146(9):701–725

King J (1996) Tsing Shan debris flow. Special project report SPR 6/96. Geotechnical Engineering Office, Hong Kong Government

Larsson S (1982) Geomorphological effects on the slopes of Langyear Valley, Spitsbergen, after a heavy rainstorm in July 1972. Geogr Ann 64A:105–125

Lin M-L, Wang K-L, Huang J-J (2005) Debris flows run off simulation and verification – case study of Chen-You-Lan Watershed, Taiwan. Nat Hazard Earth Syst Sci 5:439–445

Loat R, Meier E (2003) Wörterbuch Hochwasserschutz. Bundesamt für Wasser und Geologie, Biel

Mangeney A (2011) Landslide boost from entrainment. Nat Geosci 4:77–78

Marchi L, D'Agostino V (2004) Estimation of debris-flow magnitude in the Eastern Italian Alps. Earth Surf Process Landf 29(2):207–220

Mathewson CC, Keaton JR, Santi PM (1990) Role of bedrock ground water in the initiation of debris flows and sustained post-storm stream discharge. Bull Assoc Eng Geol 27(1):73–78

Moser M (1980) Zur Analyse von Hangbewegungen in schwachbindigen bis rolligen Lockergesteinen im alpinen Raum anlässlich von Starkniederschlägen. Interpraevent 1:121–148

Penck A, Brückner E (1909) Die Alpen im Eiszeitalter. Tauchnitz, Leipzig

Pfiffner OA, Heitzmann P, Lehner P, Frei W, Pugin A, Felber M (1997) Incision and backfilling of Alpine valleys: Pliocene, Pleistocene and Holocene processes. In: Pfiffner OA, Lehner P, Heitzmann P, Mueller S, Steck A (eds) Deep structure of the Swiss Alps: results of NRP 20. Birkhäuser, Basel, pp 265–288

Rapp A (1960) Recent development of mountain slopes in Kärkevagge and surroundings, Northern Scandinavia, Geografiska Annaler 42, Wiley-Blackwell, pp 65–200

Reitner JM (2007) Glacial dynamics at the beginning of Termination I in the Eastern Alps and their stratigraphic implications. Quat Int 164–165:64–84

Rickenmann D (1999) Empirical relationships for debris flows. Nat Hazard 19(1):47–77. doi:10.1023/A:1008064220727

Rickenmann D, Weber D, Stepanov B (2003) Erosion by debris flows in field and laboratory experiments. In: Rickenmann D, Chen C (eds) Debris-flow hazards mitigation: mechanics, prediction, and assessment. Millpress, Rotterdam, pp 883–894

Scheidl C, Rickenmann D, Chiari M (2008) The use of airborne LIDAR data for the analysis of debris flow events in Switzerland. Nat Hazard Earth Syst Sci 8:1113–1127

Scheuner T, Keusen HR, McArdell BW, Huggel C (2009) Murgangmodellierung mit dynamisch-physikalischem und GIS-basiertem Fliessmodell (in German). Wasser Energie Luft 101:15–21

Skermer NA, VanDine DF (2005) Debris flows in history. In: Jakob M, Hungr O (eds) Debris flow hazards and related phenomena. Praxis Publishing Ltd., Chichester, pp 25–51

Spreafico M, Lehmann C, Naef O (1996) Empfehlung zur Abschätzung von Feststofffrachten in Wildbächen. Teil I: Handbuch, 46p. + Anhang, und Teil II: Fachliche Grundlagen, 113p., Groupe de travail pour l'hydrologie operationelle (GHO), vol 4. Landeshydrologie und –geologie, Bern

Stiny J (1931) Die geologischen Grundlagen der Verbauung der Geschiebeherde in Gewässern. Springer, Wien

Stock JD, Dietrich WE (2003) Valley incision by debris flows: evidence of a topographic signature. Water Resour Res 39. doi:10.1029/2001WR001057

Stock JD, Dietrich WE (2006) Erosion of steepland valley by debris flows. GSA Bull 118(9–10):1125–1148. doi:10.1130/B25902.1

Stoffel M, Bollschweiler M, Beniston M (2011) Rainfall characteristics for periglacial debris flows in the Swiss Alps: past incidences – potential future evolutions. Clim Chang 105(1–2):263–280

Sukamoto YT, Otha T, Noguchi H (1982) Hydrological and geomorphological studies of debris flows on forested hill slopes in Japan. In: Recent developments in the explanation and prediction of erosion and sediment yield. Proceedings of the Exeter symposium, IAHS Publication 137. International Association of Hydrological Sciences, Paris, pp 110–122

Takahashi T (1987) High velocity flow in steep erodible channels. In: Proceedings XXII IAHR Congress, Lausanne, Technical Session A, pp 42–53

Takahashi T (1991) Debris flow, IAHR monograph series. Balkema, Rotterdam

Terzaghi K (1925) Erdbaumechanik auf Bodenphysikalischer Grundlage. Fanz Deuticke, Wien

Tognacca C, Bezzola GR, Minor HE (2000) Threshold criterion for debris-flow initiation due to channel-bed failure. In: Wieczorek G, Naeser AA (eds) Proceedings of the 2nd international conference on debris flow hazard mitigation: mechanics, prediction and assessment, Taipei, Taiwan, 16–18 Aug 2000. Balkema, Rotterdam, pp 89–97

van Husen D (2000) Geological processes during Quaternary. Mitteilungen der Österreichischen Geologischen Gesellschaft 92(1999):135–156

Varnes DJ (1978) Slope movement types and processes. In: Landslides, analysis and control, Natural Academy of Science, special report 176. Natural Academy of Science, Washington, DC, pp 11–35

Weidner S (2000) Kinematik und Mechanismus tiefgreifender alpiner Hangdefomationen unter besonderer Berücksichtigung der hydrologischen Verhältnisse. PhD-thesis, Fried-rich-Alexander-Universität Erlangen-Nürnberg

Wieczorek GF (1996) Landslides: investigation and mitigation. In: Turner AK, Schuster RL (eds) Landslides: investigation and mitigation, Special report 247. National Academic Press, Washington, DC, pp 76–90

Wieczorek GF, Glade T (2005) Climatic factors influencing occurrence of debris flows. In: Jakob M, Hungr O (eds) Debris-flow hazards and related phenomena. Springer, Chichester, pp 325–362

WP/WLI – Working Party on World Landslide Inventory (International Geotechnical Socie-ties UNESCO) (1994) A suggested method for reporting landslide causes. Bull IAEG 50:71–74

Youd TL (1973) Liquefaction, flow and associated ground failure, US Geological Survey Circular 688. US Geological Survey, Washington, DC

Zimmermann M (1990) Debris flows 1987 in Switzerland: geomorphological and meteorological aspects. International Association of Hydrological Sciences, Wallingford, pp 387–393

Zimmermann M, Mani P, Gamma P (1997) Murganggefahr und Klimaänderung – ein GIS-basierter Ansatz. vdf Hochschulverlag, Zürich

Sediment Transport Processes

Helmut Habersack and Andrea Kreisler

1 Introduction and River Scaling Concept

Sediment transport processes have recently gained importance in river engineering, torrent control and reservoir management due to an increasing discrepancy between a surplus of sediments in upstream and a deficit in downstream river sections (Habersack et al. 2010b). This development leads to problems in flood protection (channel change), river engineering (e.g. riverbed degradation), hydropower generation (e.g. reservoir sedimentation) and the ecological status of running waters (e.g. loss of instream structures).

Since sediment transport processes occur at different scales within a river catchment, analysis should take account of scale specific boundary conditions and of their interactions. One framework to assess sediment transport processes is the river scaling concept (RSC, Habersack 2000), which gives also the structure of this book chapter.

The aim of the RSC is to form a basis for the assessment of abiotic and biotic processes in a river basin by proposing a two-phase procedure (i.e. down- and upscaling; Fig. 1). During the downscaling phase an abiotic analysis is performed at each scale and interrelations are studied. The downscaling phase, leading to the hierarchical assessment of processes, is followed by an aggregation (upscaling) of the information gathered at the small scale, e.g. by including numerical sediment transport models.

H. Habersack (✉) • A. Kreisler
Christian Doppler Laboratory for Advanced Methods in River Monitoring, Modeling and Engineering, University of Natural Resources and Life Sciences, Vienna, Austria

Department for Water, Atmosphere and Environment, Institute of Water Management, Hydrology and Hydraulic Engineering, University of Natural Resources and Life Sciences, Vienna, Austria
e-mail: helmut.habersack@boku.ac.at

M. Schneuwly-Bollschweiler et al. (eds.), *Dating Torrential Processes on Fans and Cones*, Advances in Global Change Research 47, DOI 10.1007/978-94-007-4336-6_4,

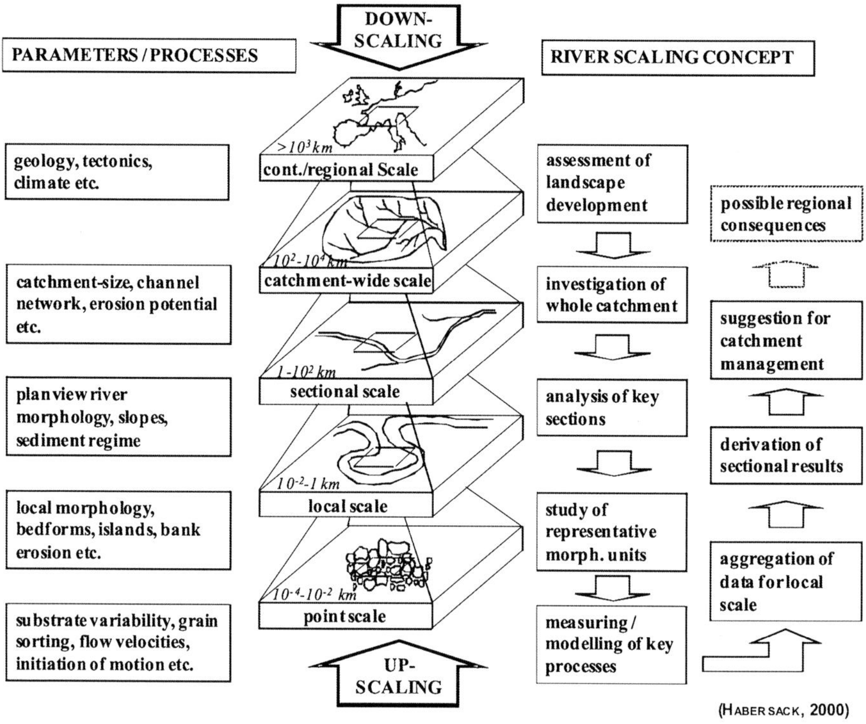

Fig. 1 The River Scaling Concept (RSC; after Habersack 2000) is a framework to assess abiotic and biotic processes in a river basin by proposing a two-phase procedure (i.e. down- and upscaling)

2 Catchment Scale

At the catchment scale tectonics, geologic processes, uplift, denudation rates, climatic change, vegetation cover and relief energy are the essential processes and boundary conditions for sediment transport. Major sources of sediment and the potential amount of mobilized material determine key locations which separate key sections. Mass movements, landslides, debris flows, ground avalanches, land cover (and change) and anthropogenic measures (e.g. land use) determine the existing sediment balance and river morphology at the sectional scale.

3 Sediment Balance and River Morphology at the Sectional Scale

On the sectional scale river reaches are investigated based on key variables. The main focus is on the balance of sediment (input – output comparison) – leading to stable, aggrading or degrading river beds (Habersack 2000). The two most important

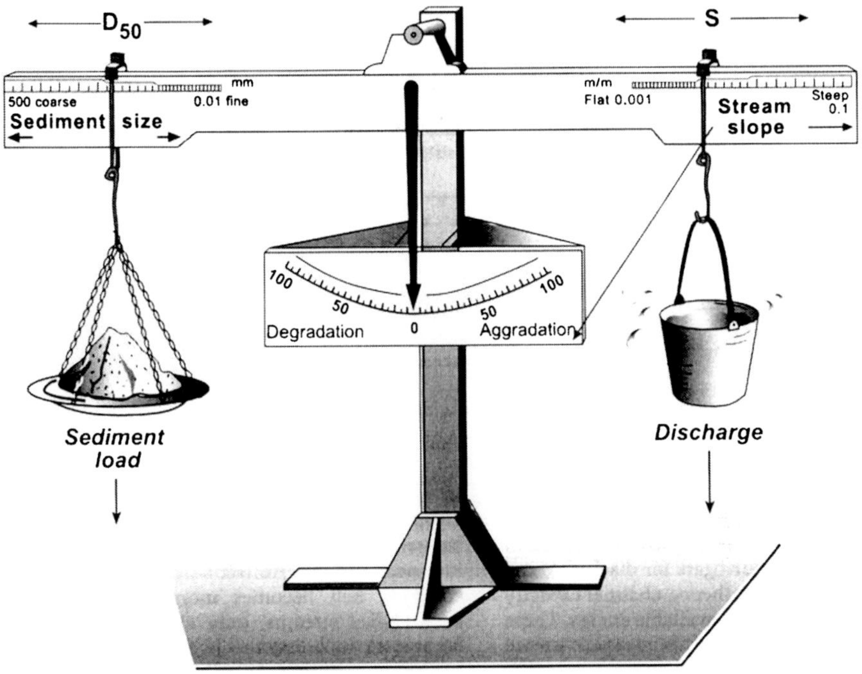

Fig. 2 The balance of a water discharge-sediment load (after Lane 1955) indicates a stable channel, where neither degradation, nor aggradation occurs

Table 1 Gives morphodynamic responses of river channels to changes in the variables water (*Qw*) and sediment (*Qs*) (After Schumm 1977)

Change	River bed morphology	Change	River bed morphology
Qs+ Qw=	Aggradation, channel instability, wider and shallower channel	Qs+ Qw−	Aggradation
Qs− Qw=	Incision, channel instability, narrower and deeper channel	Qs+ Qw+	Processes increased in intensity
Qw+ Qs=	Incision, channel instability, wider and deeper channel	Qs− Qw−	Processes increased in intensity
Qw− Qs=	Aggradation, channel instability, narrower and shallower channel	Qs− Qw+	Incision, channel instability, deeper, wider channel

variables shaping river morphodynamic are water flow and sediment transport. A generalized description of the water flow-sediment balance, indicating a stable channel balance has been proposed by Lane (1955; Fig. 2).

$$\text{Sediment Load} \times \text{Sediment SIZE} \propto \text{Stream SLOPE} \times \text{Stream DISCHARGE}$$

Morphodynamic responses of river channels to varied water flow and sediment discharge are summarized by Schumm (1977; Table 1).

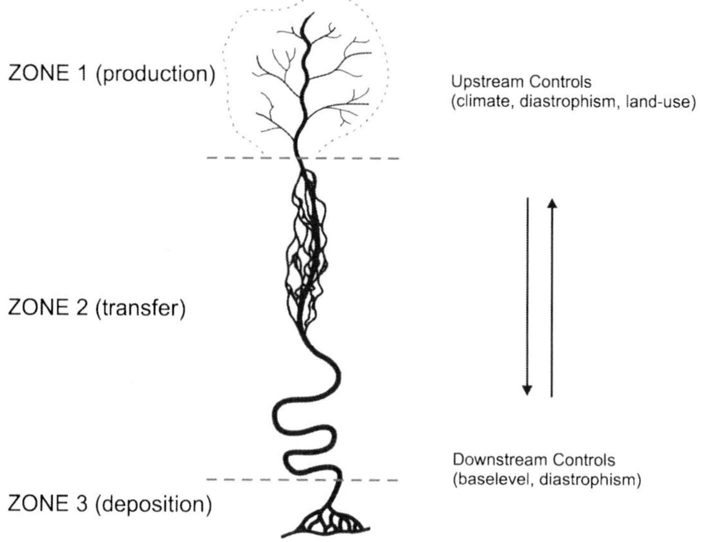

Fig. 3 Schumm (1977) divided the river reach into three zones: sediment production zone, transition zone and deposition zone

Input and output of material, change of channel geometry over time and varying sediment transport within the reach determine the sediment continuity, which is described by the Exner equation:

$$\frac{\delta z}{\delta t} = \frac{\delta q_G}{\delta s} + E_s - S_s \qquad (1)$$

where z is the change in vertical direction, t is time, s is the channel direction, q_G is bedload transport, E_s is entrained sediment and S_s is settled sediment (Zanke 2002).

A division of the river reach into sediment production, transition zone and deposition zone was suggested by Schumm (1977, Fig. 3). The drainage basin, watershed or sediment zone (River Zone 1) is the area where water and sediments are derived. It is the primary zone of sediment production. River Zone 2, referred to as the area of transition, is dominated by a balance between sediment input and sediment output. Sediment deposition occurs in Zone 3.

4 Bedforms, Erosion and Deposition at the Local Scale

Above all local river morphology is characterized by the occurring bedforms as well as erosion and deposition processes.

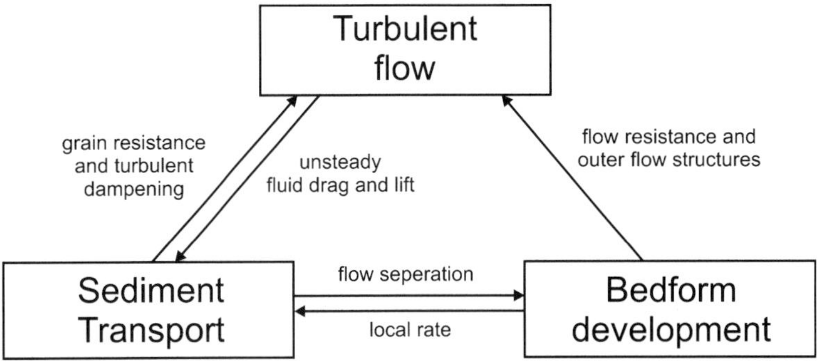

Fig. 4 Complex interaction between turbulent flow, sediment transport and bedform development (After Leeder 1983)

Bedforms have an essential effect on sediment transport and channel stability (Hassan et al. 2008). The complex interaction of bedforms, sediment transport and turbulent flow is displayed in Fig. 4 (Leeder 1983). Stabilizing bedforms are formed if sediment supply is low. They reduce the depth of the bed active layer and the mobility of grains, resulting in reduced sediment transport rates. High sediment supply conditions prevent the development of stabilizing bedforms and therefore the mobility of the grains and the depth of the active layer will increase (Hassan et al. 2008).

Simons and Richardson (1966) presented the development of ripples, dunes and antidunes above sand riverbeds (Fig. 5). Various bed shapes of mountain drainage basins, have been classified by Montgomery and Buffington (1997). The idealized longitudinal profile through the channel network, showing the distribution of these alluvial channel types is illustrated in Fig. 6.

Hassan et al. (2008) presented a bedform classification which is partly based on Lewin (1978), Church and Jones (1982) and Hassan (2005), where the bed of gravel-bed rivers is considered to be composed of storage and resistance elements. These elements are divided into microform, mesoform, macroform and megaform ranging from 10^{-2} to 10^{3} m.

In addition to the development of bedforms, riverbank erosion determines local bed morphology. The main processes in riverbank erosion are fluvial erosion at the bank toe and mass failure in the upper parts of the bank. Fluvial erosion results from the detachment of single grains or aggregates from the bank surface due to shear stresses exerted by the flow, while mass failures are triggered by gravitation (e.g. Thorne 1982; Rinaldi and Darby 2008). Bed topography change during flow events may alter the flow field along the bank and therefore induce large bank retreat (Klösch et al. 2010). In contrast, the formation of in channel berms or benches at the margins of river beds will result in a narrowing of rivers (Pizzuto 2008).

Habersack et al. (Habersack et al. 2010b) and Krapesch et al. (2010) have documented morphological effects of extreme floods causing lateral migration,

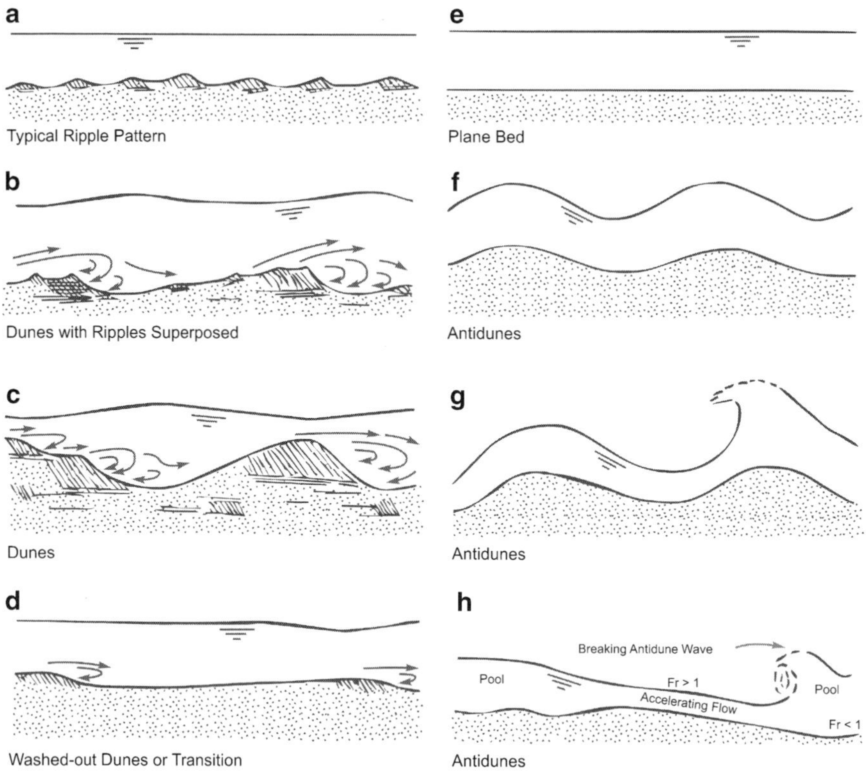

Fig. 5 The development of the bedforms ripples, dunes and antidunes above sand riverbeds has been presented by Simons and Richardson (1966)

erosion and deposition. Their analysis of width ratios (width before and after the flood) in five Austrian rivers points to strong morphodynamic activity and significant correlation between width changes and specific stream power, thus indicating that stream power could serve as a screening tool for the assessment of morphological changes (Fig. 7; Krapesch et al. 2010).

Stream power ω represents available power of the river [W/m²] and is described as (Bagnold 1956):

$$\omega = \frac{\Omega}{B} = \frac{\rho g Q S}{B} \tag{2}$$

where B = channel width, Q = discharge of the stream, S = channel gradient, g = acceleration due to gravity and ρ = density of water.

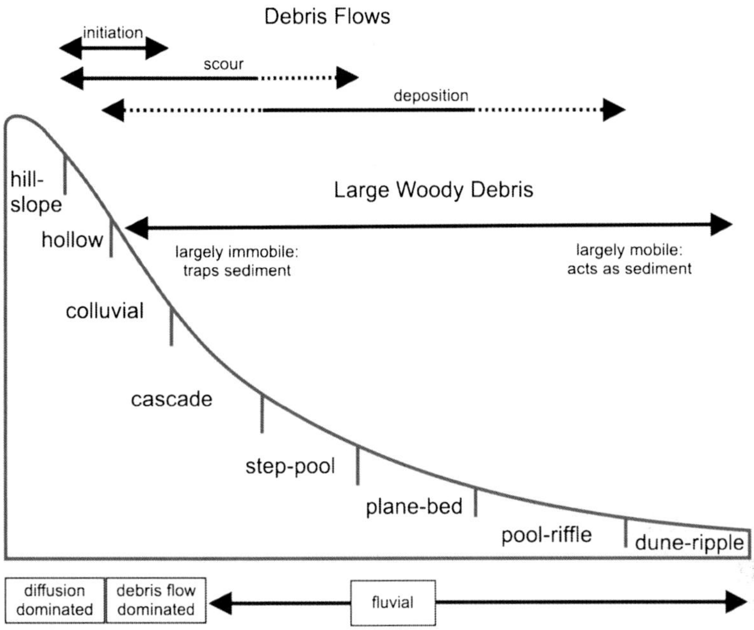

Fig. 6 Montgomery and Buffington (1997) classified various bed shapes of mountain drainage basins. The idealized longitudinal profile is shown in Fig. 6

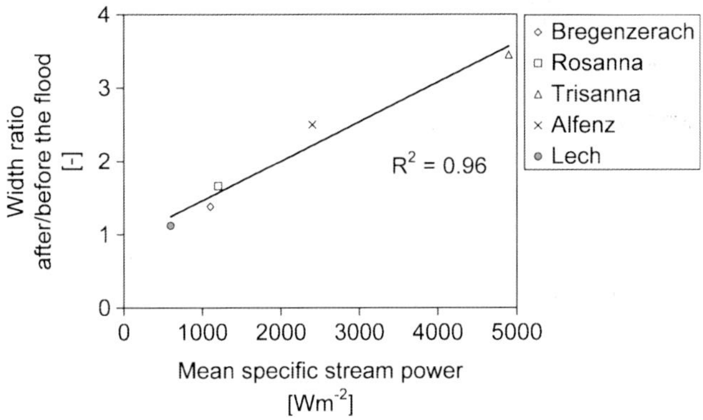

Fig. 7 An analysis (Habersack et al. 2010b; Krapesch et al. 2010) of width ratios (width before and after the flood) in five Austrian rivers shows strong morphodynamic activity and a correlation between width changes and mean specific stream power

5 Sediment Transport at the Point Scale

Figure 8 presents the definitions for open-channel flow over a sediment bed used in the following chapter. A steady, uniform flow is considered with an open-channel flow containing a mean depth H, a mean width B and a mean flow velocity U. The river bed has a mean slope S and surface roughness is characterized by the roughness height k_s, which is proportional to sediments having a mean diameter of D.

The boundary bed shear stress τ_b, a force which operates tangential at the river bed, can be described as:

$$\tau_b = \rho g R S \quad \text{or} \quad \tau_b = \rho g H S \quad \text{if } \frac{B}{H} > 30 \tag{3}$$

where R = the hydraulic radius (García 2008).

The shear velocity u_* is defined as:

$$u_* = \sqrt{\frac{\tau_b}{\rho}} \tag{4}$$

The shear velocity u_* and the boundary shear stress τ_b allow an assessment of the flow intensity. Shear stress $\tau_{(z)}$ increases linearly in the vertical direction from the water surface to the bed wall, where it reaches its maximum τ_b. The shear stress $\tau_{(z)}$, depending on the maximum shear stress and the distance of the river bed can be given by (García 2008):

$$\tau_{(z)} = \tau_b \left(1 - \frac{z}{H}\right) \tag{5}$$

5.1 *Flow Velocity Distribution in Turbulent Flow*

Wall roughness affects the velocity distribution in a turbulent flow, a phenomenon first investigated by Nikuradse (1933). He covered pipes with sand grains and

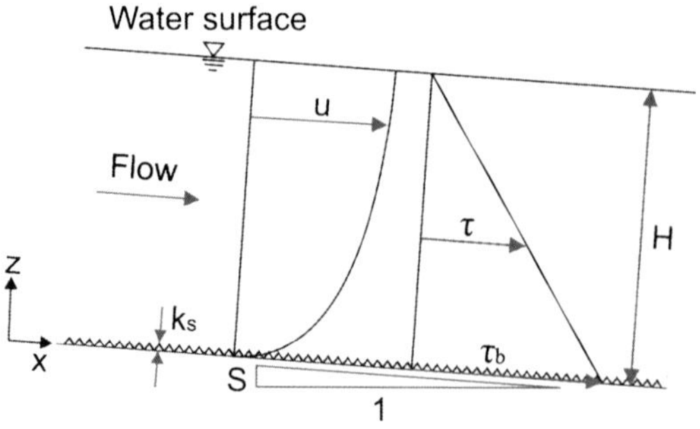

Fig. 8 Definition diagram for open-channel flow over a sediment bed (after García 2008)

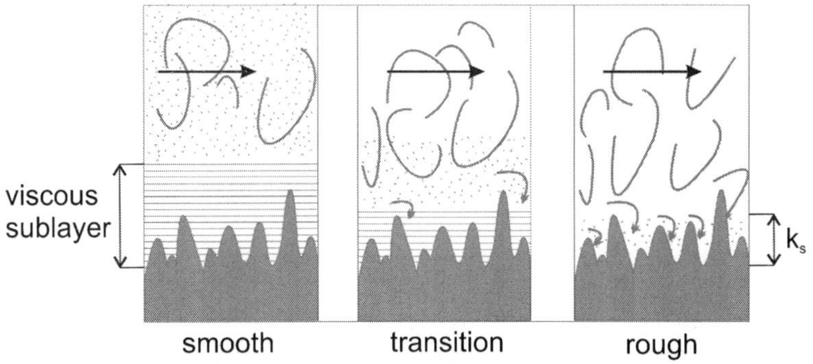

Fig. 9 There are three different hydraulic flow zones: hydraulically smooth, transition zone and hydraulic rough zone. The type of flow regime depends on the ratio of the roughness height k_s and the length scale of the viscous sublayer (after Zanke 2002)

measured velocity distributions at different Reynolds' numbers, pipe diameters and grain sizes (Van Rijn 1993) leading to the concept of equivalent sand grain roughness.

Three different hydraulic regimes exist, namely hydraulically smooth, transitional and rough flows. The type of flow regime depends on the ratio of the roughness height k_s and the length scale of the viscous sublayer (Fig. 9; Zanke 2002).

Generally, alluvial rivers possess *hydraulic rough conditions* (García 2008). At high Reynolds numbers the thickness of the viscous sublayer decreases (Zanke 2002) and roughness elements protrude fully the viscous layer, therefore the impact of the roughness elements is at a maximum (García 2008).

Above a rough surface the mean profile of turbulent flow is described as:

$$\frac{u}{u_*} = \frac{1}{\kappa} \ln\left(\frac{30z}{k_s}\right) \tag{6}$$

where u = time-averaged flow velocity at a distance z above the bed, κ = Karman's constant and is equal to about 0.4.

This equation is Prandtl's "law of the wall". It reveals that flow velocity is proportional to roughness-scaled distance (z) from the wall. The logarithmic wall layer usually applies to the lower 15–20% of the flow. However, in flow over high roughness it may extend close to the surface (Church 2008).

5.2 *Flow Resistance and Bed Roughness*

Bed roughness (k_s) mainly consists of grain roughness (k'_s), generated by skin friction forces and form roughness, (k''_s) caused by pressure forces acting on the bed forms (Van Rijn 1993). In addition, flow resistance is influenced by shape drag (e.g. roughness due to overall channel shape and meander bends; Morvan et al. 2008).

Form friction develops through the separation of the flow from the surface at bedforms. This separation causes the emergence of eddies and rollers on the lee sides of bars indicating pressure differences between the front and the rear side, so that flow resistance develops because of normal pressure acting on the bedform (Einstein 1950). Flow resistance is related to the height, steepness, shape of the bedform (Van Rijn 1993) and other elements of form roughness (e.g. vegetation).

The effect of form roughness is very important in mountain streams with irregular bed and low relative flow depth. Chiari (2008) highlighted the importance of the losses due to form roughness in mountain streams. Analysis of extreme events in Austria and Switzerland and back calculations with the SETRAC model revealed an overestimation of the observed bedload transport by a factor of ~10 on average if form roughness is neglected. The contribution of the form roughness to the total roughness in natural streams is in a range of about 50–90% (Chiari 2008).

To determine channel flow resistance Van Rijn (1993) proposes to use following relation for the grain roughness k'_s:

$$k'_s = 3d_{90} \quad \text{for } \theta \lhd 1 \quad \text{(lower regime)} \tag{7}$$

$$k'_s = 3\theta d_{90} \quad \text{for } \theta \geq 1 \quad \text{(upper regime)} \tag{8}$$

where d_{90} is the grain size of the surface bed material for which 90% of the bed is finer and θ is a mobility parameter defined as:

$$\theta = \frac{u_*^2}{(s-1)\,gd_{50}} \tag{9}$$

in which s is the relative density defined as solid density ρ_s to fluid density ρ.

The Keulegan's resistance law for rough flow (Keulegan 1938) is used to estimate grain-induced resistance. It specifies a relation of the mean velocity U and the roughness scale length k_s and is defined as:

$$\frac{U}{u_*} = \frac{1}{\kappa}\ln\left(11\frac{H}{k_s}\right) \tag{10}$$

The Gaukler-Manning-Strickler (GMS; Strickler 1923) equation is often used for engineering applications. It is an empirical equation which enables the calculation of the mean velocity in a stream:

$$U = k_{St}R^{\frac{2}{3}}S^{\frac{1}{2}} \tag{11}$$

where k_{St} is the Strickler coefficient. The Strickler coefficient compromises the total roughness and is therefore dependent on water depth. The equation of GMS is restricted to uniform, hydraulically rough, fully turbulent flows (García 2008). The Strickler coefficient k_{St} can be deduced from the equivalent sand roughness, if grain roughness dominates:

$$k_{St} = 8.3\sqrt{g}\frac{1}{\sqrt[6]{k_s}} \approx \frac{21}{\sqrt[6]{d_{50}}} \tag{12}$$

Meyer-Peter and Müller (1949) presented the following relation to define the Strickler roughness k_{St}:

$$k_{St} = \frac{26}{\sqrt[6]{d_{90}}} \tag{13}$$

The Strickler coefficient is the reciprocal value of Manning's n which is given by (Brownlie 1983):

$$n = \frac{{k_s}^{\frac{1}{6}}}{8.1g^{\frac{1}{2}}} \tag{14}$$

Equation 16 is not applicable for smooth and very rough surfaces. It is only applicable to medium-range values for the Manning parameter, in a range of $20 < 4R/k_s < 100$ (Tritthart 2005). Values for n are given in Chow (1959) and Yen (1991).

A remarkable change in the resistance of flow is observed when channels with slope gradients steeper than about 1% are considered (Rickenmann and Brauner 2003). Based on stream flow velocity observations covering a wide range of flow conditions, Rickenmann (1996) developed the following equation for the determination of the Stricker coefficient k_{St}, with the Strickler coefficient being expressed by discharge, channel slope and a characteristic grain size of the bed material:

$$k_{St} = \frac{0.97g^{0.41}Q^{0.19}}{S^{0.19}d_{90}^{0.64}} \quad \text{for } S \geq 0.008 \tag{15}$$

$$k_{St} = \frac{4.36g^{0.49}Q^{0.02}}{S^{0.03}d_{90}^{0.23}} \quad \text{for } S \leq 0.008 \tag{16}$$

5.3 *Bedload Transport*

The total load of sediment compromises bed material load and wash load. Bed material load is, according to the ISO standard ISO 4363 defined as "the part of the total sediment transport which consists of the bed material and which rate of movement is governed by the transport capacity of the channel." It is divided into

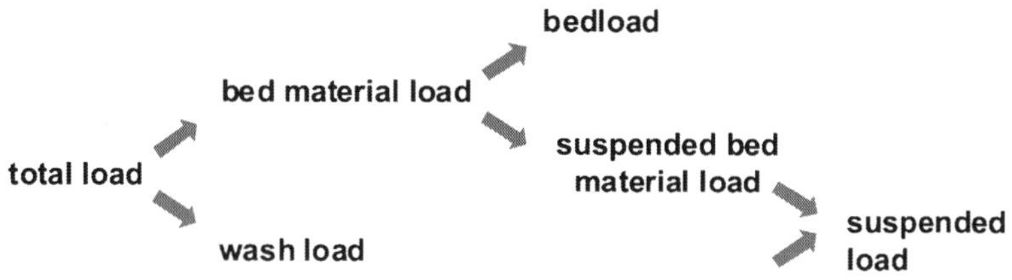

Fig. 10 Classification of the total sediment load into bed material load and wash load

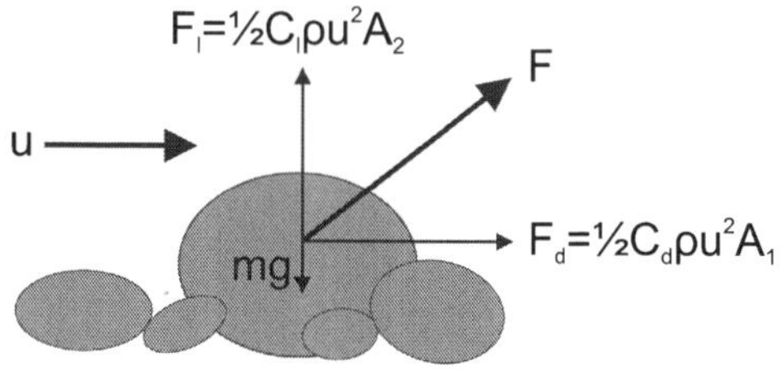

Fig. 11 Entrainment of sediment occurs when the restraining force (mg) of the particle is opposed by the driving forces (F), which obtain lift (F_l) and drag (F_d) components (after Smart and Habersack 2007)

bedload and suspended bed material load. Bedload is the sediment which is almost continuously in contact with the bed, while the suspended bed material load is maintained in suspension by turbulence in the flowing water.

Wash load is composed of particle sizes smaller than those existing in the bed material. The quantity of wash load depends on the rate with which these particles become available in the catchment (ISO 4363; Fig. 10).

Depending on discharge, particle size and flow velocity grains are transported either as bedload or as suspended load. The transition of these modes is continuous. To find a boundary between bedload transport and particles transported in suspension, Kresser (1964) defined a limiting diameter d_{gr} for a particle in relation to the average flow velocity u:

$$d_{gr} = \frac{u^2}{360g} \tag{17}$$

A rough estimation for the limiting grain diameter is given at 1 mm.

5.3.1 Initiation of Motion

Entrainment of sediment occurs when the driving forces imposed by the water flow exceeds the resisting forces of grains (Smart and Habersack 2007). Resisting forces are related to the submerged particle weight and the friction coefficient (Van Rijn 1993). Figure 11 shows the conventional tractive force approach. Here, the restraining force mg of the particle is opposed by the driving force F which obtains lift F_l and drag F_d components. The lift and drag coefficients C_l and C_d, are related to the particle cross-sectional areas A_2 und A_1 (Smart and Habersack 2007).

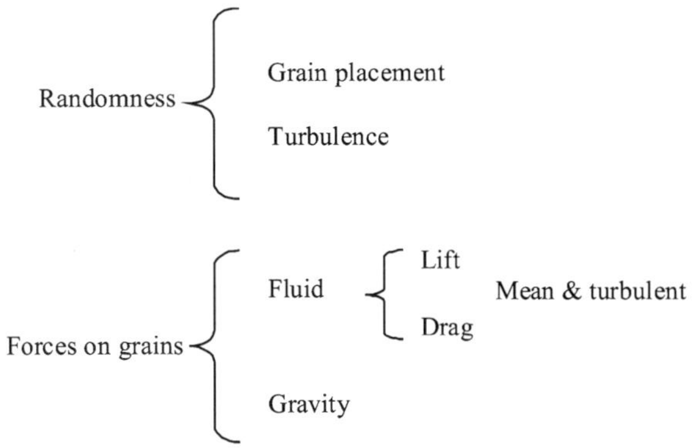

Fig. 12 Factors affecting the entrainment of particles (after García 2008)

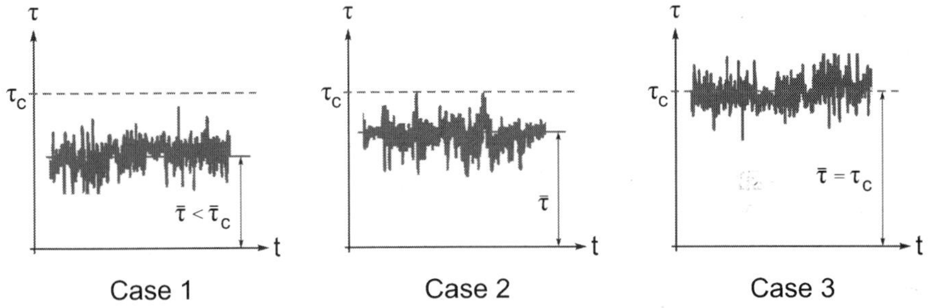

Fig. 13 Random fluctuations of shear stress prevent the definition of a single threshold of motion in turbulent flow. In this figure τ is the maximum effective shear stress and τ_c is the critical shear stress

Factors that affect the entrainment of particles are shown in Fig. 12 (García 2008). Random fluctuations of shear stress, as displayed in Fig. 13, prevent the definition of a single threshold of motion in turbulent flow. In Fig. 13 τ is the maximum effective shear stress and τ_c is the critical shear stress. Particles start moving as soon as the maximum effective shear stress τ reaches the critical shear stress τ_c (Fig. 13, case 2).

Shields (1936) presented conditions for which sediments are stable but on the verge of being entrained (Fig. 13, case 2; Günther 1971). Figure 14 shows the entrainment diagram of Shields (1936), where the Shields parameter θ_c, equal for the grain Froude number Fr* is a function of the grain Reynolds number Re*.

The Shields parameter θ_c, also known as grain Froude number Fr* is defined as:

$$\theta_c = \frac{\tau_c}{(\rho_s - \rho)gD} = Fr^* \qquad (18)$$

The Shields parameter θ_c depends on the hydraulic conditions near the river bed, particle shape and the particle position relative to other particles (Van Rijn

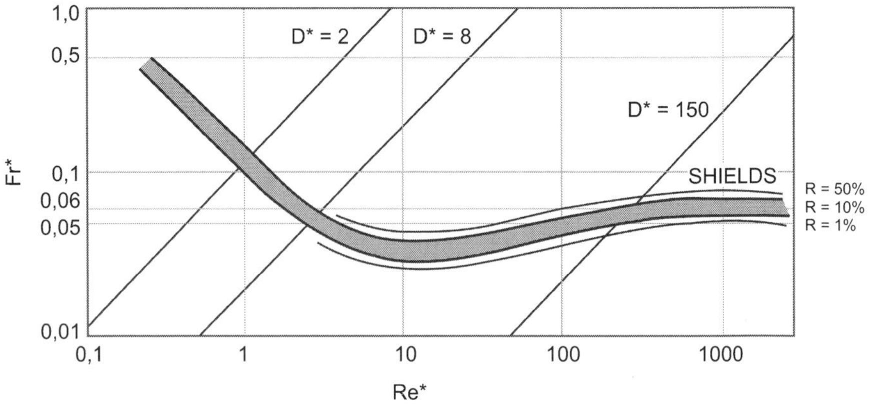

Fig. 14 Entrainment function of Shields (1936), where the grain Froude number Fr* (equal to the Shields parameter θ_c) is a function of the grain Reynolds number Re*

1993). The hydraulic conditions can be expressed by the grain Reynolds number Re*, which is given in Eq. 21. Consequently $\theta_c = f(Re^*)$ and:

$$Re^* = \frac{u_* d}{\nu} \tag{19}$$

For low grain Reynolds numbers ($Re^* < 5$) the grain Froude number is inversely related concerning the initiation of motion. The roughness height is smaller than the thickness of the viscous sublayer. Hence, particles are submerged and therefore not attached by the greater stresses in the turbulent layer (Knighton 1984).

At higher grain Reynolds numbers the roughness elements exceed the thickness of the viscous sublayer. When the roughness elements are surrounded by fully developed turbulence grain, Froude numbers reach a constant value of 0.06 (Figs. 9, 14). Here, Gessler (1971) suggested a value of 0.046.

The Shields diagram has been adapted by Zanke (1990). He assigned the critical shear stress of Shields a risk of motion of 10% (R = 10%). Zanke (2003) assumes that the critical shear stress in laminar flow is only dependent on the angle of inner friction Φ, which lies between 30° for sand and about 45° for angular stones (Zanke 2003).

In turbulent flow the increased shear stress due to fluctuations and lift forces has to be considered: The actual shear stress, which depends on the degree of its random fluctuations, is larger than the time-averaged shear stress (Fig. 13; Zanke 2003).

Further, lift forces develop because of pressure differences across the sediment particles (Smart and Habersack 2007). Coherent structures of the flow near the wall are the main causes of these lift forces. If the weight of a particle is reduced, due to this lift forces, lower shear stresses are required to initiate motion (Zanke 2003). The mechanism of pressure induced lift force is displayed in Fig. 15 (Smart and Habersack 2007).

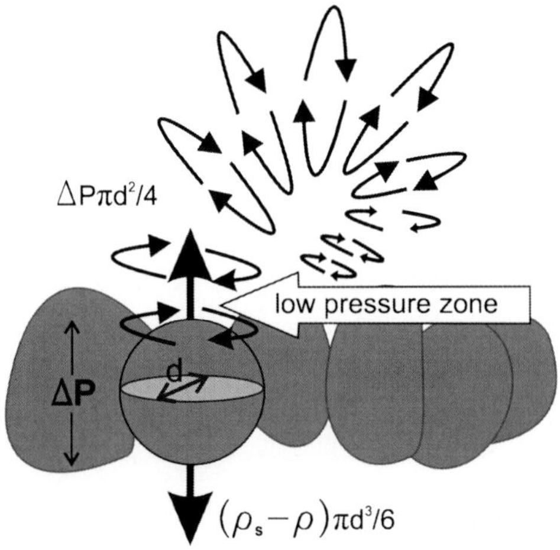

Fig. 15 Lift forces reduce the required shear stresses to initiate motion. The mechanism of pressure induced lift is displayed in Fig. 15 (after Smart and Habersack 2007)

5.3.2 Bedload Transport Process

Sediment particles start to move when the critical shear stress is exceeded, whereby the stochastic behaviour plays a central role (Habersack 2001). Movement first occurs in the form of sliding and rolling over the surface of the bed; saltation of bedload particles only occurs when shear stress increases (García 2008). García (2008) describes saltation as "the unsuspended transport of particles over a granular bed by fluid flow, in the form of consecutive hops within the near-bed region." At higher shear stresses deformation of the surface layer of the bed and movement as a grain flow or granular fluid flow may occur (García 2008).

Einstein (1950) regarded sediment transport as a problem of probability. Bedload transport is not a continuous process, but a discrete one, which is composed of phases of motion and periods of rest (Habersack 2001). These processes can be shown by a permanent monitoring of sediment particles with the application of tracer stones (Liedermann et al. 2011). McEwan et al. (2001) used techniques of discrete particle modelling and active tracers to get insights into the Lagrangian description of particle motion.

Bed material transport shows a high spatial and temporal variability, especially in gravel-bed rivers. Grain movement is intermittent. Figure 16 depicts the distribution of geophone impulses over the cross-section and over time (Habersack et al. 2010a).

Selective entrainment and abrasion are processes responsible for the characteristic downstream changes in bed material (Habersack 1997). The hypothesis of selective entrainment implies an enhancement of coarse grains at the river surface. Varying grain sizes require different critical shear stresses to be entrained. While small grains are already transported, coarse grains still do not move. The coarse fractions enhance at the river surface and develop an armour layer, thereby the

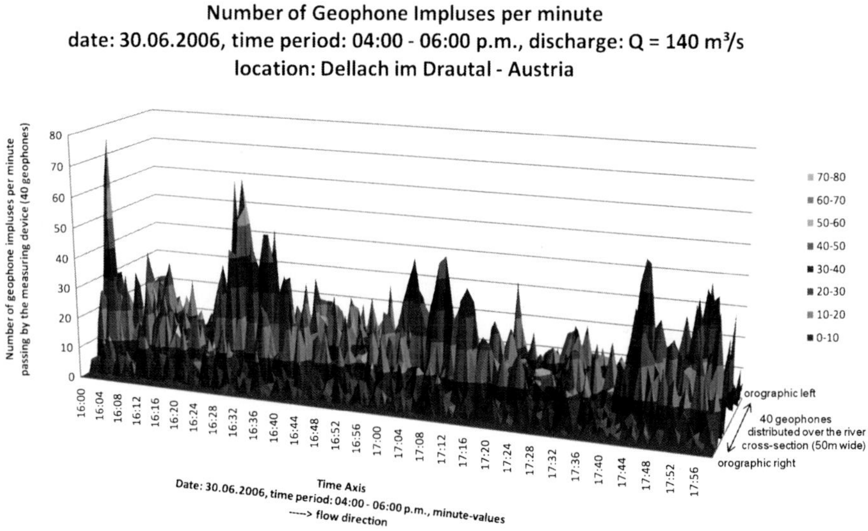

Fig. 16 Spatial and temporal variability of bedload transport, measured with a geophone device at the Drau River (Austria) (Habersack et al. 2010a)

Table 2 Five different cases of development of armour layers (Jäggi 1992)

Case	Keyword	Range	Description
A	Weak transport	$\theta < 0.05$	Fine sediment supplied from upstream, which is not identical to the bed material, moves over a still bed
B	Static armouring	$0.05 < \theta < 0.08$–0.10 no supply	Fine particles are eroded from the bed surface and a stable coarse armour layer forms, no substantial erosion
C	Mobile armouring	$0.05 < \theta < 0.08$–0.10 supply	In the same flow conditions as is case B, material more or less identical to the bed material is supplied
D	Dynamic armouring	$\theta > 0.10$	A stable armour layer can not form, but the coarser particles tend to stay longer in their positions then the fines
E	Full motion	$>>0.1$	For high flow intensities which are high for all grains of a mixture, no different behaviour is to be expected

thickness is approximately equal to the maximum grain size (d_{max}) (Habersack 1997). Jäggi (1992) differentiates between 5 ranges concerning the development of armour layers (Table 2).

The intrinsic sediment theory that smaller grains are more mobile than their coarse counterparts is applicable only at uniform material (Hunziker and Jäggi 2002) which is not the case in nature where a bed mixture contains a wide range of grain sizes. The phenomenon hiding depends upon the relative placement of individual grains at the river surface (Fig. 17). Coarse stones are more exposed to

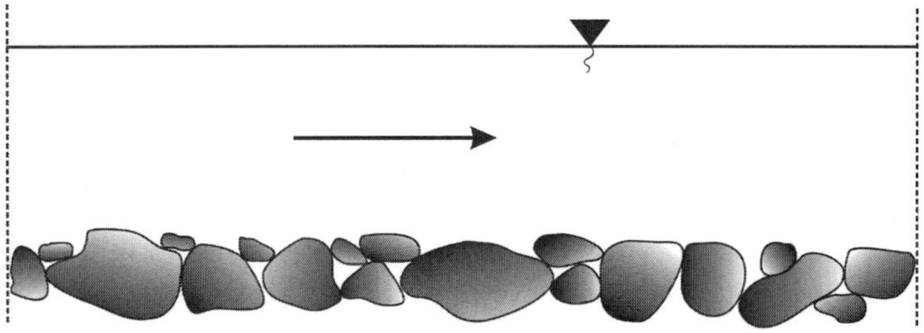

Fig. 17 The phenomenon hiding depends upon the relative placement of individual grains at the river surface. Coarse stones are more exposed to the flow than small ones, which are sheltered by the coarse ones

the flow than small ones (exposure effect). Thus, coarse stones are relatively more mobile in a grain mixture than in uniform sediment. Small stones are sheltered by the coarse ones and therefore less mobile than if surrounded by equally sized sediments (Parker and Klingeman 1982).

Abrasion of sediment particles is proportional to their weight in water and the transported length (Sternberg 1875). Sternberg described abrasion as:

$$d_w = -a_w w d_s \tag{20}$$

where w is the weight of the sediment particle, d_s the transported length and a_w a material constant, which depends on the specific weight and the resistance to abrasion.

A study of Habersack (1999) showed the contribution of both processes to downstream fining of bed material, whereby selective entrainment dominates abrasion.

5.4 *Bedload Transport Equations and Models*

The use of bedload transport equations is restricted to their application range. As shown in Habersack and Laronne (2002) these numerical approaches can differ from measurements. Field data are essential to validate and calibrate bedload transport equations (Habersack et al. 2008). Furthermore, measurements play an important role for the development of additional equations.

Bedload equations have been classified by Graf (1971) into du Boys-type equations (du Boys 1879), which relay on a shear stress relationship, Schoklitsch-type equations (Schoklitsch 1934) which are based on a discharge relationship, and Einstein-type equations (Einstein 1950) that go back to statistical considerations of bedload transport, including lift forces (Habersack and Laronne 2002).

In the following, some bedload transport equations are listed (from Habersack and Laronne 2002):

Basic equation of du Boys (1879):

$$q_{sv} = \chi \tau (\tau - \tau_c) \tag{21}$$

where q_{sv} = volumetric specific bedload discharge and χ = a characteristic sediment coefficient.

Schoklitsch equation (1934):

$$q_s = \frac{7000 S^{\frac{3}{2}}}{d_{40}^{0.5} (q - q_c)} \tag{22}$$

where q_s = specific bedload discharge in mass and q_c = critical specific discharge.

Meyer-Peter and Müller equation (1948):

$$\frac{\gamma R \left(\frac{k}{k'}\right)^{\frac{3}{2}} S}{d_m} - 0.047 (\gamma_s - \gamma) = 0.25^{\frac{1}{3}} \frac{(q'_s)^{\frac{2}{3}}}{d_m} \tag{23}$$

where k = roughness coefficient due to slope S; k' = roughness coefficient due to S' (energy loss due to grain resistance); d_m = representative grain diameter of mixture; q_s' = submerged bedload discharge mass per unit time and width.

Einstein equation (1950):

$$p = 1 - \frac{1}{\pi^{0.5}} \int_{-B_* \psi - 1/\eta_o}^{B_* \psi - 1/\eta_o} e^{-t^2} dt = \frac{A_* \varphi}{1 + A_* \varphi} \tag{24}$$

Using respective intensity values of the bedload and water discharges:

$$\varphi = \frac{q_s}{\gamma_s} \sqrt{\frac{\rho}{\rho_s - \rho} \frac{1}{gd^3}}$$

$$\psi = \frac{\rho}{\rho_s - \rho} \frac{d}{SR'_b} \tag{25}$$

where p = probability of motion; A* = 43.5; B* = 0.143; η0 = 0.5; t is a dummy variable of integration; Φ = intensity of bedload discharge; ψ = flow intensity; Rb = hydraulic radius with respect to the bed and granular boundary.

5.4.1 Bedload Transport Equations for Steep Slopes

Smart and Jäggi (1983) extended the experiments of Meyer-Peter and Müller (1948) to slopes of 20%. The relation of d_{90}/d_{30} corrects the transport capacity accordingly to the grain size distribution. This correction term increases bedload transport slightly and reveals more intensive bedload transport at wider grain size distributions:

$$q_s = 4\left(\frac{\rho_s - \rho}{\rho}\right)^{-1}\left(\frac{d_{90}}{d_{30}}\right)^{0.2} qS^{1.6}\left[1 - \frac{\theta_c\left(\frac{\rho_s-\rho}{\rho}\right)d_m}{HS}\right] \tag{26}$$

where q is the specific discharge for a river width of 1 m [$m^3\ s^{-1}$].

Rickenmann (1990, 1991) performed steep flume experiments to investigate the influence of an increasing fluid density and viscosity on the bedload transport capacity of the flow. Above a limiting grain Reynolds number of about 10 he observed higher bedload transport rates due to raised fluid density. The maximum density of the suspension was about 22.7%, which determines the transition to debris flow. The following bedload transport equation, valid for slopes ranging from 0.0004 to 0.20, has been developed:

$$q_s = 3.1\left(\frac{d_{90}}{d_{30}}\right)^{0.2}(q - q_c)\,S^{1.5}\left(\frac{\rho_s}{\rho} - 1\right)^{-1.5} \tag{27}$$

Another frequently used formula is that of Palt (2001). Flow resistance at steep slopes with low relative water depth is high, compared to flat slopes. Increased flow resistance can be considered applying the reduced energy slope instead of the bed slope S (Rickenmann et al 2006; Chiari et al. 2010).

In general all bedload transport equations contain empirical parameters which need to be estimated carefully and, if possible, should be calibrated by using field data (Habersack et al. 2011). Bedload transport formulas are also used in numerical models. Various models, differing in dimensionality and degree of sophistication have been applied over time (Tritthart et al. 2009). An integrated numerical sediment transport and morphology model was presented in Tritthart et al. (2009, 2011).

6 Conclusions

This chapter summarizes main processes and formulas related to sediment transport with a special focus on steeper sloped channels. Of course, no complete overview can be given within the available space of this chapter. For practical purposes the following conclusions can be derived:

- Sediment transport occurs at different scales (hierarchically dependent); thus a scale oriented approach, starting with the catchment scale and leading to the point scale, followed by an upscaling phase should be applied (e.g. River Scaling Concept)
- The sediment transport formulas must be selected according to the given boundary conditions (e.g. channel slope, grain size)
- Only a fractional calculation covers the existing wide range of grain sizes and interactions (e.g. hiding – exposure)
- Often there exists a difference between the potential bedload transport capacity and the effectively occurring transport (e.g. in supply limited systems)
- A minimum field data set on bedload transport should be available to calibrate and validate the selected sediment transport formulas
- Numerical sediment transport models gain increasing importance to solve hydraulic engineering problems (the dimensions necessary depend on the scale and practical questions to be answered)
- For specific questions related to sediment transport still physical model studies are necessary

Finally for calculating sediment transport and calibrating as well as validating numerical simulations of morphodynamic changes of fans and cones the documentation of past events is essential. The documentation should contain the transported sediment volumes based on geometry changes, grain sizes of transported particles and – if possible – sediment transport data from the field.

References

Bagnold RA (1956) The flow of cohesionless grains in fluids. Philos Trans R Soc Lond Ser A Math Phys Sci 249(964):235–297

Brownlie WR (1983) Flow depth in sand-bed channels. J Hydraul Eng 109(7):959–990

Chiari M (2008) Numerical modelling of bedload transport in torrents and mountain streams. PhD thesis, Institute of Mountain Risk Engineering Vienna University of Natural Resources and Applied Life Sciences

Chiari M, Friedl K, Rickenmann D (2010) A one dimensional bedload transport model for steep slopes. J Hydraul Res 48(2):152–160

Chow VT (1959) Open channel hydraulics. McGraw-Hill, New York

Church M (2008) Multiple scales in rivers. In: Habersack H, Piégay H, Rinaldi M (eds) Gravel-Bed rivers VI – from process understanding to river restoration. Elsevier, Amsterdam, pp 3–32

Church M, Jones D (1982) Channel bars in gravel bed rivers. In: Hey RD, Bathurst JD, Thorne CR (eds) Gravel Bed rivers fluvial processes, engineering and management. Wiley, Chichester, pp 291–338

du Boys MP (1879) Etudes du régime et l'action exercée par les eaux sur un lit à fond de gravière indéfiniment affouiable. Ann Ponts Chaussees 5(18):141–195

Einstein HA (1950) The bedload function for bedload transportation in open channel flows. Technical bulletin 1026. U.S. Department of Agriculture, Washington, DC

García MH (2008) Sediment transport and morphodynamics. In: García MH (ed) Sedimentation engineering: processes, measurements, modeling, and practice, Manuals and reports on engineering practice No. 110. ASCE, Reston

Gessler J (1971) Beginning and ceasing of sediment motion. In: Shen Littleton HW (ed) River mechanics. Water Resources Publications, Littleton
Graf WH (1971) Hydraulics of sediment transport. McGraw-Hill, New York
Günther A (1971) Die kritische mittlere Sohlenschubspannung bei Geschiebemischungen unter Berück-sichtigung der Deckschichtbildung und der turbulenzbedingten Sohlenschub-spannungsschwankugen. PhD thesis, Versuchsanstalt für Wasserbau, Hydrologie und Glaziologie. Zürich ETH
Habersack H (1997) Catchment-wide, sectional and local aspects in sediment transport modelling and monitoring. J Sediment Res 12(3):120–130
Habersack H (1999) Relative Bedeutung von Abrieb und selektivem Transport in einem anthropogen veränderten Fließgewässer Zeitschrift für Kulturtechnik und Landentwicklung, 40(4):145–192
Habersack H (2000) The river scaling concept (RSC): a basis for ecological assessments. Hydrobiologia 422–423:49–60
Habersack H (2001) Radio-tracking gravel particles in a large braided river in New Zealand: a field test of the stochastic theory of bed load transport proposed by Einstein. J Hydrol Process 15(3):377–391
Habersack H, Laronne J (2002) Evaluation and improvement of bed load discharge formulas based on Helley-Smith sampling in an alpine gravel bed river. J Hydraul Eng 128(5):484–499
Habersack H, Seitz H, Laronne JB (2008) Spatio temporal variability of bedload transport rate: analysis and 2D modelling approach. J Geodinamica Acta 21(1–2):67–79
Habersack H, Seitz H, Liedermann M (2010a) Integrated automatic bedload transport monitoring. In: Gray JR, Laronne JB, Marr JDG (eds) Bedload-surrogate monitoring technologies, SIR 2010-5091. U.S. Geological Survey, Reston, pp 218–235
Habersack H, Schober B, Krapesch G, Jäger E, Muhar S, Poppe M, Preis S, Weiss M, Hauer C (2010b) Neue Ansätze im integrierten Hochwassermanagement: Floodplain Evaluation Matrix FEM, flussmorphologischer Raumbedarf FMRB und räumlich differenziertes Vegetationsmanagement. VeMaFLOOD Österreichische Wasser und Abfallwirtschaft 62(1–2):15–21
Habersack H, Tritthart M, Hengl M, Lalk P, Rickenmann D, Knoblauch H, Badura H, Gabriel H (2011) River modelling – sediment transport and river morphology. ÖWAV Arbeitsbehelf
Hassan MA (2005) Characteristics of gravel bars in ephemeral streams. J Sediment Res 75:29–42
Hassan MA, Smith BJ, Hogan DL, Luzi DS, Zimmermann AE, Eaton BC (2008) Sediment storage and transport in coarse bed streams: scale considerations. In: Habersack H, Piégay H, Rinaldi M (eds) Gravel-Bed rivers VI – from process understanding to river restoration. Elsevier, Amsterdam, pp 473–497
Hunziker RP, Jäggi M (2002) Grain sorting processes. J Hydraul Eng 128(12):1060–1068
Jäggi M (1992) Sedimenthaushalt und Stabilität von Flussbauten. Mitteilungen der Versuchsanstalt für Wasserbau, Hydrologie und Glaziologie an der ETH Zürich Band 119. Versuchsanstalt für Wasserbau, Hydrologie und Glaziologie, Zürich
Keulegan GH (1938) Laws of turbulent flow in open channels. J Nat Bur Stand 21:707–741
Klösch M, Tritthart M, Habersack H (2010) Modeling of near-bank flow velocities during flow events as basis for developing bank erosion equations. In: Dittrich A, Koll K, Aberle J, Geisenhainer P (eds) River flow 2010 – proceedings of the [fifth] international conference on fluvial hydraulics, Braunschweig, Germany, 2010, pp 1301–1308
Knighton D (1984) Fluvial forms and processes. Edward Arnold, London, 218 pp
Krapesch G, Hauer C, Habersack H (2010) Scale orientated prediction of river width changes due to extreme flood hazards. Nat Hazard Earth Syst Sci 11:1–12
Kresser W (1964) Gedanken zur Geschiebe- und Schwebstoffführung der Gewässer. Österreichische Wasserwirtschaft 16, pp 6–11
Lane EW (1955) The importance of fluvial morphology in hydraulic engineering. Am Soc Civil Eng 81:1–17
Leeder MR (1983) On the interactions between turbulent flow, sediment transport and bedform mechanics in channelized flows. Mod Anc Fluv Syst 6:5–18
Lewin J (1978) Floodplain geomorphology. Prog Phys Geogr 2:408–437

Liedermann M, Tritthart M, Habersack H (2011) Particle path characteristics at the large gravel-bed river Danube: results from a tracer study and numerical modelling, Earth Surf Process Landf (submitted)
McEwan IK, Habersack H, Heald JGC (2001) Discrete particle modelling and active tracers: new techniques for studying sediment transport as a Lagrangian phenomenon. In: Mosley VMP (ed) Gravel-Bed rivers. New Zealand Hydrological Society, Wellington
Meyer-Peter E, Müller R (1948) Formulas for bed-load transport. In: Proceedings of the 2nd meeting of the IAHR International Association for Hydraulic Structures Research, Stockholm, Sweden, pp 39–64
Meyer-Peter E, Müller R (1949) Eine Formel zur Berechnung des Geschiebetriebes. Schweiz Bauztg 67(3):29–32
Montgomery DR, Buffington JM (1997) Channel-reach morphology in mountain drainage basins. Geol Soc Am Bull 109:596–611
Morvan HD, Knight D, Wright N, Tang X, Crossley A (2008) The concept of roughness in fluvial hydraulic and its formulation in 1D, 2D and 3D numerical simulation models. J Hydraul Eng 46(2):191–208
Nikuradse J (1933) Laws of flow in rough pipes, Technial memorandum 1292. National Advisory Comitee for Aeronautics, Washington, DC
Palt S (2001) Sedimenttransportprozesse im Himalaya-Karakorum und ihre Bedeutung für Wasserkraftanlagen. PhD thesis, University of Karlsruhe
Parker G, Klingeman PC (1982) On why gravel bed streams are paved. Water Resour Res 18(5):1409–1423
Pizzuto JE (2008) Streambank erosion and river width adjustment. In: Garcia M (ed) Sedimentation engineering. ASCE, Reston, pp 387–439
Rickenmann D (1990) Bedload transport capacity of slurry flows at steep slopes. Dissertation ETH Nr. 9065, Zürich Mitteilungen Nr. 103 der Versuchsanstalt für Wasserbau, Hydrologie un Glaziologie der ETH Zürich
Rickenmann D (1991) Hyperconcentrated flow and sediment transport at stepp slopes. J Hydraul Eng 117(11):1419–1439
Rickenmann D (1996) Fliessgeschwindigkeit in Wildbächen und Gebirgsflüssen. Wasser Energie Luft 88(11/12):298–304
Rickenmann D, Brauner M (2003) Ansätze zur Abschätzung des Geschiebetransportes in Wildbächen und Gebrigsflüssen (Kompendium für das Projekt ETAlp). Wien, Institut für Alpine Naturgefahren und Forstliches Ingenieurwesen, Universität für Bodenkultur, Wien
Rickenmann D, Chiari M, Friedl K (2006) Setrac – a sediment routing model for steep torrent channels. In: Ferreira R, Alves E, Leal J, Cardoso A (eds) River flow 2006. Taylor & Francis, London, pp 843–852
Rinaldi M, Darby SE (2008) Modelling river-bank erosion processes and mass failure mechanisms:progress towards fully coupled simulations. In: Habersack H, Piégay H, Rinaldi M (eds) Gravel-Bed rivers VI – from process understanding to river restoration. Elsevier, Amsterdam, pp 703–737
Schoklitsch A (1934) Der Geschiebetrieb und die Geschiebefracht. Wasserwirtschaft 39(4):1–7
Schumm SA (1977) The fluvial system. Wiley, New York
Shields A (1936) Anwendung der Ähnlichkeitsmechanik und der Turbulenzforschung auf die Geschiebebewegung, Mitteilungen der Preußischen Versuchsanstalt für Wasser-, Erd- und Schiffbau 26. Triltsch & Huther, Berlin
Simons DB, Richardson EV (1966) Resistance to flow in alluvial channels, US geological survey professional paper 422. US Government Printing Office, Washington, DC
Smart GM, Habersack H (2007) Pressure fluctuations and gravel entrainment in rivers. J Hydraul Res 45(5):661–673
Smart GM, Jäggi MNR (1983) Mitteilungen der Versuchsanstalt für Wasserbau, Hydrologie und Glaziologie der Eidgenössischen Technischen Hochschule Zürich 64. Edigenössischen Technischen Hochschule, Zürich, pp 9–188

Sternberg H (1875) Untersuchungen über Längen- und Querprofil geschiebeführender Flüsse. Zeitschrift für Bauwesen 25, pp 483–506

Strickler A (1923) Beiträge zur Frage der Geschwindigkeitsformel und der Rauhigkeitszahlen für Ströme, Kanäle und geschlossene Leitungen, Mitteilungen des Eidgenössischen Amtes für Wasserwirtschaft 16. Eidg. Amt für Wasserwirtschaft, Bern

Thorne CR (1982) Process and mechanism of riverbank erosion. Wiley, Chichester

Tritthart M (2005) Three-dimensional numerical modelling of turbulent river flow using polyhedral finite volumes. Phd thesis, Institut für Wasserbau und Ingenieurhydrologie, Fakultät für Bauingenieurwesen. Wien, Technischen Universität Wien

Tritthart M, Schober B, Liedermann M, Habersack H (2009) Development of an integrated sediment transport model and its application to the Danube River. 33rd IAHR Congress: Water Engineering for a Sustainable Environment, Vancouver, Canada

Tritthart M, Schober B, Habersack H (2011) Non-uniformity and layering in sediment transport modelling 1: flume simulations. J Hydraul Res. 49(3):325–334

Van Rijn LC (1993) Principles of sediment transport in rivers, estuaries and coastal seas. Aqua Publications, Amsterdam

Yen BC (1991) Channel flow resistance: centennial of Manning's formula. Water Resource Publications, Colorado

Zanke UCE (1990) Der Beginn der Geschiebebewegung als Wahrscheinlichkeitsproblem. Wasser & Boden 42(1):40–43

Zanke UCE (2002) Hydromechanik der Gerinne und Küstengewässer. Parey, Berlin

Zanke UCE (2003) On the influence of turbulence on the initiation of sediment motion. Int J Sediment Res 18(1):17–31

Debris-Flow Runout and Deposition on the Fan

Dieter Rickenmann and Christian Scheidl

1 Introduction

Debris flows are at the interface of several research directions dealing with natural hazards processes. It is therefore not surprising that methods for the prediction of flow and runout of debris flows have similarities to approaches originally developed for snow or debris avalanches and streamflow hydraulics (Körner 1980; Lied and Bakkehoi 1980; Perla et al. 1980; Iverson 1997). However, debris-flow volume and bulk flow behaviour may change during travel through the channel, e.g. by entrainment of loose sediment and/or incorporation of water from a tributary. At present, no generally applicable model is able to cover the range of all possible material mixtures and event scenarios. This complexity results in different torrential processes and results in a large variety of approaches to predict debris-flow mobility.

This chapter is divided into three parts. We first present classification methods to distinguish between different torrential processes. In the second part we have a closer look at the deposition patterns of "larger" debris-flow events, characterized by an overtopping of the main channel on the fan. Using data on debris-flow events from the European Alps and British Columbia, Canada, the variation of flow mobility during the runout phase is discussed for diverse process types. A particular focus is on the semi-empirical approach between deposition area and total volume which is based on the assumption of geometric similarity of deposits (Iverson et al. 1998). In the third part we give an overview of runout prediction methods for debris flows.

D. Rickenmann (✉)
Swiss Federal Research Institute WSL, CH-8903 Birmensdorf, Switzerland
e-mail: dieter.rickenmann@wsl.ch

C. Scheidl
Institute of Mountain Risk Engineering, Department of Civil Engineering and Natural Hazards, University of Natural Resources and Life Sciences-Vienna, A-1190 Vienna, Austria
e-mail: christian.scheidl@boku.ac.at

M. Schneuwly-Bollschweiler et al. (eds.), *Dating Torrential Processes on Fans and Cones*, Advances in Global Change Research 47, DOI 10.1007/978-94-007-4336-6_5,

2 Process Types Associated with Torrential Events

Torrential fans are formed by sediment accumulations due to mass movements and sediment transfer processes such as fluvial sediment transport (FST), debris floods (DFL) or debris flows (DF). The terms debris flow and debris flood are used according to the classification proposed by Hungr et al. (2001). Debris floods and fluvial sediment transport typically involve more water than sediment (by volume) in the flow and the first type has a larger sediment concentration than the latter. A debris flood in a torrent is characterized by substantial transport of coarse sediment, while hyperconcentrated flow is usually associated with substantial fine sediments in suspension. In practice, these flow processes are often classified on the basis of geomorphic investigations in the fields (e.g. Costa 1988; Hübl et al. 2002).

For hazard assessment it is important to consider the magnitude-frequency relationship of torrential events, which is likely to depend on process type. We may expect that the deposition behaviour on the fan will be influenced by the process type as well. In this section we provide some support for this assumption. In a previous study on estimating debris-flow deposition on fans (Scheidl and Rickenmann 2010), data on more than 100 torrential events in Austria, South Tyrol (Italy) and Switzerland were classified as debris flows, debris floods, and fluvial sediment transport. We use these data here to show that there is a more or less continuous transition from fluvial bedload transport to debris flows, as is also suggested from a large number of related observations from the 2005 flood events in Switzerland (Rickenmann and Koschni 2010).

Debris flows typically occur in steep and small torrent catchments. Rickenmann et al. (2008) suggested that a torrent is defined as a stream with a catchment area less than 25 km^2 and a mean channel gradient larger than 5–10%. While debris flows occur in torrent catchments, mountain rivers with fluvial bedload transport are characterized by a larger catchment area and a smaller mean channel gradient. This is confirmed by a study of Marchi and D'Agostino (2004) studying 127 debris flows throughout the North-Eastern Italian Alps, out of which 125 events exceeded a longitudinal slope of 10% with catchment areas smaller than 20 km^2. Similarly Marchi and Brochot (2000) report a maximum catchment area of 20–30 km^2 for debris flows and debris-flood events in the French Alps. For the documented debris-flow events of 1987 in Switzerland (Rickenmann and Zimmermann 1993), the maximum catchment area was 22 km^2 with a minimum mean channel slope of 19%. The data of Scheidl and Rickenmann (2010) largely confirm that debris flows and debris floods are likely to occur in catchments smaller than 25 km^2 and with a mean channel gradient steeper than 10% (Fig. 1). Figure 1 also includes two well documented debris-flow events from British Columbia, Canada. If catchment area is considered as a proxy for discharge, this parameter together with the channel gradient essentially determine the sediment concentration to be expected during high flow events (Rickenmann and Koschni 2010). Thus the proposed criterion to separate torrent catchments from mountain rivers appears to be reasonable.

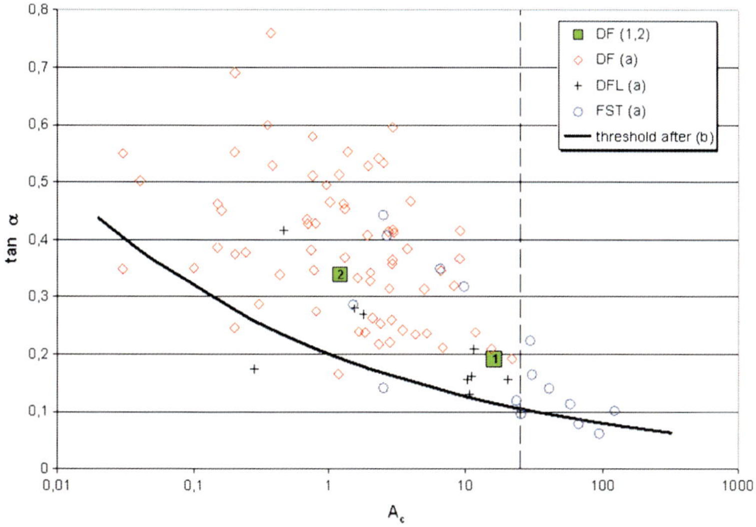

Fig. 1 Relation between mean travel angle and catchment area for the data of Scheidl and Rickemann (2010), denoted with (*a*), with the *threshold line* (*b*) of Zimmermann et al. (1997). (*1*) Hummingbird Creek (Jakob et al. 2000) and (*2*) Cathedral Gulch (Jackson et al. 1989). The *dashed line* represents the proposed limiting catchment area of 25 km^2

However, this general distinction between torrents and mountain rivers does not mean that only debris flows will occur in torrent channels. A possible classification of sediment transfer processes in torrent catchments may be based on the Melton number M_E (Melton 1965) and the average fan slope S_f (Marchi and Brochot 2000; Bardou 2002). The Melton number is an indicator for the basin ruggedness and has also been used in other studies delimiting debris-flow occurrence (Jackson et al. 1987; Crosta and Frattini 2004). M_E is determined as the maximum elevation difference in a catchment divided by the square root of the catchment area A_c. Here the difference in elevation is calculated between the elevation of the highest point within the catchment, $Elev_{max}$, and the elevation of the fan apex, $\mathrm{Elev}_{\mathrm{min}}$:

$$\mathrm{M_E} = \frac{\mathrm{Elev_{max}} - \mathrm{Elev_{min}}}{\sqrt{\mathrm{A_c}}} \tag{1}$$

Bardou (2002) proposed three different zones (A, B, and C) to distinguish between torrential process types. Figure 2 illustrates this delineation using data from Scheidl and Rickenmann (2010), based on 17 fluvial sediment transport processes, 10 debris floods and 106 debris flows, along with data from four independently documented debris-flow events in British Columbia.

The threshold line between zone A and B separates fluvial sediment transport processes from mixed or transitional processes (debris floods, hyperconcentrated

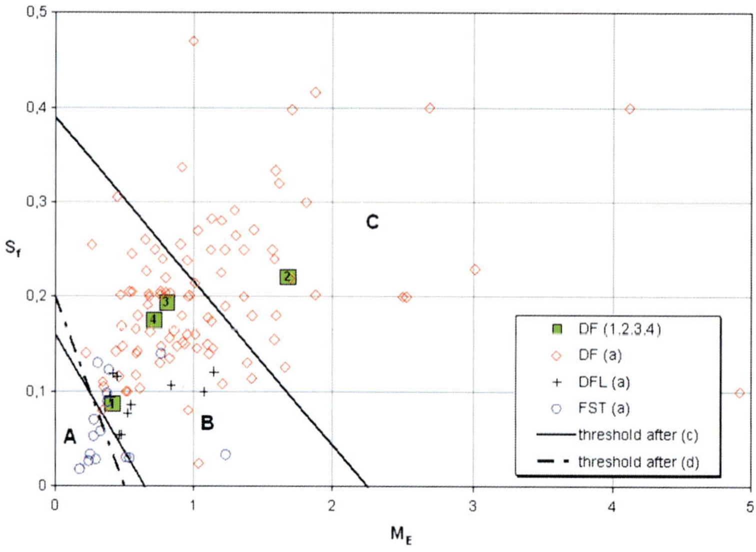

Fig. 2 Relation between average fan slope and Melton number for the data of Scheidl and Rickemann (2010), denoted with (*a*). (*1*) Hummingbird Creek (Jakob et al. 2000), (*2*) Cathedral Gulch (Both from Jackson et al. 1989), (*3*) Canyon Creek and (*4*) Hot Springs Creek (both from Jordan 1984). The threshold lines (*c*) and (*d*) are proposed by Bardou (2002) respectively Marchi and Brochot (2000). The domains *A*, *B* and *C* are discussed in the text

flows). The threshold line between zone B and C distinguishes between mixed processes and debris flows. Based on the data from Marchi and Brochot (2000), an alternative threshold line between zone A and B has been inserted. In Fig. 2 zone A includes only fluvial sediment transport data. While debris floods are represented only in zone B, debris-flow data are widely distributed within zone B and zone C. The apparent lack of a clear boundary between the latter two processes may partly reflect both their similarity and some uncertainty in classifying torrential processes. For example, Jakob et al. (2000) describe the debris-flow event in Hummingbird Creek (1 in Fig. 2) as an unusually large debris flow with a record breaking long-term antecedent precipitation, which may partly explain its proximity to zone A.

Wilford et al. (2004) studied 65 alluvial and colluvial fans in west central British Columbia, and found the following criteria to best distinguish debris flows from debris floods and floods (i.e. fluvial sediment transport process): Melton number >0.6 and watershed length (planimetric straight-line length from the fan apex to the most distant point on the watershed boundary) <2.7 km. Such a Melton number criterion is not confirmed by the data of Scheidl and Rickenmann (2010) or Rickenmann and Koschni (2010), for which many catchments with debris-flow occurrences have a Melton number <0.6.

3 Deposition Patterns of "Large" Debris Flows Outside the Channel on the Fan

Mobility on constructed fan areas may not only depend on different process types and topography but also on human infrastructures such as buildings, streets or railways. However, no fluvial sediment transport processes were considered since no distinct deposition zones could be observed.

Based on data from 42 debris-flow and five debris-flood events (Table 1), planar runout patterns were classified according to the shape of the deposition area. Altogether four types (A, B, C, D) of runout patterns could be distinguished from visual evaluation of the observed deposition zones (Fig. 3). The amount and degree of infrastructure affected by each event was qualitatively estimated, ranging from low (one building or one linear infrastructure affected) to high (several buildings and linear infrastructures affected). A null value was only applied to debris-flow events without any affected buildings, streets or railways.

Type A represents an ideal shape for uniform runout patterns (Fig. 4). Deposition is not influenced by any obstacles or distinct channels on the fan. Typically the material overtops the channel and starts to spread over the fan, influenced by the topography. It is assumed that the cross section or the conveyance of the channel near the start of deposition is too small in relation to the peak discharge of the event. The distal end of the deposition appears unconstrained, which allows the approximation of the observed runout area with a circle sector of radius L_f and angle ψ, under the assumption that the sector area equals the observed deposition area (Fig. 4). The majority of the debris-flow deposits of type A in Table 1 are related to fan situations with no or moderate potential influence by houses or infrastructure.

Types B1 and B2 as well as C1, C2 and D are associated with "disturbed" runout patterns within the observed deposition area (Fig. 5). For these types, an approximation of the planar view of the runout pattern with a simple sector (like for type A) represents a too strong simplification. For the debris-flow deposits of type B in Table 1 it appears that the deposition pattern is influenced by obstructions like buildings or mitigation measures in the lower part of the deposition zone. In our examples of types C and D in Table 1 we mainly identified linear structures crossing the deposition zone, like roads, railway tracks or receiving streams which were responsible for a pronounced lateral spreading in the lower part of the deposition area.

For the debris-flow deposition pattern of type A, the volume of the deposits can theoretically be approximated by a cylindrical circle sector (Fig. 6) with the mean deposition height $\bar{h} \propto V^{1/3}$ and the planimetric area $B \propto \psi L_f^2$:

$$V = \bar{h}B \propto V^{1/3}\psi L_f^2 \tag{2a}$$

or

$$V^{2/3} \propto \psi L_f^2 \tag{2b}$$

Table 1 Overview of the debris-flow data used in Fig. 8

Event name	Process-type	Documentation	Volume [m^3]	L_f [m]	ψ [°]	Pattern-type	Degree of infrastructure density
Bormina	DF	VAW (1992)	50,000	473	13.68	A	Null
Brichboden	DF	VAW (1992)	4,000	362	29.14	A	Null
Gerental_1	DF	VAW (1992)	7,000	364	45.40	A	Null
Gerental_2	DF	VAW (1992)	33,000	211	14.03	A	Null
Gerental_3A	DF	VAW (1992)	45,000	264	18.50	A	Null
Gerental_3B	DF	VAW (1992)	33,000	243	17.91	A	Null
Gonerli	DF	VAW (1992)	5,000	332	26.96	A	Null
Grätschliecht	DF	VAW (1992)	4,000	290	12.93	A	Null
Heugandtal	DF	VAW (1992)	7,000	340	14.03	A	Null
Mühlistein	DF	VAW (1992)	3,000	104	16.22	A	Null
Piz Caral SE	DF	VAW (1992)	3,000	142	33.40	A	Null
Piz Lugrendo NE	DF	VAW (1992)	3,000	404	50.11	A	Null
Stotzig Muttenhorn	DF	VAW (1992)	2,000	302	12.94	A	Null
Witenwasseren	DF	VAW (1992)	40,000	430	20.47	A	Null
Blauseeligraben	DF	VAW (1992)	2,000	90	45.65	A	Low
Cathedral Gulch	DF	Jackson et al. (1989)	90,000	1,000*	14.04	A	Low
Cathedral Gulch	DF	Jackson et al. (1989)	80,000	800*	14.20	A	Low
Canyon Creek	DF	Jordan (1994)	10,000	200*	49.83	A	Low
Glattbach	DF	Gertsch and Kienholz (2005)	9,000	367	41.33	A	Low
Holdenbach	DF	Gertsch and Kienholz (2005)	17,000	337	18.09	A	Low
Hot Springs Creek	DF	Jordan (1994)	60,000	800*	13.13	A	Low
Richleren	DF	VAW (1992)	9,000	805	29.81	A	Low
Ri di Cavanna	DF	VAW (1992)	7,000	217	25.72	A	Low
Ri di Gallinoso	DF	VAW (1992)	4,000	125	13.71	A	Low
Ri di Sozz	DF	VAW (1992)	3,000	198	41.76	A	Low
Schisstalbach	DF	Gertsch and Kienholz (2005)	11,000	281	22.35	A	Low

Val Meral	DF	VAW (1992)	2,000	799	14.72	A	Low
Cathedral Gulch	DF	Jackson et al. (1989)	7,000	400*	15.08	A	Medium
Nebukawa	DF	Kobayashi (1985)	1,000,000	900*	18.17	A	Medium
Sarno EP-5	DF	Oramas Dorta et al. (2007)	140,000	1,200*	30.27	A	Medium
Glyssibach	DF	NDR Consulting Zimmermann/Niederer und Pozzi Umwelt AG (2006)	60,000	387	31.11	A	High
Trachtbach	DF		6,000	140	–	B1	High
Sarno Ep-2	DF	Oramas Dorta et al. (2007)	120,000	650*	–	B2	Medium
Varuna	DF	VAW (1992)	214,000	1,011	–	B2	High
Cathedral Gulch	DF	Jackson et al. (1989)	87,000	–	9.36	C1	Medium
Rotlauibach	DF	Scheuner et al. (2009)	520,000	–	31.92	C2	Low
Saastal	DF	VAW (1992)	180,000	–	96.21	C2	Low
Cathedral Gulch	DF	Jackson et al. (1989)	136,000	–	17.83	C2	Medium
Cathedral Gulch	DF	Jackson et al. (1989)	24,000	–	31.63	C2	Medium
Grosstal	DF	VAW (1992)	20,000	–	54.12	C2	Medium
Schwendibach	DF		11,000	–	26.69	C2	Medium
Caraballeda	DF	Wieczorek et al. 2001	2,000,000	–	28.02	C2	High
Bardielbach	DFL	Schraml (2007)	6,500	671	9.79	A	Low
Eisbachgraben	DFL	Schraml (2007)	25,000	416	43.53	A	Medium
Partellstobel	DFL	Schraml (2007)	18,000	558	18.33	A	Medium
Gumpenbach	DFL	Schraml (2007)	35,000	1,328	7.61	A	Medium
Niedernsiller Mühlbach 2	DFL	Schraml (2007)	50,000	1,733	40.14	A	Medium

Unless otherwise noted, L_f values are based either on values listed in the related documentations or on measurements of original field observation data, such as maps or aerial images. L_f values denoted with a star (*) were measured from figures in the related documentations. The degree of infrastructure density refers to the relative proportion of objects such as buildings, streets or railway lines which may have influenced the deposition pattern

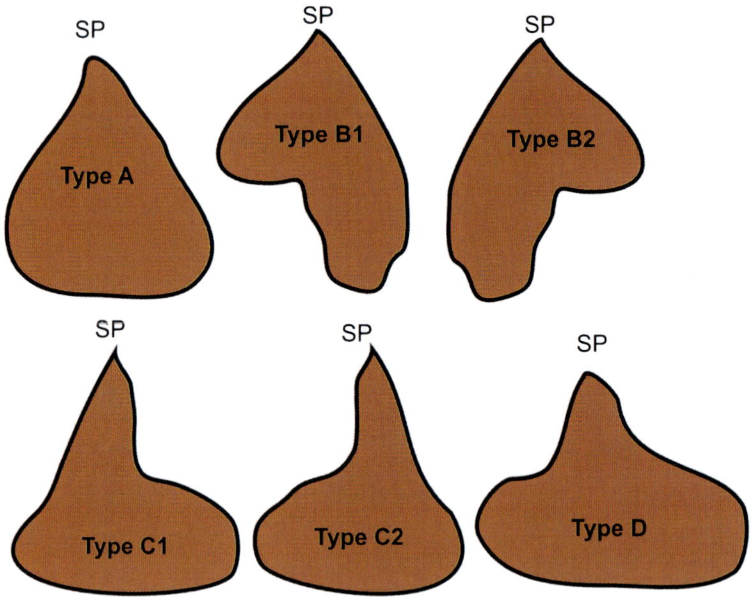

Fig. 3 Types of different recent runout patterns. SP denotes the start point of the deposition (often the fan apex)

Equation 2a is based on the assumption of geometrical similarity of the deposits. Equation 2b can be can be transformed to express the maximum runout distance L_f as a function of the deposition volume V and aperture angle ψ:

$$L_f \propto \psi^{-1/2} V^{1/3} \tag{3}$$

For debris-flow and debris-flood events of type A (Table 1), Fig. 7 shows a clear relationship between the circle sector properties (radius L_f, aperture angle ψ) and the event volume V. The data can be fitted with a power-law ($R^2 = 0.78$):

$$L_{fpred} = 65.3\left(\frac{\psi}{V^{2/3}}\right)^{-0.49} \approx 65.3\psi^{-1/2} V^{1/3} \tag{4}$$

Additionally, data of five debris-flood events, which could be classified to type A, are shown in Fig. 7. These processes show higher mobility but a similar trend of the relationship between the circle sector properties and the event volume, compared to debris-flow events.

The observed aperture angles for the development of the regression model range from 13° to 51°. Figure 7 and Eq. 4 essentially confirm that the assumption of geometric similarity is reasonable for debris-flow deposits which represent a runout pattern on the fan under quasi-natural conditions, after overtopping of the channel.

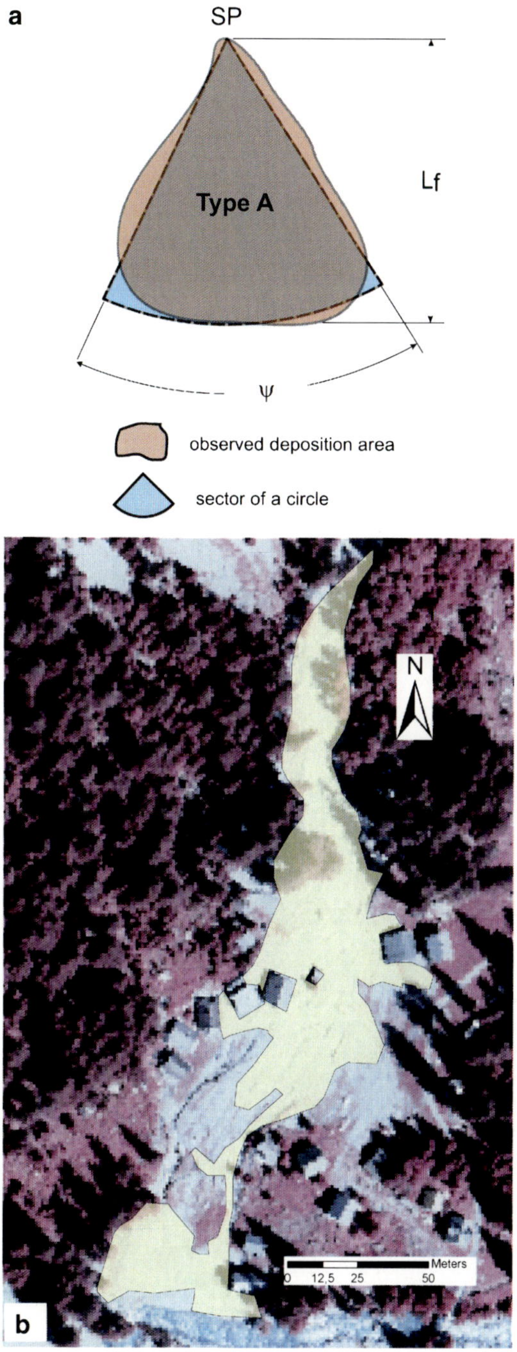

Fig. 4 (**a**) Runout pattern type A. ψ denotes the aperture angle of a fitting circle sector, with L_f, a radius equal to the distance from the starting point to the distal end of the observed deposition area. (**b**) Schisstalbach debris-flow event, Switzerland 2005 – example of deposition type A; aerial photo by Swisstopo, 29/30 August 2005 (reproduced with permission of Swisstopo, no. JA082265)

Fig. 5 (**a**) Trachtbach debris-flow event, Switzerland 2005 – example of deposition type B1; the lake of Brienz may have influenced the maximal runout distance, the aperture angle was influenced by the houses near the main channel; aerial photo by Swisstopo, 29/30 August 2005 (reproduced with permission of Swisstopo, no. JA082265). (**b**) Caraballeda debris-flow event, Venezuela 1999 – example of deposition type C2; the maximum runout was influenced by the sea (modified from Wieczorek et al. 2001)

The debris-flow and debris-flood events related to type B, C, or D would require a more detailed geometric approximation of the deposit shape and a larger number of cases for a quantitative analysis.

Fig. 5 (continued)

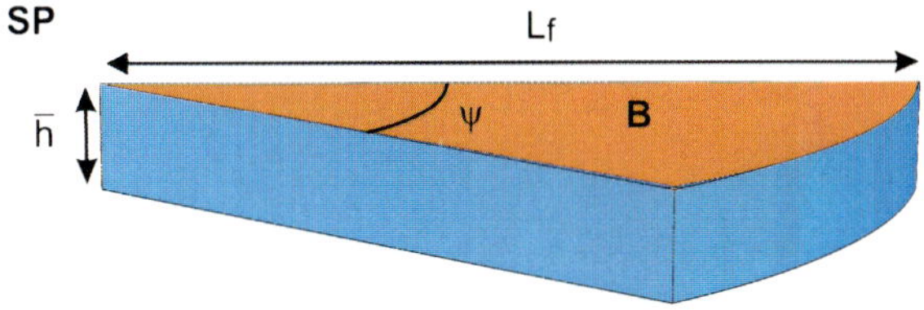

Fig. 6 Approximated cylindrical sector to describe the deposition pattern; *SP* denotes the start point of the deposition

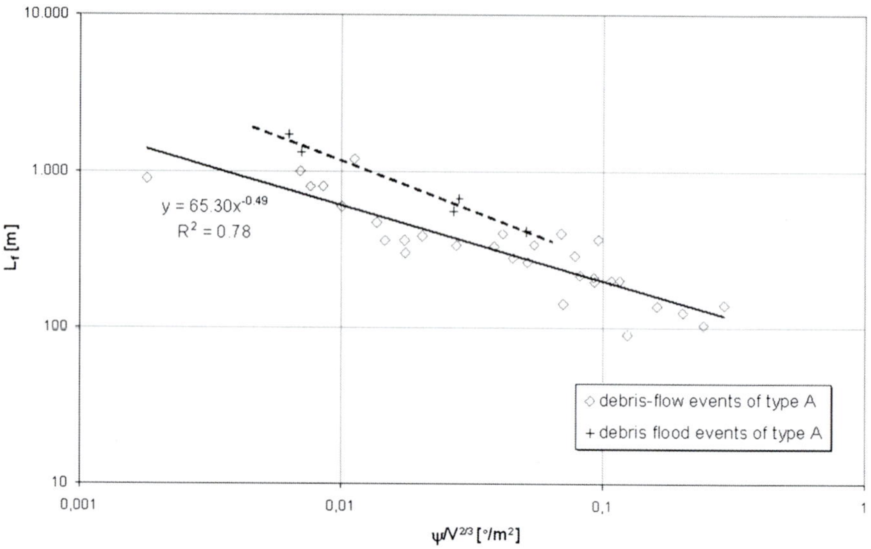

Fig. 7 Relationship between the approximated circle sector radius L_f and a factor incorporating aperture angle Ψ and event volume V, according to the geometric similarity Eq. 2b. Only debris-flow events of type A are included, as listed in Table 1

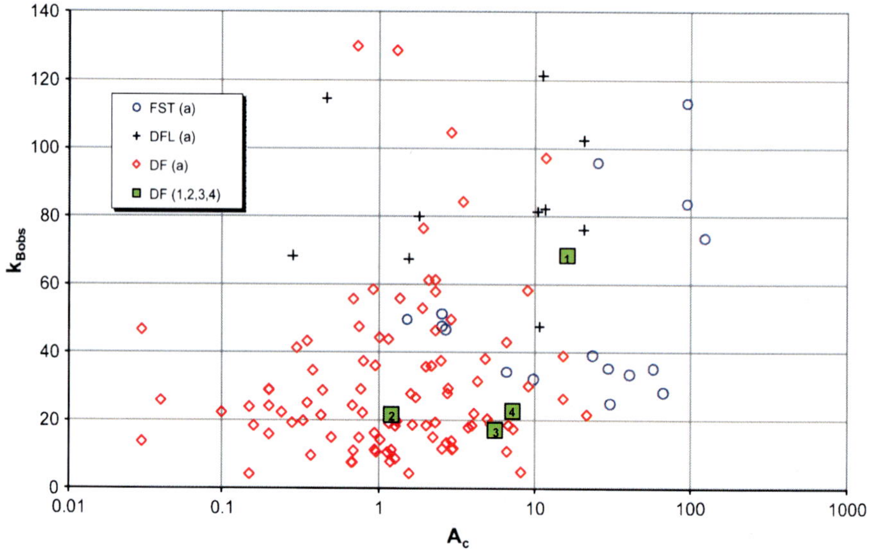

Fig. 8 Relation between k_{Bobs} as a function of catchment area A_c for the dataset of Scheidl and Rickenmann (2010) denoted with (*a*), and (*1*) Cathedral Gulch (Jackson et al. 1989), (*2*) Canyon Creek and (*3*) Hot Springs Creek (Jordan 1984) as well as (*4*) Hummingbird Creek (Jakob et al. 2000)

4 Implications of the Process Types on Runout and Mobility

Many studies document the importance of debris-flow volume on runout area on the fan (also see Sect. 2). Iverson et al. (1998) developed a GIS-based simulation model for hazard mapping and proposed that there is a close semi-empirical relation between the planimetric deposition area, *B*, of lahars or debris flows and the event volume, *V*. Assuming geometric similarity, they proposed the following power-law equation:

$$\mathrm{B} = \mathrm{k_B V^{2/3}} \tag{5}$$

where k_B is a proportionality coefficient representing an average mobility of, for example, the groups of lahar or debris-flow data. Scheidl and Rickenmann (2010) proposed a similar GIS-based simulation program for debris-flow hazard mapping. For a given volume and starting point of the deposits, a simple flow routing algorithm is used to determine the potential runout area covered by debris-flow deposits. The runout zone is then delineated by confining the simulated potential spreading area in the down slope direction using the empirically determined planimetric deposition area which is also based on Eq. 5. A further empirical relationship is used to determine the proportionality coefficient in Eq. 5 as a function of geomorphic catchment parameters. For this purpose the individual observed mobility coefficient k_{Bobs} is defined based on the observed debris-flow volume V_{obs} and observed deposition area B_{obs} of a single event as:

$$\mathrm{k_{Bobs}} = \mathrm{B_{obs} V_{obs}^{-2/3}} \tag{6}$$

Scheidl and Rickenmann (2010) then proposed that the mobility coefficient k_{Bobs} can be estimated as a function of the average fan slope S_f and the average channel slope S_c. This coefficient is thought to reflect some of the flow properties during the depositional phase of the debris-flow event.

Assuming geometric similarity, the mobility coefficient as defined by Eq. 5 has been proposed to depend on process type in several studies (Crosta et al. 2003; Griswold 2004; Yu et al. 2006; Berti and Simoni 2007; Griswold and Iverson 2008; Scheidl and Rickenmann 2008). The development of alluvial fans with water driven mass movements reflects their rheological conditions during the deposition process: The studies of Jackson et al. (1987), Marchi and Tecca (1995) and Chau et al. (2000) imply that granular flow behaviour will lead to a higher roughness and friction during depositional flow, which results in steeper fan slopes on average and in a smaller mobility. A more viscous or muddy flow behaviour, on the other hand, shows a higher mobility and results in smoother and flatter fans. A higher mobility of muddy debris flow compared to granular debris flow is also reflected by a smaller pseudo Manning's flow resistance coefficient (Rickenmann and Weber 2000). Assuming that the runout of an event can be correlated with the dominant fan-forming processes, the local mobility coefficient can be estimated based on the fan and channel slope (e.g. Scheidl and Rickenmann 2010).

If mobility coefficients k_{Bobs} are plotted as a function of catchment areas A_c (Fig. 8), there is no clear distinction between debris flows, debris floods, or fluvial sediment transport. However, the events associated with the latter two flow types show a trend for having comparatively larger values of k_{Bobs} and A_c.

5 Runout Prediction Methods for Debris Flows

In principal runout prediction methods for torrential mass movements, especially for debris flows, are either based on empirical-statistical or dynamical methods (Rickenmann 2005). Empirical equations to estimate the total travel distance (the entire horizontal path length) of debris flows and landslides were proposed by Corominas (1996), Rickenmann (1999), Legros (2002) and Toyos et al. (2008). In all these approaches the runout length is essentially a function of volume and angle of reach or the longitudinal profile of the expected flow path. The empirical model of Prochaska et al. (2008) to predict runout for non-volcanic debris flows is based on the average channel slope and on the stream profile to the drainage divide. The method is analogous to methods for snow avalanche runout estimations (Lied and Bakkehoi 1980) and does not account for a changing event volume.

Other methods estimating the total travel distance are based on a sediment budget along the flow path. Such volume balance approaches have been proposed by Cannon (1993) and Fannin and Wise (2001). The dynamical runout model "dfwalk" (Gamma 2000) uses a mass-point-model, originally developed for snow avalanches (Körner 1980; Perla et al. 1980), and requires an empirical calibration of two parameters. During the last years, the method of cellular automata modelling has been applied to debris-flow susceptibility assessments (Iovine et al. 2005; D'Ambrosio et al. 2003, 2006) and geomorphic evolution in river catchments (Coulthard et al. 2007).

For a more detailed delineation of potentially endangered areas, the runout pattern or the surface area on the debris-flow cone, downstream of the point where major deposition occurs, should be known (Rickenmann 2005). A simple GIS based empirical model for this purpose was first developed for lahars (Iverson et al. 1998). Subsequently, more observations on debris flows were used to test or modify the original approach of Iverson et al. (1998) in studies by Crosta and Agliardi (2003), Griswold (2004), Yu et al. (2006), Berti and Simoni (2007), Oramas Dorta et al. (2007), Griswold and Iverson (2008), and Scheidl and Rickenmann (2010). A probabilistic relationship between debris-flow deposition length and deposition volume to evaluate the runout for a given debris-flow volume was presented by Chen et al. (2007). Tsai (2006) studied the geometric development of debris-flow cones, and used the maximum length, width, and thickness of debris-flow cones as the characteristic parameters in the analysis of morphological similarity based on a laboratory experiment. Most of these studies underline the importance of the debris-flow volume on the resulting runout area on the fan. Many empirical equations for the total travel distance support the important role of event volume on runout distance (Rickenmann 2005).

Using a dynamical approach Takahashi and Yoshida (1979), Hungr et al. (1984) and Takahashi (1991) proposed a one-dimensional, analytical model to predict the runout due to an abrupt change of bed slope, assuming a constant flow section and a constant discharge. In principle the most accurate description of the depositional process of debris flows is provided by hydraulic simulation continuum models, like FLO-2D (O'Brien et al. 1993), RAMMS (Bartelt et al. 1999; Hose 2007; Christen et al. 2010), FlatModel (Medina et al. 2008) or MassMov2D (Beguería et al. 2009). However, these models are sometimes difficult to apply and generally require a calibration of "friction" parameters by back-calculation of past events (Naef et al. 2006; Rickenmann et al. 2006; Tecca et al. 2007; Hürlimann et al. 2008; Pirulli and Sorbino 2008; Hochschwarzer 2009). A common finding of many of these studies is that, once the flow has overtopped the channel, the runout pattern on the fan is determined to a large degree by the fan topography. For these reasons Scheidl and Rickenmann (2010) combined a multiple flow path algorithm (based on detailed fan topography) with an empirical relationship between planimetric deposition area and event volume in their TopRunDF modelling approach.

6 Conclusions

Considering classification schemes for different process types in torrential catchments, a relatively clear distinction between events dominated by debris flows or debris floods and fluvial sediment transport can be made. However, the same classification schemes do not allow for a clear distinction between debris flows and debris floods. Since process type influences depositional characteristics such as deposit thickness (or flow mobility) and steepness of the deposits, fan and catchment properties may support empirical runout prediction.

Analyses of deposition patterns on the fan of larger debris-flow events support the assumption of geometric similarity of the deposits, as proposed by Iverson et al. (1998) and used in other runout prediction approaches. Exceptions may be expected in "disturbed" or overbuilt deposition areas where both a specification of a starting point of deposition and the runout flow may be influenced not only by topography but also by human infrastructures. More detailed studies are required to better quantify the effect of log jams, narrow outlets or flow obstacles along the flow path on the deposition pattern.

A short review of debris-flow runout prediction methods leads to the following conclusions: (i) most approaches are essentially of empirical nature, and therefore better predictions may be expected for situations which are similar to the underlying data, (ii) the runout length or the depositional area depend on event volume, similar to the case for debris or snow avalanches, and (iii) the fan topography is an important element controlling the distribution of the deposits on the fan for larger events overtopping the main channel.

In general, continuum based dynamic runout prediction methods provide the most complete characterization of debris-flow processes and their behaviour. As in other two-dimensional runout prediction methods, the simulated inundated areas appear to be mainly controlled by the topography of the fan. However, numerical models need to be further evaluated with regard to appropriate values for the rheologic or friction parameters, which are often assumed or back-calculated from field observations.

Acknowledgements The study has been funded by the Austrian Science Fund (FWF) project no. L 180-N10 on 'Runout prediction of debris flows'. The Swiss Federal Office for Environment supported the analysis of the Swiss debris-flow events of 2005. Markus Zimmermann provided the original field data concerning the 1987 debris flows in the Swiss Alps.

References

Bardou E (2002) Methodologie de diagnostic des laves torrentielles sur un bassin versant alpin. PhD thesis, École Polytechnique Fédérale de Lausanne, p 382

Bartelt P, Salm B, Gruber U (1999) Calculating dense-snow avalanche runout using a Voellmyfluid model with active/passive longitudinal straining. J Glaciol 45:212–254

Beguería S, Asch TWJV, Malet J-P, Gröndahl S (2009) A GIS-based numerical model for simulating the kinematics of mud and debris flows over complex terrain. Nat Hazard Earth Syst Sci 9:1897–1909

Berti M, Simoni A (2007) Prediction of debris flow inundation areas using empirical mobility relationships. Geomorphology 90:144–161

Cannon SH (1993) An empirical model for the volume-change behavior of debris flows. In: Shen HW, Su ST, Wen F (eds) Hydraulic engineering'93, vol 2. American Society of Civil Engineers, New York, pp 1768–1773

Chau KT, Chan LC, Wai WH (2000) Shape of deposition fan and runout distance of debris-flow: Effects of granular and contents. In: Wieczorek, Naeser (eds) Debris-Flow Hazards Mitigation: Mechanics, Prediction, and Assessment, pp 387–395

Chen J, Chang S, Tsang Y, Shieh C (2007) Empirical relationships for deposited length of debris-flow: a case study in Taiwan. In: Chen-lung C, Major JJ (eds) 4th International conference on debris-flow hazards mitigation: mechanics, prediction, and assessment, Chengdu, China. Millpress, Rotterdam

Christen M, Bartelt P, Kowalski J (2010) Back calculation of the In den Arelen avalanche with RAMMS: interpretation of model results. Ann Glaciol 51:161–168

Corominas J (1996) The angle of reach as a mobility index for small and large landslides. Can Geotech J 33:260–271

Costa JE (1988) Rheologic, geomorphic, and sedimentologic differentiation of water floods, hyperconcentrated flows, and debris flows. In: Baker VR, Patton PC (eds) Flood Geomorphology, pp 113–122

Coulthard TJ, Hicks DM, van de Wiel MJ (2007) Cellular modelling of river catchments and reaches: advantages, limitations and prospects. Geomorphology 90:192–207

Crosta G, Agliardi F (2003) A methodology for physically-based rockfall hazard assessment. Nat Hazard Earth Syst Sci 3:407–422

Crosta G, Cucchiaro S, Frattini P (2003) Validation of semi-empirical relationships for the definition of debris-flow behavior in granular materials. In: Rickenmann D, Chen (eds) Debris-Flow Hazards Mitigation: Mechanics, Prediction, and Assessment, pp. 821–833

Crosta GB, Frattini P (2004) Controls on modern alluvial fan processes in the Central Alps, Northern Italy. Earth Surf Process Landf 29:267–293

D'Ambrosio D, Gregorio SD, Iovine G, Lupiano V, Rongo R, Spataro W (2003) First simulations of the Sarno debris flows through Cellular Automata modelling. Geomorphology 54:91–117

D'Ambrosio D, Spataro W, Iovine G (2006) Parallel genetic algorithms for optimising cellular automata models of natural complex phenomena: an application to debris flows. Comput Geosci 32:861–875

Fannin RJ, Wise MP (2001) An empirical-statistical model for debris flow travel distance. Can Geotech J 38:982–994

Gamma P (2000) dfwalk – Ein Murgang-Simulationsprogramm zur Gefahrenzonierung. Geographica Bernensia G66. Geographisches Institut der Universität Bern, Bern, p 144

Gertsch E, Kienholz H (2005) Ereignisdokumentation StorMe, Unwetter 22./23. August 2005, Gemeinde Lütschental. Angewandte Geomorphologie und Naturrisiken AGNAT, Geographisches Institut Universität Bern, Bern

Griswold JP (2004) Mobility statistics and hazard mapping for non-volcanic debris flows and rock avalanches. Master's thesis, Portland State University, p 200

Griswold JP, Iverson RM (2008) Mobility statistics and automated hazard mapping for debris flows and rock avalanches, U.S. Geological Survey scientific investigations report 5276. U.S. Geological Survey, Reston, p 59

Hochschwarzer M (2009) Vergleich von Simulationsmodellen zur Reichweitenabschätzung alpiner Murgänge am Beispiel Südtiroler Ereignisse. Master's thesis, University of Applied Life Sciences and Natural Resources, p 135

Hose A (2007) Einfluss der Topographie auf die Murgangsimulation im Ablagerungsbereich und Implikationen auf die Gefahrenbeurteilung. Master thesis, Universität Stuttgart, Institut für Geographie, p 130

Hübl J, Kienholz H, Loipersberger A (2002) DOMODIS – Documentation of Mountain Disasters, State of Discussion in the European Mountain Areas. International Research Society INTERPRAEVENT, Schriftenreihe 1, Handbuch 1, p 38

Hungr O, Morgan G, Kellerhals R (1984) Quantitative analysis of debris torrent hazards for design of remedial measures. Can Geotech J 21:663–677

Hungr O, Evans S, Bovis MJ, Hutchinson JN (2001) A review of the classification of landslides of the flow type. Environ Eng Geosci 7:221–238

Hürlimann M, Rickenmann D, Medina V, Bateman A (2008) Evaluation of approaches to calculate debris-flow parameters for hazard assessment. Eng Geol 102:152–163

Iovine G, D'Ambrosio D, Gregorio SD (2005) Applying genetic algorithms for calibrating a hexagonal cellular automata model for the simulation of debris flows characterised by strong inertial effects. Geomorphology 66:287–303

Iverson RM (1997) The physics of debris flows. Rev Geophys 35(3):245–296

Iverson RM, Schilling SP, Vallance JW (1998) Objective delineation of lahar-inundation hazard zones. Geol Soc Am Bull 110:972–984

Jackson L, Kostaschuk R, McDonald G (1987) Identification of debris flow hazard on alluvial fans in the Canadian Rocky Mountains. Geol Soc Am Rev Eng Geol 7:115–124

Jackson L, Hungr O, Gardner J, Mackay C (1989) Cathedral Mountain debris flows, Canada. Bull Int Assoc Eng Geol 40:35–54

Jakob M, Anderson D, Fuller T, Hungr O, Ayotte D (2000) An unusually large debris flow at Hummingbird Creek, Mara Lake, British Columbia. Can Geotech J 37:1109–1125

Jordan RP (1994) Debris flows in the southern coast mountains, British Columbia: dynamic behaviour and physical properties. PhD thesis, The University of British Columbia, p 272

Kobayashi Y (1985) A catastrophic debris avalanche induced by the 1923 great Kanto earthquake. J Nat Disaster Sci 7:1–9

Körner H (1980) Modelle zur Berechnung der Bergsturz- und Lawinenberechnung. In: Internationales Symposium "Interpraevent", vol 2, Klagenfurt, Austria, pp 15–55

Legros F (2002) The mobility of long-runout landslides. Eng Geol 63:301–331

Lied K, Bakkehoi S (1980) Empirical calculations of snow-avalanche run-out distance, based on topographic parameters. J Glaciol 26(94):165–177
Marchi L, Brochot S (2000) Les cônes de díjection torrentiels dans les Alpes francaises. Morphomètrie et processus de transport solide torrential. Rev Gèogr Alpine 88(3):23–38
Marchi L, D'Agostino V (2004) Estimation of debris-flow magnitude in the Eastern Italian Alps. Earth Surf Process Landf 29:207–220
Marchi L, Tecca P (1995) Alluvial fans of the Eastern Italian Alps: morphology and depositional processes. Geodinamica Acta 8:20–27
Medina V, Hürlimann M, Bateman A (2008) Application of FLATModel, a 2D finite volume code, to debris flows in the northeastern part of the Iberian Peninsula. Landslides 5:127–142
Melton MA (1965) The geomorphic and paleoclimatic significance of alluvial deposits in Southern Arizona. J Geol 73:1–38
Naef D, Rickenmann D, Rutschmann P, McArdell BW (2006) Comparison of flow resistance relations for debris flows using a one-dimensional finite element simulation model. Nat Hazard Earth Syst Sci 6:155–165
O'Brien JS, Julien PY, Fullerton W (1993) Two-dimensional water flood and mudflood simulation. J Hydraul Eng 119:244–260
Oramas Dorta D, Toyos G, Oppenheimer C, Pareschi M, Zanchetta G (2007) Empirical modelling of the May 1998 small debris flows in Sarno (Italy) using LAHARZ. Nat Hazard 40:381–396
Perla R, Cheng T, McClung D (1980) A two parameter model of snow avalanche motion. J Glaciol 26:197–208
Pirulli M, Sorbino G (2008) Assessing potential debris flow runout: a comparison of two simulation models. Nat Hazard Earth Syst Sci 8:961–971
NDR Consulting Zimmermann/Niederer und Pozzi Umwelt AG (2006) Lokale lösungsorientierte Ereignisanalyse, Glyssibach, Bericht zum Vorprojekt. Tech. rep., Tiefbauamt des Kantons Bern
Prochaska AB, Santi PM, Higgins J, Cannon SH (2008) Debris-flow runout predictions based on the average channel slope (ACS). Eng Geol 98:29–40
Rickenmann D (1999) Empirical relationships for debris flows. Nat Hazard 19:47–77
Rickenmann D, Weber D (2000) Flow resistance of natural and experimental debris-flows in torrent channels. In: Wieczorek, Naeser (eds) Debris-Flow Hazards Mitigation: Mechanics, Prediction, and Assessment, pp 245–254
Rickenmann D (2005) Runout prediction methods. In: Jakob M, Hungr O (eds) Debris-flow hazards and related phenomena. Praxis/Springer, Berlin/Heidelberg, pp 305–324
Rickenmann D, Koschni A (2010) Sediment loads due to fluvial transport and debris flows during the 2005 flood events in Switzerland. Hydrol Process 24:993–1007. doi:10.1002/hyp. 7536
Rickenmann D, Laigle D, McArdell BW, Hübl J (2006) Comparison of 2D debris-flow simulation models with field events. Comput Geosci 10:241–264
Rickenmann D, Hunzinger L, Koschni A (2008) Hochwasser und Sedimenttransport während des Unwetters vom August 2005 in der Schweiz. In: Mikos M, Huebl J, Koboltschnig G (eds) Schutz des Lebensraumes vor Hochwasser, Muren, Massenbewegungen und Lawinen. Interpraevent 26–30 May 2008, Dornbirn, Vorarlberg, Austria. Conference proceedings, vol 1. Internat. Research Society Interpraevent, Klagenfurt, pp 465–476
Rickenmann D, Zimmermann M (1993) The 1987 debris flows in Switzerland: documentation and analysis. Geomorphology 8:175–189
Scheidl C, Rickenmann D (2008) Depositional characteristics and runout of Alpine debris flows. In: Conference proceedings Interpraevent'08, vol 1, Dornbirn, Austria, pp 477–488
Scheidl C, Rickenmann D (2010) Empirical prediction of debris-flow mobility and deposition on fans. Earth Surf Process Landf 35:157–173
Scheuner T, Keusen HR, McArdell BW, Huggel C (2009) Murgangmodellierung mit dynamisch physikalischem und GIS-basiertem Fliessmodell, Fallbeispiel Rotlauigraben, Guttannen, August 2005. Wasser Energie Luft 101:15–21
Schraml C (2007) Ablagerung von Feststoffen auf Wildbachkegeln. Master's thesis, University of Natural Resources and Life Sciences, Vienna
Takahashi T (1991) Debris flow. A.A. Balkema, Rotterdam/Brookfield

Takahashi T, Yoshida H (1979) Study on the deposition of debris flows, part 1-deposition due to abrupt change of bed slope. Annu Disaster Prev Res Inst Kyoto Univ 22:315–328
Tecca P, Genevois R, Deganutti A, Armento M (2007) Numerical modelling of two debris flows in the Dolomites (Northeastern Italian Alps). In: Chen-lung C, Major JJ (eds) 4th International conference on debris-flow hazards mitigation: mechanics, prediction, and assessment, Chengdu, China. Millpress, Rotterdam
Toyos G, Gunasekera R, Zanchetta G, Oppenheimer C, Sulpizio R, Favalli M, Pareschi MT (2008) GIS-assisted modelling for debris flow hazard assessment based on the events of May 1998 in the area of Sarno, Southern Italy: II. Velocity and dynamic pressure. Earth Surf Process Landf 33:1693–1708
Tsai Y-F (2006) Three-dimensional topography of debris-flow fan. J Hydraul Eng 132:307–318
VAW (1992) Murgänge 1987, Dokumentation und Analyse. Bericht Nr. 97.6 der Versuchsanstalt für Wasserbau, Hydrologie und Glaziologie, ETH Zürich (unpublished report)
Wieczorek GF, Larsen MC, Eaton LS, Morgan BA, Blair JL (2001) Debris-flow and flooding hazards associated with the December 1999 storm in coastal Venezuela and strategies for mitigation. U.S. Geological Survey, Reston
Wilford D, Sakals M, Innes J, Sidle R, Bergerud W (2004) Recognition of debris flow, debris flood and flood hazard through watershed morphometrics. Landslides 1:61–66
Yu F-C, Chen C-Y, Chen T-C, Hung F-Y, Lin S-C (2006) A GIS process for delimitating areas potentially endangered by debris flow. Nat Hazard 37:169–189
Zimmermann M, Mani P, Gamma P (1997) Murganggefahr und Klimaänderung – ein GIS-basierter Ansatz. vdf Hochschulverlag AG an der ETH Zürich: 162

Events on Fans and Cones: Recurrence Interval and Magnitude

Matthias Jakob

1 Introduction

Debris flows, their volcanic counterparts lahars, and debris floods that may occur as part of a debris-flow event, are the mass movement events that result in most losses worldwide. Hundreds of deaths occur every year, while horrific multi-thousand death tolls are announced every decade or so. Debris-flow related economic and life losses are not limited to emerging and developing countries but regularly occur in the developed world as well. In the few month leading up to the writing of this chapter, debris flows in Madeira (Portugal), Italy and Brazil lead to well over 100 fatalities and created losses well over 100 Million USD. Consequently, debris flows have been the focus of intensive scientific investigation with hundreds of papers appearing annually on various aspects of debris-flow research. Several dedicated conferences have been held whose sole focus is debris flows. In 2005 a book on debris flows and related processes was published (Jakob and Hungr 2005) to offer a more systematic review of the state-of-the-art. The book was published in 2005 and thus reflects mostly knowledge up to 2003 or 2004. With that it is outdated in some fields. Furthermore, it only touched on various topics in chapter form while it is clear to the practitioner that some of the chapters in the Jakob and Hungr book would have deserved to be written up as individual volumes themselves.

One of the key issues in determining the risk in an area subject to debris flows or debris floods is to determine a sound long-term relationship between frequency and magnitude. In most cases the best information for such relationship lies buried in the fan or cone. In absence of well-established frequency-magnitude relationships, modelling of flows at given return periods is associated with error that propagates if quantitative or semi-quantitative risk assessments follow the study of

M. Jakob (✉)
BGC Engineering Inc., 500-1045 Howe Street, Vancouver, BC V6Z2A9, Canada
e-mail: mjakob@bgcengineering.ca

M. Schneuwly-Bollschweiler et al. (eds.), *Dating Torrential Processes on Fans and Cones*, Advances in Global Change Research 47, DOI 10.1007/978-94-007-4336-6_6,

debris-flow intensity. Despite this well-known realization, there is a striking paucity of papers that address individual-fan/-cone frequency-magnitude relationships. The Fourth International Conference on Debris-Flow Hazards Mitigation: Mechanics, Prediction and Assessment in Chengdu, China, serves as a good example. Fifteen papers are listed under the heading hazard and risk, and at first glance most of these papers ought to include some aspects of frequency-magnitude analyses since, by definition, those form the basis of hazard and risk. However, none of the papers that deal with individual fans or cones provide enough material to verify how frequency-magnitude relationships were developed.

Excellent frequency-magnitude data exist from a few monitored sites such as Japan (Suwa, personal communication, 2006), Italy (Marchi et al. 2002; Tecca et al. 2003), and Switzerland (Hürlimann et al. 2003).

2 Debris Flow Frequency

2.1 Methodical Assessment

Debris-flow frequency is of interest to academics working on long-term landscape development. From a practical point of view, debris-flow frequency is a prerequisite to assess risk. Risk is the ultimate variable against which any mitigation measures should be weighed. Risk combines the hazard (expressed as frequency and debris-flow intensity) and the consequences (expressed, for example, as the potential for loss-of-life, economic damage, damage to the reputation of a company whose operations may be affected by debris flows and environmental losses).

Several chapters in this volume (Schneuwly-Bollschweiler et al. 2012) address the types of methods that can be applied to reconstruct the chronology of past events and thus provide an indication (though no certainty) of the average return period for future events. These include directly observations (Marchi and Tecca 2012, *thisvolume*), archival records, particularly in areas that have a long settlement history (D'Agostino 2012, *this volume*), air photos and satellite images (Staley and Wasklewicz 2012, *this volume*), field observations of so-called "silent witnesses" though these tend to not disclose their respective age (Kaitna and Huebl 2012, *this volume*), the continuously expanding field of dendrogeomorphology (Schneuwly-Bollschweiler and Stoffel 2012, *this volume*) supplemented by the broader field of vegetation analysis (Baroni et al. 2012, *this volume*), varve chronologies in lacustrine sediments (Gilli et al. 2012, *this volume*), cosmologic nuclides (Ivy-Ochs et al. 2012, *this volume*), radiocarbon dating (Chiverell and Jakob 2012, *this volume*), and luminescence dating (Lang 2012, *this volume*). Given the detailed treatment in other chapter, this one will not attempt to provide additional insight in the individual methods.

It will be very rare to be able to use a combination of methods, either because the local conditions do not allow their application, because they are prohibitively expensive, or because they cover a time range that is inappropriate for a risk assessment.

For example, we can consider a road or railway through the northern Chilean and Argentine Andes, an area characterized by extremely arid conditions with no or very little organic materials. The area is also sparsely or not settled at all. This condition will not allow any direct observations because no infrastructure crossed this particular stretch of land prior to the planning of the new route. For the same reason, archival records do not exist. Dendrogeomorphology and vegetation analysis are useless since there is no vegetation to analyse. Cosmogenic and luminescence dating do not provide the time horizons of interest (chronology of several hundred years). The absence of lakes or former lakes along the infrastructure corridor negates the possibility of dating via varve chronologies. This leaves few methods. Silent witnesses may be abundant in the form of well-preserved boulder lobes and levees, large rafted rocks or truncated fan or cone surfaces that have exposed numerous individual debris-flow events; however, these are relatively meaningless unless their age can be determined in absolute terms. Fortunately, there is an increasing amount of satellite imagery and air photography available that can be used to map and date debris-flow events, at least for the duration of the record. If no debris-flow event can be discerned in the period of satellite or air photograph record, this will provide at least a minimum estimate of debris-flow frequency, which although unsatisfactory, is better than nothing. This case is in stark contrast to, for an example, an Austrian village, where historical records of debris-flow event reach back several hundred years, and where old conifers on the fan or cone and along the channel allow accurate dating of the debris flows and even a reconstruction of the flow cross-sections in bedrock reaches of the channel. Furthermore, the abundance of organic materials in the watershed and the fan or cone sediments allows an extension of the dendrochronologic record to a few thousand years through radiometric dating of the enclosed organics and buried organic horizons. In the latter case, an exact dating for a long period is only a question of available funds, but not a limitation of methods.

2.2 *Limitations of Frequency Analysis*

There are several reasons why data censoring and data limitations increase the difficulties in applying the methods of flood frequency analysis to debris flows. In hydrology, appropriate data sets are selected to which theoretical frequencies are fitted. Inferences are then made about the underlying population. Statistical criteria such as randomness and independence of successive events as well as homogeneity and data stationarity form the basis for applicability of statistical techniques. In particularly independence and homogeneity assumptions are grossly violated in many debris-flow watersheds. This suggests that a sophisticated statistical treatment of debris-flow time series is unwarranted in these cases. Similarly, use of annual maximum series (AMS), as frequently used in hydrology, are inappropriate for debris flows since events rarely occur every year and, unlike river flows, are usually not continuously monitored. Partial duration series (PDS) are preferable because a time interval suitable for debris-flow occurrence could be determined by choosing an appropriate magnitude base level. However, PDS requires independence of magnitude exceedance and assumes an underlying distribution. Independence may only be warranted in transport-limited (supply-unlimited) basins (Jakob 1996). Researchers have failed to agree on an underlying theoretical frequency distribution that depend on the process responsible for the frequency of debris flows (i.e. Ohmori and Hirano 1988; Johnson et al. 1991). Costa (personal communication, 1996) suggests that uncertainties in the magnitude estimates are greater than the assumptions inherent in fitting underlying distributions, and suggests the application of the Weibull formula [$T = (n + 1)/m$] with T being the return period, n being the number of years of record and m being the rank. This formula, for example, has found application in Jakob and Friele (2010) in their work to reconstruct debris-flow frequency for a large fan in southern British Columbia. Jakob (2012) provides a discussion of methods applied to establish frequency-magnitude relationships with several case studies. Finally, debris flows can be initiated by a large variety of processes thus violating the homogeneity criterion. This criterion, however, seems to be of lesser importance given that floods are also created by different processes.

This discussion highlights an important relation between frequency and magnitude of debris flows that is further discussed below. It also demonstrates that the difference between flood-frequency analysis and debris-flow frequency analysis is primarily one of degree rather than kind, particularly for transport-limited basins where sufficient data are available.

A few nations have developed national guidelines or standards to map and treat natural hazards. For example, Switzerland has published recommendations for planning and geohazards. In this well-researched document (see Bründl 2012, *this volume*), so-called "Gefahrenkarten" (Danger Maps) are produced that classify geohazards into five zones (high hazard, medium hazard, low hazard, remaining hazard zone and no or negligible hazard). "Danger" or hazard levels are defined though a matrix that includes the probability of occurrence as well as the hazard intensity. Similar to the hazard categories, five frequency (or return period) categories have been introduced (<30 years, 30–100 years, 100–300 years, >300 years). A different matrix specifies which kind of developments should be protected against what kind of intensity event. On the other end of the spectrum, settlements, businesses, industry, camping and sports arenas need to be protected fully against any intensity up to the 100-year return period, low intensity for the 100–300 year return period event and moderate intensity for the >300-year return period event. Risk-based thinking underlies this matrix because the consequences for a geohazard impacting permanently used facilities are significantly higher than for temporarily or transiently used routes.

The examples above imply that for most cases, i.e. urban and industrial developments, the methods to determine frequencies must allow a frequency analysis that encompasses the entire range of return periods. In the case of a settlement, this requires that a record will need to extend to at least some 500 years (the >300 years leaves a very long time window for analysis!). This time frame will likely bring the practitioner to the limits of some dating techniques such as dendrochronology (see Stoffel 2010 for the most extensive dendrochronologic cone frequency investigation undertaken to date, but whose frequency-magnitude relationship could only be extended with reasonable confidence to 150 years). Similarly archival information will be limited in most cases and may require radiocarbon dating to supplement the dendrochronologic, lichenometric (i.e. Helsen et al. 2002) or historical record.

In other countries, there are no specified frequency categories because the decision on mitigation and zoning may hinge on the outcome of a quantitative risk assessment whose data input may span several millennia. A good example is the Cheekye fan in southern British Columbia, where there is a desire to develop portions of the fan with urban housing and some industry. A three-fold study was conducted in 2007 and 2008 that included an in-depth frequency-magnitude assessment (Jakob and Friele 2010), a numerical modelling exercise and a quantitative risk assessment (BGC 2009) that included risk of loss-of-life calculations for existing and future development and comparison to internationally defined risk tolerance thresholds. Such analysis, particularly for high consequence developments encourages a full exploitation of dating methods for the entire Holocene record. Information from over 100 test trenches over 5 m deep, dendrochronology and direct observations were assembled to provide a reasonable frequency-magnitude relationship, however, at a cost of many hundreds of thousands of dollars.

3 Debris-Flow Magnitude

3.1 Methodical Assessment

Debris-flow magnitude can be expressed as the total volume of an event that travels beyond a certain point of interest, the peak discharge reaching an element at risk or the area inundated. All three variables are interrelated. As previously discussed (Jakob 2005), the method of which magnitude variable to choose depends on the objective of the hazard assessment. Debris-flow volumes are important as input to runout models and to design a containment structure. Peak discharge needs to be estimated to size conveyance structures (culverts, bridges, infrastructure crossings). The area inundated must be known for a range of return periods for urban planning and infrastructure layout and design. Jakob (2005) and many other authors prior and since have provided information on how to determine debris-flow volumes using a variety of methods. In general these can be classified in (a) methods using watershed morphometric parameters, (b) methods using detailed geomorphic mapping from remote sensing and field traverses in estimating the amount of material stored in channels, (c) yield rate estimates and (d) fan or cone studies. Method (a) and (b) are only suited to provide an estimate of the volume of the next debris flow, or perhaps a "median" volume. They are less suited to determine "design magnitudes" that are defined as the magnitude corresponding to a frequency for which a given mitigation structure is designed. A multivariate morphometric analysis can only then be applied successfully if the analysis is based on correlation of historical data of very large flows to basin parameters. It cannot produce a frequency-magnitude relationship. Method (c) in which runoff frequency – magnitude relationships are equated with sediment yield can produce debris-flow frequency-magnitude relationships, but those are associated with large errors due to the non-linearity of sediment supply mechanisms. Above a certain return period, the debris-flow trigger may no longer be single debris slides or debris avalanches, but a larger landslide. If a certain hydroclimatic threshold is exceeded, it is also possible that the majority of surface sediments above a given slope angle detach and evolve into debris flows. There are hundreds of examples for this worldwide, especially in the subtropics and tropics. Furthermore, the coincidence of a strong seismic event and a wet season could lead to a non-linear quasi-simultaneous increase in point source failures that would unite to yield an unexpectedly high total debris-flow volume.

Method (d) provides the surest tool to determine debris-flow volumes because it provides a chronology of events and associated thicknesses and areas inundated. Lamentably, such studies require significant capital expenditures and time. Depending on the fan or cone size, tens of test trenches, often to the full depth of a standard excavator reach (5–6 m), are required to extract a sufficiently complete chronology of events. For safety reason, the trenches need to be shored or access be obtained through lowering of steel cages. This is required to allow the practitioner to access the test trench walls and extract organic material for dating. Each debris-flow layer needs to be carefully mapped and ideally bulk samples taken for grain size analyses. If debris-flow deposits are separated by soils (particularly in the more distal, non-erosive portions of the fan or cone) organic material found in these soils can also be used to identify individual debris-flow deposits and correlated them between test pits (i.e. Jakob and Weatherly 2005). Three-dimensional representation of the fan or cone architecture can then be achieved via specialized software, or through pen and paper area integration of dated debris-flow deposits.

3.2 Limitation of Magnitude Analysis

While fan/cone-based investigation methods yield reliable frequency-magnitude data, there are several factors that may severely impede their applications. For example, frequency-magnitude curves may be truncated towards the high frequency – low magnitude range. Thin disconnected stringers of such debris flows that cover only small portions of a fan or cone have been observed following wildfires in the southwestern United States (Cannon, personal communication, 2010) that are unlikely to be recorded in even a detailed fan or cone study. This volume truncation of high frequency flows, however, is not considered a major problem for engineering application because design events (usually of return periods between 100 and

500 years) will likely be recorded in the fan or cone stratigraphy. For example, Scott (1989) investigated the recurrence intervals and peak discharge of very large lahars in The Toutle-Cowlitz River system at Mount St. Helens and associated the occurrence of lahars to periods of volcanic activity over a 40,000–50,000 year period. Scott recognized the problem of data censoring of smaller flows that are not recorded but stressed the importance of hazardous large events for planning purposes. For this reason, he focused on lahars that travelled at least 20 km and concluded that a major lahar with a travel distance in the order of 50 km and peak discharge of 10,000 m^3/s is likely to occur at recurrence intervals around 100 years. Similarly high frequency – low volume debris flows have little influence on quantitative risk assessments because such flows are characterized by low flow depth, low flow velocities (thus low impact force) and low portion of the fan surface inundated. This implies that the probability of death of individuals or groups is very low and will thus not affect the shape of F-N (cumulative frequency - number of potential fatalities) curves or PDI (probability of death of individuals) estimates.

The reconstruction of debris-flow magnitudes on fans or cones can also be constrained by data censoring. Deposits of earlier events are invariably reworked or at least partially eroded by the debris-flow creek or trunk stream. Masking by more recent events is not a problem since the test trenching aims to discover such obliterated events.

In summary, several methods to determine debris-flow magnitude and especially volume have been presented in the past 30 years. No method better allows the reconstruction of long-term debris-flow magnitudes than detailed three-dimensional fan or cone studies. These are labour and cost-intensive but are usually only a fraction, the cost of a vital infrastructure outage, or a key building.

4 Frequency-Magnitude Relationships

4.1 Methodical Assessment

Few studies have provided conclusive data on which to base frequency-magnitude relationships for debris flows. Innes (1985) emphasized that theoretically-derived frequency-magnitude relationships cannot be transferred to mass movement processes. He suggests that at a given point in time, the frequency of flows adjusts to the rate of debris supply, which has also been suggested by Okuda et al. (1980a, b). Innes concludes that the frequency of debris flows should decline exponentially with magnitude, which may hold for transport-limited basin but is questionable for weathering-limited basins because the amount of erodible material determines the shape of the frequency distribution (Jakob 1996). Johnson et al. (1991) developed a predictive model to estimate frequency and magnitude of debris flows in an area where stratigraphic data of earlier events were lacking. A specific debris yield was obtained by dividing the recorded yield by the basin area and by the length of record in years. The means and standard deviations of specific debris yields were based on the log-normal distribution, which was assumed to underlie the physical process. Expected annual debris-flow volumes were computed using a power-law regression model based on various morphometric variables and wildfire recurrence intervals. Sufficient data on frequency and magnitude were available to construct a probability graph. This approach is probably the most desirable but requires high quality data that, in this case, were only available because most debris-flow deposits ended in artificial catchment basins that are emptied after each event.

As one of the three (together with varve chronology and sclerochronology) annual dating methods, dendrochronology plays a crucial part in reconstructing debris-flow activity in the past. This can be accomplished for single debris-flow fans or cones (i.e. Stoffel et al. 2010; Bollschweiler et al. 2010), or on a regional scale (i.e. Strunk 1995; Jakob 1996; May and Gresswell 2004; Bollschweiler et al. 2010). This is important to answer questions of landform evolution and especially to develop reliable frequency-magnitude (F-M) relationships of debris flows that form the basis for quantitative risk assessments (QRAs).

Several countries are now calling for landslide QRAs for their existing or new developments. The development of reliable F-M relationships requires a significant effort. (Stoffel et al. 2010) provides an example where 2,246 tree ring series were analysed from 1,102 old-growth trees (see also Schneuwly-Bollschweiler and Stoffel 2012, *this volume*), although recent work (Stoffel and Bollschweiler 2008) has suggested that a much smaller subset of data may provide results that are still usable for frequency reconstructions. Frequency estimates can and should be extended and completed from other data sources such as interviews with locals, newspaper searches, air photograph interpretation, lake sediment analysis, vegetation succession (i.e. Bollschweiler et al. 2008), lichenometry (Helsen et al. 2002; Wilkerson and Schmidt 2009) and radiometric dating (i.e. Jakob and Weatherly 2005). Dendrochronology offers the advantage of being precise and allowing a more or less continuous record over a few hundred years if performed by qualified personnel.

Dendrochronology can also support the reconstruction of debris-flow or flash-flood volume and peak discharge (Ballesteros et al. 2010). Volumes can be deciphered by mapping trees affected by the same event that may provide at least areas affected particularly for fans or cones in densely vegetated areas that preserve the spatial pattern of debris-flow inundation (Bollschweiler et al. 2007; Stoffel et al. 2010). Areas are related to volumes and can thus be used as a volume surrogate if subsurface investigations cannot be conducted due to budget limitation, land ownership or lack of access for heavy machinery. Peak discharge can be back-calculated if tree scars along bedrock controlled sections are dated, the cross-section measured and the flow velocities calculated from empirical equations (Jakob 2005). If risk assessments need to exceed the dating range for dendrochronology because, for example, the development has a risk potential that requires modelling of several thousand year return period events, radiometric methods (standard radiocarbon and AMS dating) will need to extend the dendrochronologic record (Jakob and Friele 2010).

4.2 Limitations of Frequency-Magnitude Analysis

The previous elaborations point towards the requirements for a practitioner to develop reliable F-M relationships. It is obvious that without a combination of methods in determining both the F and M variables, reliable long-term relationships cannot be determined. Since these F-M relationships build the foundation for hazard or risk zoning, numerical modelling and quantitative or qualitative risk assessments, their correctness is of fundamental importance. In most cases private consulting firms establish the F-M relationships. Since the skill set required to perform these analyses is not unique, and a number of companies in each country compete for such assignments, costs play a significant role in selecting the winning proposal and adhering to a pre-determined budget that is allocated to such studies. Herein lays a problem in that budgets may be estimated too low by the consultant or the contracting agency to fulfil the study objectives satisfactorily. In the author's opinion, an average fan or cone study for a 1 km^2 fan/cone complex may cost well over USD $ 100,000 but, in the case of particularly large or complex fans or cones with multiple use watersheds, may be significantly higher.

F-M relationships may plot as straight lines on a log-log scale, or they may have breaking points, particularly for long time series. Such breaking points could occur due to a significant change in climate (Jomelli et al. 2004; Jomelli et al. 2007; Jakob and Lambert 2009) sediment availability, watershed-scale environmental change such as insect infestations, mining, wildfires, and large mass movements or by a different triggering mechanism for debris flows. These breaks or knick points in the frequency-magnitude relationship are of fundamental importance as their ignorance can lead to severe over- or underestimation of low frequency (high return period) events. For example, on Cheekye River, it was found that such knick point exists in the low frequency scale when debris flows are no longer triggered by shallow debris avalanches and the coalescence of debris flows from multiple tributary channels, but rather from rock avalanches that either evolve into debris flows or that dam the main stream and lead to outbreak floods (Jakob and Friele 2010). A linear extrapolation of the rainfall generated debris-flow F-M curve would have resulted in a significant overestimation of the 10,000-year return period event volume (8 Mm^3 vs. approx. 3 Mm^3), see Fig. 1. In the former case debris-flow mitigation would have been practically and economically impossible; in the latter case mitigation can be considered at a feasible cost. Similar differences in debris-flow triggering processes that affect the frequency-magnitude relationships have been identified elsewhere (i.e. Lugon and Monbaron 1998; Jakob and Weatherly 2005).

5 Discussion and Conclusions

Qualitative, semi-quantitative and quantitative risk assessments of debris flows all require a sound analysis of the F-M relationships of debris flows. The more elements are at risk (population, infrastructure, urban housing, hazardous installations), the

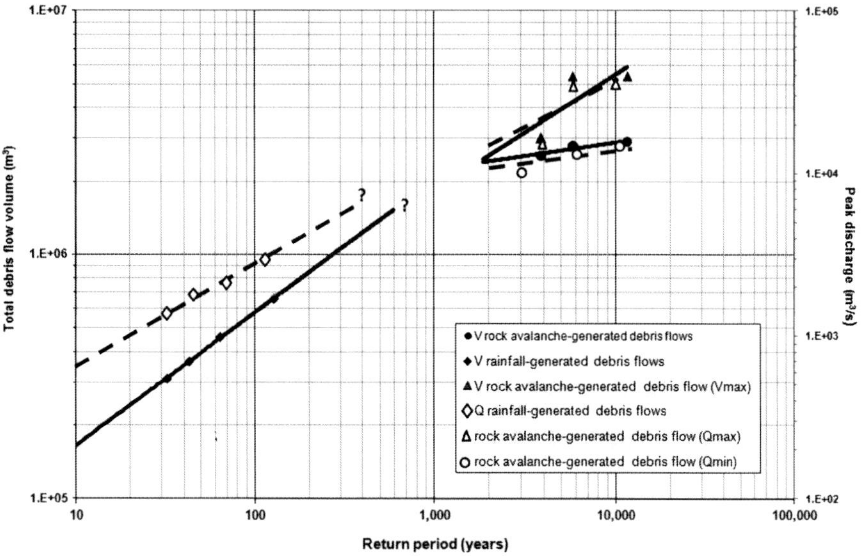

Fig. 1 Frequency-magnitude relationship for debris flows on Cheekye fan, British Columbia. Note the pronounced break in slope that is due to a different data population for very large debris flows (rock avalanche transitioning to debris flows) than the lower magnitude debris flows that are largely triggered by shallow slope failures during heavy rainfall events

higher the need for a multidisciplinary approach in which a series of methods should be applied for comparison and to supplement each other. For high consequence cases, short time series (tens of years to a few hundred years) are insufficient and linear extrapolation may be a dangerous practice. The practitioner who is managing the assignment must be highly critical and literally dig deeper to decipher the historic archive stored in the natural environment. Alternative methods exist such as demonstrated by Stoffel (2010) in the form of very detailed dendrochronologic study, but those can be even more labour-intensive. Due to the competitive nature of many hazard or risk consulting projects that create price pressures, it is also important that the authorities that are writing the requests for proposals are fully aware of the limitations of F-M analysis. Only then are they equipped with the tools to better evaluate proposals by technical merit.

The non-linearity of F-M relationships may be founded in a number of geophysical events that are extrinsic or intrinsic to the watershed in question. The extrinsic factors include climate change and associated changes in the frequency and magnitude of hydroclimatic events (i.e. Van Steijn 1996; Bollschweiler et al. 2008), changes in the wildfire frequency, either by changes in drought frequency, thunderstorm frequency or changes in wildfire management. The intrinsic factors, that can be linked to the extrinsic ones, include changes in watershed management (logging, animal grazing, forestation), construction activities such as mining, forest road construction, ski area development, or be associated with cataclysmic events such as large landslides or rock avalanches that can influence the sediment availability and thus the magnitude of future flows.

Once a reliable F-M relationship has been created, and which is ideally linked to a sound error analysis, the practitioner can begin to calibrate a runout model based on the observed runout characteristics that develop as a by-product of the frequency-magnitude analysis. In this sense, test trenching or drilling of the fan or cone surface provide an invaluable resource. The material characteristics (bulk sample grain size distribution) and the degree of down-fan thinning of sediment layers provide important information on the material rheologies. Comparison of different events from a stratigraphic and material property point-of-view also allows a study of the variability in rheologies that would need to be incorporated in the runout model. For this reason, F-M analyses should not be viewed in isolation. Moreover, a poorly developed frequency-magnitude analysis will invariably lead to erroneous model assumptions that will propagate into the risk quantification. In the best case, the implications on risk are minor and only unsatisfactory from a scientific viewpoint. In the worst case, the neglect of developing a reliable F-M relationship can lead to faulty danger (hazard) maps on which important planning, mitigation and development decisions are based. Ultimately, an erroneous F-M analysis could lead to unexpected economic and life losses.

Within jurisdictions that are overseeing debris-flow risk assessments, a few test studies should be conducted, perhaps in association with research institutions, in which very detailed frequency-magnitude relationships be developed using several methods. This could be done on fans or cones that have already been studied by practitioners and then compared to previous results. Such studies would answer the question of just how much effort should be expended for fan or cone studies. Given that a single life in the developed world may be monetized to $5–$ 10 million, and given that a catastrophic debris flow may lead to tens or even hundreds of fatalities, spending a few hundred thousand dollars more likely constitutes a cost-effective investment in an increasingly risk-averse society. Moreover, the better developed a F-M curve; the less uncertainty (factor-of-safety) will need to be factored into the design of structural debris-flow mitigation measures, which can easily offset the extra cost for a detailed F-M assessment.

Acknowledgements The author gratefully acknowledges Kris Holm, Michelle Schneuwly-Bollschweiler and Markus Stoffel for reviewing this chapter.

References

Ballesteros-Canovas JA, Stoffel M, Bollschweiler M, Bodoque del Pozo JM, Díez-Herrero A (2010) Flash-flood impacts cause changes in wood anatomy of Alnus glutinosa, Fraxinus angustifolia and Quercus pyrenaica. Tree Physiology 30:773–781

Baroni G, Gentili R, Armiraglio S (2012) Vegetation analysis on composite debris cones. In: Schneuwly-Bollschweiler M, Stoffel M, Rudolf-Miklau M (eds) Dating torrential processes on fans and cones – methods and their application for hazard and risk assessment, Advances in global change research. Springer, Dordrecht/Heidelberg/London/New York

BGC Engineering Consultants Inc. (2009) A quantitative risk assessment for Cheekye fan. Report to Cheekeye River Development Corporation

Bollschweiler M, Stoffel M, Ehmisch M, Monbaron M (2007) Reconstructing spatio-temporal patterns of debris-flow activity using dendrogeomorphological methods. Geomorphology 87:337–351

Bollschweiler M, Stoffel M, Schneuwly DM (2008) Dynamics in debris-flow activity on a forested cone – A case study using different dendroecological approaches. Catena 72(1):67–78

Bollschweiler M, Stoffel M, Schneuwly DM (2010) Using event and minimum age dating for the assessment of hazards on a debris-flow cone. In: Stoffel M, Bollschweiler M, Butler DR, Luckman BH (eds) Tree rings and natural hazards: A state-of-the-art. Springer, Berlin, Heidelberg, New York

Bründl M (2012) Dealing with natural hazard risks in Switzerland – the influence of hazard mapping on risk-based decision making. In: Schneuwly-Bollschweiler M, Stoffel M, Rudolf-Miklau M (eds) Dating torrential processes on fans and cones – methods and their application for hazard and risk assessment, Advances in global change research. Springer, Dordrecht/Heidelberg/London/New York

Chiverell R, Jakob M (2012) Radiocarbon dating: alluvial fan/debris cone evolution and hazards. In: Schneuwly-Bollschweiler M, Stoffel M, Rudolf-Miklau M (eds) Dating torrential processes on fans and cones – methods and their application for hazard and risk assessment, Advances in global change research. Springer, Dordrecht/Heidelberg/London/New York

D'Agostino V (2012) Assessment of past torrential events through historical sources. In: Schneuwly-Bollschweiler M, Stoffel M, Rudolf-Miklau M (eds) Dating torrential processes on fans and cones – methods and their application for hazard and risk assessment, Advances in global change research. Springer, Dordrecht/Heidelberg/London/New York

Gilli A, Anselmetti FS, Glur L, Wirth SB (2012) Lake sediments as archives of recurrence rates and intensities of past flood events. In: Schneuwly-Bollschweiler M, Stoffel M, Rudolf-Miklau M (eds) Dating torrential processes on fans and cones – methods and their application for hazard and risk assessment, Advances in global change research. Springer, Dordrecht/Heidelberg/London/New York

Helsen MM, Koop PJM, van Steijn H (2002) Magnitude-frequency relationship for debris flows on the fan of the Chalance torrent, Valgaudemar (French Alps). Earth Surf Processes Landforms 27:1299–1307

Hürlimann M, Rickenmann D, Graf C (2003) Field monitoring and modeling in the Swiss Alps. Can Geotech J 40:161–175

Innes JL (1985) Magnitude – frequency relations of debris flows in northwest Europe. Geogr Ann 67 A(1–2):23–32

Ivy-Ochs S, Dühnforth M, Densmore AL, Alfimov V (2012) Dating fan deposits with cosmogenic nuclides. In: Schneuwly-Bollschweiler M, Stoffel M, Rudolf-Miklau M (eds) Dating torrential processes on fans and cones – methods and their application for hazard and risk assessment, Advances in global change research. Springer, Dordrecht/Heidelberg/London/New York

Jakob M (1996) Morphometric and geotechnical controls on debris flow frequency and magnitude in Southwestern British Columbia. PhD dissertation, University of British Columbia

Jakob M (2005) Debris flow hazard assessments. In: Jakob M, Hungr O (eds) Debris flows and debris avalanches – a practically oriented overview of the state-of-the-art. Praxis and Springer, Heidelberg

Jakob M (2012) The fallacy of frequency: statistical techniques for debris-flow frequency-magnitude analysis. In: 11th International Symposium on Landslides. Banff ISL. June 3–8 2012. In print

Jakob M, Friele AP (2010) Landslide hazards and risks from volcanic debris flows at Mount Garibaldi, British Columbia. Geomorphology 114:382–395

Jakob M, Hungr O (eds) (2005) Debris-flow hazards and related phenomena. Springer, Berlin, Heidelberg, New York

Jakob M, Lambert S (2009) Climate change effects on landslides along the south-west coast of British Columbia. Geomorphology 107:275–284

Jakob M, Weatherly H (2005) Debris flow hazard and risk assessment. Jones Creek, Washington. In: Hungr O, Fell R, Couture R, Eberhardt E (eds) Landslide risk management. Proceedings, pp 533–542

Johnson PA, McCuen RH, Hromadga Tv (1991) Magnitude and frequency of debris flows. J Hydrol 123:68–82

Jomelli V, Pech VP, Chochillon C, Brunstein D (2004) Geomorphic variations of debris flows and recent climatic change in the French Alps. Clim Chang 64(1–2):77–102

Jomelli V, Brunstein D, Grancher D, Pech P (2007) Is the response of hill slope debris flows to recent climate change univocal? A case study in the Massif des Ecrins (French Alps). Clim Chang 85(1–2):119–137

Kaitna R, Huebl J (2012) Silent witnesses for torrential processes. In: Schneuwly-Bollschweiler M, Stoffel M, Rudolf-Miklau M (eds) Dating torrential processes on fans and cones – methods and their application for hazard and risk assessment, Advances in global change research. Springer, Dordrecht/Heidelberg/London/New York

Lang A (2012) Luminescence dating of alluvial fans and cones. In: Schneuwly-Bollschweiler M, Stoffel M, Rudolf-Miklau M (eds) Dating torrential processes on fans and cones – methods and their application for hazard and risk assessment, Advances in global change research. Springer, Dordrecht/Heidelberg/London/New York

Lugon R and Monbaron M (1998) Stabilité des terrains meubles en zone de pergélisol et changements climatiques. Deux études de cas en Valais: Le Ritigraben (Mattertal) et la moraine du Dolent (Val Ferret). vdf Huchschulverlag, Zürich

Marchi L, Tecca PR (2012) Debris-flow monitoring in Italy. In: Schneuwly-Bollschweiler M, Stoffel M, Rudolf-Miklau M (eds) Dating torrential processes on fans and cones – methods and their application for hazard and risk assessment, Advances in global change research. Springer, Dordrecht/Heidelberg/London/New York

Marchi L, Arattano M, Deganutti AM (2002) Ten years of debris-flow monitoring in the Moscardo Torrent (Italian Alps). Geomorphology 46:1–17

May CL, Gresswell RE (2004) Spatial and temporal patterns of debris-flow deposition in the Oregon Coast Range, USA. Geomorphology 57:135–149

Ohmori H, Hirano M (1988) Magnitude, frequency and geomorphological significance of rock mud flows, landcreep and the collapse of steep slopes. Zeitschrift für Geomorphologie, Neue Folge, Supplementband 67:55–65

Okuda S, Suwa H, Okunishi K, Yokoyama K, Nakano M (1980a) Observations on the motion of a debris flow and its geomorphological effects. Zeitschrift für Geomorphologie Supplementband 35:142–163

Okuda S, Okunishi K. and Suwa H (1980b) Observation of debris flow at Kamikamihori Valley of Mt. Yakedake. In: Okuda S (ed) Excursion guide book, 3rd meeting of IGU commission on field experiments in geomorphology, Kyoto, pp 116–139

Schneuwly-Bollschweiler M, Stoffel M (2012) Dendrogeomorphology – tracking past events with tree rings. In: Schneuwly-Bollschweiler M, Stoffel M, Rudolf-Miklau M (eds) Dating torrential processes on fans and cones – methods and their application for hazard and risk assessment, Advances in global change research. Springer, Dordrecht/Heidelberg/London/New York

Schneuwly-Bollschweiler M, Stoffel M, Rudolf-Miklau M (2012) Dating torrential processes on fans and cones – methods and their application for hazard and risk assessment, Advances in global change research. Springer, Dordrecht/Heidelberg/London/New York

Scott KM (1989) Magnitude and frequency of lahars and lahar runout flows in the Toutle-Cowlitz River system. U.S. Geological Survey Open-file Report 90–385, 106 p

Staley DM, Wasklewicz TA (2012) The use of airborne laser swath mapping on fans and cones: an example from the Colorado Front Range. In: Schneuwly-Bollschweiler M, Stoffel M, Rudolf-Miklau M (eds) Dating torrential processes on fans and cones – methods and their application for hazard and risk assessment, Advances in global change research. Springer, Dordrecht/Heidelberg/London/New York

Stoffel M (2010) Magnitude-frequency relationships of debris flows – a case study based on field surveys and tree-ring records. Geomorphology 116:67–76

Stoffel M, Bollschweiler M (2008) Tree-ring analysis in natural hazards research – an overview. Nat Haz Earth Syst Sci 8:187–202

Stoffel M, Bollschweiler M, Widmer S, Sorg A (2010) Spatio-temporal variability in debris-flow activity: a tree-ring study at Geisstriftbach (Swiss Alps) extending back to AD 1736. Swiss J Geosci 103:283–292

Strunk H (1995) Dendrogeomorphologische Methoden zur Emittelung der Murfrequenz und Beispiele ihrer Anwendung. Roderer, Regensburg

Tecca PR, Galgaro A, Genevois R, Deganutti AM (2003) Development of a remotely controlled debris-flow monitoring system in the Dolomites (Acquabona, Italy). Hydrol Processes 17:1771–1784

Van Steijn H (1996) Debris-flow magnitude-frequency relationships for mountainous regions of central and northwest Europe. Geomorphology 15:259–273

Wilkerson F, Schmidt G (2009) Dendrogeomorphic applications to debris flows in Glacier National Park, Montana, USA. In: Stoffel M, Bollschweiler M, Butler DR, Luckman BH (eds) Tree rings and natural hazards. Springer, Berlin/Heidelberg/New York

Part II
Dating Past Events

Silent Witnesses for Torrential Processes

Roland Kaitna and Johannes Huebl

1 Introduction

Torrential processes like debris flows, debris floods, and intensive bedload transport represent a serious hazard for settlements and infrastructure located on Alpine fans. Protection works and mitigation measures are mostly based on the magnitude of a design event with a defined recurrence interval. However, engineering hazard assessment of torrential processes is often limited to rough estimation of possible future events since the data base is small and statistically reliable relationships between event frequency and magnitude are rare. Some information can be found in historical records of local communities. For this reason some effort has been done recently to collect this information from public administration offices and chronics in Austria (Huebl et al. 2010). Although the data is heterogeneous and its applicability for standard engineering extreme-value analysis is limited, information on historical events represents a helpful tool for hazard assessment in Alpine watershed and should be gathered if available.

Field investigation of the natural setting of a watershed and the associated hazard potential for human assets is and will be necessary in engineering practise. A useful tool to identify the process, the magnitude and frequency of past events (backward directed approach, Kienholz 1998) and subsequently to assess possible future events is the identification and interpretation of "silent witnesses". Silent witnesses are geomorphic and biologic evidence of past mass movement events found in the field. The term has been coined in the first half of the last century by Emil Straube, a member of the Austrian Service for Torrent and Avalanche Control, and has been used in scientific context since Aulitzky (1992). Investigation of a torrential fan is of

R. Kaitna • J. Huebl (✉)
Department of Civil Engineering and Natural Hazards, Institute of Mountain Risk Engineering, University of Natural Resources and Life Sciences, Vienna, A-1190 Vienna, Austria
e-mail: roland.kaitna@boku.ac.at; johannes.huebl@boku.ac.at

M. Schneuwly-Bollschweiler et al. (eds.), *Dating Torrential Processes on Fans and Cones*, Advances in Global Change Research 47, DOI 10.1007/978-94-007-4336-6_7,

particular interest, since it stores the history of developments within the watershed. Parameters like inclination, area, and surface topography are directly connected to the processes in the catchment. Glacial activity can have a strong influence on the development of torrential fans. Resulting geomorphic evidence represents additional information for hazard assessment and therefore should not be neglected.

Information from silent witnesses enables engineers and scientists to increase the data base of past events and helps to improve the understanding of the reaction of the watershed system to particular climate and weather conditions.

To give a systematic overview of possible silent witnesses we describe some large-scale geomorphic pattern which can be identified by remote sensing techniques and aerial pictures in Sect. 2. Section 3 lists small-scale geomorphic features and vegetation evidence of recent mass-flow events which can be identified by field investigations in the headwaters, transit reach and on the fan.

2 Large-Scale Geomorphic Evidence of Torrential Processes

Torrential watersheds that produce debris flows, debris floods and intensive bedload transport are steep headwater channels of mountain rivers and their tributaries. Geomorphic processes on the hillslope and in the channel are closely connected and mostly triggered by local rainfall events of high intensity and short duration (e.g. Costa 1984; Takahashi 1991; Marchi et al. 2002). Sediment production and transport represent a hazard for human settlements but nevertheless are an important factor for bedload equilibrium in larger rivers. This has been often neglected in engineering projects in the past. Several classification schemes (cf. chapter 'Processes of sediment supply to alluvial fans and debris cones') were suggested to describe the development and behaviour of torrential watersheds. We want to mention a few of these to stake out some general trends, which are useful to consider with regard to silent witnesses found in the field.

Based on the presence of a depositional fan Salzer (1886) described torrent catchments in high Alpine regions as a sequence of an initiation zone, a transit zone and a deposition zone (=fan or cone). Alpine torrential catchments in lower regions are characterized by an initiation zone and an elongated reach of sediment storage respectively re-erosion. Such catchments are often characterized by the non-existence of a distinct fan or cone.

In his pioneering work Stiny (1910) identified the main sediment sources for mass wasting processes. He differentiated between torrents which transfer 'old' sediment that has been deposited over geologic time scales (e.g. glacial deposition during cold periods) and torrents where most of the transported sediment has been produced 'recently' by physical and chemical weathering. The first type of torrent (german '*Altschuttwildbach*') tends to produce episodic and mostly large debris-flow and debris-flood events. As long as material is available the limiting factor is the rainfall input. The second type (german '*Jungschuttwildbach*') can be regarded as sediment limited, since the loose material is produced by weathering, and is transported and stored in or close to the channel. Periodic mass flow events can be subsequently triggered by rainfall events as soon as enough sediment is available. Hence, in the context of hazard assessment it is important to gain knowledge on the long term history of a watershed and therefore to look for traces of glacial activity and glacial deposits. A more process based classification scheme has been suggested by Aulitzky (1980). In his "two-fold classification of torrents", Aulitzky differentiates between debris-flow torrents, debris-flood torrents, bedload torrents and flood creeks. Subsequently dominant erosion processes in the watershed are assessed and a checklist of geomorphic and biologic evidence (silent witnesses) is used as hazard indicators to identify expected type and magnitude of torrential processes. A further classification scheme is given by Bunza et al. (1982). They connect large-scale erosion and deposition features with geology and the dynamic of the system to identify major sediment sources. This is especially useful for obtaining a first assessment of process types to be expected.

2.1 *Watershed*

For an initial hazard assessment it is helpful to identify large-scale evidence of mass-wasting processes by using geologic maps, digital elevation models and/or aerial photographs. Large-scale depth erosion ('file' erosion) and erosion of talus material below competent bedrock ('wedge' erosion) suggest a deep layer of soil or colluvial material which results from high transport capacity of the channel and high sediment availability for torrential processes (see Bunza et al. 1982). Unexpected topographic pattern and anomalies on the hillslope as well as in the channel network may indicate mass movements and erosion processes that can affect torrential activity. An example for this is the Auenbach (Austria) showing a degraded channel system due to several active landslides in the head-watershed (Fig. 1a) serving as sediment sources for intensive bedload transport in the channel.

From a hydrologic perspective the shape of a catchment can influence runoff concentration (an elongated catchment tends to produce an attenuated hydrograph compared to a more circular one). A high channel density indicates a substrate of low permeability and potentially high surface runoff. In combination with information of relief energy and sediment availability a first assessment of the hazard potential of a torrent can be done. A detailed analysis of channel networks as indicators for torrential activity can be found in Zeller (1995). Deep seated mass creep may be identified by a wide and well defined head scarp and convex toe of the hillslope. Especially in soft rocks (e.g. phyllitic material) depth creep as well as sagging is often observed and is potentially connected to an unloading of the slopes after de-glaciations. Due to a very slow movement vegetation has time to recover the head scarps. The geomorphic evidence of double-ridges (Fig. 2) indicates a deep seated mass-creep at its upper end. The toe of a deep seated mass-creep can modify the course and the width of the channel which subsequently can lead to an increased erosion of the banks and destruction of engineering mitigation measures. Consequently bank failure is the main source for sediment transport.

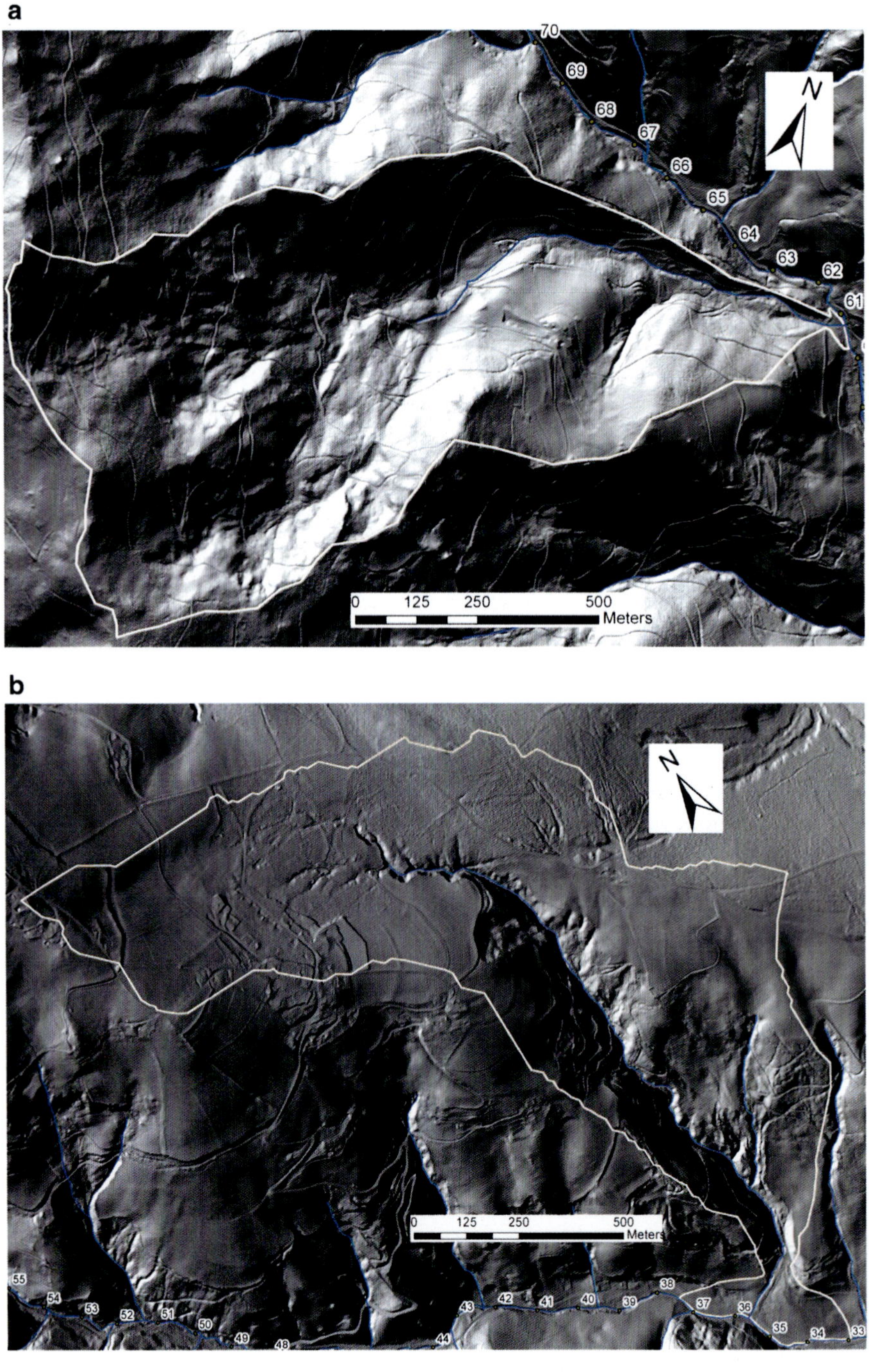

Fig. 1 (**a**) Disturbed surface topography and lack of a distinct channel system due to active landslides of a tributary of the Auenbach (Austria). (**b**) Linear erosion pattern of a well developed head-water channel of another tributary of the same torrent

Fig. 2 Double ridge at the upper limit of the Loehnersbach watershed in Austria (photo: courtesy H.W. Weinmeister)

2.2 Fan

The topography of the fan reflects the development and erosion processes within the watershed. Already the slope of the fan may give a clue of the sediment concentration of past and expected torrential processes. Alluvial fans are formed by fluvial action, whereas colluvial fans are formed by landslide activity, including debris flows. In some regions, also the term debris-flow cone is used. Fans which have been formed at least partially by debris-flow activity show generally a higher gradient than alluvial fans. As pointed out by Jakob (2005) there is no unique threshold value. Fans with a mean slope of $<4°$ are unlikely to be formed by debris flows, but this can vary depending on the geologic setting and material involved. Muddy debris-flows and flows of volcanic origin (lahars) can produce fan gradients as low as 1–2°. Marchi and Tecca (1995), Bardou (2002) and Scheidl and Rickenmann (2010) identified the dominating process in several alpine watersheds

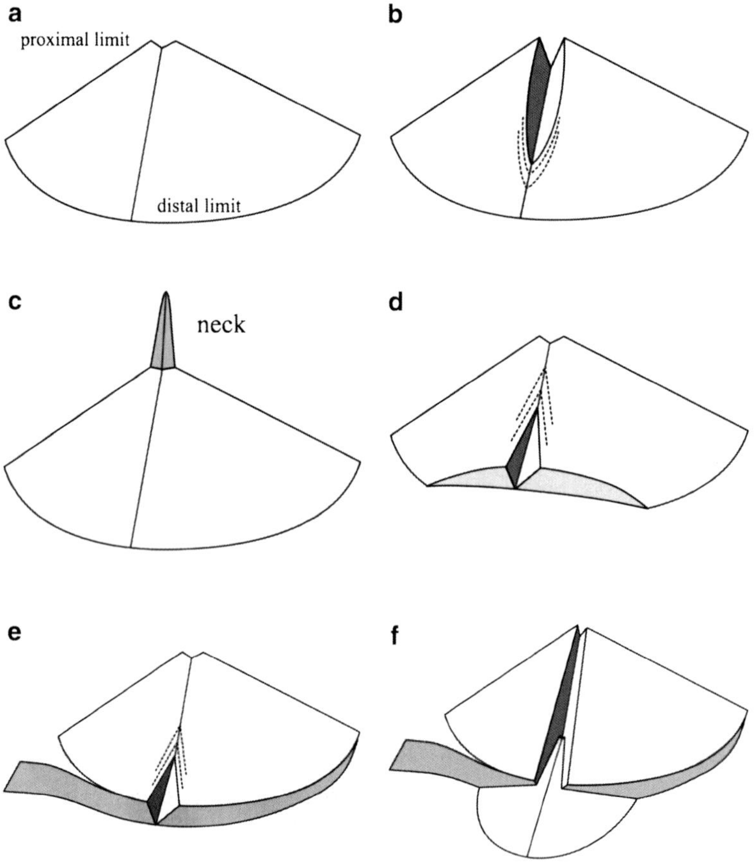

Fig. 3 (**a**) Fan in dynamic equilibrium. (**b**) Downward erosion due to system changes in the watershed. (**c**) Deposition above the proximal limits of the fan due to increased sediment transfer to the fan. (**d**) Retrogressive erosion of the distal limits of the fan due to lateral change of the base level. (**e**) Retrogressive erosion of the distal limits of the fan due to vertical change of the base level. (**f**) Situation (**e**) with additional formation of a secondary fan

using the inclination of the fan and Melton number (specific elevation difference in the watershed divided by the square root of the catchment area above the fan) as a proxy for relief energy.

The longitudinal profile of alluvial fans is often observed to be concave up. Any large-scale convexity may indicate mass flow activities by landslides and earthflows, as recently observed at the Gschliefgraben in Austria. The morphology of the fan and its proximal and distal limits are useful indications of possible system changes within the watershed which result in different stages of development of the fan development. Figure 3 sketches observations made in the Austrian Alps.

In case of ongoing intensive sediment discharge beyond the distal limits of the fan and a limited transport capacity of the receiving river, morphologic pattern like induced meandering may be observed.

For the identification of the dominant process as well as hazard assessment it is useful to delineate active and in-active areas on the fan (Blair and McPherson 1994). This is especially true for fans with deeply incised channels (Fig. 3f). An example is the alluvial fan of the Blaubach, which is one of the largest fans in Austria. Its distal limits have been eroded by the receiving river (Krimmler Ache) resulting in a vertical drop between the fan surface and the base level of 50 m. Subsequently the late glacial fan has been strongly incised by the torrent (Blaubach). Several secondary fans have been developed and eroded by the receiving river, leaving reaches of different levels. Due the deep incision of the Blaubach most parts of the large primary fan are inactive and only the secondary fan is affected by torrential events.

3 Small-Scale Evidence of Torrential Processes

3.1 Headwatershed/Hillslope

Mass wasting processes on the hillslope are strongly connected to torrential processes, as they may represent an important source of sediment for hazard processes. For this reason it is necessary to focus field work not only to the channel but to expand investigation to the adjacent hillslope, vegetation and the erosion processes in the head-watershed. Channel incision, reaches of sediment storage, recent bank failures, shallow and deep-seated landslides are important sources for transport processes in the torrent. Especially in high Alpine areas sediment transfer from scree slopes below rock-faces can add significant amounts of solids. In combination with topographic information like relief energy and an assessment of the hydrologic response of a watershed to intensive precipitation events, the general disposition of the headwatershed to produce hazard processes can be evaluated.

The biological signs of mass movements include inhomogeneous vegetation cover, bended trees, tension cracks in the soil mantle and possibly cut-off root networks (e.g. Aulitzky 1992; Huebl et al. 2003). A forest stand including a high fraction of stems deflecting from the vertical ("drunken forest") indicates an area of recent creeping or sliding activity as recently observed at an earthflow area in Upper Austria (Fig. 4a, b). Widely bended trees result from long term, deep-seated, rotational movement on a distinct failure plane. In some cases it is possible to estimate the age of last movement, by comparison of the habitus of younger trees in the vicinity (Fig. 4c). A closer look at the type of vegetation can give additional information on soil moisture conditions which is important to assess slope stability. There are typical plants that indicate wet soils (e.g. *Caltha palustris, Ranunculus aconitifolius, Mentha longifolia*) with constantly high pore water pressures. A sound overview of vegetation units and indicators of soil condition in Alpine regions is given by Markhart et al. (2004).

Fig. 4 (**a**) Tension crack and strained root network. (**b**) "Drunken" forest at the Gschliefgraben earthflow in Upper Austria. (**c**) Bended tree on an rotational landslide area; note that the young tree is not affected (photos: R. Kaitna)

3.2 Transit Zone/Channel

Traces in the transport reach of the channel can be used to deduce the type of process and important process parameters like maximum flow depth, cross section area and flow velocity. Due to the dynamic nature of most channels and adjacent channel banks, traces rarely persist over long time spans.

An exception represent the erosion marks left from debris flows in bedrock reaches and on artificial structures like check dams. Visible wear and plucking of the bedrock underline the high frictional and collisional stresses related to the passage of debris flows (Fig. 5a). Stock and Dietrich (2003) and Hsu et al. (2008) hypothesize that this effect may be the main agent of channel cutting into bedrock over geologic time scales.

Mudlines and sediment deposits along the channel (Fig. 5b) are reliable traces of the maximum flow depth during floods and debris flow which can be used to back-calculate the maximum cross-section and therefore also peak discharge if a velocity estimate is available. Lack of severe damage next to the channel, scour marks in the channel as well as traces above the typical flood limit indicate debris-flood activity. Additional hints are enhanced sediment deposits at the outer side of channel bends. Great damage or even complete removal of vegetation can be observed after the passage of a debris flow within the flow path, whereas often little or no damage at the edges of the flow or on gentle gradients can be found (Costa 1988). If trees in the flow path have not been removed and/or the bark of the trunks is only peeled, debris-flow velocity was probably low and/or only small grain sizes were present (Costa 1984). Deposition of sediment by debris floods often results in similar accumulation heights, but does not necessarily cause severe mechanical damage to the buried objects (Fig. 6).

Parabolic (or often termed U-shaped) channel cross-sections indicate recent debris flow activity in a torrent (Fig. 8a). Based on the principle of increased flow depth on the outside of channel bends a method to assess the mean debris-flow velocity from the super-elevation of sediment deposits or mudlines (Fig. 7) has been suggested already by Haiden (1925). The forced vortex equation from open channel hydraulics (cf. Chow 1959) applied for debris flows is mostly written as

$$\Delta h = k \frac{w v^2}{R_c g \cos \theta_c}$$

where Δh = elevation difference between inner and outer bend, w = channel width, v = mean flow velocity, R_c = channel radius, g = acceleration due to gravity, θ_c = channel slope, and k = a correction factor. The choice of the correction factor is associated with large uncertainty since literature suggestions range between 1 and 10 (e.g. Hungr et al. 1984; Chen 1987; Bulmer et al. 2002).

Similarly, assuming complete conversion of kinetic energy to potential energy the impacts on trees, the gradient of mudlines on the upstream and downstream side

Fig. 5 Lattenbach torrent (Austria, August 2007). (**a**) Strongly degraded flowsection of a check dam due to wear of frequent debris flows. (**b**) Clear mudlines indicating peak flow after the passage of a debris flow (photos: R. Kaitna)

Fig. 6 Intensive sediment deposition in the village of Pfunds (Austria, August 2005) did not destroy a fragile glass-wood construction. Although 60,000 m^3 of sediment has been deposited on the fan within a short time, the event has been classified as debris flood (photo: R.Kaitna)

of trunks and the runup against obstacles perpendicular to the flow may be used to roughly estimate its velocity using

$$\Delta h = \alpha \frac{v^2}{2g}$$

with α = a correction factor for debris flows. Unfortunately no specific values for the correction factor can be given. Both methods need further investigations.

3.3 *Deposition Zone/Fan*

The proximal limit of the fan represents a significant knick-point in the longitudinal profile of a torrent. This transition from mostly steep and confined conditions to less steep and unconfined enhances spreading of debris flow material and deposition. If not modified my human activity (e.g. plowing) an irregular (hummocky) surface topography is a hint of past debris-flow activity. If a debris-flow deposits stalls within

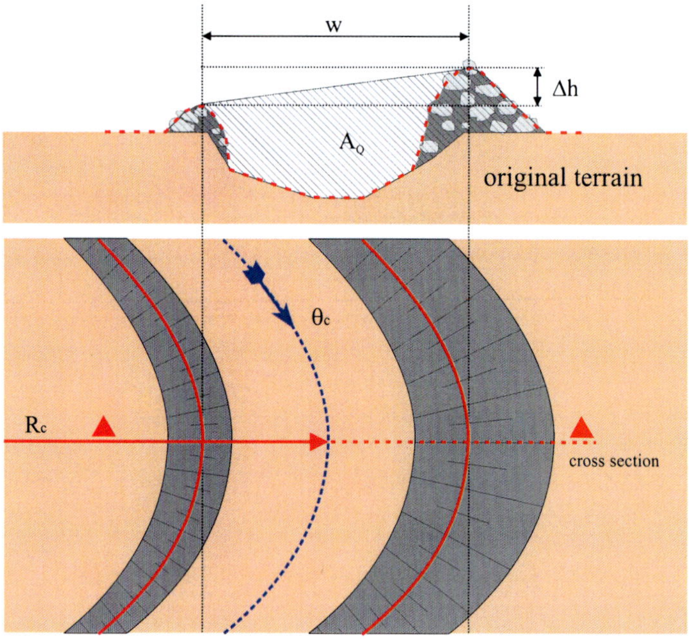

Fig. 7 Sketch of super-elevation of deposits in channel bends after debris-flow passage (Huebl 2008)

the original channel and therefore blocks it, subsequent surges may be deposited outside of the channel and/or a new channel may be formed by consecutive erosion processes (e.g. debris cone as described by Schirmer 1988).

Selective deposition in the course of bedload transport typically leads to a decrease of grain sizes from the fan apex to the distal limit of the fan and to well recognizable sorting pattern or imbrication (e.g. Miall 1996). Deposits of debris flows leave randomly distributed gravel and boulders in a matrix of finer material. Frequently the phenomena of "inverse grading", i.e. large grains overlaying smaller material, has been reported (Fig. 8a). Sporadic large, angular boulders ("megaclasts") that cannot be attributed to glacial activity, rock fall or other mass wasting processes often results from debris flow activity (Fig. 12a). Uncovered boulders that relict from past events may be dated using the method of lichenometry (cf. Jomelli 2012, *this volume*).

Well-defined boulder trains and levées can be found in the transition reach as well as in the deposition area of fans prone to debris flows (Fig. 8a).

Terminal lobes of debris flows often are composed of large boulders lacking a significant amount of fines. They represent a zone of high frictional resistance and therefore tend to deposit first and may then be overrun and/or bypassed by the more liquid body and tail of the flow. Additionally this material pushing from

Fig. 8 Traces of a debris-flow event in Lassen National Park, USA. (**a**) Elevated lateral levées and *U-shaped channel* cross section at a debris-flow transit reach. (**b**) Large boulders overlaying finer sediment (photos: R.Kaitna)

behind and/or subsequent surges can produce pressure ridges perpendicular to flow direction (Fig. 9a). When steep-fronted lobes stop within the channel the subsequent flow may erode a new channel. As already mentioned, several of these events may

Fig. 9 (**a**) Pressure ridges perpendicular to the flow direction. (**b**) Hummocky micro-relief and vegetation covering debris-flow deposits (photos: J. Huebl)

produce a hummocky micro-relief (Fig. 9b). Based on the age of the vegetation cover (e.g. forest stand) or dendrogeomorphic investigations one may assess the frequency and spatial extent of flow events on the fan (cf. Schneuwly-Bollschweiler and Stoffel 2012, *this volume*).

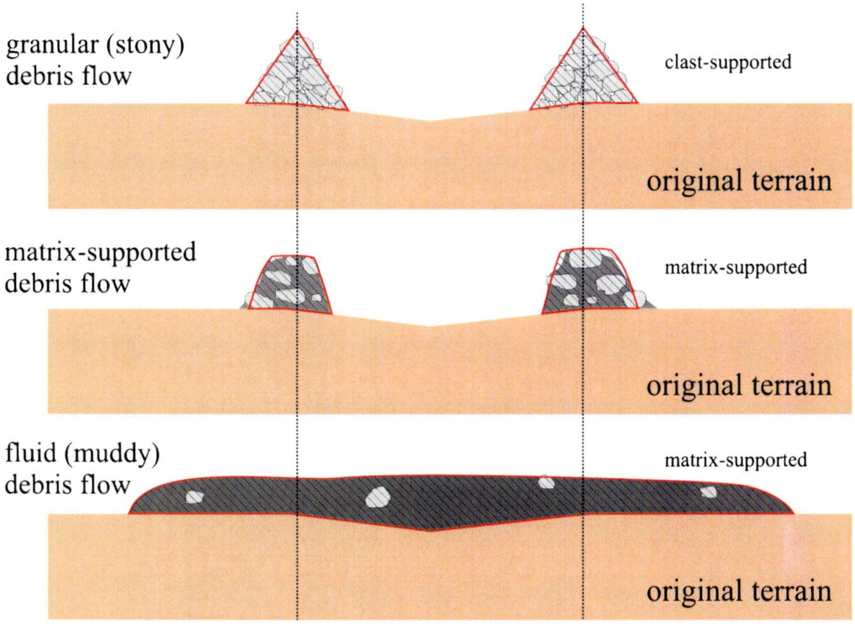

Fig. 10 Sketch reproducing the typical cross-sections of debris-flow deposits as outlined by Bardou et al. (2003); not to scale

The deposits of debris flows consisting of a significant fraction of fines mostly show a sharp limit separating inundated areas and completely unaffected areas. In the case of more granular debris flows the liquid matrix has been observed to drain from the coarse deposits.

Bardou et al. (2003) used debris flow traces to link geomorphic features of the deposits with mechanical properties of the material. This approach, which is based on investigations of debris-flows material in laboratory and on the "rheophysical classification" by Ancey (1999), may be useful for selection of the appropriate simulation tool for hazard assessment. It is assumed that the matrix, which is the fine fraction of the complete grain size distribution, has a great influence on debris flow mechanical behaviour and this is finally also expressed in the cross-sectional shape of the levées (Fig. 10): a curved cross-section often results from muddy debris flows containing a high fraction of fine and cohesive material, as well as coarse material. The deposit appearance in the field is smooth with clear limits and very cohesive once dry. This matrix supported debris flow is expected to have a visco-plastic flow behaviour and may be modelled by a conventional rheologic model (e.g. O'Brien et al. 1993). The typical stopping slope is as low as 2°. Granular debris-flows, where the dominant energy dissipation is due to friction and collision between large particles, typically produce levées with a sharp cross-section shape. The deposits lack fine material, are rough with no clear limits in the field and are non-cohesive when dry. The high flow resistance is indicated by stopping slopes

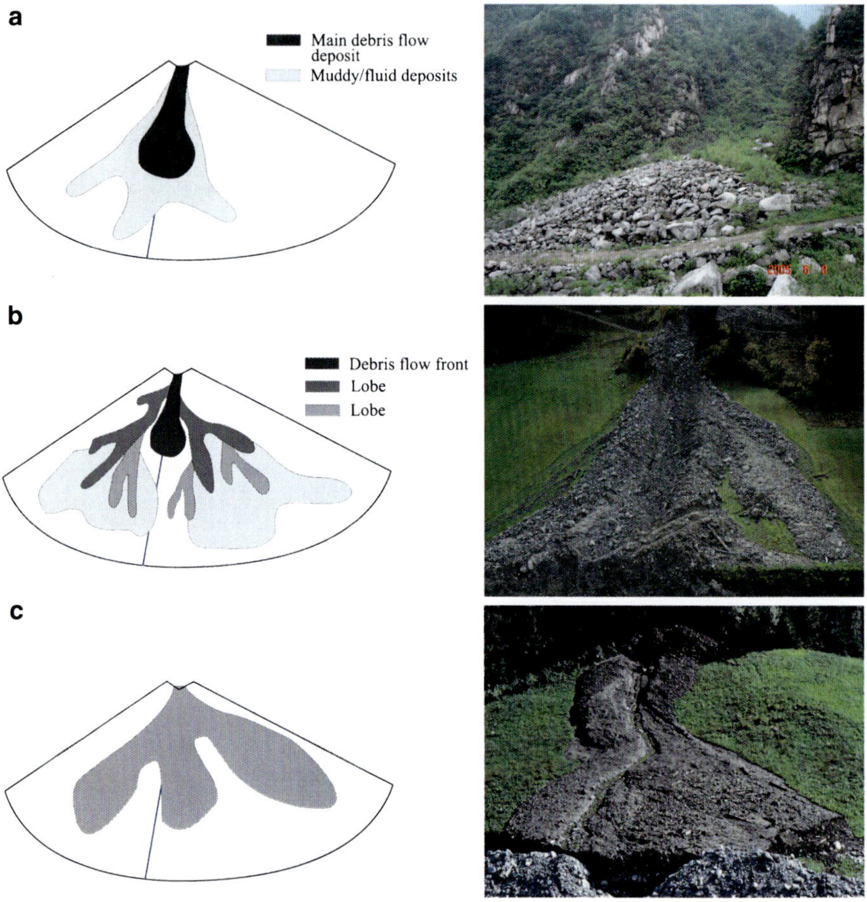

Fig. 11 (**a**) Granular/stony debris flow. (**b**) Matrix-supported debris flow. (**c**) Fluid/muddy debris flow (Huebl 2008; photos: J. Huebl)

of up to 10°. Flat and wide depositional cross-section shapes indicate more fluid debris flows with frictional-viscous flow behaviour. These terrace-like deposits can be found at slopes lower than 1° and are cohesive when dry due to a high content of sand and silt. Bardou et al. (2003) suggest that type 2 and type 3 debris flows should be modelled with a Coulomb like or collisional flow model. It is noted that these models are in development (e.g. Hungr and McDougall 2009; Scheuner et al. 2009) and up to now mainly used in academics rather than engineering practice.

Similar to the classification of Bardou et al. (2003) we identify different types of debris flows based on the shape of the final deposits on the fan (Fig. 11): granular (or stony) debris flows rarely produce multiple lobes as observed for debris flows composed of a significant amount of fine material (matrix supported). The shape of fluid/muddy debris flows is similar, but flat with increased lateral spreading (Huebl 2008).

Fig. 12 (**a**) Large boulder indicating exceptional transport capacity of debris flows. (**b**) Below ground entrance due to historic sediment deposition (photos: J. Huebl)

Relicts of intensive past sediment deposition can be frequently found on the inhabited reaches of European alpine fans. Close ground level windows and entrances below ground (Fig. 12b) indicate that the level was built up by infrequent mass movement processes while depositions have not been removed. Deposition heights of up to 2 m can be deduced in the village of Inzing, Austria (Skermer and VanDine 2005).

4 Conclusion

Identification and interpretation of geomorphic and biologic evidence can yield valuable hints for engineering hazard assessment of torrential processes. Information on landslide activity, sediment availability, type of process, event parameters like peak flow and velocity, as well as magnitude and frequency can be qualitatively and sometimes even quantitatively estimated. Time is an important factor for some of these silent witnesses since traces of the process are often cleared in the course of recovery works in well developed regions or vanish with time e.g. due to rain washing away mud traces or re-working of deposits. Other evidence of past torrential activity may persist of longer time periods and can be found even after decades. However, detailed event documentation is preferably carried out soon after the process.

It is noted that all information gathered by the interpretation of silent witnesses should be brought into the context of watershed disposition. A drastic system change, e.g. due to change of landuse and/or sediment availability may alter the hazard potential significantly and may bias magnitude and frequency estimations based on silent witnesses. For all conclusions from these field interpretations it is of great importance to assess its informative value and quantitative limitations.

Acknowledgements The authors gratefully acknowledge Dr. Christian Scheidl for his GIS support and critical comments on the manuscript.

References

Ancey C (1999) Rhéologie des laves torrentielles. Final scientific report PNRN 1998–99, CEMAGREF

Aulitzky H (1980) Preliminary two-fold classification of torrents. In: Proceedings of international symposium interpraevent, Bad Ischl, Austria, Band 4, pp 285–310

Aulitzky H (1992) Die Sprache der "Stummen Zeugen" (The language of "silent wittnesses"). In: Proceedings of international symposium interpraevent, Bern, Switzerland, Band 6, pp 139–173

Bardou E (2002) Methodologie de diagnostic des laves torrentielles sur un basin versant alpin. PhD thesis, École Polytechnique Fédéral de Laussane, 382 p

Bardou E, Ancey C, Bonnard C, Vulliet L (2003) Classification of debris-flow deposits for hazard assessment in alpine areas. In: Rickenmann D, Chen LC (eds) Debris-flow hazards mitigation: mechanics, prediction, and assessment. Millpress, Rotterdam, pp 799–808

Blair TC, McPherson JG (1994) Alluvial fan processes and forms. In: Abrahams AD (ed) Geomorpholgy of desert environments. Chapman & Hall, London/Glasgow/New York/Tokyo/Melbourne, pp 354–402

Bulmer MH, Barnouin-Jha OS, Peitersen MN, Bourke M (2002) An empirical approach to studying debris flows: implications for planetary modeling studies. J Geophys Res 107:1–16

Bunza G, Karl J, Mangelsdorf J (1982) Geologisch-morphologische Grundlage der Wildbachkunde. Schriftenreihe des Bayerischen Landesamt fuer Wasserwirtschaft 17

Chen C (1987) Comprehensive review of debris flow modeling concepts in Japan. In: Costa JE, Wieczorek GF (eds) Debris flows/avalanches: process, recognition, and mitigation. Rev Eng Geol 7:13–29

Chow VT (1959) Open channel hydraulics. McGraw-Hill, New York, p 680

Costa JE (1984) Physical geomorphology of debris flows. In: Costa JE, Fleisher PJ (eds) Developments and applications of geomorphology. Springer, Berlin, pp 268–317

Costa JE (1988) Rheologic, geomorphic, and sedimentologic differentiation of water floods, hyperconcentrated flows, and debris flows. In: Baker VR, Kochel RC, Patton DC (eds) Flood geomorphology. Wiley, New York, pp 113–122

Haiden A (1925) Ueber die Berechnung der Murstromdurchflussmengen [On the calculation of debris flow discharge] Wildbach- und Lawinenverbau 2, Wien

Hsu L, Dietrich WE, Sklar LS (2008) Experimental study of bedrock erosion by granular flows. J Geophys Res-Earth Surf 113(F2):F02001

Huebl J (2008) Ereignisdokumentation Prozesse: Wasser. Universitaetskurs Ereignisdokumentation. Institut fuer Alpine Naturgefahren. Universitaet fuer Bodenkultur. Wien [unpublished report]

Huebl J, Bunza G, Hafner K, Klaus W (2003) Stummer Zeugen Katalog. In: Projektteam ETAlp (Bundesministerium fuer Land- und Forstwirtschaft, Umwelt und Wasserwirtschaft), Kompendium zu ETAlp-Erosion, Transport in alpinen Systemen, 48; Bundesministerium fuer Land- und Forstwirtschaft, Umwelt und Wasserwirtschaft. Wien

Huebl J, Totschnig R, Sitter F, Mayer B, Schneider A (2010) Historische Ereignisse – Band 1–3: Auswertung von Wildbach Schadereignissen in Oesterreich. IAN Report 111, Institut fuer Alpine Naturgefahren. Universitaet fuer Bodenkultur. Wien [unpublished report]

Hungr O, McDougall S (2009) Two numerical models for landslide dynamic analysis. Comput Geosci 35:978–992

Hungr O, Morgan G, Kellerhals R (1984) Quantitative analysis of debris torrent hazards for design of remedial measures. Can Geotech J 21:663–677

Jakob M (2005) Debris-flow hazard analysis. In: Jakob M, Hung O (eds) Debris-flow hazards and related phenomena. Springer, Berlin, pp 411–444

Jomelli V (2012) Lichenometric dating of debris avalanche deposits with an example from the French Alps. In: Schneuwly-Bollschweiler M, Stoffel M, Rudolf-Miklau M (eds) Dating torrential processes on fans and cones – methods and their application for hazard and risk assessment. Advances in Global Change Research. Springer, Dordrecht/Heidelberg/London/New York

Kienholz H (1998) Early warning systems related to mountain hazards. In: Proceedings of the international conference on early warning systems for the reduction of natural disasters EWC'98, 7–11 Sept 1998, Potsdam

Marchi L, Tecca PR (1995) Alluvial fans of the Eastern Italien Alps: morphometry and depositional processes. Geodinamica Acta (Paris) 8(1):20–27

Marchi L, Arattano M, Deganutti A (2002) Ten years of debris-flow monitoring in the Moscardo Torrent (Italian Alps). Geomorphology 46:1–17

Markhart G, Kohl B, Sotier B, Schauer T, Bunza G, Stern R (2004) Provisorische Geländeanleitung zur Abschätzung des Oberflächenabflussbeiwertes auf alpinen Boden-/Vegetationseinheiten bei konvektiven Starkregen [A simple code of practice for assessment of surface runoff coefficients for alpine soil-/vegetation units in torrential rain]. BFW-Dokumentation; Schriftenreihe des Bundesamtes und Forschungszentrums fuer Wald, No. 3. Wien

Miall AD (1996) The geology of fluvial deposits. Springer, Berlin

O'Brien JS, Julien PY, Fullerton WT (1993) Two-dimensional water flood and mudflow simulation. J Hydraul Eng 119(2):244–261

Salzer J (1886) Ueber den Stand der Wildbachverbauung in Oesterreich. Verlag des krainisch-küstenlaendischen Forstvereins. Wien

Scheidl C, Rickenmann D (2010) Empirical prediction of debris-flow mobility and deposition on fans. Earth Surf Process Landforms 35:157–173

Scheuner T, Keusen HR, McArdell BW, Huggel C (2009) Murgangmodellierung mit dynamisch-physikalischem und GIS-basiertem Fliessmodell. Wasser Energie Luft 101(1):15–21

Schirmer W (1988) Murstappel bei St. Virgil/Suedtirol, Exkursionsfuehrer. Deutscher Arbeitskreis fuer Geomorphologie 15

Schneuwly-Bollschweiler M, Stoffel M (2012) Dendrogeomorphology – tracking past events with tree rings. In: Schneuwly-Bollschweiler M, Stoffel M, Rudolf-Miklau M (eds) Dating torrential processes on fans and cones – methods and their application for hazard and risk assessment. Advances in Global Change Research. Springer, Dordrecht/Heidelberg/London/New York

Skermer NA, VanDine DF (2005) Debris flows in history. In: Jakob M, Hung O (eds) Debris-flow hazards and related phenomena. Springer, Berlin, pp 25–52

Stiny J (1910) Die Muren [Debris flows]. Wagner, Innsbruck

Stock JD, Dietrich WE (2003) Valley incision by debris flows: evidence of a topographic signature. Water Resour Res 39(4):1089. doi:10.1029/2001WR001057

Takahashi T (1991) Debris flow, IAHR monograph series. Balkema Publishers, Rotterdam

Zeller J (1995) Das Gewässernetz, ein quantitativer Indikator fuer den Charakter von Kleineinzugsgebieten. Wildbach- und Lawinenverbau, 59/128

Assessment of Past Torrential Events Through Historical Sources

Vincenzo D'Agostino

1 Why Is Historical Information So Crucial for 'Torrents'?

The noun "torrent" to indicate a stream or a small river is quite uncommon outside Europe. The word comes from the Latin adjective "*torrens*" which means rushing, violent, fast. In torrents a substantial difference of magnitude exists between the low water periods – including a seasonal lack of surface water in ephemeral streams – and flood events capable of rearranging the river-bed topography heavily. The extremely low frequencies of important flow rates might cause the reduction of hazard perception and leads towards an incorrect land use and an unsafe management of the catchment. The collection of historical information of past events must be the starting point to deal with inhabitant security, land planning, and watershed management. Moreover, there are at least two other basic reasons which strengthen this need for analysis of those 'torrential' rivers located in steep lands.

First, torrents are often found in remote areas and drain small watersheds (<100 km^2), thus inducing several operational difficulties for a long duration and good quality monitoring of rains and discharges. Past event reconstruction can help to assess the rainfall-runoff behaviour of the catchment and contemporary land transformations.

The second reason descends from the complex water-sediment-slopes interactions which are active in mountain basins. In a large river system the estimation of a scenario flood and its related hazard goes through a detailed hydrologic analysis, where flood generation is driven by rainfall and snowmelt. This statement is not fully valid for a mountain torrent because floods descend from a composition of geomorphic and hydrologic processes (Jakob and Jordan 2001). Geomorphic

V. D'Agostino (✉)
Department Land, Environment, Agriculture and Forestry, TeSAF, University of Padova, IT-35020 Legnaro (PD), Italy
e-mail: vincenzo.dagostino@unipd.it

M. Schneuwly-Bollschweiler et al. (eds.), *Dating Torrential Processes on Fans and Cones*, Advances in Global Change Research 47, DOI 10.1007/978-94-007-4336-6_8,

processes collaborate to trigger flood events where the solid component might be dominant, as occurs in cases of hyperconcentrated flows, debris flows, and rock avalanches. The bulked maximum flow rates can reach values one or two orders of magnitude higher than regional water discharges associated with a centennial return period. Peak values are heavily affected by sediment supply conditions, collapses of landslides next to the banks, temporary clogging of cross-sections (woody debris jams, sediment entrapment at the inlet of critical river crossing) or even by glacial lake outburst. In a few words a large set of anomalous or apparently anomalous floods exists in small mountain watersheds: past documentation highlights that such floods could be obliterated in the risk analysis and in the connected topics.

The explicit paradigm that forms the basis for historical research (HR) is that flooding disasters occurring today in a given territory represent mostly physical-stochastic processes which have already occurred in the past and also have a certain site-specific repetitiveness: geologic instability depends, to a large extent, on environmental and unchangeable factors, hydrologic responses are driven by the meteo–climatic trend and by the land evolution over years.

2 Sources of Information

The historical research (HR) on flooding allows: (i) the finding of information that is not directly, or rarely, traceable in the field, (ii) providing confirmation of documentary sources which are not exhaustive or oral (Cerato 1999). By following this definition, HR should not to be merely addressed in far-off times, because each investigation about a significant event before the present can be deemed as an HR. Obviously countries of ancient civilization offer a longer period of records regarding natural hazards and require more effort of homogenization and interpretation of documents taken over different years. In this case the critic patchwork of extended information is an integral part of a consistent HR.

Sources of information are numerous and partly depend on the geographic location of the investigated area and on the technical-administrative apparatus that was operative at the time of the events. Including all type of historical knowledge, it is possible to mention these main sources:

- Official documents and cartography from historical state archives, public libraries, municipalities, offices for torrent control and hydraulic works, geologic surveys, forestry services, civil defence, environmental sciences museum, agencies for communication roads (highways, railways), research centres, universities. In the Trento (Italy) archives, mid-nineteenth century maps exist indicating a subdivision in landed estates of a "Defence Consortia". Each owner must contribute to the expenses for work construction in proportion to the degree of hazard which was estimated for his property (Cerato 1999).
- Unpublished documents gathered from archives, local authorities, and written evidence or transcription of verbal evidence.

- Interviews of elderly people; they can provide detailed information to characterize event dynamics. If possible, the collection of several witnesses is advisable in order to establish a coherent overview and eliminate inconsistencies.
- Technical reports describing ante-event and/or post-event conditions of the watershed. There exists an emblematic "Memorial of technical-building carried out by the province commission for water regulation in Tyrol because of 1882 flooding" that was published in Innsbruck in 1892. The disaster that occurred from 16th to 20th September affected a vast area of the eastern alpine arc (Tyrol County and Carinthia) and caused the loss of more than 400 lives. The floods occurred during a period of slump and of tremendous exploitation of forestry resources.
- Experts' reports of damage and priority works after the events.
- Designs of soil bioengineering and hydraulic works for the watershed management and the recovery of the public safety after the event. The first organized actions at the basin scale to control sediment transport in the European Alps date back to the sixteenth century. The motivation which brought about the building of torrent control works is often closely linked to the dynamics of the last event. The presence itself of intensive landslide settlement and ancient check dams sequences for torrent consolidation is the first evidence of torrential activity of the basin (Cerato 1999; Marchi et al. 2010).
- Old drawings, photos (since 1870), aerial photos (since 1930), and, more recently, videos preserved by public agencies, military corps, TV companies, and individuals; inhabitants who experienced the events have often the most enlightening images. Fast diffusion of digital cameras and video-cameras is letting this valuable form of information grow greatly. Some photographic information might be also provided by parish registers or associations devoted to the mountain protection. Old paintings, picture postcards or tourist guides may help to reconstruct land morphology, landscape evolution, ancient soil use and vegetation cover. Antique postcards must be analyzed with caution, because they frequently present distortions related to their acquiring, colouring, printing and artistic arrangement (Sawyer and Butler 2006). The ancient toponym of the torrent itself may indicate its dangerous activity. An example much used in Italy was the term "Ruins of ..." to express a largely gullied, and prone-to-collapse mountain stream.
- Newspapers or newspaper collections. The increasing awareness of the HR importance is encouraging some public agencies for torrent control works to set up regional archives containing all published documentation (Provincia Autonoma di Trento 2006).
- Technical journal, scientific papers, books (Fig. 1). Some flood events impact on a territory in such an intensive way that many papers appear with different aims of record keeping (Rickenmann and Zimmermann 1993; García and López 2005; Marchi and Cavalli 2007).
- Memorial stones and inscription plates which record the event and its characteristics (Fig. 1).

Fig. 1 Post-event photos of the event occurred in the Tinne torrent (Chiusa, Bolzano, Italy) on 9th August 1921: a debris flow mobilized about 800,000 m^3 of sediment. Image composition has been created from Junger (1921). The *small image* on the *top right corner* (taken in 2010) represents one of many indicators scattered around the town and marking the maximum level reached by the flow

– Monitoring system data, which have been operating for several years; this information may not appear exactly historical but is getting effective for highly active torrents (D'Agostino and Marchi 2001; Marchi and Tecca 2006).
– Cadastres of regional authorities specifically dedicated to event documentation. In the European Alpine Regions this source has been improved over the last 10 years thanks to well focused research projects like DOMODIS – Documentation of Mountain Disasters (Hübl et al. 2002) and DIS-ALP – Disaster Information Systems of Alpine Regions (Kienholz et al. 2006). An excellent example of this activity is the report published by the Autonomous Province of Bolzano (Italy) (Macconi and Formaggioni 2010), which presents the 2009 event documentation and a summary of 12 years of post-event surveys. This systematic register represents a modern approach to assess the future historical documentation.

An interesting report related to the use of historical documents to prevent slope instabilities, floods and debris flow has been published by Tropeano and Turconi (2004). The paper illustrates a wide sample of documentary investigations carried out since the Middle Age in Northeastern Italy, Savoyard torrents (France), Switzerland, Austria, Great Britain, Japan, Northern California, Appalachian Mts. in Virginia, Utah, Hawaii, and China. Notable is also the contribution of Skermer and VanDine (2005), dealing with debris flow in history for European Alps, Asia and

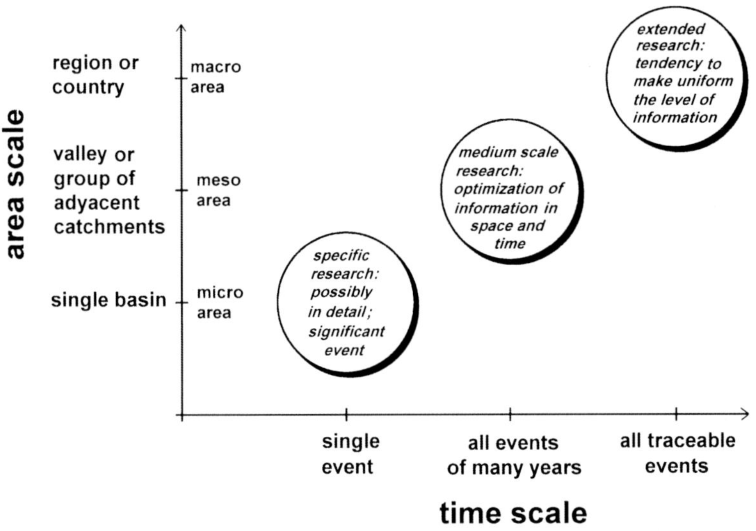

Fig. 2 Sketch of time and area coordinates to identify the type of historical research

semi-arid regions (South America, Western North America, Kazakhstan, Spain). Several authors state limitation of the historical method to assess events, arguing that the reading of sources must be tempered by the context of socio-economic and life conditions of population: a debris flow during wars, famines, or plague is not probably a social priority. As a consequence, historical analyses could be affected by biases originating from different sensibility to similar floods during different historical ages.

Documentary activity can be oriented to different scales in time and space (Marchi and Tecca 2006). A semi-quantitative graph is proposed in Fig. 2 showing a further refinement of this concept. Three gradations of time scale (a single event, a multiple years period, all documentable time) can be coupled with three area scales of investigations (single catchment, neighbourhood basins or part of a mountain valley, a region) generating nine options of time-area investigation. The three circles in Fig. 2 represent the trend of the cognitive typology on the average. When a 'torrential' event is highly significant or very extreme, often a *specific HR* (single event, single catchment) can be carried out in depth and is capable to represent by itself a complete analysis to assess the behaviour of a basin. On the other extreme the *extended HR* (all events, region; Fig. 2) research aims to collect all information about a country or a vast region. The work is very exacting in terms of searching all possible documentary sources and of making the reading of collected historical data homogeneous. When historical information goes back to before the eighteenth century, it becomes more muddled and less precise in the process description. Such research is powerful to appreciate the degree of activity of an extensive territory and the time evolution of this activity in the light of variation of land use, anthropogenic conditioning and climate. A *medium scale HR* (Fig. 2) is common because it allows the optimization of information for maximum number of years which warrant the same level of reliability. Furthermore the meso-scale offers a precise distribution of land vulnerability for contemporary events. This scale is frequently adopted by researchers, when they focus on a specific phenomenon and its magnitude. Thanks to this data they can develop empirical predictive equations (Kronfellner-Kraus 1984; D'Agostino et al. 1996; Rickenmann 1999), explore time-series on number of events (Marchi and Tecca 2006) and their probability of occurrence versus magnitude (D'Agostino and Marchi 2001), and recognize typical patterns of occurrence (Zimmermann et al. 1997). From many medium scale HRs it is then easier to up-scale previously collected data to a macro-area (D'Agostino et al. 1996; Marchi and D'Agostino 2004).

The HR tries basically to answers the following questions:

- What type of event occurred and when?
- How did it occur and what was its magnitude?
- Which consequences affected people, land, and goods?

The first question concerns a typological characterization of the phenomenon and dating and possibly the time and/or the times of concurrent processes.

The second question responds to the events dynamics, which involves: the cause of triggering, propagation and cessation, duration, velocity and depth of the flow in high quality documentation, time interval between the storm initiation (sometimes a waterfall activation) and the passage of the phenomenon at a certain torrent location; data quantifying magnitude and affected areas; and effectiveness of protection measures which have been operating.

The last topic integrates event dynamics and emphasizes damages, land transformation, vulnerability of mitigation measures, and finally the socio-economic impact of the flood event.

In the next paragraph details are given about process identification, magnitude-frequency assessment and related hazards through HR, these being the most tricky items in the activity.

3 Process Interpretation

3.1 Typology and Dynamics of Alluvial Events

The process types of torrential events in mountain creeks are classified in four categories involving rapid movement of water and sediments, and whether or not they are aggregate with woody debris. Categories are listed as follows:

1. Events which are characterized by water floods and at most by bedload transport (BL); clean water flood in mountain creeks are rare except for rocky channels. During this event the sediment rate changes suddenly when the armour layer of the bed and bed forms clusters (steps, bars, rapids) are broken (Warburton 1992).

Above this threshold of collapse, the bed load transport changes from supply limited to transport capacity driven conditions (D'Agostino and Lenzi 1999). Volumetric concentration of sediment may reach maximum values of 10–15%; the sediment transport is grain selective; deposits are clast supported, well sorted and have some stratification which reflects the flow competence.

2. Floods which entail debris flow (DFW). Pulsating sediment-water surges move all together incorporating bed material regardless of its grain-size. The sediment concentration by volume may vary from 30% to 70%. The massive flow is non-Newtonian and its rheology depends on the proportion between fine and granular components; debris-flow deposits are quite often matrix supported and minimally sorted and weakly stratified (floating texture). Mudflow can also indicate a debris flow, when fine components are abundant.
3. Floods belonging to a transition between BL and DFW are termed debris flood (DFD) (Aulitzky 1982; Hungr et al. 2001). The flow behaviour is Newtonian and nearly maintains the water kinematics. DFD does not cause dense and blocking surge as DF does. Aulitzky (1982) describes DFD "spreading as flat blankets on torrent cones and fans". Deposits are poorly sorted, often fine grained, with possible large boulders standing at the surface or partially embedded like random stopped 'missiles' because they arrived at a more gentler slope.
4. A landslide with flood-like characteristics is finally the debris avalanche (DAV): a very rapid to extremely rapid large bulk of mostly dry granular flow (Angeli et al. 1996), which usually triggers on open slopes converging or not in confined channels. DAV develops by falling or collapsing of rock bodies which progressively break during downhill movement or by rapid destabilization – due to water flow, ice melting, large rocks falling, earthquake – of loose debris deposits.

Some authors use the term hyperconcentrated flow as DFD synonym (Costa 1984), others insert the hyperconcentrated flow as a stand-alone non-Newtonian category between DFD and DFW (Jakob and Jordan 2001) or simply between BL and DFW (Pierson 2005). A hyperconcentrated flow is marked out by the transportation of "large quantities of coarse sediment (sand and possibly gravel) at high concentration in intermittent dynamic suspension" (Pierson 2005). In short we can say that in the transitional phase between BL and DFW the flow behaviour is extremely sensitive to the grain size distribution and grain density. Under their influence, the flow may be or not to be Newtonian.

Now, after recalling the definitions and going back to the historical documentation of past BL, DFD, DFW, DAV events, it is easy to understand that a clear back-analysis distinguishing DFW and DFD events represents a big challenge. To facilitate the interpretation, it is convenient to place the Newtonian and rich-in-sediment-flows into the DFD category and the non-Newtonian hyperconcentrated flows and mud flows in the DFW events.

The recognition of BL versus DFD might also be doubtful in old documentation and demands a critical reading of the chronicles if proper images are not available.

According to personal experience in reconstructing past events (D'Agostino et al. 1996) or in post-event surveys, a tendency to overload the event estimation with sediment concentration exists, when the burial with sediments of fans is large (>10 ha), deposits exceed 100,000 m^3, or the event hit a meso/macro area causing serious destruction. This is the case of the disaster that occurred on 16th December 1999 in twenty watersheds of north-eastern Venezuela (Caribbean Sea coast). About 15,000 people died and nearly 90,000 houses were destroyed or damaged (García and López 2005). Post-event field observations revealed that the disaster was heavily conditioned by a huge amount of runoff, the typical pattern of debris-flow deposit was not evident in many basins and the sediments extended as nearly uniform "flat blankets" over accreted fans, thus showing a typical DFD event. Deep gulling and bed degradation along fans and upstream of the fan apex for long reaches showed that the event was supply-limited if compared to the stream power provided by running waters.

A practice rule aiding the interpretation from HR of sediment transport mode is offered by the concept of competent mean torrent slope. So, reworking formulas to predict volumetric concentration of bedload transport (Smart and Jaeggi 1983) and debris flow (Ou and Mizuyama 1994), a BL event is expected for stream bed gradients lower than 18%, DFD exists in the field slope of 18–28%, and DFW begins to occur above 24% (D'Agostino 2010). The slope limits are not rigid and can be easily violated if sudden volumetric releases (glacial outburst, breaks of woody debris jams, check dam collapse) occur just upstream of – or not too far from – the deposition area. Furthermore, when a deep incision takes place in those torrents which are energetically marked as potential DFW channels (slope above 28%), great attention has to be paid to the decision whether it is a debris flow or not. In fact, a mature debris flow that moves at the equilibrium condition along a channel is moderately erosive, unless the bed bottom is covered by uniformly loose and unstable debris or consolidation check dams collapse.

The typological distribution of historical events in a mountain macro-area frequently shows the prevalence of BL and DFW. After data collection of landslides and stream floods in the last 150 years in Northern Italy Tropeano and Turconi (2004) determined that 36% were DFW, 34% BL, 9% rock falling or sliding, 2% soil slips, and the remaining 19% landslides.

3.2 *Frequency, Magnitude, Hazard*

The hazard related to one or more type of process is assessable at a first reliable level by means of the HR. Aulitzky (1973) proposed a check-list to assess a torrent index. One question underlines the role of historical documentation in finding out which area of a village suffered damage in the past. Assuming that older houses were built in the safest places, the estimation of a higher torrent hazard is reasonable if chronicles report wide damage even for most ancient settlements of the village.

The HR can further support the hazard assessment if quantitative estimation of frequency and intensity of the triggering rainfall event is not available or is not significant. The statement is more in keeping with DFW events, because most damages depend on the sediment contribution and this latter is less directly related to the runoff volume (Rickenmann and Koschni 2010) and thus to the probability of occurrence of a certain amount of rainfall. The total sediment which is delivered to a fan area – and here partially or completely deposited after a flood event – results from the combination among following several variables: type of process (BL, DFW, DFD, DAV), efficiency of the works operating against the process, land topography, event duration, possible runoff volume, sediment supply conditions. The relative weight of the last four factors is process-dependent and keeps the scientific community busy. At scale of medium (>30 years) and rare (>100 years) recurrence interval rainfall events, we can state:

1. BL and DFD usually deliver larger sediment volumes to the basin outlet if compared to DFW events (D'Agostino et al. 1996). Generally BL/DFD floods last longer and have greater runoff volumes than DFW; so, even though their sediment concentration is quite lower than that of DFW, they produce overall sediment volumes larger than those of flash and short-duration DFWs. This behaviour is then obliterated in the field because most DFW sediments are inclined to halt close to the confluence, while BL sediments are discharged and diluted in the valley river.
2. Post-event surveys of deposited BL volumes are not frequent (e.g. avulsion or sediment blockage by some channel constraints), because the process is more dispersive (see above). DFW deposits, on the contrary, become the basic data to assess the event magnitude and to rank its effects on fan areas (Jakob 2005).
3. DFWs are heavily conditioned by the basin disposition (Zimmermann et al. 1997) to supply sediments; this involves geotechnical characteristics of hillslopes and hydrographic network, degree of sediment availability and level of channel recharge (Bovis and Jakob 1999; Jakob et al. 2005).

The assessment of frequency and magnitude of events by means of the HR is intrinsically approximated. People tend to obliterate or simply not to record small and/or low impact floods causing a background noise to the occurrence reconstruction through the history and then reducing the historical degree of activity. Such filtering is not homogeneous because the sensibility changes with ages, geographical location, and world progress. Marchi and Tecca (2006) analyzed factors influencing the building of DFW time series from archive data at regional (Northeastern Italy) and single basin scales (Moscardo torrent and Acquabona channel, Italy). They deduced for both scales an increasing number of DFW events in the last decades (1980–2000) and concluded that higher frequency can not be ascribed clearly to changes of rainfall intensity because the contemporary larger availability of documentation could justify the apparent increase.

The overall sediment volume deposited by BL/DFD/DFWs includes an inaccuracy, which decreases from the past toward the present. Consistent data on sediment volumes begin in the second half of nineteenth century for the European Alps. They initiate to be more precise when removal of sediments by means of powerful excavators has been conducted and the construction of wide deposition basins closed by filtering-retention check dams (D'Agostino 2010) was widespread (since

1960–1970). Another source of uncertainness is a correct field identification of deposition area: partial deposits, placed upstream of the fan apex, could have been neglected due to a difficult access to the torrent. Alluvial events following each other in a short time do not allow separating partial contributions.

The use of old maps and photos makes the back-reconstruction of flooded area (*B*) easier. In case of DFW the following scale invariant relationship exists relating *B* with the volume *V* (metric units) of deposition (Iverson et al. 1998):

$$B = kV^{2/3} \tag{1}$$

where *k* is a quasi-rheological parameter which varies in the European Alps between the lowest values of 6 (Crosta et al. 2003) and 14 (D'Agostino et al. 2010) and highest values of 33 (Berti and Simoni 2007) and 45 (Scheidl and Rickenmann 2010). The *k* parameter increases for DFD and BL. A sample related to Austrian torrential events gives $k = 56$ (Scheidl and Rickenmann 2010). Through the elaboration of DFD/DFW data published by López et al. (2000) and García and López (2005) for the Venezuela event it can be obtained $k = 47$. If the affected surface (*B*) is known, the estimation of *V* is feasible with calibrating *k* locally through present debris-flow deposits, or via literature data. The future historical data on volumes will reduce these approximations and benefit from comparative topographical surveys (ante and post event) by means of laser-scanning techniques (Scheidl et al. 2008).

Magnitude-frequency analysis provides information about the trend and the response from sediment sources areas. Supply unlimited basins, with widespread landslides along the main channel and at the basin head, have quasi-regular DFW occurrence, ranging, as an example, from 0.5–3 events/year (Moscardo torrent, Udine Province, Italy) for highly erosive situations (Marchi and Tecca 2006) to 0.06–0.03 events/year (Leimbach, Canton BE, Switzerland) (Zimmermann et al. 1997), when material is excavated along the flow path. These supply unlimited torrents release their sediment recharge frequently and they cause events with small to medium magnitude ($V < 30{,}000\ m^3$) interrupted by some episodic maxima. It is not a contradiction that such mountain streams are less dangerous, because the public perception of hazard remains constantly high and then the monitoring and protection measures are usually more active. The opposite situation deals with catchments showing extremely rare DFWs (<0.01 events/year). This results from limiting topographic constraints, which deliver DFWs to the fan only for extreme volumetric and kinematic conditions (Zimmermann et al. 1997), and/or due to the circumstance that the basin has a very slow sediment recharge rate and this intrinsic threshold to trigger debris flow matures every 200–500 years.

Finally, the assessment of DFW volumes through the HR is the basis to develop predictive empirical equations at different spatial scales. The research carried out by D'Agostino et al. (1996) (Province of Trento) and then extended to the larger area of northeastern Italian Alps (Marchi and D'Agostino 2004) is one example of a fruitful result. The investigation took advantage of a careful selection of good quality data

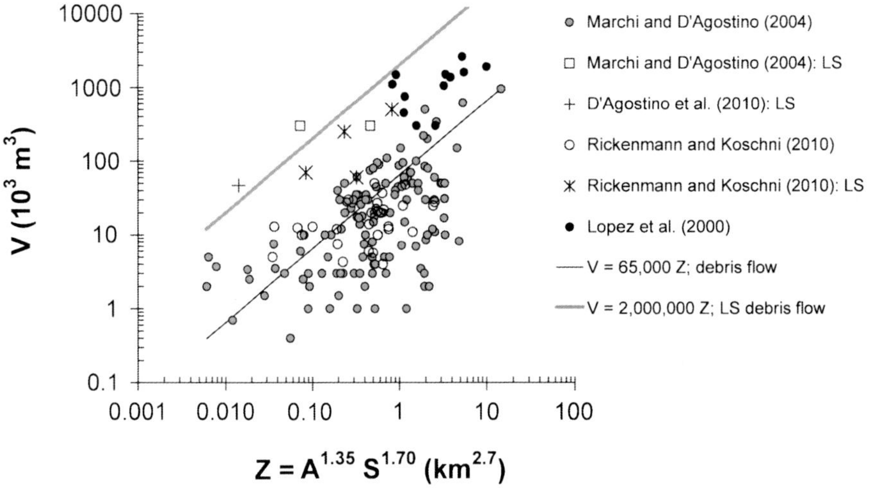

Fig. 3 Collection of published data on debris-flow volumes: morpho-energetic parameter Z ($Z = A^{1.35} S^{1.7}$; A: basin area, S: mean channel slope) versus deposited volume V. Symbol *LS* indicates debris-flow events conditioned by the occurrence of large landslide. The equations provide the volumes in cubic metres

and explored a period (1882–1999) containing the two most severe floods (1882 and 1966) of the last century. A predictive equation was obtained as follows:

$$V = C(A^{1.35}S^{1.70}) = CZ \tag{2}$$

where A is the basin area (km^2) and S (m/m) is the mean slope of the main stream (from the channel head to the fan apex), the group $Z = A^{1.35} S^{1.70}$ is a morpho-energetic variable resulting from the multiple regression, and C is the parameter of calibration equal to 65,000 (Fig. 3). Equation 2 was also tested by Scheidl and Rickenmann (2010) for 1987 Swiss DFWs including the severe August 2005 event (DFW, BL-DFD). They obtained that V values still depend on Z but according to a lower parametric value ($C \cong 16{,}400$). The extreme volume data of Marchi and D'Agostino (2004), D'Agostino et al. (2010), Scheidl and Rickenmann (2010) influenced by large landslides or by a deep channel gulling and those of López et al. (2000) (Venezuela event) depict the same trend of Eq. 2 but with a C value amplified by two orders of magnitude (Fig. 3). The confirmation of Z meaningfulness may help in adjusting C (similarly to k) for a macro-area and for a certain recurrence period (empirical-historical frequency).

3.2.1 An Example of Quantitative Outcomes from Historical Sources

A mountain catchment in Italy (Rotolon torrent, Recoaro, Vicenza Province) makes historical documentation available since 1573 allowing the reconstruction of a

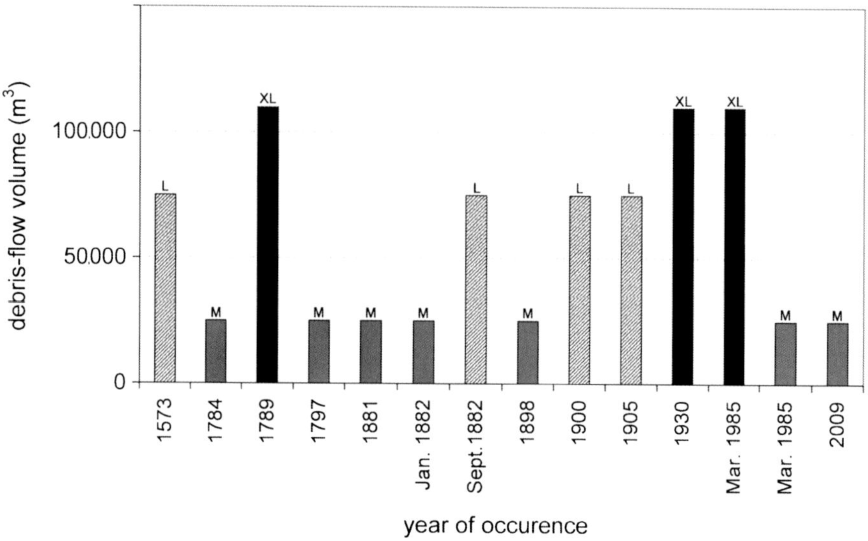

Fig. 4 Historical sequence of debris-flow events in the Rotolon torrent (Vicenza, Italy): time of occurrence in the x-axis (distorted sequential scale), semi-quantitative volumes of deposition in y-axis

quasi-quantitative volumetric sequence of DFW events over 436 years. The torrent name comes from the Cimbri's language through the composition of *rot* (red) with *lön* (landslide). For oldest events the magnitude has been evaluated thanks to the description of sediment depositions and damage reported in the historical documentation and the comparison of these effects with those of recent events. In Fig. 4 the time of DFW occurrence against magnitude is represented: extra large (XL) volumes indicate magnitudes greater than 100,000 m^3; large volumes (L) correspond to magnitude in the range 50,000–100,000 m^3; medium volumes (M) mean values in the range between 10,000 and 50,000 m^3; small volumes (S) certainly occurred but they are not reported owing to null impact on the safety of inhabitants. A similar approach to the volume classification was followed by Brochot et al. (2002). The Rotolon time series expresses a basin behaviour which is conditioned by a unique wide instable area at its head ("the Red Landslide", Fig. 5), located at the toe of calcareous rocky cliffs and very sensitive to the raising and springing of subsurface waters. The estimated magnitudes are here directly originated by the volume collapsed in the landslide area. That results from historical information clearly. Unstable slopes – composed by marl, sandstone, argillite – are rich in silt and clay; a grain size analysis of a matrix sample collected after the last event (May 2, 2009; $V = 45{,}500$ m^3) contained a weight proportion of 18% of silt and of 21% of clay. This makes debris-flow material highly mobile when it enters the main channel and mixes with water flows. As a confirmation, the mobility parameter (Eq. 1) resulted in $k = 51$. The Rotolon example highlights (Fig. 4) the empirical recurrences of 150 years (436/3) for XL debris flows, of 60 years (436/7) for event larger or equal to L, and of 30 years (436/14) for those equal or bigger than

Fig. 5 Head of the basin of the Rotolon torrent and main channel of debris-flow propagation (photo taken after the 2009 event)

M. A refinement should likely be necessary accounting for the two dependent 1985 DFWs, but this simple reconstruction could anyway be meaningful to depict the basin behaviour.

4 Benefits for Practitioners

Research, analysis and critical comparison of historical sources on past torrential events are extremely useful both for public and private practitioners planning and designing active and passive measures for land safety and hazard mitigation. As a general rule each technical report should initiate with a sort of inventory of historic events. The concurrence of difference experts is also advisable. In fact, the cooperation between the technical component (e.g. geologists, engineers, dendrochronologists) and the humanistic component (e.g. philologists, archivists, journalists) warrants the consistence and the multiple view of the inventory. This last concept may appear obvious but is not yet fully applied in most event reports which are either technical or literally.

The historic inventory offers several direct benefits to the practitioners; the main advantages are listed as follows:

- Magnitude, typology and consequences of historic torrential events support decisions about functionality and sizes of torrent control works under design.
- Soil use regulations are better informed and then matched with the series of torrential disasters occurred in watershed and fan.
- The knowledge of efficiency of previous torrent control works enlightens the designer in avoiding same basic mistakes (design hypotheses, choice of materials, building method, work typology), in improving the overall quality of new works and in figuring out their mitigation effects.

- The knowledge of vulnerability of past torrent control works on the occasion of most hazardous events underlines the dual behaviour of the protection works: they can also amplify the damages under very unfavorable circumstances. The practitioner takes here precious information when hazard mapping in extreme hypotheses is required by the law and when he must clarify the functional limits of his management proposal. Furthermore, the historic vulnerability makes practitioners more confident with the expected lifetime of different torrent control works and with the consequent need of periodic ordinary and extraordinary maintenances.
- The dissemination and the defence of the planned actions with respect to the stakeholders (population, environmental organizations, tourists, etc. . . .) is greatly helped by the historic evidence of the torrential activity. Technical arguments and calculations are less immediate to communicate, particularly when the landscape impact of designed works is strong or soil use restrictions are heavy. Protection measures might appear out of proportion towards the expected torrential hazard if the memory of an extreme historic event is not recalled.

5 Conclusions

Historical research (HR) on the subject of past activity of mountain torrents suffers a certain approximation because society changes over time the way and the intensity in which it records flood events. This lack can be partly filled if these changes are interpreted by the investigator, the investigation extends to a significant time span in the past and the HR is carried out searching for multiple information sources and analyzing the post-event measures. Several scientific investigations through HR contributed to improve qualitatively and quantitatively our knowledge of torrential processes. This progress confirmed and increased the awareness that it is extremely productive to record evidence of past flooding events.

A new generation of historical documentation is in progress thanks to the intensification of post-event surveys, and technologic-scientific progress in data acquisition, processing and archiving. The history of historical research teaches us that most efforts have to be addressed in a distinct univocal recognition of process typology, and to pay attention also to low magnitude events. These were often disregarded in the past documentation but are becoming more influential on the alpine land along with the increasing anthropic pressure on mountain valleys.

Acknowledgements The author gratefully acknowledges Enrico Pozza for the Rotolon torrent documentation. Remarks on the December 1999 disaster in Venezuela have been formulated arguing with Gary Parker during a field visit. Heartfelt thanks are addressed to the Editors and to David Butler for their precise revision. Research was granted by PRIN 2007- MIUR and by Strategic Project 2008 GEO-RISKS, University of Padova (Italy).

References

Angeli MG, Gasparetto P, Menotti RM, Pasuto A, Silvano S, Soldati M (1996) Rock avalanche. In: Dikau R, Brunsden D, Schrott L, Ibsen ML (eds) Landslide recognition. Wiley, Chichester, pp 190–201

Aulitzky H (1973) Vorläufige Wildbach-Gefährlichkeits-Klassifikation für Schwemmkegel. In: 100 Jahre Hochschule für Bodenkultur in Wien – Fachveranstaltungen, Band IV, Studienrichtung Forst – und Holzwirtschaft, Teil 2, pp 114–117 (in German)

Aulitzky H (1982) Preliminary two-fold classification of torrents. Mitteilungen der forstlichen Bundesversuchsanstalt Wien 144:243–256

Berti M, Simoni A (2007) Prediction of debris flow inundation areas using empirical mobility relationships. Geomorphology 90:144–161

Bovis MJ, Jakob M (1999) The role of debris supply conditions predicting in predicting debris flow activity. Earth Surf Process Landf 24:1039–1054

Brochot S, Marchi L, Lang M (2002) L'estimation des volumes des laves torrentielles: méthodes disponibles et application au torrent du Poucet (Savoie). Bull Eng Geol Environ 61:389–402 (in French)

Cerato M (1999) Contributi per una storia delle sistemazioni idrauliche forestali in Valsugana. Provincia Autonoma di Trento, Azienda Speciale di Sistemazione Montana (ed), Trento (in Italian)

Costa JE (1984) Physical geomorphology of debris flow. In: Costa JE, Fleisher PJ (eds) Developments and application of geomorphology. Springer, Berlin, pp 268–317

Crosta GB, Cucchiaro S, Frattini P (2003) Validation of semi-empirical relationships for the definition of debris-flow behaviour in granular materials. In: Rickenmann D, Chen C-L (eds) Proceedings of the third international DFHM conference, Davos, Switzerland. Millpress Science Publishers, Rotterdam, pp 821–832

D'Agostino V (2010) Filtering-retention check dam design in mountain torrents. In: Garcia CC, Lenzi MA (eds) Check dams, morphological adjustments and erosion control in torrential streams. Nova, Hauppage

D'Agostino V, Lenzi MA (1999) Bedload transport in the instrumented catchment of the Rio Cordon. Part II: analysis of the bedload rate. Catena 36:191–204

D'Agostino V, Marchi L (2001) Debris flow magnitude in the Eastern Italian Alps: data collection and analysis. Phys Chem Earth C 26:657–663

D'Agostino V, Cerato M, Coali R (1996) Il trasporto solido di eventi estremi nei torrenti del Trentino Orientale. In: Proceedings of INTERPRAEVENT 1996, Garmisch-Partenkirchen, Germany, Band 1, pp 377–386 (in Italian)

D'Agostino V, Cesca M, Marchi L (2010) Field and laboratory investigation of runout distances of debris flow in the Dolomites (Eastern Italian Alps). Geomorphology 115:294–304

García R, López JL (2005) Debris flow of December 1999 in Venezuela. In: Jakob M, Hungr O (eds) Debris-flow hazards and related phenomena. Springer/Praxis, Berlin/Chichester, pp 519–538

Hübl J, Kienholz H, Loipersberger A, Balteanu D, Corominas J, Egli T, Glade T, Hegg C, Schrott L, Sperling M (2002) DOMODIS – documentation of mountain disasters state of discussion in the European Mountain Areas. In: Hübl J, Kienholz H, Loipersberger A (eds) International Forschungsgesellschaft INTERPRAEVENT, Handbuch 1. International Research Society, Klagenfurt.

Hungr O, Evans SG, Bovis MG, Hutchinson JN (2001) A review of the classification of landslides of the flow type. Environ Eng Geosci 7:221–238

Iverson RM, Schilling SP, Vallance JW (1998) Objective delineation of lahar-hazard zones downstream from volcanoes. Geol Soc Am Bull 110:972–984

Jakob M (2005) A size classification for debris flow. Eng Geol 79:151–161

Jakob M, Jordan P (2001) Design flood estimates in mountain streams – the need for a geomorphic approach. Can J Civ Eng 28:425–439

Jakob M, Bovis M, Oden M (2005) The significance of channel recharge rates for estimating debris-flow magnitude and frequency. Earth Surf Process Landf 30:755–766

Junger F (1921) Klausen. Zum Gedenken an das große Unglück am 9. August 1921. Der Schlern, Bozen (in German)

Kienholz H, Perret S, Schmid F (2006) Documentazione degli eventi naturali – Guida per la documentazione sul terreno. In: Piattaforma Pericoli Naturali Convenzione delle Alpi (ed), Innsbruck/Bern (in Italian)

Kronfellner-Kraus G (1984) Extreme Festofffrachten und Grabenbildungen von Wildbächen. In: Proceedings of INTERPRAEVENT 1984, Villach, Austria, Band 2, pp 109–118

López JL, Pérez D, García R, Shucheng Z (2000) Evaluacion hidro-geomorfólogica del los aludes torrenciales de Decembre de 1999 en Venezuela. In: Proceedings of The debris flow disaster of December 1999 in Venezuela, Universidad Central de Venezuela, Caracas

Macconi P, Formaggioni O (2010) ED30 Report 2009 – Relazione riassuntiva Documentazione Eventi 2009. Provincia Autonoma di Bolzano, Ripartizione 30 – Opere idrauliche – Azienda speciale per la regolazione dei corsi d'acqua e la difesa del suolo (ed), Bolzano (in Italian)

Marchi L, Cavalli M (2007) Procedures for documentation of historical debris flow: application to the Chieppena torrent (Italian Alps). Environ Manag 40:493–503

Marchi L, D'Agostino M (2004) Estimation of debris-flow magnitude in the Eastern Italian Alps. Earth Surf Process Landf 29:207–220

Marchi L, Tecca PR (2006) Some observations and the use of data from historical documents in debris-flow studies. Nat Hazard 38:301–320

Marchi L, Cavalli M, D'Agostino V (2010) Hydrogeomorphic processes and torrent control works on a large alluvial fan in the eastern Italian Alps. Nat Hazard Earth Syst Sci 10:547–558

Ou G, Mizuyama T (1994) Predicting the average sediment concentration of debris flow. J Jpn Soc Eros Control Eng 47:9–13 (in Japanese)

Pierson CP (2005) Hyperconcentrated flow – transitional process between water flow and debris flow. In: Jakob M, Hungr O (eds) Debris-flow hazards and related phenomena. Springer/Praxis, Berlin/Chichester, pp 159–202

Provincia Autonoma di Trento (2006) Progetto ARCA Archivio storico degli eventi calamitosi del territorio della Provincia Autonoma di Trento – fonti cronachistiche ed archivistiche –Rapporto conclusivo. In: GEB Strategie per il territorio (ed), Trento (in Italian)

Rickenmann D (1999) Empirical relationships for debris flows. Nat Hazard 19:47–77

Rickenmann D, Koschni A (2010) Sediment loads due to fluvial transport and debris flows during the 2005 flood events in Switzerland. Hydrol Process 24:993–1007

Rickenmann D, Zimmermann M (1993) The 1987 debris flow in Switzerland: documentation and analysis. Geomorphology 8:175–189

Sawyer CF, Butler DR (2006) The use of historical picture postcards as photographic sources for examining environmental change: promises and problems. Geocarto Int 21:73–80

Scheidl C, Rickenmann D (2010) Empirical prediction of debris-flow mobility and deposition on fans. Earth Surf Process Landf 35:157–173

Scheidl C, Rickenmann D, Chiari M (2008) The use of airbone LiDAR data for the analysis of debris flow events in Switzerland. Nat Hazard Earth Syst Sci 8:1113–1127

Skermer NA, VanDine DF (2005) Debris flow in history. In: Jakob M, Hungr O (eds) Debris-flow hazards and related phenomena. Springer/Praxis, Berlin/Chichester, pp 25–51

Smart GM, Jaeggi MNR (1983) Sediment transport on steep slopes. Mitt. Versuchsanstalt für Wasserbau Hydrologie und Glaziologie, vol 64, ETH, Zürich

Tropeano D, Turconi L (2004) Using historical documents for landslides, debris flow and stream flood prevention. Application in Northern Italy. Nat Hazard 31:663–679

Warburton J (1992) Observation of bed load transport and channel bed changes in a proglacial mountain stream. Arct Alp Res 24:195–203

Zimmermann M, Mani P, Romang H (1997) Magnitude-frequency aspects of alpine debris flows. Eclogae Geol Helv 90:415–420

The Use of Airborne Laser Swath Mapping on Fans and Cones: An Example from the Colorado Front Range

Dennis M. Staley and Thad A. Wasklewicz

1 Introduction

Debris fan systems are common features of mountainous environments, particularly in formerly glaciated valleys in the Arctic, Antarctic, and high mountains of the lower latitudes (Rapp 1960a, b; Albjar et al. 1979; Church et al. 1979; White 1981; Caine 1983; Perez 1993). The prevalence of these landforms in these physiographic settings is attributed to rapid weathering associated with the presence of oversteepened valley sides, stress unloading of rock walls following deglaciation, and the severity of the climate (Matsuoka and Sakai 1999; Ballantyne 2002; Curry and Morris 2004). These factors are conducive to the development of fan deposits by rockfall, debris-flow and avalanche activity, each of which represent a significant hazards to fan areas.

A majority of alpine debris fan and talus research attempts to relate fan and source area morphometry to the processes which form and sculpt the deposit (Rapp 1960a, b; Caine 1967, 1969; Kirkby and Statham 1975; Statham 1976; Carson 1977; White 1981; Perez 1989; Francou and Mante 1990; Akerman 1994; van Steijn et al. 1995, 2002; Perez 1998; Cavalli and Marchi 2008; Trevisani et al. 2009). Direct measurement of the geomorphic processes that produce and modify these features is difficult since they often operate at intermittent or relatively long temporal scales (Schumm and Lichty 1965; Schumm 1991). Instead of direct measurement, researchers often rely upon measurement of form as a method of understanding process rates and mechanics. Despite a large body of literature

D.M. Staley (✉)
Landslide Hazards Program, United States Geological Survey, Golden, CO, USA
e-mail: dstaley@usgs.gov

T.A. Wasklewicz
Department of Geography, East Carolina University, Greenville, NC, USA
e-mail: wasklewiczt@ecu.edu

M. Schneuwly-Bollschweiler et al. (eds.), *Dating Torrential Processes on Fans and Cones*, Advances in Global Change Research 47, DOI 10.1007/978-94-007-4336-6_9,

devoted to the study of alpine debris fan morphometry, very few studies have made measurements of surface form coincident with the spatial scales at which the geomorphic processes are reflected as features on the surface (Cavalli and Marchi 2008; Trevisani et al. 2009). Most measurements of debris fan form have been made randomly or along talus profiles or transects with average point spacing between 10 and 50 m (Caine 1969; Francou and Mante 1990; Perez 1998). Field reconnaissance and morphometric analysis suggest distinct surficial forms are evident at finer spatial scales on alpine debris fans (Caine 1969; Bones 1973; White 1981; Akerman 1994; Blikra and Nemec 1998; Perez 1998). There is a need to generate measurements of surficial features and processes at fine spatial scales to more accurately quantify the resultant forms of sediment transport processes influencing alpine debris fans.

Airborne methods, such as photogrammetry, synthetic aperture radar (SAR) or interferometric SAR (InSAR) and laser range finding methods, such as airborne laser swath mapping (ALSM) have proven to be useful in the analysis of alluvial fan forms and processes (Coe et al. 1997; Catani et al. 2003, 2005; Staley et al. 2006; Frankel and Dolan 2007; Cavalli and Marchi 2008; Volker et al. 2007; Wasklewicz et al. 2008; Roering et al. 2009). These data sources are able to provide much greater detail regarding the shape and surficial form of the analyzed fans than coarse scale topographic maps (e.g. USGS 7.5-min quadrangles at 1:24,000).

Prior to (and even after) the increased availability of InSAR and airborne laser scanning data, high-resolution topographic information related to alluvial fan processes was typically constrained to photogrammetric acquisition. Stereo-matched low altitude photography allowed researchers to analyze orthoimagery and associated terrain data with resolutions typically on the order of 2 m and finer (Coe et al. 1997; Lane et al. 2000). Thorough reviews of the photogrammetric techniques and applications in geomorphic research can be found in Lane et al. (1993, 2000). Photogrammetric data have been used to produce inventories of alluvial fans (Crosta and Frattini 2004; Godt and Coe 2007), analyze alluvial fan surficial forms and processes (Crosta and Frattini 2004), and track changes associated with torrential processes using pre- and post-event imagery (Coe et al. 1997; Crosta and Frattini 2004; Roering et al. 2009).

Improvements to, and reductions in the cost of radar and laser technology allowed researchers to incorporate SAR, InSAR and laser scanning data into geomorphic research. These methods allowed researchers to study fans in greater detail than traditional photogrammetric techniques. While photogrammetric analysis permitted high resolution analysis of surfaces, the amount of effort required process these data constrained photogrammetrically derived high resolution measurements to relatively small areas. InSAR and airborne laser scanning allowed for high resolution measurements of surface form over much larger areal extents.

Interferometric synthetic aperture radar (InSAR) refers to a form of radar data processing that allows for the characterization of topography using air or spacecraft-mounted SAR systems (Massonnet and Feigl 1998). SAR systems use an antenna to produce a narrow band of electromagnetic energy directed at the surface. SAR systems both transmit and receive electromagnetic waves at wavelengths commonly ranging from 3 cm (*X band*) to 24 cm (*C band*) (Massonnet and Feigl 1998), though shorter wavelength sensors in the *Ku band* (roughly 1.7 cm) are recently being used (Okada et al. 2007). At these wavelengths, the energy is able to travel through the atmosphere with minimal degradation. Distance from the sensor to the Earth's surface is calculated based upon the round-trip travel time of the radar wave(s). Inferometry refers to the process of improving data accuracy by processing information from two or more SAR images of the same location. The multiple images can be obtained by (1) using the same SAR antenna during sequential flightpaths or orbits (multipass InSAR), or (2) using two (or more) SAR antennae mounted perpendicular to the flight path or orbit on the same platform (single pass InSAR) (Catani et al. 2005). Depending upon aircraft height and atmospheric conditions, InSAR is able to produce digital terrain models ranging in resolution from ~30-m (Roering et al. 2009) to less than 1 m (Okada et al. 2007). Readers seeking a more thorough review of InSAR methods are referred to Gens and van Genderen (1996), Massonnet and Fiegl (1998) and Rosen et al. (2000).

Application of InSAR data to the quantification of alluvial fan and form has been relatively limited. Farr and Chadwick (1996) use InSAR to differentiate between salt-dominated weathering processes and desert varnish and pavement development of alluvial fans in China and the southwestern United States. Catani et al. (2003) use spaceborne InSAR data to derive terrain measurements useful for the identification, prediction and assessment of flood hazards on alluvial fans. Despite the limited use in alluvial fan research, InSAR is a promising tool for the quantification of alluvial fan processes and forms, particular as sensor technology permits the generation of higher resolution terrain data (Okada et al. 2007) and repeated data acquisition allow for the temporal analysis of topographic change (Rosen et al. 2000; Roering et al. 2009).

Airborne laser swath mapping (ALSM, commonly referred to as airborne LiDAR) provides a novel means of mapping and analyzing the surficial features of debris fans and cones at high spatial resolutions over large spatial extents. A detailed explanation of the history and mechanics of laser range finding data for topographic analysis is beyond the scope of this chapter. Readers seeking more details on ALSM and other forms of laser range finding should consult Shan and Toth (2009) and Vosselman and Maas (2010).

ALSM is an optical remote sensing technology that measures properties of scattered light to find range and other information of a distant target (Shrestha et al. 1999; Carter et al. 2001). For ALSM data collection, an airplane is equipped with a precise kinematic global positioning system (GPS) receiver to position the aircraft and an accurate inertial navigation system to monitor the orientation (roll, yaw, and pitch) of the aircraft. A pulsed laser ranging system, secured on the underside of the aircraft's fuselage, produces measurements of the distance to the Earth's surface by calculating the roundtrip time of the laser pulse. The aircraft flies at a speed of 150–200 km per hour at an altitude of 500–1,000 m (dependent upon topography and atmospheric conditions) above the desired site. Current generation airborne laser scanners emit laser pulses in excess of 100,000 pulses per second (Slatton et al. 2007). Combining the positioning information, aircraft orientation, and range data from the laser sensor permits the creation of digital terrain models with horizontal accuracies of 15–20 cm and vertical accuracies of 5–10 cm (Slatton et al. 2007).

ALSM data are commonly used to produce DTMs with 1-m resolution (Shrestha et al. 1999). These high resolution data permit the evaluation of the surface form of alluvial fans at unprecedented spatial resolution over large areas. The use of this method of data acquisition for the analysis of fan system morphometry has increased dramatically in recent years, with applications related to fan systems in both arid (Staley et al. 2006; Frankel and Dolan; 2007; Volker et al. 2007; Wasklewicz et al. 2008) and alpine (Staley 2006; Cavalli and Marchi 2008; Trevisani et al. 2009) environments. ALSM data have been used to quantitatively identify the spatial patterns of debris-flow deposits (Staley et al. 2006) and distinguish debris-flow features from flood deposits (Volker et al. 2007; Wasklewicz et al. 2008) on alluvial fans. In addition, Frankel and Dolan (2007) demonstrated that the surface roughness of fans decreases over time, and measures of roughness may be used as a dating method in Death Valley, California. Cavalli and Marchi (2008) demonstrated the effectiveness of roughness indices derived from ALSM data to classify debris flow and non-debris-flow features on a single alpine alluvial fan. Trevisani et al. (2009) utilized variograms to classify the surface of alpine fans into categories similar to Cavalli and Marchi (2008). While these are novel approaches to quantifying alluvial fan and debris fan morphometry, they do not investigate the difference between the different types of processes that influence the fan surface, and the relationship of the form of the deposit to the form of the source areas.

The remainder of this chapter provides a case study that analyzes the morphometric characteristics of debris fans and source areas, and their relationship to the types of processes that contribute to the form of the fan surface. Here, we incorporate ALSM methods to analyze the surface form of debris fans and sources areas in two Alpine study areas in the Colorado Front Range (U.S.). Our analysis quantifies relationships between dominant geomorphic process and expressions of surface form (area, gradient, and roughness). We do this in two ways: (1) assessing differences in fan and source area morphometry between dominant processes, and, (2) by attempting to predict the degree of influence by debris flow based upon morphometry. Here, we pay particular attention to the morphometric characterization of debris-flow dominated fans and source areas, as they represent a greater hazard in terms of inundation area and travel distances for downslope populations than rockfall or slush flows. The quantitative relationships identified in this analysis may be used to rapidly assess the types of hazards occurring on alpine debris fans from high resolution ALSM data.

2 Study Site

This study analyzed ALSM data from 49 debris fans located in Stevens Gulch and Arapahoe Basin, which are located in the Colorado Front Range (Fig. 1). The Front Range is an exposed core of a large anticline uplifted during the Laramide

Orogeny, beginning ca. 75 million years B.P. (Kellogg et al. 2008) Both study areas are underlain by igneous and metamorphic bedrock formed as early as 1.7 billion years B.P. (Kellogg et al. 2008).

The location and form of the debris fan systems reflect the legacy of multiple Pleistocene and Holocene glacial advances, and Holocene geomorphic processes such as rockfall, debris flow, slush flow, grain flow, and talus shift (Rapp 1960a, b; Caine 1969; White 1981; van Steijn et al. 1995, 2002; Perez 1998; Cavalli and Marchi 2008; Trevisani et al. 2009). Debris fans have formed under well-developed couloirs and gullies, most of which are located along zones of structural weaknesses (Williams 1951; White 1981). Minimal couloir development occurred where bedrock was less resistant to weathering processes and valley sides are dominated by talus sheet systems.

Stevens Gulch (SG) is located immediately east of the Continental Divide near Loveland Pass (Fig. 1a). Data from 33 debris fan systems in Stevens Gulch are incorporated into this analysis. Elevations within the alpine portion of Stevens Gulch range from 3,500 m at the base of the talus slopes, to 4,350 m at the summit of Grays Peak, the highest point on the Continental Divide in the conterminous United States. Debris fans are located along the eastern and southern scarp of the valley. A gradational contact exists in Stevens Gulch between Middle Proterozoic Silver Plume granite in the northern (lower) portion of the valley and Precambrian metasedimentary biotite-gneiss in the southern (higher) portion of the valley (Kellogg et al. 2008). All 33 debris fan systems are situated above timberline.

Arapahoe Basin is located immediately west of the Continental Divide, just south of Loveland Pass (Fig. 1b). This study area is characterized by 16 debris fans. The six fans within Arapahoe Basin Ski Area glacial cirque are distributed at the base of the northwest-southeast trending ridge. A small cirque north of Arapahoe Basin contains ten well developed debris fans. All analyzed debris fan systems in Arapahoe Basin are underlain by Precambrian biotite gneiss (Kellogg et al. 2008) and are situated above treeline.

Within the study area, processes responsible for erosion, transport, and deposition of material include rockfall, debris flow, and slush flow. Rockfall is defined as the freefall, bouncing, and rolling of material shed directly from rockwalls (primary rockfall), or from debris stored on ledges or couloirs in the source area (secondary rockfall) (Whalley 1984; Matsuoka and Sakai 1999; Dorren 2003). Rockfall has been identified on fan surfaces through analysis of particle size distributions (Bones 1973; Kirkby and Statham 1975; Church et al. 1979; White 1981; Perez 1989; Blikra and Nemec 1998; Jomelli and Francou 2000) and fan profiles (Rapp 1960a, b; Caine 1967, 1983; Kirkby and Statham 1975; White 1981; Blikra and Nemec 1998; Jomelli and Francou 2000). Debris flows are defined as rapid mass movements of poorly sorted granular solids, water, and air (Varnes 1978). Debris flow produces surface features such as channels, levees, and terminal lobes that are easily identified on fan surfaces. Slush flows represent wet avalanches composed of water-saturated snow, rock debris and mud that are typically associated with spring snowmelt (Caine 1969; Jomelli and Bertran 2001; Decaulne and Saemundsson 2006). Slushflows may resemble highly viscous debris flows or low intensity, low viscosity flows resulting

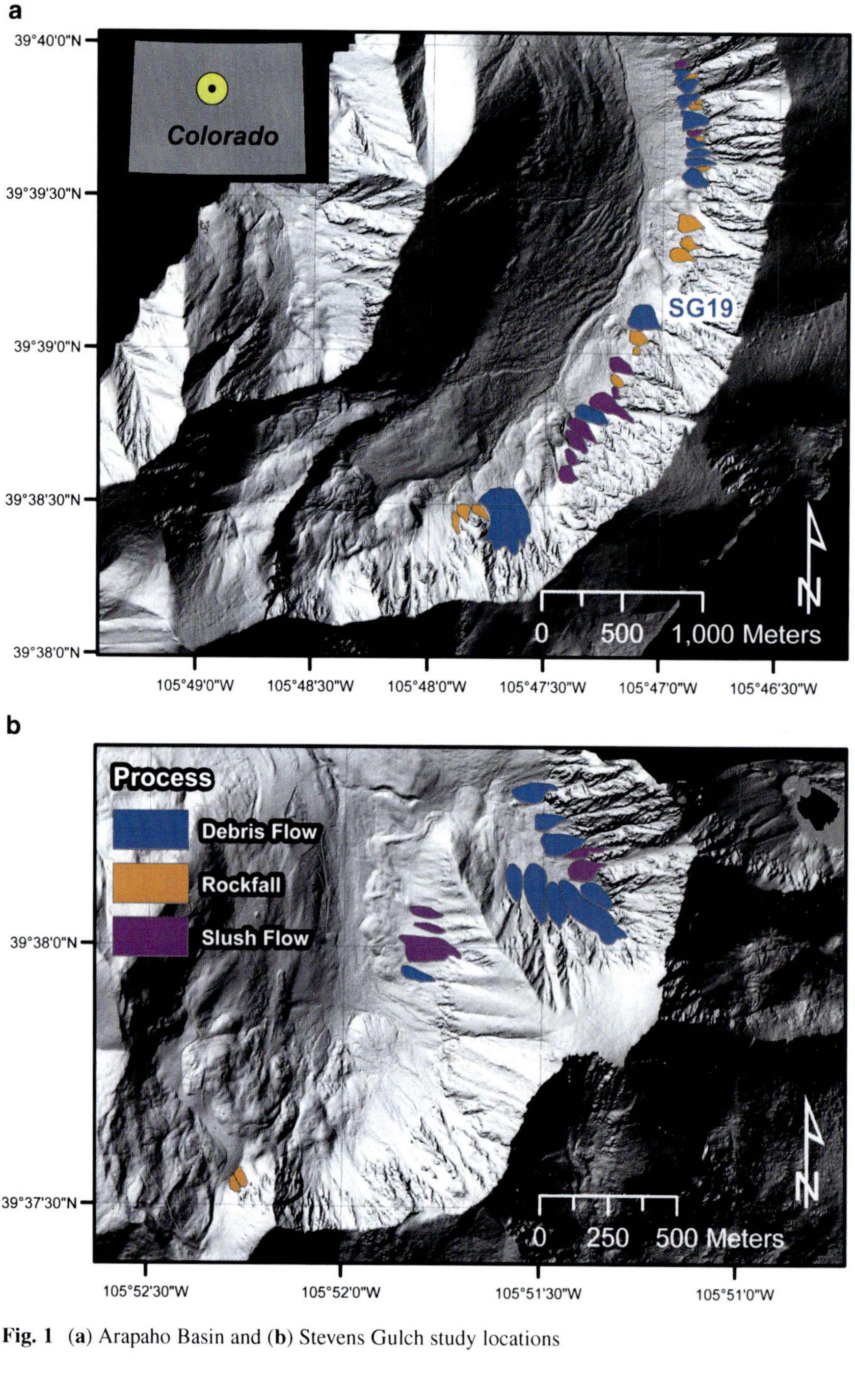

Fig. 1 (**a**) Arapaho Basin and (**b**) Stevens Gulch study locations

from ice breakups in small streams (Nyberg 1989; Bull et al. 1995; Decaulne and Saemundsson 2006). Most slushflows in the study areas resemble the former and runout only a short distance on the talus surface. Other processes, such as snow avalanches (Gardner 1970; Luckman 1977; White 1981; Jomelli and Francou 2000; Jomelli and Bertran 2001), solifluction (van Steijn et al. 1995), and frost coated clast flows (van Steijn et al. 1995, 2002) have been described in the literature, but the resultant forms associated with these processes are not evident on the fan surfaces included in this analysis.

3 Methods

3.1 ALSM Acquisition

ALSM data were generated by the NSF supported National Center for Airborne Laser Mapping (NCALM) at the University of Florida's Department of Civil & Coastal Engineering. The ALSM data collection occurred on 29 September 2005, using the following parameters: a scan angle of 20° (1/2 FOV); a scan frequency of 28 Hz; and a pulse rate of 33,333 Hz. NCALM conducted all of the post-processing of the raw ALSM data point cloud. Data were received as ASCII files that contain X, Y, and Z triplets processed from laser shots. Laser shots were obtained by using a kriging algorithm on separate flight strips that overlapped at greater than 30%. Grid nodes were combined with a weighted scheme based on strip geometry to produce a best estimate of elevation on a 1-m grid. The 1-m grid nodes contained in the ASCII file format were converted into 1-m raster grids with the aid of ArcGIS 9.1. The 1-m raster grids were imported into ArcInfo Workstation for calculation of morphometric parameters and spatial analysis. ALSM data were analyzed for the 49 alpine debris fan systems described in Sect. 2.

3.2 Derivation of Surface Morphometry

Geomorphic maps of the debris fans (Fig. 2) were derived from shaded relief and curvature maps (following the methods of Staley et al. 2006) using ALSM data. Here, planimetric and profile curvature were classified as concave, rectilinear and convex (Blaszczynski 1997; Staley et al. 2006) to create surface polygons. Each polygon was then assigned a process type based upon field observations. For each fan, the proportion of area occupied by each process type was calculated, and the process type that occupied the greatest proportion of surface area was defined as the dominant process (Fig. 3). The spatial extent of each source area was determined using the Watershed and Stream Delineation Tool developed by the Department of Civil Engineering at Texas A&M University (Olivera and Valenzuela 2006).

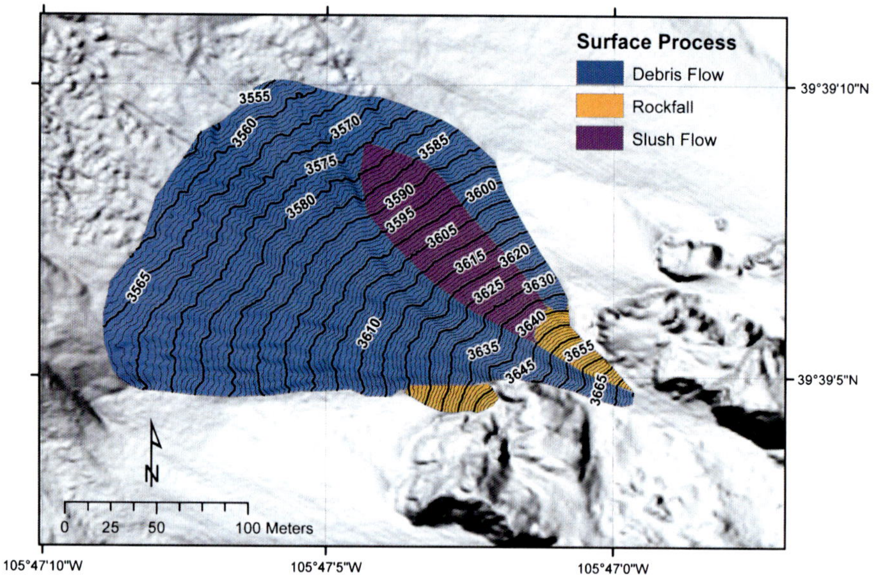

Fig. 2 Geomorphic map of the processes influencing debris fan SG19

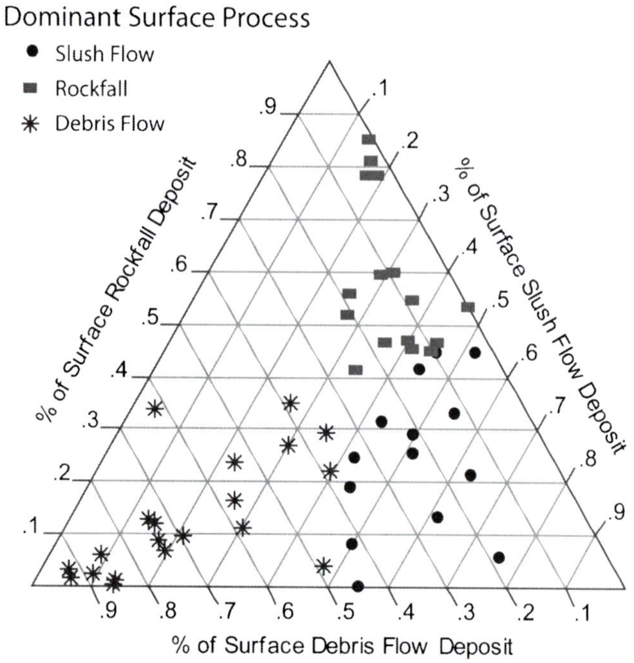

Fig. 3 Ternary diagram identifying the dominant process type and percentage of debris flow, rockfall and slush flow features on the surface of the 49 analyzed debris fans

Surface morphometry was analyzed using 2 dimensional (2D) and 2.5 dimensional (2.5D) measures. 2D measures included fan area, basin area, and the total relief of the basin. Areal calculations were determined from the geomorphic maps of the fans and source areas. Total relief was calculated from ALSM data and geomorphic maps by subtracting the elevation of the fan apex from the maximum elevation within the source area. The 2.5D measures included gradient and surface roughness. These measures were measured at length scales of 1- and 9-m using a moving window approach. These length scales were selected as they represent both the fine-scale features (1-m) and coarser-scale form (9-m) of the fan surface. The moving window calculates, on a cell-by-cell basis, the value of each parameter based upon the value of the surrounding cells. The length scale of measurement is defined by the radius of a circle centred on the cell. Gradient is defined as the change in elevation over a specified distance (represented in degrees). The roughness of the surface was quantified using the standard deviation of relative relief values at the length scales identified above. Relative relief was calculated using the methods outlined by Volker et al. (2007). Rough surfaces exhibit a higher standard deviation than smooth surfaces.

Detailed measurements of the identified terrain parameters were made for each 1-m pixel in all 49 debris fans and source areas. Areal calculations, total relief and mean values for gradient and roughness were derived for both fan surface and source areas. ANOVA, Student's t-test and linear regression analyses to statistically evaluate the morphometric differences between debris fans and source areas dominated by debris-flow processes and those dominated by rockfall and slush flow activity. Data were evaluated for normality using a Shapiro-Wilk W test and for homoscedasticity using both Leveen's and Bartlett's tests of equal variances. A simple log transformation allowed use of these tests when the original data were not able to meet the assumptions of normality and homoscedasticity, (e.g. basin and fan areas). P-values smaller than 0.05 were considered sufficient to reject the null hypotheses of the various statistical tests.

4 Results

4.1 Morphometry of Alpine Debris Fans

Geomorphic mapping of surficial features from hillshade and field validation identified 20 debris fans dominated by debris-flow processes, 16 fans dominated by rockfall processes, and 14 fans dominated by slush flow processes (Fig. 3). Assessment of the relationships between area, gradient and roughness and the dominant process exhibited on the fan surface identified some significant relationships and statistical predictability.

ANOVA results indicated significant differences in the areal extent of fans of the different process types (Table 1). The areal extent of debris-flow fans was

Table 1 Summary of mean parameter values for each fan type and ANOVA results

	Dominant fan process			ANOVA results	
Morphometric parameter	Debris flow	Rockfall	Slushflow	F-ratio	P
N	20	16	14		
Log(Fan area) (m^2)	9.2	8.2	8.8	9.62	0.0003
Fan gradient 1-m (°)	29.1	34.3	32.4	28.20	<0.0001
Fan gradient 9-m (°)	28.3	34.1	32.2	30.30	<0.0001
Fan roughness 1-m (°)	0.1	0.1	0.1	3.07	0.06
Fan roughness 9-m (°)	0.5	0.4	0.4	3.97	0.02
Log(Basin area) (m^2)	10.6	9.0	9.8	13.22	<0.0001
Basin relief (m)	342.6	190.3	278.0	13.87	<0.0001
Basin gradient 9-m (°)	40.8	44.4	43.4	3.92	0.02

significantly greater than both rockfall and slush flow dominated fans. A weak, though statistically significant positive correlation between fan area and the percent of the fan surface influenced by debris flow was identified (Fig. 4a).

Gradient differed significantly between process types at 1- and 9-m length-scales. ANOVA results yielded significant differences in gradient between debris flow, rockfall and slush flow dominated fans (Table 1). Debris-flow fans were consistently lower gradient than both rockfall and slush flow dominated fans, while rockfall dominated fans were steepest at both length-scales. At the 1-m length-scale, a strong negative relationship was identified between fan gradient and percent of surface influenced by debris-flow processes (Fig. 4b). A strong negative relationship was also identified between fan gradient and percent of surface influenced by debris-flow processes at the 9-m length-scale (Fig. 4c).

Surface roughness was also analyzed at both 1- and 9-m length-scales. ANOVA revealed significant differences between debris-flow and non-debris-flow processes only at the 9-m length scale (Table 1). At this length-scale, debris-flow dominated fans tended to exhibit greater surface roughness than either rockfall or slush flow, though rockfall and slush flow fans did not differ significantly from one another. Weak positive relationships were identified between proportion of surface area influenced by debris flow and roughness at both length-scales (Fig. 4e, f).

4.2 *Morphometry of Source Areas*

Analysis of source areas indicated that differences in total basin relief and basin area produced significant differences in the types of processes that were evident on fan surfaces, and therefore, fan morphometry. Debris-flow producing source areas were significantly larger than those that produced a dominance of slush flow or rockfall processes (Table 1). ANOVA revealed significant differences in basin area between all dominant process types, with debris-flow dominated sources the largest and

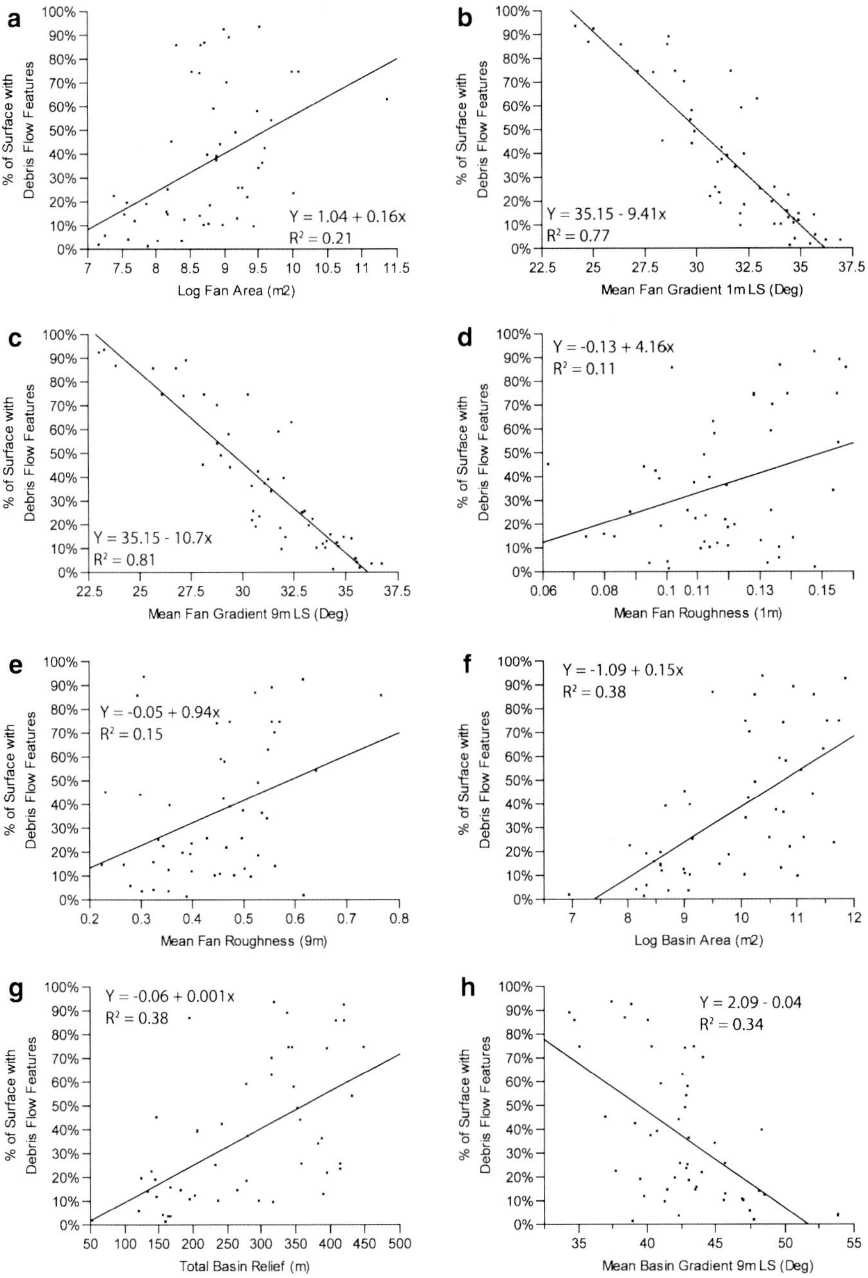

Fig. 4 Simple linear regression results assessing the relationship between morphometric parameter and fan area influenced by debris flow

rockfall dominated sources areas located in the smallest basins. A linear regression model suggested a weak positive relationship between basin area and percentage of fan surface influenced by debris flow (Fig. 4f).

Debris-flow dominated fans were characterized by source areas with significantly greater total relief than those of slush flow and rockfall dominated fans (Table 1). In addition, slush flow dominated fan source areas tended to have significantly more total relief than rockfall dominated fans. Linear regression suggested a weak positive correlation between total source area relief and the percentage of the fan surface influenced by debris flow (Fig. 4g).

No significant relationship between source area gradient and dominant process types were detected at the 1-m length-scale. At the 9-m length-scale, debris-flow dominated fans exhibited source areas with significantly lower gradients than those of rockfall dominated fans, and did not differ significantly from slush flow dominated fans (Table 1). No significant differences were identified for source area gradients at the 9-m length-scale between slush flow dominated fans and rockfall dominated fans. Linear regression suggested a very weak negative correlation between source area gradient (9-m length-scale) and the percentage of the fan surface influenced by debris flow (Fig. 4h).

4.3 Relationships Between Source Area and Fan Morphometry

Simple linear regressions were used to detect significant relations between source area morphometry (area, relief and gradient) and fan morphometry (area, gradient and roughness). Here, we report only those models that are statistically strong ($r^2 > 0.35$). As such, we only report on the relationships associated with the source area morphometric parameters of total relief, the log of basin area, and mean gradient at the 9-m length-scales. No models were able to provide a reasonable correlation between any of the source area parameters and fan roughness at either length-scale.

The log-transformed basin area was correlated with the log-transformed fan area measurement (Fig. 5a). Here, a strong positive relationship was identified between measures. A weaker, inverse relationships was identified between basin area and fan gradient (9-m length-scale).

Correlations were also identified between total basin relief and both fan area and fan gradient at the 9-m length-scale (Fig. 5). A relatively strong positive relationship existed between total source area relief and fan area, while a weaker ($r^2 = 0.37$), though still statistically significant relationship was identified between total source area relief and fan gradient. Additionally, a weak positive relationship ($r^2 = 0.39$) was identified between source area gradient and fan gradient (both measured at the 9-m length-scale).

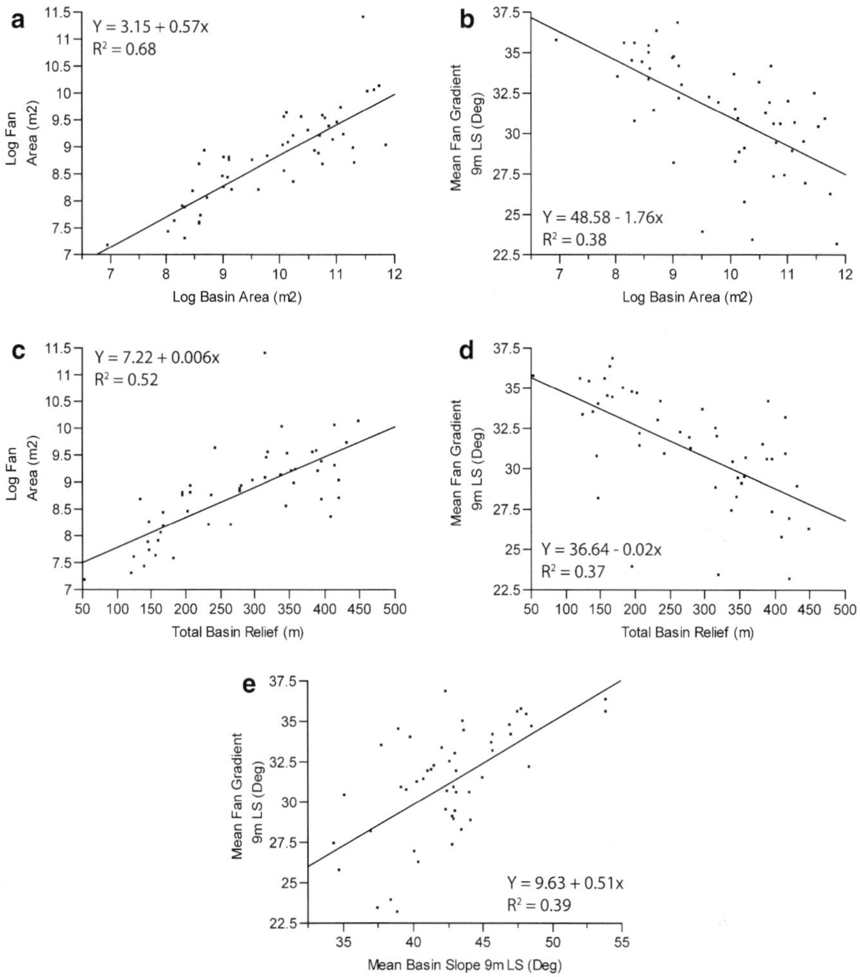

Fig. 5 Simple linear regression analyses of basin and fan morphometry

5 Discussion

This study quantified the relations between geomorphic process types and the surface form of both alpine debris fans and their source areas. Distinct differences in both fan and source area morphometry have been identified for those systems dominated by debris-flow, rockfall, or slush flow processes.

Debris-flow fans tended to be larger, have shallower gradients, and are characterized by higher surface roughness values (particularly at longer length-scales) when compared to rockfall or slush flow dominated fans. Debris-flow source areas also tended to have the greatest areas, largest total relief, and lowest gradient of the three types of processes analyzed in this study. The larger source areas associated with debris-flow dominated fan systems were able to collect more water during precipitation events, but are also able to collect more snow blown by prevailing winds. Greater moisture availability increased the likelihood of debris-flow initiation processes in the source area or on the fan surface. Once flow was initiated, the rheology of debris flows allowed them to runout over significantly longer distances and shallower gradients than the slush flow processes operating in the study areas. As such, debris-flow systems were characterized by larger fans with lower mean gradients than other dominant process-types. Higher surface roughness values reflect the greater topographic complexity associated with debris-flow deposits relative to rockfall or slush flow deposits. Channel erosion and deposition of levees and terminal lobes produced features with higher roughness values at all analyzed length-scales.

Slush flow dominated fans tended to reside between debris-flow and rockfall dominated fans in regards to area and gradient, but be indistinguishable from rockfall fans in regards to surface roughness. Source areas for slush flow dominated systems were larger and more dissected than rockfall systems, but smaller and less dissected than debris-flow systems. Collection of snow, rain and debris were important in producing slush flow processes. Relative to debris-flow dominated systems, the slush flow systems were unable to consistently produce the runoff required to trigger debris-flow activity. Instead, snowmelt increased the pore-pressures in late-spring patches of snow. Slush flows were initiated when the pore-pressures exceeded the critical shear strength of the snowpack. These flows were not able to maintain mobility for as long a distance, or produce features with as high of a degree of topographic complexity as debris flows.

Rockfall dominated fans and their source areas tend to be the smallest and steepest of all three process types. Material shed from smaller, steeper basins has a greater likelihood of being deposited at the base of the source area, where rockfall in larger basins is often deposited on ledges or steeper channels, where the deposits are more likely to be mobilized by subsequent debris flows or slush flows. The smaller fan areas reflect the tendency for smaller, more frequent rockfall mobilized clasts to be deposited close to the fan apex. While long distance runout is possible and has been widely recognized, the relatively low frequency of this phenomenon is reflected in the relatively small size of rockfall dominated debris fans.

The general relationships identified between source area morphometry and fan surface morphometry reflect the influence of surface form on the generation of different types of geomorphic processes. Basin area, relief, and gradient represent important boundary conditions which are critical in the determining of the type, extent, and frequency of the different processes responsible for the development and evolution of alpine debris fans. These findings reflect the importance of studying both how surface form reflects the process types responsible for the development of the fans, but also how surface form influences subsequent geomorphic processes.

The rapidly increasing human population will result in significant influences on previously undeveloped mountainous regions worldwide. Settlement, as well as transportation networks and utility corridors will all be exposed to the hazards of mountainous terrain, including debris flow, rockfall, and other mass-movement processes. In the future, airborne laser swath mapping data will become less-costly and more readily available throughout the world. These data provide an important source of topographic information for the analysis of potential debris-flow and other mass-movement related hazards. The methods documented in this case study present a means of rapidly and effectively assessing the potential hazards associated with the geomorphic processes that create and sculpt alpine debris fans. The relations defined here are specific to alpine fans in two valleys within the Colorado Front Range. While the exact relations identified in this study may not be directly transferable to other environments, the documented methods of analysis should be applicable to fans in both humid and arid environments.

6 Conclusions

The relations between geomorphic process types and surface form are significant not only to the study of alpine slope processes (with particular emphasis on talus systems), but are also applicable to the discipline of geomorphology as a whole. In this study, we have quantified the relationship between process type and form on alpine debris fan systems at high-spatial resolutions and over large spatial extents. We have found that there are statistically significant differences in the size, gradient and roughness of fan surfaces and contributing areas between systems dominated by different geomorphic processes. New insight has been gained into how geomorphic processes are reflected in, and are in turn influenced by, the surface form of various components of the alpine debris fan system.

High resolution airborne data (ALSM) have provided highly accurate terrain characterization over large spatial extents. These data were used to derive measurements of surface form, develop the statistical models and test differences between process types. The analysis has proved that it is possible to statistically differentiate the processes influence alluvial fan surfaces. These quantitative relationships provide a means to rapidly assess the extent and types of hazards associated with alpine fan systems, and plan for the reduction or mitigation of the effects of these hazards.

While we have used ALSM data in the case study in the Colorado Front Range, other airborne data sources, such as low altitude aerial photography and high-resolution InSAR data also would provide the level of detail necessary to conduct surface analyses of process types on alluvial fan systems. In addition, a multi-temporal dataset would provide insight into how these surfaces and relationships between process type and surface form evolve over time.

Acknowledgements This material is based upon work supported by the National Science Foundation under Grant No. 05–02343 (Doctoral Dissertation Research: Process – form linkages in alpine talus deposits, Front Range, Colorado, USA) and Grant No. 02–39749 (CAREER: Alluvial fan form quantification to advance geographic science and education). Any opinions, finding, or conclusions or recommendations expressed in this material are those of the authors and do not necessarily reflect the views of the National Science Foundation. The authors gratefully acknowledge Jeff Coe and Jason Kean for reviews that have dramatically improved this manuscript. We also would like to acknowledge Chris Smith for assistance in the field.

References

Akerman HJ (1994) Notes on talus morphology and processes in Spitsbergen. Geogr Ann Ser A Phys Geogr 66:267–285
Albjar G, Rehn J, Stromquist L (1979) Notes on talus formation in different climates. Geogr Ann Ser A Phys Geogr 61(3–4):179–185
Ballantyne CK (2002) Paraglacial geomorphology. Quat Sci Rev 21:1935–2017
Blaszczynski JS (1997) Landform characterization with geographic information systems. Photogramm Eng Remote Sens 63(2):183–191
Blikra LH, Nemec W (1998) Postglacial colluvium in western Norway: depositional processes, facies and paleoclimatic record. Sedimentology 45:909–959
Bones JG (1973) Processes and sediment size arrangement on high arctic talus, southwest Devon Island, NWT, Canada. Arct Alp Res 5(1):29–40
Bull WB, Schlyter P, Brogaard S (1995) Lichenometric analysis of the Karkerieppe slush-avalanche fan, Karkevagge, Sweden. Geogr Ann Ser A Phys Geogr 77(4):231–240
Caine N (1967) The texture of talus in Tasmania. J Sediment Petrol 37(3):796–803
Caine N (1969) A model for alpine talus slope development by slush avalanching. J Geol 77(1): 92–100
Caine N (1983) The mountains of northeastern Tasmania. AA Balkema, Rotterdam, 201 pp
Carson MA (1977) Angle of repose, angles of shearing resistance and angles of talus slopes. Earth Surf Process Landf 2(4):363–380
Carter W, Shrestha R, Tuell G, Bloomquist D, Sartori M (2001) Airborne laser mapping shines new light on Earth's topography. EOS Trans Am Geophys Union 82(46):549, 550, 555
Catani F, Farina P, Moretti S (2003) Spaceborne radar interferometry: a promising tool for hydrological analysis in mountain alluvial fan environments. In: Erosion prediction in ungaged basins: integrating methods and technique, IAHS publication No. 279. IAHS, Wallingford, pp 241–248
Catani F, Farina P, Moretti S, Nico G, Strozzi T (2005) On the application of SAR interferometry to geomorphological studies: estimation of landform attributes and mass movements. Geomorphology 66:119–131
Cavalli M, Marchi L (2008) Characterisation of the surface morphology of an alpine alluvial fan using airborne LiDAR. Nat Hazard Earth Syst Sci 8:323–333
Church M, Stock RF, Ryder M (1979) Contemporary sedimentary environments on Baffin Island, NWT Canada: debris slope accumulations. Arct Alp Res 11(4):371–402
Coe JA, Glancy PA, Whitney JW (1997) Volumetric analysis and hydrologic characterization of a modern debris flow near Yucca Mountain, Nevada. Geomorphology 20:11–28
Crosta GB, Frattini P (2004) Controls on modern alluvial fan processes in the central Alps, northern Italy. Earth Surf Process Landf 29:267–293
Curry AM, Morris CJ (2004) Lateglacial and Holocene talus slope development and rockwall retreat on Mynydd Du, UK. Geomorphology 58:85–106
Decaulne A, Saemundsson P (2006) Meteorological conditions during slushflow release and their geomorphic impact in the northwestern Icelandic Westfjords. Geomorphology 80(1–2):80–93
Dorren LKA (2003) A review of rockfall mechanics and modeling approaches. Prog Phys Geogr 27(1):69–87
Farr TG, Chadwick OA (1996) Geomorphic processes and remote sensing signatures of alluvial fans in the Kun Lun Mountains, China. J Geophys Res 101(E10):23091–23100

Francou B, Mante C (1990) Analysis of the segmentation in the profile of alpine talus slopes. Permafr Periglac Process 1(1):53–60

Frankel KL, Dolan JF (2007) Characterizing arid region alluvial fan surface roughness with airborne laser swath mapping digital topgraphic data. J Geophys Res 112:F02025

Gardner J (1970) Geomorphic significance of avalanches in the Lake Louise area, Alberta, Canada. Arct Alp Res 2(2):135–144

Gens R, van Genderen JL (1996) Review article SAR interferometry – issues, techniques, applications. Int J Remote Sens 17(10):1803–1835

Godt JW, Coe JA (2007) Alpine debris flows triggered by a 28 July 1999 thunderstorm in the central Front Range, Colorado. Geomorphology 84:80–97

Jomelli V, Bertran P (2001) Wet snow avalanche deposits in the French Alps: structure and sedimentology. Geogr Ann Ser A Phys Geogr 83(1–2):15–28

Jomelli V, Francou B (2000) Comparing the characteristics of rockfall talus and snow avalanche landforms in an alpine environment using a new methodological approach, Massif de Ecrins, French Alps. Geomorphology 35(3–4):181–192

Kellogg KS, Shroba RR, Bryant B, Premo WR (2008) Geologic map of the Denver West 30′ × 60′ Quadrangle, North-Central Colorado: U.S. Geological Survey Scientific Investigations Map 3000, scale 1:100,000, 48 pp

Kirkby MJ, Statham I (1975) Surface stone movement and scree formation. J Geol 83(3):349–362

Lane SN, Richards KS, Chandler JH (1993) Developments in photogrammetry: the geomorphic potential. Prog Phys Geogr 17(3):306–328

Lane SN, James TD, Crowell MD (2000) Application of digital photogrammetry for complex topography for geomorphological research. Photogramm Rec 16(95):793–821

Luckman BH (1977) The geomorphic activity of snow avalanches. Geogr Ann Ser A Phys Geogr 59(1–2):31–48

Massonnet D, Feigl K (1998) Radar interferometry and its application to changes in the Earth's surface. Rev Geophys 36(4):441–500

Matsuoka N, Sakai H (1999) Rockfall activity from an alpine cliff during thawing periods. Geomorphology 28:309–328

Nyberg R (1989) Observations of slush flows and their geomorphological effects in the Swedish mountain area. Geogr Ann Ser A Phys Geogr 71(3–4):185–198

Okada Y, Hirao C, Horiuchi T, Hara Y, Yedidia JS, Azarbayejani A, Oishi N (2007) Highly Accurate DSM Reconstruction Using Ku-band Airborne InSAR. IEEE Int Geosci Remote Sens Soc Symp (IGARSS):5049–5052. http://ieeexplore.ieee.org/xpl/articleDetails.jsp?arnumber=4423996

Olivera F, Valenzuela M (2006) Watershed and stream delineation tool 2002. Available from http://ceprofs.tamu.edu/folivera/GISTools/wsdt/home.htm. Cited 15 Mar 2006

Perez F (1989) Talus fabric and particle morphology on Lassen Peak, California. Geogr Ann Ser A Phys Geogr 71(1–2):43–57

Perez F (1993) Talus movement in the high Equatorial Andes: a synthesis of ten years of data. Permafr Periglac Process 4:199–215

Perez F (1998) Talus fabric, clast morphology, and botanical indicators of slope processes on the Chaos Crags (California Cascades), USA. Geogr Phys Quat 52(1):47–68

Rapp A (1960a) Recent development of mountain slopes in Karkevagge and surroundings, northern Scandinavia. Geogr Ann Ser A Phys Geogr 42(2–3):65–200

Rapp A (1960b) Talus slopes and mountain walls at Tempelfjorden, Spitsbergen, Norsk Polarinstitutt Skrifter 119. Oslo University Press, Oslo, 96 pp

Roering JJ, Stimley LL, Mackey BH, Schmidt DA (2009) Using DInSAR, airborne LiDAR, and archival air photos to quantify landsliding and sediment transport. Geophys Res Lett 36:L19402

Rosen PA, Hensley S, Joughin IR, Li FK, Madsen SN, Rodriguez E, Goldstein RM (2000) Synthetic aperture radar interferometry. Proc IEEE 88:333–382

Schumm SA (1991) To interpret the Earth, ten ways to be wrong. Cambridge University Press, New York, 133 pp

Schumm SA, Lichty RW (1965) Time, space and causality in geomorphology. Am J Sci 263: 110–119
Shan J, Toth KC (2009) Topographic laser ranging and scanning – principles and processing. CRC Press, Boca Raton, 590 pp
Shrestha RL, Carter WE, Lee M, Finer P, Sartori M (1999) Airborne laser swath mapping: accuracy assessment for surveying and mapping applications. J Am Congr Surv Mapp 59(2):83–94
Slatton KC, Carter WE, Shrestha RL, Dietrich W (2007) Airborne laser swath mapping: achieving the resolution and accuracy required for geosurficial research. Geophys Res Lett 35:L23S10
Staley DM (2006) Process and form on Alpine Talus Cone Systems, Front Range, Colorado, USA. Unpublished PhD dissertation, University of Memphis, Memphis, 214 pp
Staley DM, Wasklewicz TA, Blaszczynski JS (2006) Surficial patterns of debris flow deposition on alluvial fans in Death Valley, CA using airborne laser swath mapping data. Geomorphology 74:152–163
Statham I (1976) A scree slope rockfall model. Earth Surf Process 1:43–62
Trevisani S, Cavalli M, Marchi L (2009) Variogram maps from LiDAR data as fingerprints of surface morphology on scree slopes. Nat Hazard Earth Syst Sci 9:129–133
van Steijn H (2002) Long-term landform evolution; evidence from talus studies. Earth Surf Process Landf 27(11):1189–1199
van Steijn H, Bertran P, Francou B, Hetu B, Texier J-P (1995) Models for the genetic and environmental interpretation of stratified slope deposits: review. Permafr Periglac Process 6:125–146
van Steijn H, Boelhouwers J, Harris S, Hetu B (2002) Recent research on the nature, origin and climatic relations of blocky and stratified slope deposits. Prog Phys Geogr 26(4):551–575
Varnes DJ (1978) Slope movement and types and processes. In: Schuster RL, Krizek RJ (eds) Landslides: analysis and control, National Research Council, Transportation Research Board, special report 247. National Academy of Sciences, Washington, DC, pp 76–90
Volker HX, Wasklewicz TA, Ellis MA (2007) A topographic fingerprint to distinguish Holocene alluvial fan formative processes. Geomorphology 88:34–45
Vosselman G, Maas HG (2010) Airborne and terrestrial laser scanning. Whittles Publishing, Dunbeath, 336 pp
Wasklewicz TA, Mihir MA, Whitworth J (2008) Surface variability of alluvial fans generated by disparate processes, eastern Death Valley, CA. Prof Geogr 60(2):207–223
Whalley WB (1984) Rockfalls. In: Brundsen D, Prior DB (eds) Slope instability. Wiley, New York, pp 217–255
White SE (1981) Alpine mass movement forms (non-catastrophic): classification, description and significance. Arct Alp Res 13(2):127–137
Williams FJ (1951) The geology of the Stevens Mine. Unpublished Masters thesis, Department of Geology, Colorado School of Mines, Golden, CO, USA

Dendrogeomorphology – Tracking Past Events with Tree Rings

Michelle Schneuwly-Bollschweiler and Markus Stoffel

1 Introduction

Tree rings were first used for the pure dating of wood, for instance in archaeology. However, in the 1970s, researcher started drawing environmental information from tree rings as their growth reflects the influences to which a tree is exposed during its live. The term "dendrogeomorphology" was first introduced in the 1970s by Alestalo (1971) and refers to the study of geomorphic processes with tree rings. The method was then further developed by Shroder (1978, 1980) and Braam et al. (1987a, b).

In the beginnings of dendrogeomorphic research, applications of tree rings to geomorphic questions focused mainly on the simple dating of processes. In the recent past, however, research in dendrogeomorphology and its application have developed to a discipline able to provide various kinds of temporal and spatial information on geomorphic processes and natural hazards (Stoffel et al. 2010a). Besides the reconstruction of past event frequencies, tree-ring studies may provide spatial data on the lateral spread and on runout distances of events on fans and cones. Through additional field investigations magnitudes of past events can be assessed and tree rings may also give indications on the event triggers when coupled with meteorological data.

In the following, tree-ring formation, the impact of torrential activity on trees and their growth reactions to external disturbance will be described. Methods used in dendrogeomorphology are then described and their applications on fans and cones are shown. Finally, the limitations of the method will be explained.

M. Schneuwly-Bollschweiler (✉) • M. Stoffel
Laboratory of Dendrogeomorphology (dendrolab.ch), Institute of Geological Sciences, University of Berne, CH-3012 Berne, Switzerland

Chair for Climate and Climatic Impacts, Environmental Sciences, University of Geneva, CH-1227 Carouge-Geneva, Switzerland
e-mail: michelle.schneuwly@dendrolab.ch; markus.stoffel@dendrolab.ch

M. Schneuwly-Bollschweiler et al. (eds.), *Dating Torrential Processes on Fans and Cones*, Advances in Global Change Research 47, DOI 10.1007/978-94-007-4336-6_10,

2 Tree-Ring Formation and Impact of Torrential Activity on Trees

Dendrochronology – and therefore also dendrogeomorphology – is based on the principle that trees growing in regions with distinct seasons form annual growth rings. While cells are produced during the growing season, growth ceases during the dormant season. Tree growth and the size of each tree ring is influenced by intrinsic factors (e.g. genetics, age) on one hand and by extrinsic factors including water availability, temperature, light, soil properties and competition on the other hand. Trees growing at the same site therefore normally record the same external influences in their tree-ring series.

Apart from these biotic factors common to all trees growing at a site, mechanical impacts such as from torrential activity may cause changes in the growth pattern of individual trees at the location. This impact will be recorded in the tree-ring series and allows the later dating of events. In the following, we will provide an overview on mechanical disturbances in trees caused by torrential activity and on related growth reactions. Examples on impacted trees and associated growth reactions are illustrated in Figs. 1 and 2.

Boulders, wood or other solid material transported in the water during a flood or debris flow may cause injuries to the stem (Fig. 1). At the location of the injury, tree growth becomes interrupted and the tree will therefore start to overgrow the wound from the lateral edges of the injured segment through the production of callus tissue (Larson 1994; Sachs 1991). In addition, tangential rows of traumatic resin ducts (TRD) are formed in several conifer species. Given that the injury occurred during the growing season, resin production will normally start directly after the impact and the first TRD can be observed about 3 weeks after the moment of the injury (Luchi et al. 2005; Ruel et al. 1998). TRD represent therefore a very valuable and precise tool for the dating of past torrential events (Bollschweiler et al. 2008b) with seasonal and even monthly precision (Kaczka et al. 2010; Stoffel and Beniston 2006).

Unilateral pressure, for instance through material transported in the flow or deep erosion, may incline the stem axis of a tree (Fig. 2a). Such trees will try to regain their vertical position by forming reaction wood in order to compensate for the inclination (Braam et al. 1987a; Timell 1986). Conifers form compression wood composed of larger cells with thicker cell walls on the lower side of the stem whereas broadleaved trees form tension wood on the upper side of the trunk (Pilate et al. 2004). In the tree-ring series, a sudden onset of eccentric growth can be observed. The growth reaction will be most pronounced at the location of the tilting (Mattheck 1993).

Material transported in a flow and deposited on the fan or cone may also bury the stem base of trees (Fig. 2b). Through the resulting shortage in nutriment supply and water, tree growth will be limited and a sudden and abrupt suppression in the yearly increment of tree rings will be observed (Friedman et al. 2005; Hupp et al. 1987). Provided that the material is fine-grained, rich in nutriments, water supply not disrupted and that the depth of deposited material is not too important, an increase in

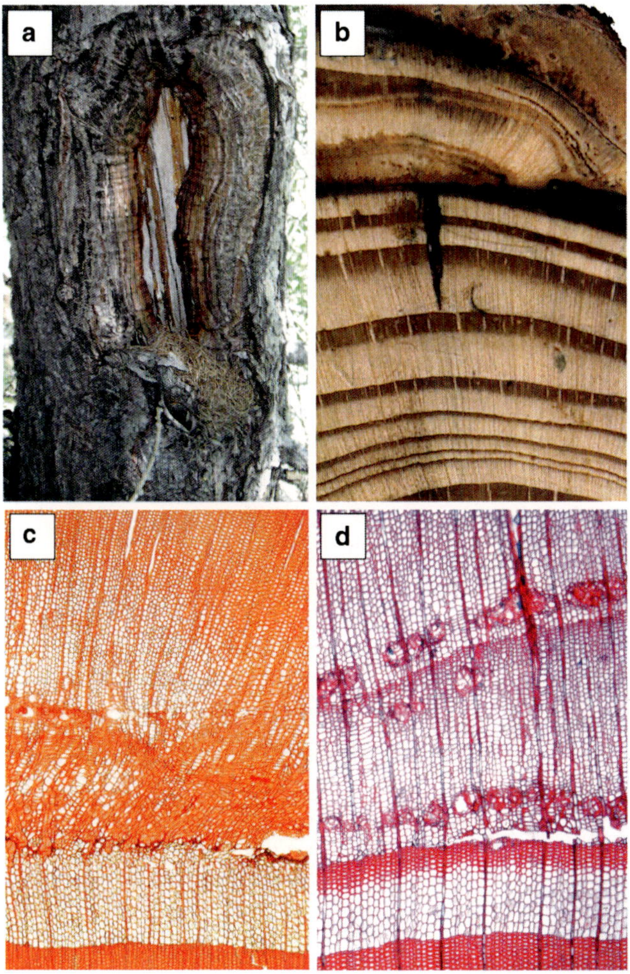

Fig. 1 Images of trees injured by torrential activity and associated growth reactions. (**a**) Injury as seen on the stem surface. (**b**) The tree starts overgrowing the open injury from the edges through the formation of callus tissue as illustrated in (**c**). (**d**) On both sides of the injury, tangential rows of traumatic resin ducts are formed

growth can occasionally be observed (Strunk 1995). If burial depth is too important, trees will die from water and nutrient shortage.

Boulders transported in the flow may decapitate smaller trees. Lateral branches will take over the lead forming the new apex (Fig. 2c). In addition, the loss of the crown often results in sudden growth suppression (Butler and Malanson 1985).

The gradual and sudden denudation of tree roots through erosive events results in different growth reactions in stem and roots (LaMarche 1961) (Fig. 2d). If only part of the root is exposed, it will continue to fulfil its function; however, anatomical changes can be observed in the cell structure of the root (Bodoque et al. 2005;

Fig. 2 Trees impacted by torrential processes and related growth reactions. (**a**) Inclination of the tree through unilateral pressure. (**b**) Deposition of material around the stem base. (**c**) In a decapitated tree, lateral branches form new leaders. (**d**) Tree roots exposed through erosion in a channel. (**e**) Large events may clear entire surfaces creating better growth conditions for survivors. (**f**) Compression wood. (**g**) Abrupt growth suppression. (**h**) Abrupt growth increase

Hitz et al. 2008). If the root is exposed completely, it will die off leading to a shortage in water and nutrient supply. Yearly increment will therefore be smaller given that an important amount of the roots system has been affected.

Devastating events may also partly eliminate forest stands (Fig. 2e). Uninjured survivor trees will benefit from less competition and improved light conditions, more water and nutrients. Tree will react to these improve conditions with a sudden growth increase.

3 Methods Used in Dendrogeomorphic Research

3.1 Fieldwork

A first prerequisite for any dendrogeomorphic study is the presence of trees predating the last event and being affected by past hydrogeomorphic processes. In the previous section, it has been shown that trees record mechanical impacts in their tree-ring series allowing an accurate dating of the moment of the impact. However, the visual inspection of the tree rings is not sufficient for the determination of the process causing the growth reaction. Therefore, a detailed identification of the geomorphic process, the related features on the cone or fan surface and all other possible impacts at the study site is an essential part of the investigation. Topographic sheets, geologic maps and aerial photographs may support the definition of geomorphic processes and affected sectors. Technical reports, archival data or newspaper reports provide additional sources of information on the occurrence of previous events at the study site.

In the field, geomorphic features related to past activity need to be identified before a map of all geomorphic deposits and forms can be realized. In open forests, LiDAR data with field verification can serve as a basis for the geomorphic map. High-precision GPS can be used where forest cover and topography allow reasonable satellite cover. An alternative method is the use of tape or laser measure, compass and inclinometer. The geomorphic map resulting from the different approaches outlined above can then be used as a basis for the sampling strategy of disturbed trees (Stoffel and Bollschweiler 2009).

The choice of trees to sample will depend on their position in deposits and on their external aspect. Trees showing obvious signs of past torrential activity are sampled either with cross sections, wedges or increment cores. Cross sections cut with a hand or chain saw provide a complete image of the entire tree-ring sequence at sampling height. However, this destructive method is only exceptionally feasible as forest may have a protective function. Therefore, the sampling has normally to be based on small wedges cut from injuries or increment cores sampled with an increment borer (Grissino-Mayer 2003). Several increment cores are normally extracted per tree at the height of the visible growth defect. The position of the sampled tree is marked on the geomorphic map and additional information noted including (i) description of the external growth defect, (ii) position of all samples taken, (iii) tree species, (iv) diameter at breast height, (v) information on diseases of the tree and (vi) information on neighbouring trees before pictures are taken of the tree and of all visible growth defects. Samples taken are then labelled and stored for transport. A more detailed information on the sampling of trees can be found in Stoffel and Bollschweiler (2008).

3.2 Laboratory Analysis and Definition of Event Years

In the laboratory, samples are dried and sanded before tree rings are counted and measured using a digital positioning table linked to a stereomicroscope and a time-series analysis program. Measured growth curves of all samples are then compared to a local reference chronology representing normal growth conditions at the study site (Cook and Kairiukstis 1990; Vaganov et al. 2006) to identify false or missing tree rings and to correct faulty tree-ring series (Schweingruber et al. 1990) as illustrated in Fig. 3.

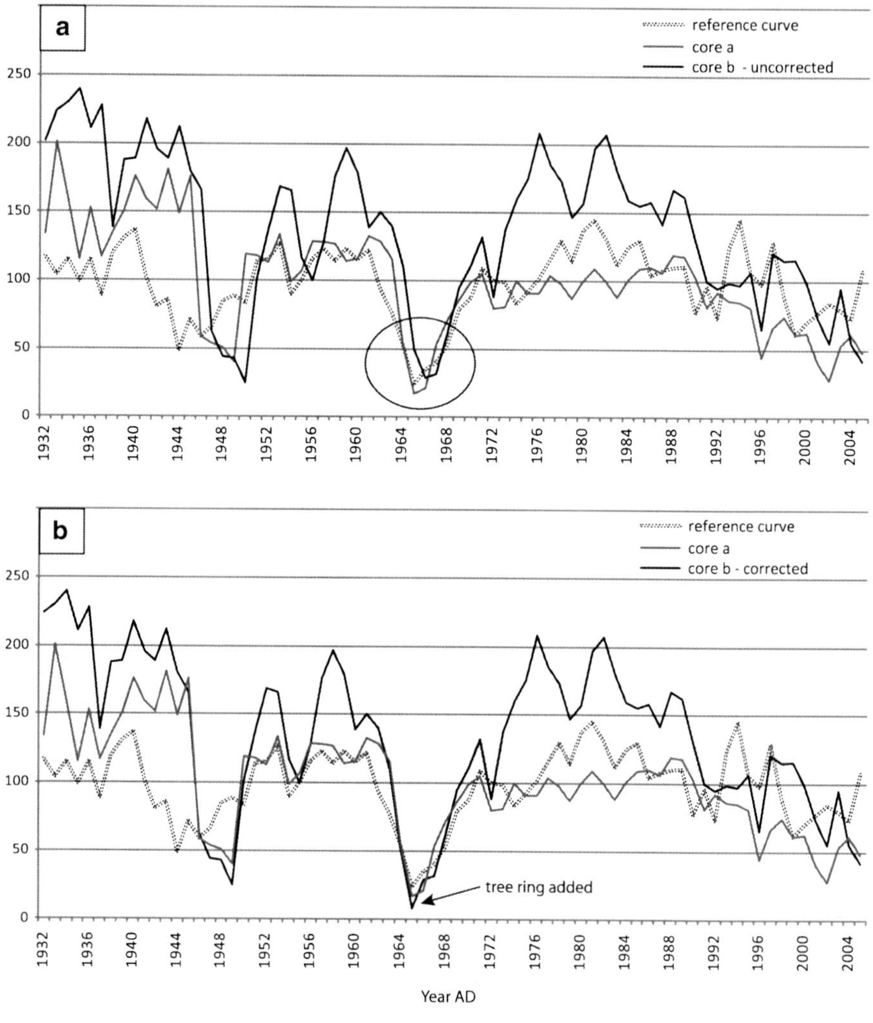

Fig. 3 (**a**) Uncorrected growth curves of a tree impacted by torrential activity as compared to the reference chronology. The *circle* indicates the location of a missing tree ring. (**b**) A tree ring has been added in 1965 and both growth curves are now in concert with the reference chronology

The comparison between growth curves of affected trees with the reference chronology also serves the identification of abrupt growth suppression or increase after an event. In addition, visual inspect of samples under the stereomicroscope helps the determination of other growth changes such as compression wood, callus tissue overgrowing injuries and TRD. All growth changes related to past torrential activity are noted before master plots representing all growth changes in a specific year are created. Finally, a map showing the location of trees with growth defects in a specific year will be produced.

As torrential processes generally affect several trees on a fan or cone or along the channel, a single growth defect in a year is not considered sufficient evidence for an event. Therefore, the definition of event years is based either on quantitative or semi-quantitative approaches separating "signal" from "noise". Quantitative approaches use fixed thresholds as for example described in the "tree-ring response index" suggested by Butler et al. (1987) for the analysis of high-magnitude snow avalanches. For torrential activity, the use of fixed thresholds (as for example 10% or 15% of all trees living in a specific year) for the definition of events is problematic as smaller events may only affect trees along the channel or at isolated breakout locations. Therefore, the spatial distribution of trees showing growth changes is of crucial importance in the reconstruction of past torrential activity.

4 Where Can Tree Rings Help the Tracking of Past Events on Fans and Cones?

4.1 Frequency

A major key to the assessment of hazards and risks related to torrential disasters is knowledge on the occurrence and return periods of events. Archival data is normally incomplete and biased towards events that caused damage to human infrastructure or the loss of live. In contrast, data on smaller or more isolated events is normally scarce or completely missing. For the understanding of the entire history of a torrent and its possible future evolution, the reconstruction of past events therefore represents an important parameter for hazard assessment also in the context of climate change.

As was shown in the chapter above, tree rings record the impact of events in their tree-ring series and are therefore able to provide detailed data on the frequency of events in the past with a yearly and sometimes even monthly resolution (Kaczka et al. 2010; Stoffel et al. 2008). Pioneering studies using tree rings for the reconstruction of event frequency of torrents were performed at Mount Shasta in California (USA) by Hupp (1984) and Hupp et al. (1987). Other studies in Northern America followed (May and Gresswell 2004; Wilkerson and Schmid 2003). In the European Alps Strunk (1988, 1989, 1991, 1995) was the first researcher to apply dendrogeomorphic methods to torrential processes. After Strunks first studies and in the recent past, several researchers investigated torrential frequencies for many catchments in Europe (e.g. Bollschweiler and Stoffel 2007; Mayer et al. 2010; Ruiz-Villanueva et al. 2010; Stoffel and Beniston 2006).

By way of example, Fig. 4 illustrates the frequency of debris-flow events in a catchment in the Swiss Alps reconstructed with tree rings. Before the dendrogeomorphic investigations, three events were recorded in local archives. An additional 50 events could be identified in the tree-ring series of 232 conifer trees between 1730 and 2006 (Stoffel et al. 2010b).

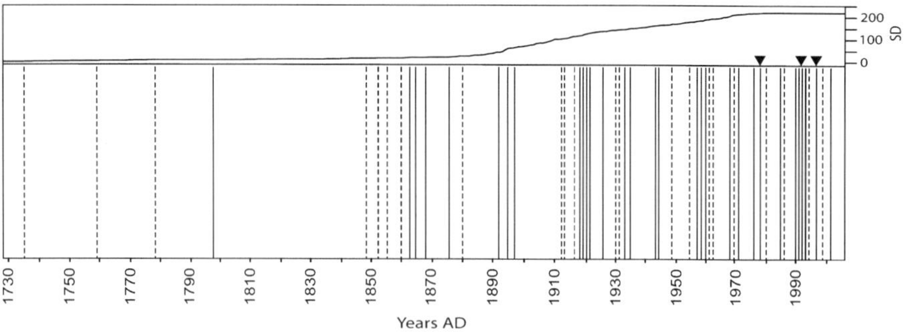

Fig. 4 Debris-flow frequency of the Geisstriftbach in the Swiss Alps where each *bar* represents an event. *Bold lines* indicate events identified based on a large number of growth disturbances whereas events with *dashed lines* were reconstructed with less evidence. The *line above* the frequency shows the sample depth. *Triangles* indicate events noted in local archives (adapted from Stoffel et al. 2010b)

It is important to stress that tree-ring based reconstructions of past torrential activity provide minimum frequencies. Small-scale events remaining within the torrent do not always affect trees. This limitation represents, however, only a minor restriction for the hazard assessment as only low-magnitude events will be neglected.

The frequency of events in a torrent cannot be assumed a constant but may vary considerably over time. Material availability, the influence of climatic changes, forest fires or land-use changes may influence the activity of a torrent considerably. Therefore, we first have to understand the significance of these parameters on past and contemporary activity in individual catchments before any prediction about future evolutions can be made. First studies on the response of catchments have been performed in the French and Swiss Alps (Bollschweiler and Stoffel 2010; Jomelli et al. 2004, 2007; Stoffel et al. 2008). Nevertheless, further studies are needed to gain a better understanding of changes in the frequency of torrential processes.

4.2 Lateral Spread and Runout Distance

Data on the timing of past events in torrents coupled with spatial analysis allows determination of the spatial behaviour of past events. Especially the precise positioning of trees reacting simultaneously to an event on geomorphic maps provides valuable data on flow paths, lateral spread and runout distance of individual incidents. If the spatial behaviour of past events is reconstructed for a fan or cone, preferred flow paths and their temporal activity or inactivity can be determined as for example suggested by Bollschweiler et al. (2008a) and illustrated in Fig. 5.

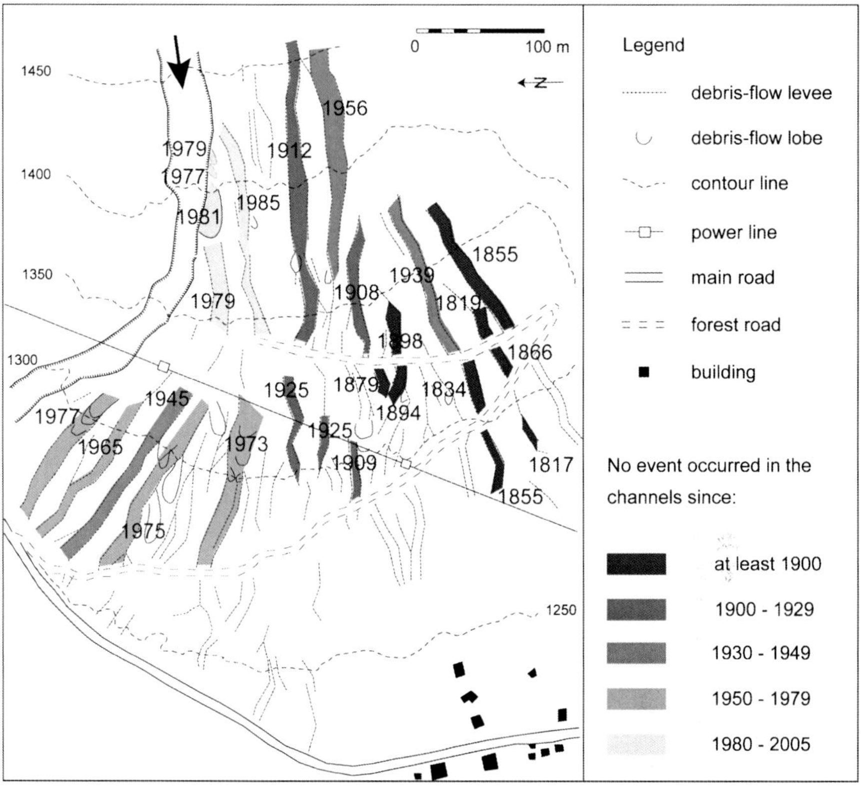

Fig. 5 Reconstruction of the last possible moment of activity in the Grosse Grabe torrent based on tree rings. Note that activity has ceased earlier in sectors located at greater distances from the current channel (*arrow*) than in sectors in the vicinity of the current channel (adapted from Bollschweiler et al. 2008a)

Based on growth disturbances identified in affected trees and on germination dates of trees re-colonizing flow paths after they were abandoned by debris-flow activity, the last possible moment of events in the channels present on the current-day cone surface were determined. Figure 5 also shows that the sectors more distant from the currently active channel evince older dates of inactivity than those in the vicinity of the currently active channel.

Similarly, the coupling of data on spatial and temporal activity allows the determination of breakout locations along the channel or river bed and the corresponding years of their activity as for example shown by Mayer et al. (2010). The dating of deposits with tree rings is a further approach that improves the understanding of depositional processes on fan or cone surfaces (Stoffel et al. 2008).

4.3 Magnitude

To estimate hazards and risks on a fan or cone and to compile hazard maps, the frequency of events should be complemented with a determination of magnitudes. As direct measurements of event magnitudes are not normally available for older events, indirect indicators have to be consulted.

An estimation of magnitudes can be derived from volumetric data of deposits on the current-day cone surface, grain size distributions of boulders, and other surrogates such as snout elevations, tree survival in deposits, or the lateral spread of surges (Stoffel 2010). Definition of magnitude is sometimes also possible through an analysis of the height and dimension of visible growth defects on tree stems and the spatial distribution of trees affected. For instance, Sigafoos (1964) or Bégin (2001) used the height of scars in trees or tilted stems to determine minimum estimates of peak flood heights in rivers and streams. Based on the amount of material deposited, grain sizes of deposits, the spread of the event on the cone, the seasonality and rainfall intensities of events, Stoffel (2010) defined five classes of debris-flow magnitudes of 62 events in a periglacial catchment in the Swiss Alps since 1864. Although the definition of paleo-magnitude will continue to remain approximative, tree rings can greatly help the understanding and estimation of past magnitudes (Ballesteros et al 2011a, b) and therefore represent a valuable tool for the acquisition of data which are normally only scarcely available.

4.4 Identification of Triggers

Torrential processes are normally triggered through a sudden and large input of water (see Kaitna et al. 2012, *this volume*). Triggers can easily be defined for current events, but data on the factors responsible for the release of past events is far from being trivial to assess. Nevertheless, knowledge on the triggering factors of past events is important for hazard assessment and the elaboration of mitigation strategies such as warning systems. The coupling of tree-ring based reconstructions with meteorological records may help the understanding of past triggering mechanisms as well as current activity and may also be used for the prediction of future activity in a changing climate. An example for a detailed study of past debris-flow occurrence based on tree rings, the assessment of triggers and a prediction of the future evolution of activity in the periglacial catchment is provided in the case study in the next chapter.

The meteorological influence on the occurrence of past events has been assessed in several other studies. For instance, Bégin (2001) observed that heavy snowfalls in winter had a significant influence on spring ice-flood events in Québec between 1930 and 1980. Szymczak et al. (2010) determined rainfall events for the triggering of debris flows in a catchment in the Swiss Alps. Even though the reconstruction of amounts of rainfall responsible for the triggering of events remains approximative due to the absence of local rain gauge stations missing data on the precise moment of triggering, the coupling of tree rings with meteorological records provides valuable data on hydrometeorological controls of past events.

5 Limitation of the Method

Besides the huge potential of dendrogeomorphic methods for the tracking of past events on fans and cones – reconstruction of frequency, lateral spread and reach, magnitudes and trigger of events –, the limitations of the method should not be peculated here. First of all, tree availability is the most important restrictions in the reconstruction of past geomorphic activity. On fans and cones located above tree line or deforested by natural or anthropogenic activity (e.g. logging, construction works, grazing, frowing), tree-ring methods cannot be applied for obvious reasons. The ages of trees affected by past activity also considerably influences the quality and length of the reconstruction. While conifers growing on fans and cones not affected by anthropogenic activities often attain several hundred years, broadleaved species do normally not grow much older than some decades. Past logging activity further limits tree ages and therefore the temporal reach of the reconstruction.

When doing tree-ring studies, it is important to consider that trees at the location are only affected by the process under study as a determination of the process causing the growth disturbance is not normally possible through the analysis of the tree ring record alone. Similarly, anthropogenic disturbance on trees should be avoided as, for example, logging activity may injure neighbouring trees.

6 Conclusions

Despite the limitations shown above, trees and tree rings represent a very valuable natural archive for the tracking of past torrential activity over the past decades to centuries. When the method is applied appropriately, not only the frequency of past events but also spatial information such as the spread, runout distance, breakout locations or preferred flow path can be obtained. The coupling with meteorological data may help the understanding of triggering mechanisms of past events and is therefore important for the elaboration of warning systems.

References

Alestalo J (1971) Dendrochronological interpretation of geomorphic processes. Fennia 105:1–139

Ballesteros JA, Bodoque JM, Díez A, Sánchez M, Stoffel M (2011a) Calibration of floodplain roughness and estimation of palaeoflood discharge based on tree-ring evidence and hydraulic modeling. J Hydrol 403:103–115

Ballesteros JA, Eguibar M, Bodoque JM, Díez A, Stoffel M, Gutiérrez I (2011b) Estimating flash flood discharge in an ungauged mountain catchment with 2D hydraulic models and dendrogeomorphic paleostage indicators. Hydrol Process 25:970–979

Bégin Y (2001) Tree-ring dating of extreme lake levels at the subarctic-boreal interface. Quat Res 55(2):133–139

Bodoque JM, Diez-Herrero A, Martin-Duque JF, Rubiales JM, Godfrey A, Pedraza J, Carrasco RM, Sanz MA (2005) Sheet erosion rates determined by using dendrogeomorphological analysis of exposed tree roots: two examples from Central Spain. Catena 64(1):81–102

Bollschweiler M, Stoffel M (2007) Debris flows on forested cones – reconstruction and comparison of frequencies in two catchments in Val Ferret, Switzerland. Nat Hazard Earth Syst Sci 7(2):207–218

Bollschweiler M, Stoffel M (2010) Changes and trends in debris-flow frequency since A.D. 1850 – results from the Swiss Alps. Holocene 20(6):907–916

Bollschweiler M, Stoffel M, Schneuwly DM (2008a) Dynamics in debris-flow activity on a forested cone – A case study using different dendroecological approaches. Catena 72(1):67–78

Bollschweiler M, Stoffel M, Schneuwly DM, Bourqui K (2008b) Traumatic resin ducts in *Larix decidua* stems impacted by debris flows. Tree Physiol 28(2):255–263

Braam RR, Weiss EEJ, Burrough PA (1987a) Dendrogeomorphological analysis of mass movement – a technical note on the research method. Catena 14(6):585–589

Braam RR, Weiss EEJ, Burrough PA (1987b) Spatial and temporal analysis of mass movement using dendrochronology. Catena 14(6):573–584

Butler DR, Malanson GP (1985) A reconstruction of snow-avalanche characteristics in Montana, USA, using vegetation indicators. J Glaciol 31(108):185–187

Butler DR, Malanson GP, Oelfke JG (1987) Tree-ring analysis and natural hazard chronologies – minimum sample sizes and index values. Prof Geogr 39(1):41–47

Cook ER, Kairiukstis LA (1990) Methods of dendrochronology – applications in the environmental sciences. Kluwer, London

Friedman JM, Vincent KR, Shafroth PB (2005) Dating floodplain sediments using tree-ring response to burial. Earth Surf Process Landf 30(9):1077–1091

Grissino-Mayer HD (2003) A manual and tutorial for the proper use of an increment borer. Tree Ring Res 59(2):63–79

Hitz OM, Gärtner H, Heinrich I, Monbaron M (2008) Wood anatomical changes in roots of European ash (*Fraxinus excelsior* L.) after exposure. Dendrochronologia 25(3):145–152

Hupp CR (1984) Dendrogeomorphic evidence of debris flow frequency and magnitude at Mount Shasta, California. Environ Geol Water Sci 6(2):121–128

Hupp CR, Osterkamp WR, Thornton JL (1987) Dendrogeomorphic evidence and dating of recent debris flows on Mount Shasta, northern California, U.S. Geological Survey professional paper 1396B. United States Government Printing Office, Washington, DC, pp 1–39

Jomelli V, Pech VP, Chochillon C, Brunstein D (2004) Geomorphic variations of debris flows and recent climatic change in the French Alps. Clim Chang 64(1–2):77–102

Jomelli V, Brunstein D, Grancher D, Pech P (2007) Is the response of hill slope debris flows to recent climate change univocal? A case study in the Massif des Ecrins (French Alps). Clim Chang 85(1–2):119–137

Kaczka RJ, Deslauriers A, Morin H (2010) High-precision dating of debris-flow events within the growing season. In: Stoffel M, Bollschweiler M, Butler DR, Luckman BH (eds) Tree rings and natural hazards – a state-of-the-art. Springer, Berlin/Heidelberg/New York

Kaitna R, Schneuwly-Bollschweiler M, Sausgruber T, Moser M, Stoffel M, Rudolf-Miklau F (2012) Susceptibility and triggers for debris flows: emergence, loading, release and entrainment. In: Schneuwly-Bollschweiler M, Stoffel M, Rudolf-Miklau M (eds) Dating torrential processes on fans and cones – methods and their application for hazard and risk assessment. Advances in Global Change Research. Springer, Dordrecht/Heidelberg/London/New York

LaMarche V (1961) Rate of slope erosion in the White Mountains, California. Geol Soc Am Bull 72(10):1579–1580

Larson PR (1994) The vascular cambium. Development and structure. Springer, Berlin/Heidelberg/New York

Luchi N, Ma R, Capretti P, Bonello P (2005) Systemic induction of traumatic resin ducts and resin flow in Austrian pine by wounding and inoculation with *Sphaeropsis sapinea* and *Diplodia scrobiculata*. Planta 221(1):75–84

Mattheck C (1993) Design in der Natur. Rombach Wissenschaft, Freiburg

May CL, Gresswell RE (2004) Spatial and temporal patterns of debris-flow deposition in the Oregon Coast Range, USA. Geomorphology 57(3–4):135–149

Mayer B, Stoffel M, Bollschweiler M, Hübl J, Rudolf-Miklau F (2010) Frequency and spread of debris floods on fans: a dendrogeomorphic case-study from a dolomite catchment in the Austrian Alps. Geomorphology 118:199–206

Pilate G, Chabbert B, Cathala B, Yoshinaga A, Leple JC, Laurans F, Lapierre C, Ruel K (2004) Lignification and tension wood. C R Biol 327(9–10):889–901

Ruel JJ, Ayres MP, Lorio PL (1998) Loblolly pine responds to mechanical wounding with increased resin flow. Can J For Res-Revue Canadienne De Recherche Forestiere 28(4):596–602

Ruiz-Villanueva V, Díez-Herrero A, Stoffel M, Bollschweiler M, Bodoque JM, Ballesteros JA (2010) Dendrogeomorphic analysis of flash floods in a small ungauged mountain catchment (Central Spain). Geomorphology 118(3–4):383–392

Sachs T (1991) Pattern formation in plant tissue. Cambridge University Press, Cambridge

Schweingruber F, Eckstein D, Serre-Bachet F, Bräker OU (1990) Identification, presentation and interpretation of event years and pointer years in dendrochronology. Dendrochronologia 8:9–39

Shroder JF (1978) Dendro-geomorphological analysis of mass movement on Table Cliffs Plateau, Utah. Quat Res 9(2):168–185

Shroder JF (1980) Dendrogeomorphology; review and new dating techniques of tree-ring dating. Prog Phys Geogr 4:161–188

Sigafoos RH (1964) Botanical evidence of floods and floodplain deposition, U.S. Geological Survey professional paper 485-A. U.S. Geological Survey, Washington, DC

Stoffel M (2010) Magnitude-frequency relationships of debris flows – a case study based on field survey and tree-ring records. Geomorphology 116:67–76

Stoffel M, Beniston M (2006) On the incidence of debris flows from the early Little Ice Age to a future greenhouse climate: a case study from the Swiss Alps. Geophys Res Lett 33(16):L16404

Stoffel M, Bollschweiler M (2008) Tree-ring analysis in natural hazards research – an overview. Nat Hazard Earth Syst Sci 8(2):187–202

Stoffel M, Bollschweiler M (2009) What tree rings can tell about earth-surface processes: teaching the principles of dendrogeomorphology. Geogr Compass 3(3):1013–1037

Stoffel M, Conus D, Grichting MA, Lièvre I, Maitre G (2008) Unraveling the patterns of late Holocene debris-flow activity on a cone in the Swiss Alps: chronology, environment and implications for the future. Glob Planet Chang 60(3–4):222–234

Stoffel M, Bollschweiler M, Butler DR, Luckman BH (2010a) Tree rings and natural hazards – a state-of-the-art. Springer, Dordrecht

Stoffel M, Bollschweiler M, Widmer S, Sorg A (2010b) Spatio-temporal variability in debris-flow activity: a tree-ring study at Geisstriftbach (Swiss Alps) extending back to AD 1736. Swiss J Geosci 103(2):283–292

Strunk H (1988) Episodische Murschübe in den Pragser Dolomiten – semiquantitative Erfassung von Frequenz und Transportmenge. Z Geomorphol 83:71–81

Strunk H (1989) Dendrogeomorphology of debris flows. Dendrochronologia 7:15–25

Strunk H (1991) Frequency distribution of debris flow in the Alps since the "Little Ice Age". Z Geomorphol 83:71–81

Strunk H (1995) Dendrogeomorphologische Methoden zur Ermittlung der Murfrequenz und Beispiele ihrer Anwendung. Roderer, Regensburg

Szymczak S, Bollschweiler M, Stoffel M, Dikau R (2010) Debris-flow activity and snow avalanches in a steep watershed of the Valais Alps (Switzerland): dendrogeomorphic event reconstruction and identification of triggers. Geomorphology 116:107–114

Timell TE (1986) Compression wood in gymnosperms. Springer, Berlin/Heidelberg/New York
Vaganov EA, Hughes MK, Shashkin AV (2006) Growth dynamics of conifer tree rings. Images of past and future environments. Springer, Berlin/Heidelberg/New York
Wilkerson FD, Schmid GL (2003) Debris flows in Glacier National Park, Montana: geomorphology and hazards. Geomorphology 55(1–4):317–328

Tree-Ring Based Record of Debris-Flow Dynamics and Triggering Rain Storms at Ritigraben (Swiss Alps) since AD 1570

Markus Stoffel

1 Introduction

The Ritigraben torrent is located on the west-facing slope of the Mattertal valley (Valais, 46°11′N., 7°49′E.) and spans a vertical range of 2,000 m from its confluence with the Vispa River at 1,080 m asl to the summit of the Seetalhorn (3,100 m asl). A rock glacier occupies most of the headwater basin (1.4 km^2) between 2,500 and 2,800 m asl (Fig. 1a), constituting the main starting zone of debris flows. On its downward course to the Mattervispa river, the Ritigraben torrent passes a forested cone (32 ha; 4.3×10^6 m^3; Fig. 1b) on a structural terrace (1,500–1,800 m asl), where debris-flow material affects trees composed of European larch (*Larix decidua* Mill.), Norway spruce (*Picea abies* (L.) Karst.) and Cembran pine (*Pinus cembra* L.). Virtually all trees on the cone show growth disturbances (GD) related to past debris-flow activity. Based on the geomorphic map and an outer inspection of the trunks, we sampled trees obviously disturbed by past debris flows and selected undisturbed reference trees from a forest stand located next to the cone. In total, 1,204 trees were sampled (2,450 increment cores): 539 *L. decidua*, 429 *P. abies* and 134 *P. cembra* trees (2,246 cores) from the debris-flow cone and 102 trees (204 cores) from undisturbed reference sites. Samples were analyzed following the standard procedures described in Stoffel et al. (2010) and Schneuwly-Bollschweiler and Stoffel (2012, *this volume*).

M. Stoffel (✉)
Laboratory of Dendrogeomorphology (dendrolab.ch), Institute of Geological Sciences, University of Berne, CH-3012 Berne, Switzerland

Chair for Climate and Climatic Impacts, Institute for Environmental Sciences, University of Geneva, CH-1227 Carouge-Geneva, Switzerland
e-mail: markus.stoffel@dendrolab.ch

M. Schneuwly-Bollschweiler et al. (eds.), *Dating Torrential Processes on Fans and Cones*, Advances in Global Change Research 47, DOI 10.1007/978-94-007-4336-6_11,

Fig. 1 (**a**) Photo of the Ritigraben debris-flow system from its source to the confluence (catchment area: 1.36 km^2, channel lengths: 3.5 km). (**b**) Detailed view of the intermediate debris-flow cone (32 ha) and its mixed conifer stand

2 Debris-Flow Deposits on the Ritigraben Cone

Geomorphic mapping permitted identification of 769 features related to past debris-flow activity on the intermediate cone (Stoffel et al. 2008). The features inventoried in the study area included 291 lobes, 465 levees and 13 well-conserved paleochannels (Fig. 2). Based on tree-ring records of disturbed trees growing in or next to the deposits, 86% of the lobes identified on the present-day surface could be dated. A majority of the dated material was deposited over the past century and signs of pre-twentieth century events are often recognizable in the tree-ring record of survivor trees, but the material causing the GD in trees has been completely overridden or eroded by more recent debris-flow activity. The detailed mapping of geomorphic features was also helpful in the identification of signs of activity in currently abandoned channels, thus allowing to document changes in the spatial occurrence and reach of flows in time (Stoffel 2010).

3 Debris-Flow Frequency and Seasonality

The study of the 2,246 tree-ring sequences allowed reconstruction of 124 events since AD 1570 (Fig. 3), thus enlarging the history of events recorded in local archives by more than one order of magnitude (124 instead of 10 events) and by adding 350 years to the debris-flow history of the Ritigraben torrent. For the period documented in local archives (1922–2009), the frequency of events could be augmented from 10 to 35, therefore pointing to the incompleteness of archives even for the recent past.

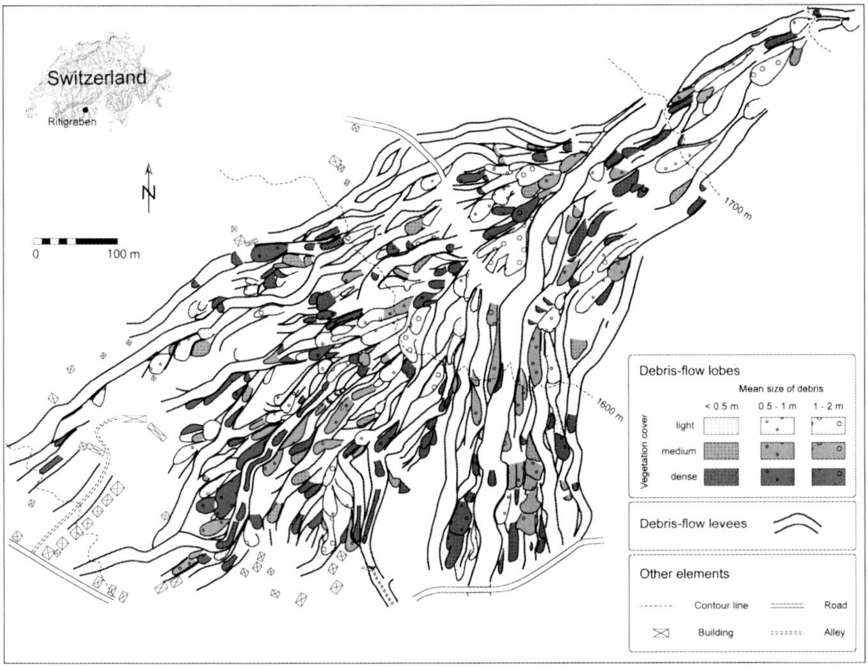

Fig. 2 Detailed geomorphic map of the intermediate debris-flow cone with lobes, levees and channels. The key indicates the density of the vegetation cover and the mean size of debris

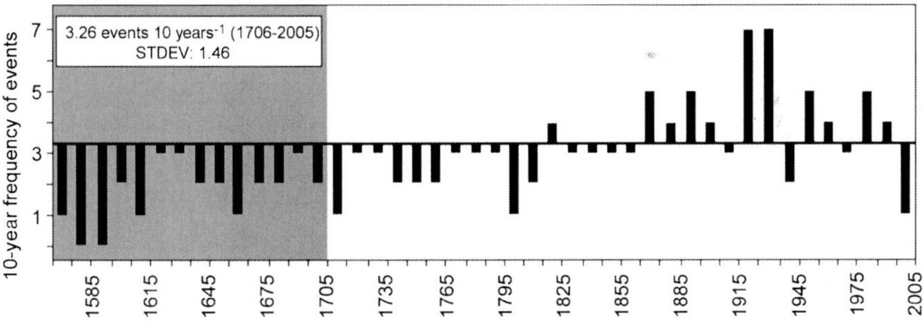

Fig. 3 Reconstructed 10-year frequencies of debris-flow events between AD 1566 and 2005. Data are presented as variations from the mean decadal frequency of debris flows of the last 300 years (AD 1706–2005)

Based on the tree-ring records, we believe that cool summers with frequent snowfalls in the upper catchment of the Ritigraben torrent regularly prevented the release of debris flows between the 1570s and 1860s (Stoffel and Beniston 2006); the warming trend combined with greater precipitation totals in summer and autumn between 1864 and 1895 (Pfister 1999) provided conditions that were increasingly favourable for the release of events. Enhanced debris-flow activity continued well into the twentieth century and reconstructions show a clustering of events between 1916 and 1935 when warm-wet conditions prevailed during summer in the Swiss Alps (Pfister 1999). In contrast, very low activity is observed for the recent past (1996–2005) with only one debris flow recorded on August 27, 2002. Since sediment availability is not a limiting factor, this temporal absence of debris-flow activity is due to an absence of triggering events, which shifted from June and July to August and September over the twentieth century (Stoffel et al. 2008, 2011).

4 Debris Sources and Magnitude-Frequency Relations of Flows

Magnitude-frequency (M–F) relations were assessed for 62 debris flows since AD 1863, i.e. for the period where meteorological records are available. The magnitude of individual events is given as *S*, *M*, *L*, *XL*, and derived from volumetric data of deposits, grain size distributions of boulders, and a series of surrogates (snout elevations, tree survival, lateral spread of surges; Stoffel 2010). *Class S* and *M* debris flows ($<5 \times 10^3$ m^3) encompass a typical size of events and have mean recurrence intervals of 5.4 (SD: 3.2) and 7.4 years (SD: 6.7), respectively. *Class XL* events ($10^4 - 5 \times 10^4$ m^3) are, in contrast, only identified three times over the past 150 years, and major erosional activity on the cone was restricted to two of these events in 1948 and 1993 (Fig. 4). A comparison of results with hydro-meteorological records shows that *class L* and *XL* events are typically triggered by advective storms (rainfall >50 mm) in August and September, when the active layer of the permafrost in the source area is largest. It also becomes obvious that over the past 150 years, climate has exerted control on material released from the source area and prevented triggering of *class XL* events before 1922.

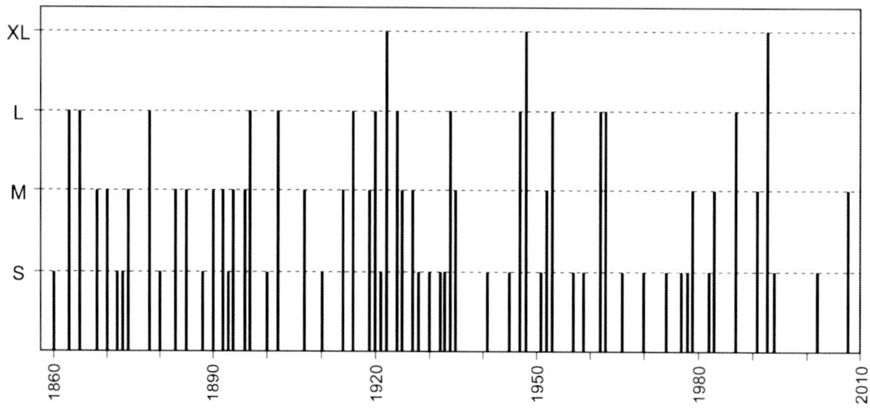

Fig. 4 Reconstructed time series of debris-flow magnitudes, in five classes, for the period from 1858 to 2008. Note the clustering of important events in the early decades of the twentieth century and the absence of *class XL* debris flows before 1922

In addition, we compared debris production and volumetric changes at the rock-glacier front (Lugon and Stoffel 2010) with debris-flow activity recorded on the cone and tried to find potential couplings and feedbacks between debris sources, channel processes and debris sinks. Acceleration in rock-glacier movements is observed in the (1950s and) 1960s, followed by a decrease in flow rates by the 1970s, before movements increase again after the early 1990s. At a decadal scale, measured changes in rock-glacier movements at Ritigraben are in concert with changes in atmospheric temperatures in the Alps. Geodetic data indicates displacement rates of up to 0.6–0.9 m year^{-1} since the beginning of systematic measurements in 1995 and annual horizontal displacement rates of the rock-glacier body have remained quite small. Sediment delivery from the rock-glacier front alone could not therefore be sufficient for the alimentation of the 16 debris flows reconstructed on the cone since 1958. On the contrary, debris accumulated at the foot of the rock glacier, landslide and rockfall activity as well as the partial collapse of oversteepened channel walls have to be seen as important sediment sources of debris flows and represent 65–90% of the material arriving on the Ritigraben cone. Noteworthy, there does not seem to exist a direct coupling between displacement rates of and sediment delivery by the rock glacier and the frequency of *class S* and *M* debris flows. In contrast, a direct link between source and sink processes exists in the case of active-layer failures. Here, failure processes at the snout of the rock glacier and debris-flow events in the channel occur simultaneously and are triggered by heavy rainfalls.

5 Triggers and Climatic Impacts on Debris-Flow Activity

Based on observational data, Stoffel et al. (2005, 2011) assessed changes in rainfall characteristics and their impact on the triggering of debris flows over the past ~150 year. No trends are visible in the frequency of heavy rainfalls, but we observe a cluster of advective storms in late summer and early autumn since the late 1980s (Fig. 5). At the same time, dendrogeomorphic data point to a reduction of convective rainfall capable of triggering events and a lowering of the overall debris-flow activity since the mid-1990s. These changes in triggering meteorological conditions may be mirroring the observed changes in persistent high-pressure systems over the Alps. In addition to the long-term changes in debris-flow systems, we observe changing responses of the watershed to the amount of rain at different moments within the debris-flow season (i.e., June through September). Differences in system response clearly reflect the state of the permafrost body in the source area of debris flows, which allows for very small debris-flow events after limited rainfall inputs (<20 mm) in June and July. The same quantities of rain will not result in debris flows in August or September, when a large active layer of the permafrost body is capable of absorbing water without producing events.

With the projected amplitude of climatic change, changes will probably occur in the seasonality, return intervals and volumes of debris flows. Regional climate model

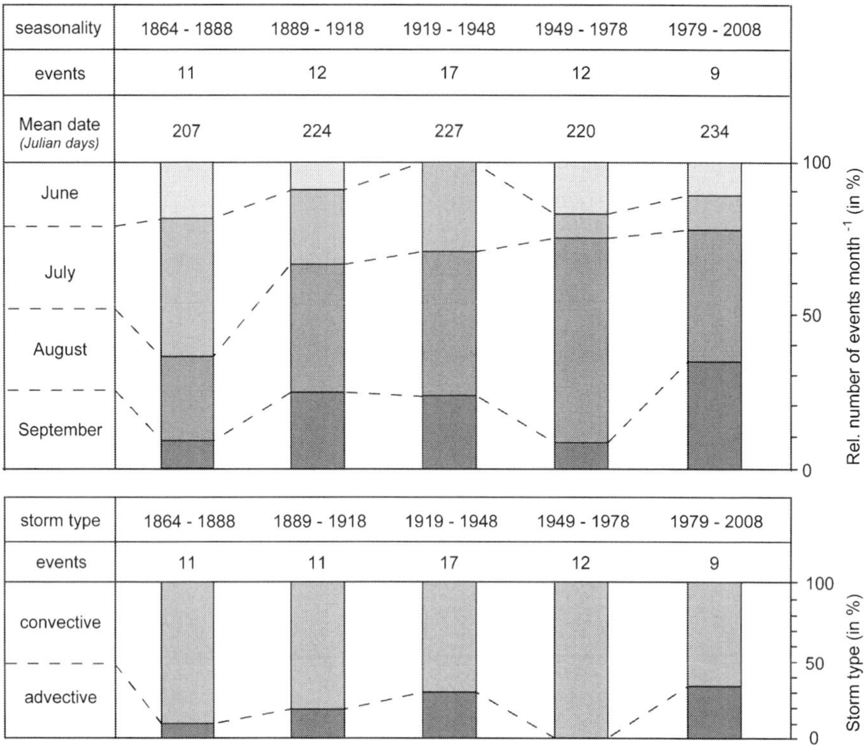

precipitation totals recorded during convective / advective rainfall events (in mm)					
mean	30.9 / 48.7	20.2 / 30.7	20.9 / 66.7	29.6 / NA	26.1 / 70.0
max	49.3 / 48.7	30.0 / 30.7	30.2 / 179.4	52.2 / NA	43.3 / 115.8
min	16.5 / 48.7	10.0 / 30.7	19.0 / 35.1	13.1 / NA	11.7 / 46.9

Fig. 5 Changes in seasonality of debris-flow events (*top*), storm types triggering debris flows (*centre*) and rainfall totals recorded during convective and advective storms with subsequent debris-flow releases (*bottom*)

projections suggest a decrease in heavy summer rainfall events which will likely result in a reduction of the overall frequency of debris flows, leaving more time for debris to accumulate in the channel. Such increases of channel accumulation rates along with the projected destabilization of the rock-glacier body will ultimately exert control on the volumes of material released from the source area. It is thus possible that extremely large *"class XXL"* events, beyond historical experience and with volumes surpassing 5×10^4 m^3 at the level of the debris-flow cone, may be observed in the future originating from this periglacial environment.

References

Lugon R, Stoffel M (2010) Rock-glacier dynamics and magnitude-frequency relations of debris flows in a high-elevation watershed: Ritigraben, Swiss Alps. Glob Planet Chang 73(3–4): 202–210

Pfister C (1999) Wetternachhersage – 500 Jahre Klimavariationen und Naturkatastrophen. Paul Haupt, Bern/Stuttgart/Wien, 325 pp

Schneuwly-Bollschweiler M, Stoffel M (2012) Dendrogeomorphology – tracking past events with tree rings. In: Schneuwly-Bollschweiler M, Stoffel M, Rudolf-Miklau M (eds) Dating torrential processes on fans and cones – methods and their application for hazard and risk assessment, Advances in Global Change Research. Springer, Dordrecht/Heidelberg/London/New York

Stoffel M (2010) Magnitude-frequency relationships of debris flows – a case study based on field surveys and tree-ring records. Geomorphology 116:67–76

Stoffel M, Beniston M (2006) On the incidence of debris flows from the early Little Ice Age to a future greenhouse climate: a case study from the Swiss Alps. Geophys Res Lett 33:L16404

Stoffel M, Lièvre I, Conus D, Grichting MA, Raetzo H, Gärtner HW, Monbaron M (2005) 400 years of debris-flow activity and triggering weather conditions: Ritigraben, Valais, Switzerland. Arct Antarct Alp Res 37:387–395

Stoffel M, Conus D, Grichting MA, Lièvre I, Maître G (2008) Unraveling the patterns of late Holocene debris-flow activity on a cone in the Swiss Alps: chronology, environment and implications for the future. Glob Planet Chang 60:222–234

Stoffel M, Bollschweiler M, Butler DR, Luckman BH (2010) Tree rings and natural hazards: a state-of-the-art. Springer, Heidelberg/Berlin/New York, 505 pp

Stoffel M, Bollschweiler M, Beniston M (2011) Rainfall characteristics for periglacial debris flows in the Swiss Alps: past incidences – potential future evolutions. Clim Chang 105(1–2):263–280

Vegetation Analysis on Composite Debris Cones

Carlo Baroni, Rodolfo Gentili, and Stefano Armiraglio

1 Introduction

Vegetation cover has been widely recognized for indicating the status and dynamics of physical components of the environment (Ellenberg 1988; Ozenda 2002; Pignatti et al. 2005; Huc 2008). Phytogeomorphology, the study of interdependence of vegetation and landforms (Howard and Mitchell 1985), has become a central topic in environmental studies (Hession et al. 2010; Marston 2010). According to a geo-ecologic approach (Troll 1971), this subject combines aspects of geomorphology and vegetation ecology with climate, soil science, natural hazards, and environmental management at different spatial scales (Kozlowska and Raczkowska 2002; Walker et al. 2004). Interactions between landforms and vegetation cover are particularly evident in mountain areas where ecological factors change along altitudinal gradients within a limited geographical area (Ellenberg 1988). In particular, landform heterogeneity in extreme plant habitats gives rise to environmental diversity (Kruckeberg 2002; Burga et al. 2004; Baroni et al. 2007) and biodiversity gradients (Gentili et al. 2010).

C. Baroni (✉)
Dipartimento di Scienze della Terra, Università di Pisa, Via S. Maria n. 53, I-56126 Pisa, Italy
e-mail: baroni@dst.unipi.it

R. Gentili
Dipartimento di Scienze dell'Ambiente e del Territorio, Università degli Studi di Milano-Bicocca, Piazza della Scienza n. 1, I-20126 Milan, Italy
e-mail: rodolfo.gentili@unimib.it

S. Armiraglio
Museo Civico di Scienze Naturali di Brescia, via Ozanam n. 4, I-25128 Brescia, Italy
e-mail: botanica@comune.brescia.it

M. Schneuwly-Bollschweiler et al. (eds.), *Dating Torrential Processes on Fans and Cones*, Advances in Global Change Research 47, DOI 10.1007/978-94-007-4336-6_12,

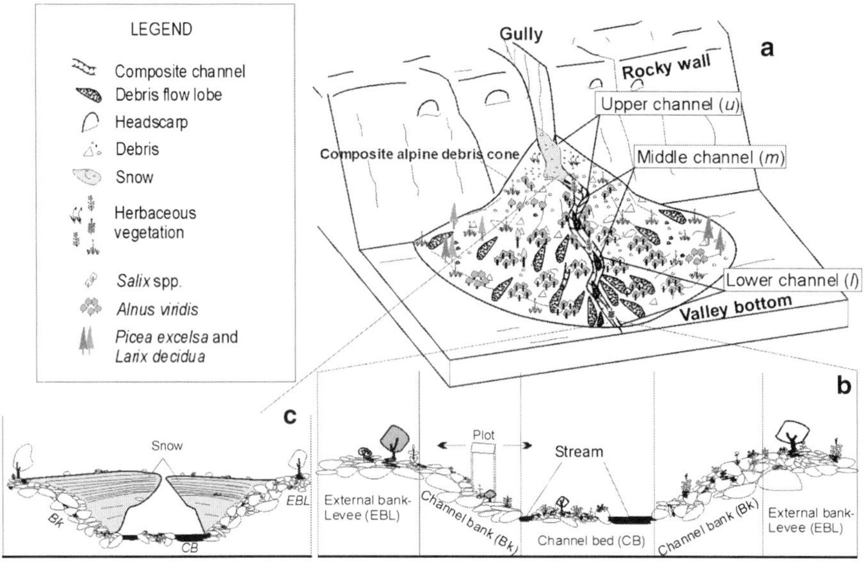

Fig. 1 (**a**) Geomorphologic and vegetation features of a composite alpine debris cone. (**b**) Transversal section of a channel showing its main subunits (modified from Gentili et al. 2010). (**c**) Schematic section of channel/avalanche track with still unmelted snow within the channel. The disturbance of snow avalanche and snow duration on vegetation colonization and establishment is different between the track (zone of transition) and the runout zone (zone of deposition). In the zone of transition where the moving snow runs downhill, plants are not likely to grow because debris in channel beds and on channel banks is frequently disturbed

Landscape features, such as snow avalanche channels and cones, debris-flow channels and lobes, and debris cones, are widespread along valley slopes in mountain regions. Both erosional and depositional landforms often coexist and develop on the same slope, reflecting the complex interaction of various seasonal geomorphic processes. Composite alpine cones (known as "*Murkegel*" in German literature) have such an origin, and their structure retains useful information for understanding the mechanisms controlling their dynamics and evolution, as well as the frequency of various geomorphic processes throughout their histories (Fig. 1a, b).

Vegetation colonizes debris deposits according to particular ecological parameters such as surface stability, water availability, and microclimate. Colonization processes both directly and indirectly depend on the surface characteristics and internal structure of geomorphic features. Time also plays a key role within these factors in favouring the evolution of mature vegetation cover (in biostasy-like environments).

Studies of vegetation on composite cones have strong potential for interpreting the dynamics, relative age, and evolution of landforms due to the combined action of various geomorphic processes.

2 Composite Alpine Debris Cones and Channels

Most of the debris cones and alluvial fans that are presently found in the alpine region began developing after the retreat of the Pleistocene glaciers that occupied the valleys during the Last Glacial Maximum. The streams that created the cones and fans at their initial stages were presumably associated with paraglacial dynamics (Ballantyne 2002 and references therein). Composite channels usually feed debris cones with the exception of those cones exclusively generated by gravity. If the contribution of rock debris and falls is significant, the apical sector of the composite cones shows a steeper slope (up to 25–30°), meaning that the steeper the slope, the greater the intake of rock debris and falls. In this case, a break of slopes separates the apical sector from the medium-distal portion of the cone where debris-flow lobes juxtapose and overlap. Shallower slopes (less than 20°) are found where debris and rock fall deposits overlap coalescent debris-flow lobes. Rock falls with boulders up to several metres in diameter can spread all over the cone surface. The oldest isolated blocks and boulders show abundant lichen cover and significant weathering.

On the contrary, fresh, perched, unstable clasts or irregularly-sized blocks, as well as turfs, roots, and wood debris, scattered on weathered boulders clearly delimit areas of avalanche activity (Jomelli 1999; Jomelli and Francou 2000).

Composite channels are polygenic landforms of the composite alpine debris cones. The channels generally cross cones in the central portion and are connected to deep gullies, cutting through rocky walls behind the cones, eventually originating at the glacial foreland, in the highest portions of the basins.

Geomorphic processes vary in intensity and frequency within channels depending on the season and altitude. The channels can converge running water, collect avalanches, and produce debris flows. Therefore, landforms and deposits of different origin, sedimentologic structure and texture are clearly distinguishable along the channel corridors, overlapping each other to different extents (Figs. 1a, b and 2a, b). Abandoned channels can be locally recognized as relict geomorphic features.

As a consequence of eventual erosion phases, channels can deeply cut the cone-head, thus generating converging erosive scarps. In this case, debris-flow lobes deposit down the valley of the erosional scarp and generate entrenched cones.

Channel depths can range from less than 1 m in the distal portion of the cones to several metres (more than 10 m) in the upper cone head. The channels typically have concave profiles and become shallower down the valley with widths ranging from tens of metres (in the upper sectors) to a few metres (in the lower sectors). Such features demonstrate that erosion caused by the synergetic action of snow avalanches and running water is mainly concentrated in the upper channel and erosion efficacy is reduced with increasing distance from the cone head. The upper and middle portions of the channels are also affected by the debris and rock fall deposits. Deposits due to debris flow (i.e., levees and lobes) accumulate in the middle and lower portions of the channels.

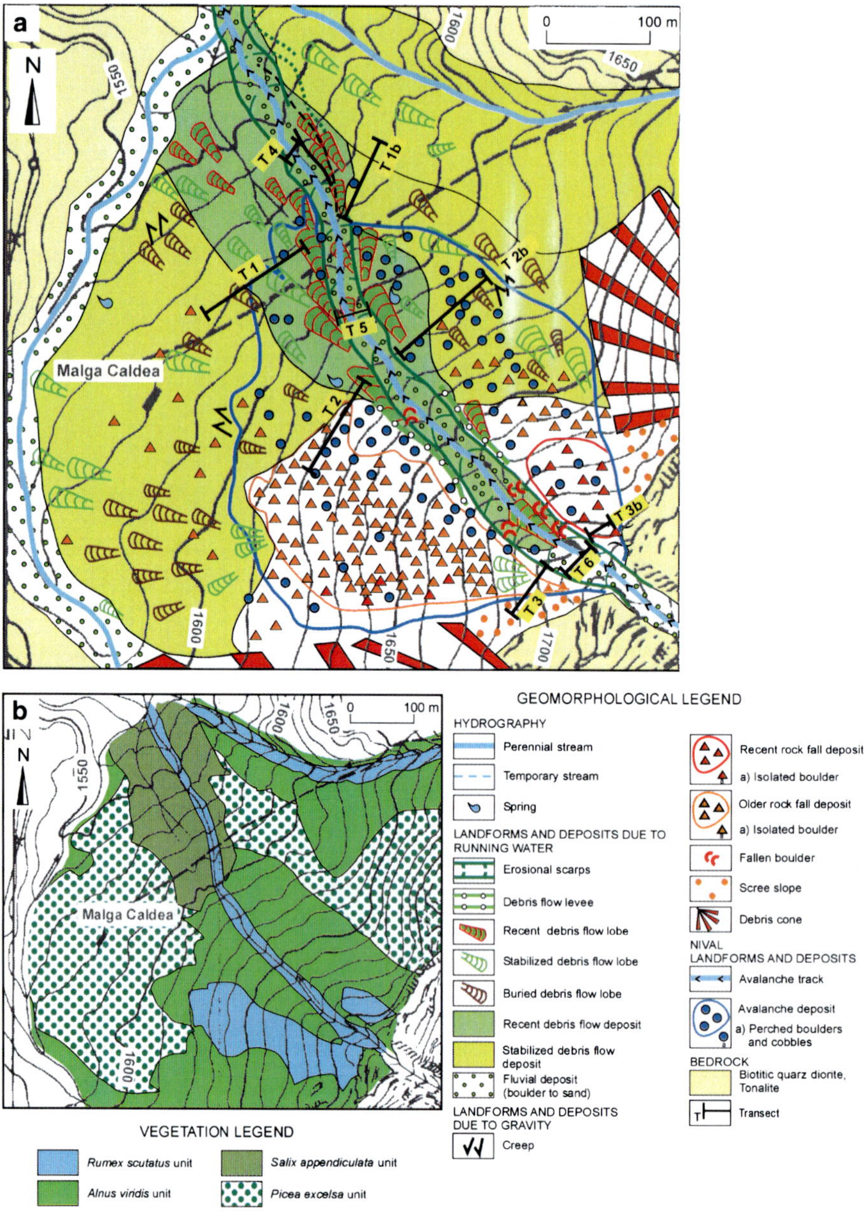

Fig. 2 Maps of (**a**) landform, (**b**) vegetation and (**c**) transects of the Malga Caldea cone (Valle dell'Avio, Central Alps, Italy; modified from Baroni et al. 2007)

In the transverse section, the channels consist of three distinct subunits: channel beds (CB), channel banks (Bk) and external banklevees (EBL). Snow accumulates in the channels during winter and spring and persists in some places until late summer (Fig. 1c). Erosion in the active channel beds often causes collapses of channel banks.

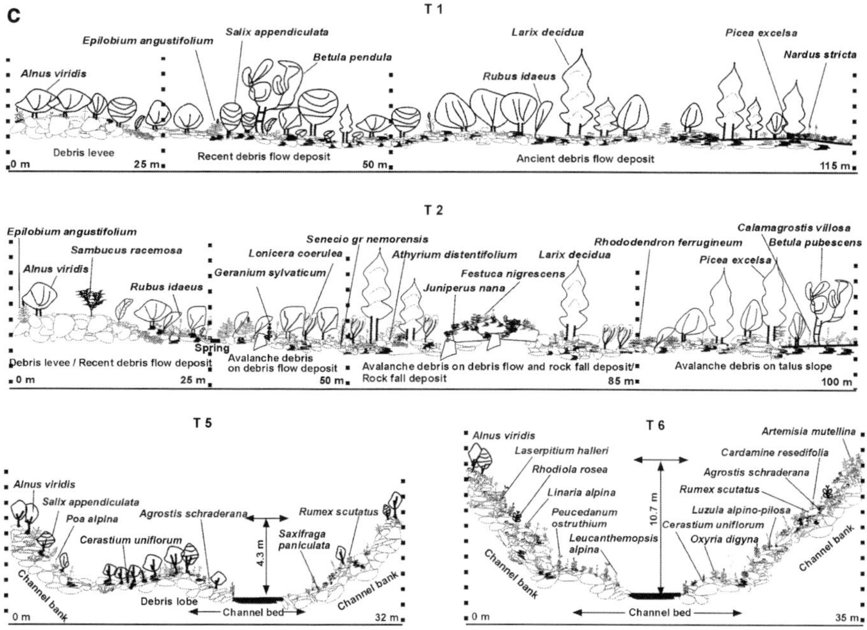

Fig. 2 (continued)

External bank-levees are mainly found in the middle and lower parts of the channels. The clast sets forming the levees (i.e., boulders with a scarce sandy matrix) are clearly imbricated (Fig. 2b).

On the cone surface, we can recognize areas of varying degrees of stability. Moving outward from the active channel, we can identify recent debris-flow levees and/or lobes, stabilized debris-flow lobes and deposits and buried debris-flow deposits. In a broad sense, this evidence reflects a progressive increase in relative age from the centre of the channel toward the margins of the cones mostly in their medium-lower sectors. Furthermore, soils that develop on the different portions of the cone increasingly thicken as a function of the cone's relative age.

3 Method of Studying Plant Communities

The general distribution of natural vegetation, especially on the Alps, is distributed in altitudinal belts (Fig. 3) and is strongly conditioned by the treeline and snowline position.

At the highest elevation, in the alpine belt, the mature communities are natural grassland while deciduous and coniferous forests characterize the lower belts. Climate control is a determinant for mature plant communities, while local factors (such as slope dip, aspect, and stability, as well as nature of bedrock, grain size and moisture of soils) considerably influence pioneer communities.

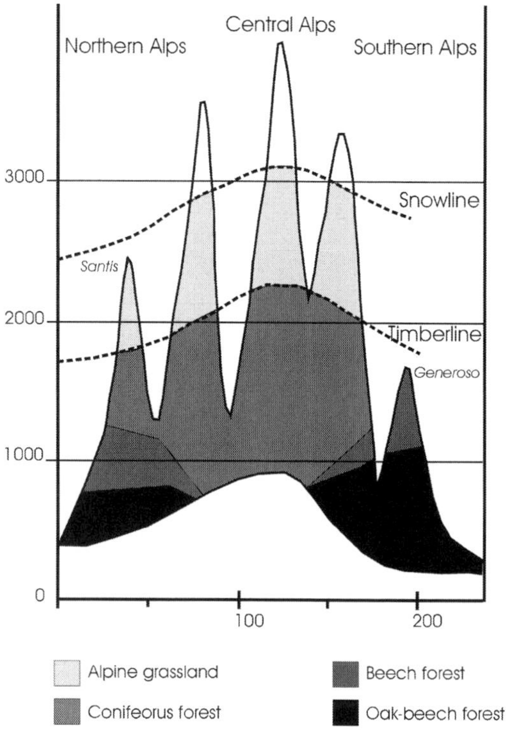

Fig. 3 Schematic cross-section through the natural vegetation belts of the Alps (modified and simplified from Ellenberg 1988)

Detailed field surveys are necessary for defining plant communities associated with geomorphic features. Aerial photographs and satellite image analysis can also be applied to produce vegetation maps.

Vegetation transects are necessary for defining vegetation patterns. Transects are made by a sequence of vegetation plots surveyed along a line ranging from a few metres to hundreds of metres in length.

Transects can be used to characterize single units or to identify differences between units of a single cone. A vegetation plot is defined on a square of standard measure (range from 0.25 to 5 m^2) where all the species are identified and listed. Percent cover of each species can also be determined in the same plot.

All the plots are organized in a matrix; all the species listed along the transect are inserted in the rows and columns represent single plots.

The matrix is classified with multivariate analysis and cluster analysis, and the plots are thereafter grouped by floristic similarity. Each group of plots represents a vegetation unit (Fig. 4). Vegetation units can be summarized in a synthetic table where the frequency of the species surveyed is reported (Table 1).

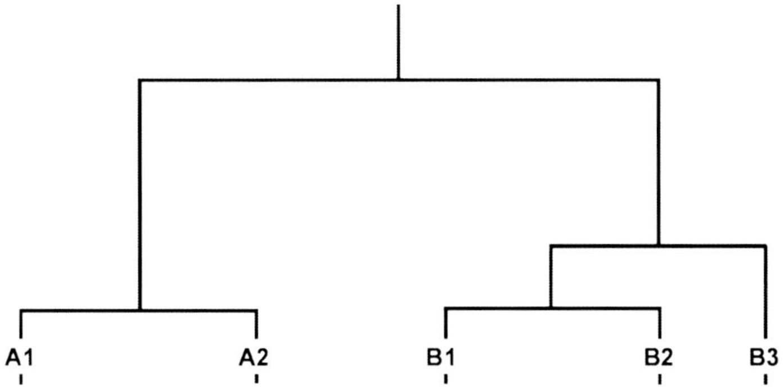

Fig. 4 Simplified dendrogram obtained by cluster analysis of 245 plots from debris cones of the Rhaetian Alps (simplified from Baroni et al. 2007). Each cluster in the dendrogram is a vegetation unit

Table 1 Synthetic vegetation table in which frequency of species in each vegetation unit is reported (simplified from Baroni et al. 2007). Numbers in bold identify more frequent species

	Channel		Debris flow		Debris and rock fall
Landforms units	A1	A2	B1	B2	B3
Vegetation units	*Rumex scutatus*	*Salix appendiculata*	*Alnus viridis*	*Picea excelsa*	*Festuca nigrescens*
Average altitude (m s.l.m.)	1,569	1,563	1,575	1,565	1,694
Number of plots	51	57	79	28	14
Average species number	8	6	10	7	11
Rumex scutatus L.	**75**	16	1		**64**
Agrostis schraderana Becherer	**48**	**51**	8	7	**64**
Salix appendiculata Vill.	18	**72**	23	21	
Alnus viridis (Chaix) DC.	9	**88**	**62**	21	7
Rubus idaeus L.	25	2	**82**	18	**50**
Viola biflora L.	9	2	**57**	14	**43**
Phegopteris polypodioides Fée			**51**	29	
Oxalis acetosella L.	2	4	**48**	**46**	
Picea excelsa (Lam.) Link	2	9	19	**64**	
Festuca nigrescens s.l.	23	5	6	4	**79**
Rhodiola rosea L.	18	9	8	7	**57**
Chaerophyllum hirsutum L.	15		5		**43**
Sedum annuum L.	11	2	1		**50**
Geranium sylvaticum L.	9	2	22		**43**
Peucedanum ostruthium (L.) Koch	26	7	30		**43**

4 Geomorphic Processes and Plant Communities

Plant community distribution on the cones is very heterogeneous with discontinuous and scattered vegetation cover that correlate with the direction, intensity and overlapping of the geomorphic processes. In particular, geomorphic processes create habitats where specific plant communities can develop. These processes act as disturbance factors that induce partial or total vegetation destruction and limit plant biomass production (Grime et al. 1996). The disturbance is measured as a function of intensity and frequency of the events (Crawley 1997). Thus, on geomorphic active landforms, plant communities consist of species with similar ecological requirements. Therefore, geomorphic processes are a decisive selective factor.

On active landforms, species tend to adapt to disturbance if they do not go extinct. Species mainly consist of pioneer herbaceous plants with large radical structures; their reproduction is both sexual and clonal. According to Austin (1981), the complex of plant communities on a composite debris cone defines a toposequence because each plant community is related to a predominant geomorphic process acting on a single portion of the cone

On alpine composite debris cones and channels, all the active geomorphic processes generate highly disturbed micro-habitats (Fig. 5). Additionally, water availability, avalanches occurrence (particularly in spring), flooding processes and debris-flow accumulation (in summer and autumn) can determine micro-habitats. Plant communities grow at lower levels of disturbance and are conditioned by other abiotic factors, such as temperature, grain size, soil moisture and nutrients, sunlight, slope dipping and aspect (Zimmermann and Eggenberg 1994; Brancaleoni et al. 2003).

Debris flows give rise to more disturbed and heterogeneous habitats in the erosion area (channel) where *Oxyria digyna*, *Agrostis schraderana* and other plants have adapted to continuous erosion.

The debris and the rockfall deposits support various coexisting microenvironments: (1) on the surfaces of the blocks and boulders there are lithophytic plants (such as *Sedum annuum*, *S. atratum* and *Cardamine resedifolia*); (2) the finely grained debris (dm- and cm-sized clasts with discontinuous soil cover) is colonized by species of shrublands and/or mesophilous grassland (*Alnus viridis*, *Agrostis schraderana*, *Peucedanum ostruthium*); (3) the sandy pockets and soil that develop between the slabs and boulders are associated with species of grassland (*Festuca nigrescens*, *F. scabriculmis, Campanula scheuchzeri*).

Snow avalanches create disturbance and their intensity is higher in the proximal portion of composites cones, while in the distal portions, avalanche intensity decreases and the frequency of disturbance becomes a selective factor.

Snow avalanches and enduring heavy snow cover exert strong mechanical stresses on plants that are often bent. This is particularly true in the case of shrubby and arboreal species, such as *Alnus viridis*, *Betula pendula*, *Larix decidua*, *Picea excelsa*, *Prunus padus* and *Salix appendiculata*. Trunks and branches of these species are often leaning or broken, or they show a typical sabre shape.

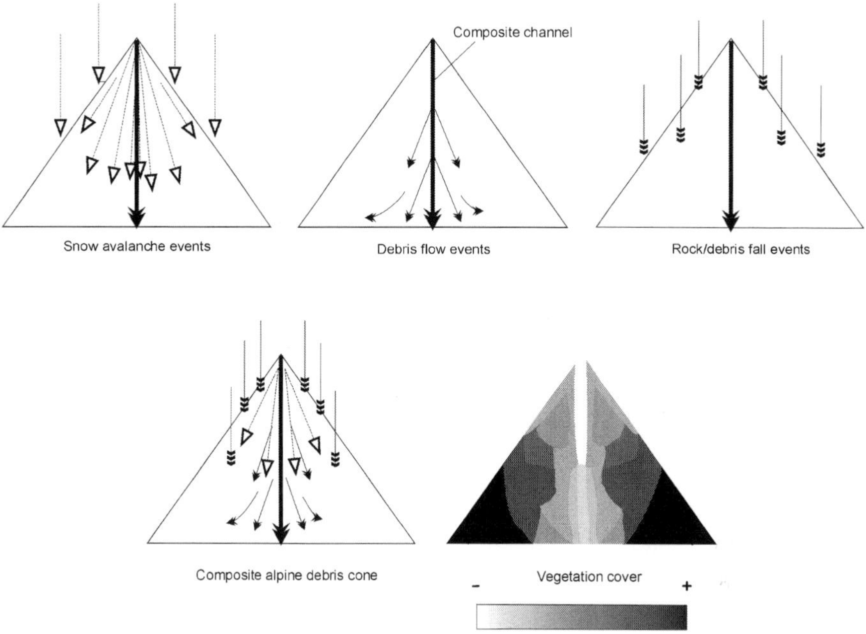

Fig. 5 Schematic representation of the action range of geomorphic processes on alpine composite debris cones. Geomorphic processes strongly affect vegetation cover according to their intensity and frequency on cone

According to our results (Baroni et al. 2007), the joint action of various geomorphic processes acting in the same sector of a cone produces dissimilar habitats where different plant communities can grow. Therefore, not only is each form associated with a specific morphogenetic process (snow avalanche, debris flow, and rockfall) characterized by its own specific vegetation type, but the overlapping of deposits of distinct origins also gives rise to specific habitats with clear influences on the distribution of plant communities (distinct from communities produced by single processes).

This evidence allows for the identification of **landform–vegetation units** that express consistent correlations among landforms, geomorphic processes and diagnostic plant communities. These communities act as phytoindicators of specific geomorphic units derived from the activity of a single geomorphic process or from a combination of different geomorphic processes.

By way of example, the comparison between landform and vegetation units in the Rhaetian Alps revealed the following relationships (Fig. 6):

1. Channel beds and channel banks were typically associated with herbaceous vegetation units (*Rumex scutatus* units), while phanerophytic pioneer vegetation (see *Salix appendiculata* units) were common in the external bank-levees.

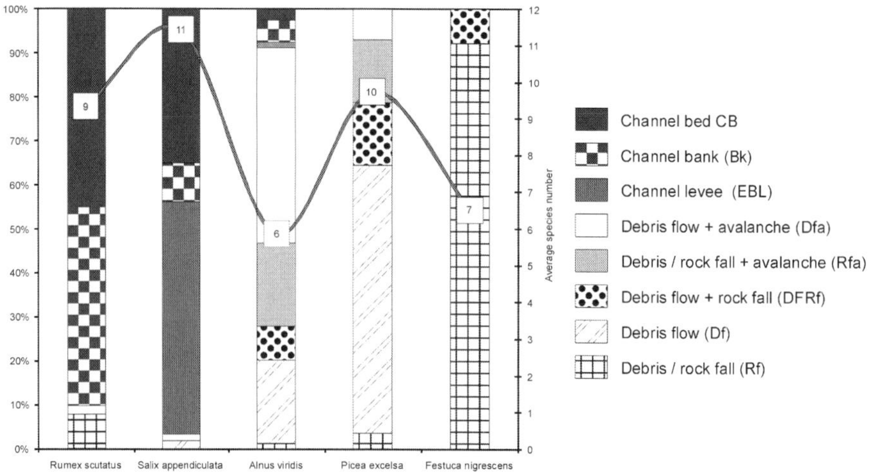

Fig. 6 Frequency of the landform types (*on the right*) according to the vegetation units and sub-units (indicated at the *base*). *C* Channel bed (avalanche track), *Cb* Channel bank, *DfL* Debris-flow levee and recent debris flow, *Df* Debris-flow deposits, *ADf* Avalanche debris on debris-flow deposits, *ADfRf* Avalanche debris on debris-flow deposits (with sparse debris talus) and rock fall deposits, *ARf* Avalanche debris on talus slope *Rf* Debris/rock fall deposits (modified from Baroni et al. 2007)

2. Debris-flow deposits were associated with phanerophytic communities (*Picea excelsa* units) and avalanche deposits were associated with *Alnus viridis* communities. Finally, debris and rockfall deposits are correlated with herbaceous mesophilous vegetation units (*Festuca nigrescens* units).

5 Vegetation Dynamics

The stability of landform surfaces in biostasy-like environments drives the development of mature plant communities.

On active geomorphic landforms, pioneer plant communities can eventually grow. On dormant, inactive landforms, pioneer vegetation is progressively replaced by mature plant communities through successional phases that start when the geomorphic processes become inactive. In other terms, the development of vegetation on landform features (both constructional and erosional landforms) is a function of the length of time since new features have been available for colonization. In this temporal sequence (chronosequence), the effect of competition between species and other intrinsic factor of the community tends to prevail over the effect of the geomorphic disturbance. Competitive species can create environmental stress for other species by reducing resource availability such as nutrients and light. In respect to the disturbance, stress is a continuous phenomenon that does not destroy biomass but reduces photosynthetic production (Grime et al. 1996). During the successional phases, the most competitive (*Alnus viridis, Rubus idaeus*) and stress-tolerant species (*Picea excelsa*) become dominant over disturbance-tolerant species (*Rumex scutatus, Oxyria digyna*) (Caccianiga et al. 2006). In the absence of disturbance, each chronosequence converges to the natural potential vegetation (Fig. 4).

As a consequence, we can recognize a relative chronology of landforms and deposits in accordance with a hierarchic successional trend of plant communities. For example, an alpine belt from pioneer herbaceous types with scattered individual on recent deposit may succeed to a structured grassland, while a subalpine and montane belt from pioneer herbaceous vegetation may change to a shrubland rather than woodland.

Vegetation dynamics can be studied through the analyses of plant community succession by studying permanent plots or considering only plant communities of a toposequence in which local factors are identical and the same geomorphic process has acted in a different time. On composite cones, debris-flow events periodically can generate new deposits and shift the ecosystem of the most disturbed area (i.e. channels). In the latter example, debris flows can damage or destroy vegetation, and thus trigger the development of new, pioneer vegetation, while mature vegetation may begin to develop in the oldest (relict) channels. These processes may be considered spatial sequences of vegetation units reflecting a chrono-/toposequence of deposits on which they develop (Harris and Gustaffson 1993), also known as "space for time substitution" or simply chronosequence (Johnson and Miyanishi 2008).

This practice of equating a few spatially separated sites with different periods of time since a specific disturbance should not be accepted in absence of evidence that their potential vegetation dynamics have been similar during those periods (Austin 1981). Recent studies suggest applying this method of studying vegetation dynamics only at local scales (Del Moral 2007) and to submit the results of these studies to statistic tests (Johnson and Miyanishi 2008).

6 Channels as Migration Route for Alpine Plant Species

Channels act as biological corridors that connect nival and high mountain belts along both altitudinal and horizontal gradients. Snow avalanche events and nival persistence in late spring and early summer increases spatial heterogeneity, enhances species richness below the treeline and induces a specific floristic composition (Körner 1999). Enduring snow cover, low temperature of streams and cold air currents can create microthermal conditions and suitable niches for the colonization of high altitude species at lower elevations. Species typical of deglaciated areas at glacier fronts (*Cerastium uniflorum, Oxyria digyna* and *Poa laxa*) can be identified as pioneer elements at very low elevations in these topographic/geomorphic traps. These species are very sensitive indicators of climate change.

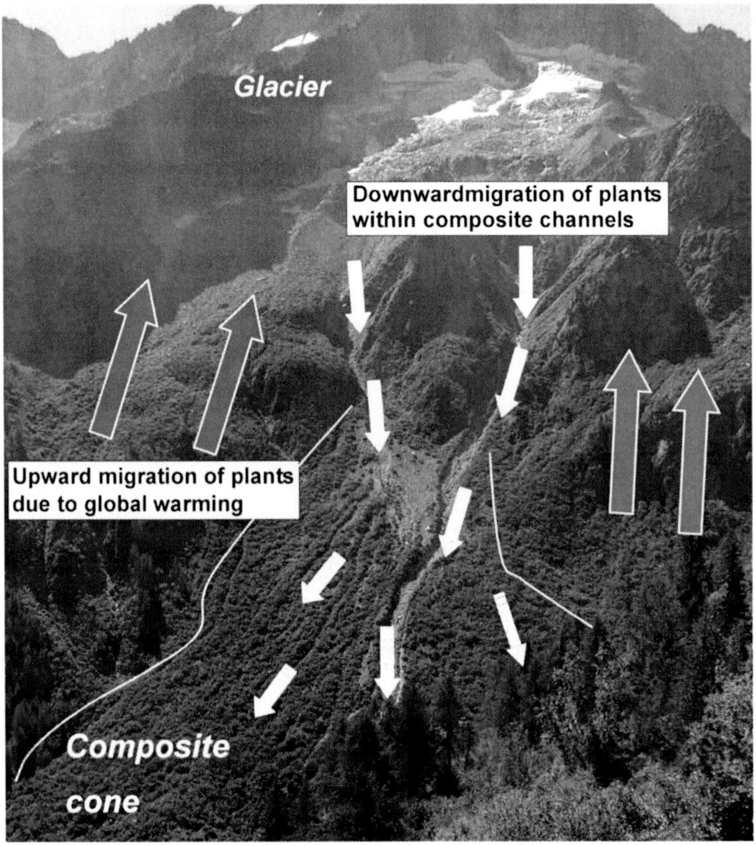

Fig. 7 Composite channels act as potential biodiversity refuges and migration routes for high mountain plants during a phase of global warming (Val Presena, Adamello Group, Italy)

Active geomorphic processes within channels act as vectors carrying seeds, pollen, spores, and (occasionally) juvenile or adult plants downhill. Alpine composite channels are the most potentially exclusive mountain feature where species migrate from higher to lower altitudes favoured by seed/propagule dispersal of geomorphic processes and favourable microclimatic conditions (Fig. 7). For these reasons, they should be constantly monitored for their roles as refuges for high mountain plant diversity and bioindicators of climatic changes to protect and conserve rare alpine plant species.

7 Conclusions

Vegetation units in the toposequences can indicate relevant variations of abiotic factors that affect the stability of the cones such as water availability (i.e. presence of underground water), superficial grain size, texture, and structure of landform units.

The distribution of disturbance-tolerant, competitive and stress-tolerant species in plant communities give a measure of the stability of the cones on which these communities develop. As a consequence studies on distribution of plant communities and toposequences may be important to classify geomorphic processes acting in small parts of the cones. Therefore, vegetation types can be used as bioindicators for the zonation of geomorphic hazard on alpine slopes due to snow avalanches, debris flows, debris/rock falls and landslides occurrences.

Studies of the vegetation combined with other subjects (lichenometry, dendochronology, pedology) can support the reconstruction of the frequency of occurrence of debris-flow events and a relative chronology of debris-flow deposits.

Alpine composite cones of alpine regions represent a tension area between mountain and nival belts where geomorphic processes and vegetation dynamics overlap each other and deeply interact. Understanding the numerous and complex interactions between vegetation and landforms in the cones at different scale levels (local, landscape, and regional) can disclose mechanisms for monitoring and also for predicting the effects of climate change at the local scale.

8 Use for Practitioners

Vegetation provides comprehensive information of geomorphic processes acting along slopes in alpine areas. In fact, the ordination of the mosaic of vegetation types on alpine composite cones can allow the identification of landforms and deposits with different degree of stability (Table 2). For this reason vegetation analysis can be applied for preventing geomorphic hazard, mostly in the alpine regions, by means of the construction of a specific vegetation index for each vegetation type based on the link between species composition, species dominance, damages to species, nature and origin of superficial deposits and local environmental factors.

Therefore, vegetation analysis provides information on the nature of the main geomorphic hazard acting at present on a specific cone, or on specific portions of a certain cone. In particular, the degree of vegetation cover is proportional to the disturbance; furthermore, the vegetation structure indicates the intensity of the disturbance (partial or total vegetation destruction). For example, below the treeline,

Table 2 Simplified example of scale of hazard based on vegetation types for the Central Alps area

	Degree of risk		
Vegetation type	Debris flow	Debris/rock fall	Snow avalanche
Rumex scutatus	High	High	High
Salix appendiculata	Medium	Low	High
Alnus viridis	Medium	Medium	High
Festuca nigrescens	Medium/low	High	Medium
Picea excelsa	Low	Medium	Low

the presence of pioneer herbaceous species indicates high disturbance, generally generated by debris flows; shrubs communities (such as *Alnus viridis*) can indicate less disturbance caused by debris flows or debris/rockfall, but an increase in stress caused by snow avalanches, which are generally less destructive than debris flow and debris/rock fall. In this case the probability of occurrence of debris flows and debris/rock fall will be lower than the probability of occurrence of snow avalanches.

On the other hand, it is clear that the in portions of the cones where the trees become dominant, the risk of debris flows decreases while rock falls and snow avalanches can occasionally occur, particularly in the vicinity of the highest limit of shrubs communities (or at the proximal border of the shrubs communities).

Vegetation transects identify the limits between vegetation types. Floristic changes among plots allow the discrimination of the areas where two or more geomorphic processes are acting and at which degree of intensity. In areas at the transition between vegetation types, the probability of the occurrence of each process could be higher, although their intensity may be lower than in the areas where a single type of vegetation is prevalent.

After the identification of vegetation types and transition areas through detailed survey and mapping and using GIS facilities, a vegetation-based zonation of the geomorphic hazard is at hand. Such result may support decision makers in the development of action plan after the evaluation of vulnerability.

For the mapping process, vegetation plots along transect series on alpine composite cones can have a highest discriminating power in comparison to other vegetation survey methods such as line-transect, merely physiognomic, vegetation relevé (one per unit), etc. In fact plot transects facilitate the identification of each kind of vegetation/deposit change.

On the other hand, vegetation-based methods are strongly associated and affected by the local flora according to biogeographic history of a region. For this reason a distinct vegetation index for hazard assessment needs to be planned and formulated for homogeneous phytogeographic regions.

Acknowledgements This work was carried out as part of a COFIN-MIUR 2009 project, "Effects of global warming in the glacial and periglacial environments of the Maritime and Cozie and Central Alps" (P.R. Federici).

References

Austin MP (1981) Permanent quadrats: an interface for theory and practice. Vegetatio 46:1–10

Ballantyne CK (2002) Paraglacial geomorphology. Quat Sci Rev 21:1935–2017

Baroni C, Armiraglio S, Gentili R, Carton A (2007) Landforms-vegetation units for reconstructing the geomorphologic evolution of composite alpine debris cones (Valle dell'Avio, Adamello Group, Italy). Geomorphology 84:59–79

Brancaleoni L, Strelin J, Gerdol R (2003) Relationships between geomorphology and vegetation patterns in subantarctic Andean tundra of Tierra del Fuego. Polar Biol 26:404–410

Burga CA, Frauenfelder R, Ruffet J, Hoelzle M, Kääb A (2004) Vegetation on alpine rock glacier surfaces: a contribution to abundance and dynamics on extreme plant habitats. Flora 199:505–515

Caccianiga M, Luzzaro A, Pierce S, Ceriani RM, Cerabolini B (2006) The functional basis of a primary succession resolved by CSR classification. Oikos 112:10–20

Crawley MJ (1997) Life history and environment. In: Crawley MJ (ed) Plant ecology. Blackwell Science, Boston, p 717

Del Moral R (2007) Limits to convergence of vegetation during early primary succession. J Veg Sci 18:479–488

Ellenberg H (1988) Vegetation ecology of Central Europe. Cambridge University Press, Cambridge

Gentili R, Armiraglio S, Rossi G, Sgorbati S, Baroni C (2010) Floristic patterns, ecological gradients and biodiversità in the composite channels (Central Alps, Italy). Flora 205:388–398

Grime JP, Hodgson JG, Hunt R (1996) Comparative plant ecology. A functional approach to common British species. Unwin Hyman, London

Harris SA, Gustafson CA (1993) Debris flow in an area of continuous permafrost, St. Elias Range, Yukon Territory. Z Geomorphol N S 37:41–56

Hession WC, Curran JC, Resler LM (2010) Geomorphology and vegetation: interactions, dependencies, and feedback loops. Geomorphology 116:203–205

Howard JA, Mitchell CW (1985) Phytogeomorphology. Wiley, New York, 222 pp

Huc S (2008) Scree mobility above the treeline in the eastern Pyrenees (France): morphodynamic and biological indicators. Géomorphol Relief Process Environ 2008(2):99–112

Johnson EA, Miyanishi K (2008) Testing the assumptions of chronosequences in succession. Ecol Lett 11:419–431

Jomelli V (1999) Dépôts d'avalanches dans les Alpes françaises: géometrie, sédimentologie et géodynamique depuis le Petit Âge glaciaire. Geogr Phys Quat 53:199–209

Jomelli V, Francou B (2000) Comparing the characteristics of rockfall talus and snow avalanche landforms in an Alpine environment using a new methodological approach: Massif des Ecrin, French Alps. Geomorphology 35:181–192

Körner C (1999) Alpine plant life. Functional plant ecology of high mountain ecosystems. Springer, Berlin

Kozlowska A, Raczkowska Z (2002) Vegetation as a tool in the characterisation of geomorphological forms and processes: an example from the Abisko Mountains. Geogr Ann 84:233–244

Kruckeberg AR (2002) Geology and plant life. The effects of landforms and rocky types on plants. University of Washington Press, Washington, DC

Marston RA (2010) Geomorphology and vegetation on hillslopes: interactions, dependencies, and feedback loops. Geomorphology 116:206–217

Ozenda P (2002) Prospectives pour una geobiologie des montagnes. Presses Polytechniques et Universitaire Romandes, Lausanne

Pignatti S, Menegoni P, Pietrosanti S (2005) Bioindicazione attraverso le piante vascolari. Valori di indicazione secondo Ellenberg per le piante della flora d'Italia. Braun Blanquetia 39:3–97

Troll C (1971) Landscape ecology (geoecology) and biogeocenology. A terminological study. Geoforum 8:43–48

Walker DA, Epstein HE, Gould WA, Kelley AM, Kade AN, Knudson JA, Krantz WB, Michaelson G, Peterson RA, Ping CL, Raynolds MK, Romanovsky VE, Shur Y (2004) Frost-boil ecosystems: complex interactions between landforms, soils, vegetation and climate. Permafr Periglac Process 15:171–188

Zimmermann K, Eggenberg S (1994) Comparison of vegetation and geomorphology: problems and approach. Suppl Rev Valdôtaine d'Hist Nat 48:357–365

Using Age of Colonizing Douglas-Fir for the Dating of Young Geomorphic Surfaces—A Case Study

Thomas C. Pierson

1 Introduction

Dating of many types of young (<500 year), dynamic, geomorphic landforms (e.g. mass-movement erosional tracks and deposits, alluvial terraces, flood plains, etc.) for purposes of hazard assessment and mitigation commonly requires greater dating precision than is available through radiocarbon dating or other methods. Ages of trees growing on landform surfaces have been used in a number of studies to estimate the time of landform creation or surface clearing, but the time lag between surface formation or disturbance and the reestablishment of trees can vary from 1 to more than 200 years (Desloges and Ryder 1990; Frenzen et al. 1988, 2005; Larsen and Bliss 1998; McCarthy and Luckman 1993; Sigafoos and Hendricks 1969; Winter et al. 2002). Appropriate lag times for selected tree species and for particular climatic and altitudinal ranges must be determined for the method to be useful.

This article summarizes a study carried out to determine the average time interval from formation of a stable barren landform surface until the first Douglas-fir (*Pseudotsuga menziesii*) seedlings survive and grow to breast height (BH) in lowland river valley settings (<750 m a.s.l.) of the southwestern Washington and northwestern Oregon Cascades (Pierson 2007). Ideally, if the oldest tree growing on a terrace or flood plain could be found, cored, and dated, and if that tree belonged to the first group (cohort) of trees to colonize the new geomorphic surface, the true age of the surface could be determined with considerable precision. In order to sample the oldest tree(s) on the surface for dating, the largest-circumference trees were sought, with the assumption that within a relatively small area the oldest trees should be the largest.

T.C. Pierson (✉)
Volcano Science Center, Cascades Volcano Observatory, U.S. Geological Survey, 1300 SE Cardinal Court, Suite 100, Vancouver, WA 98683, USA
e-mail: tpierson@usgs.gov

M. Schneuwly-Bollschweiler et al. (eds.), *Dating Torrential Processes on Fans and Cones*, Advances in Global Change Research 47, DOI 10.1007/978-94-007-4336-6_13,

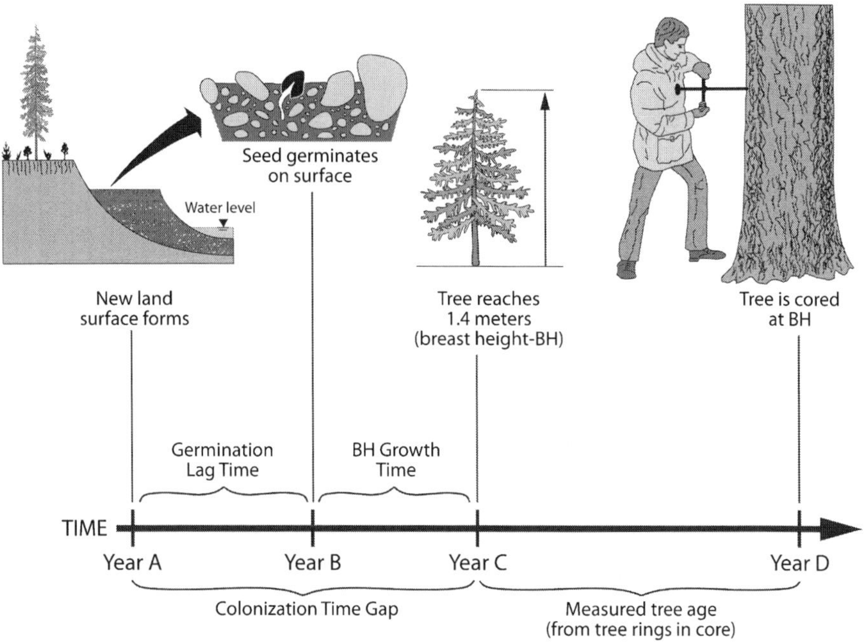

Fig. 1 Schematic timeline showing conceptual basis for this study. Timeline goes from time of landform creation until a tree growing on that landform is cored at breast height (*BH*) for dating purposes

This study determined and compared the time corrections to be added to tree-ring ages obtained from increment cores from the truly oldest tree(s) and the largest trees growing on fresh landform surfaces in the study area (Fig. 1). Those corrections are (1) the time interval between stabilization of the new landform surface and seedling germination (germination lag time or GLT); and (2) the interval between seedling germination and growth to sampling height if the trees are not cored at ground level (breast height growth time or BHGT, when trees are cored at about 1.4 m above the ground surface) (see Savage et al. 1996). The two time intervals together are termed the colonization time gap (CTG).

2 Regional Setting and Sampling Strategy

Eight study sites were chosen on geologically recent (<60 year) landform surfaces of known age in river valleys draining three Cascade volcanoes—Mount Rainier, Mount St. Helens, and Mount Hood (Fig. 2). The single study site at Mount Rainier is a stream terrace along Kautz Creek that was swept in 1947 by a debris flow, which left a thick, compact, more or less smooth-surfaced deposit of muddy sandy

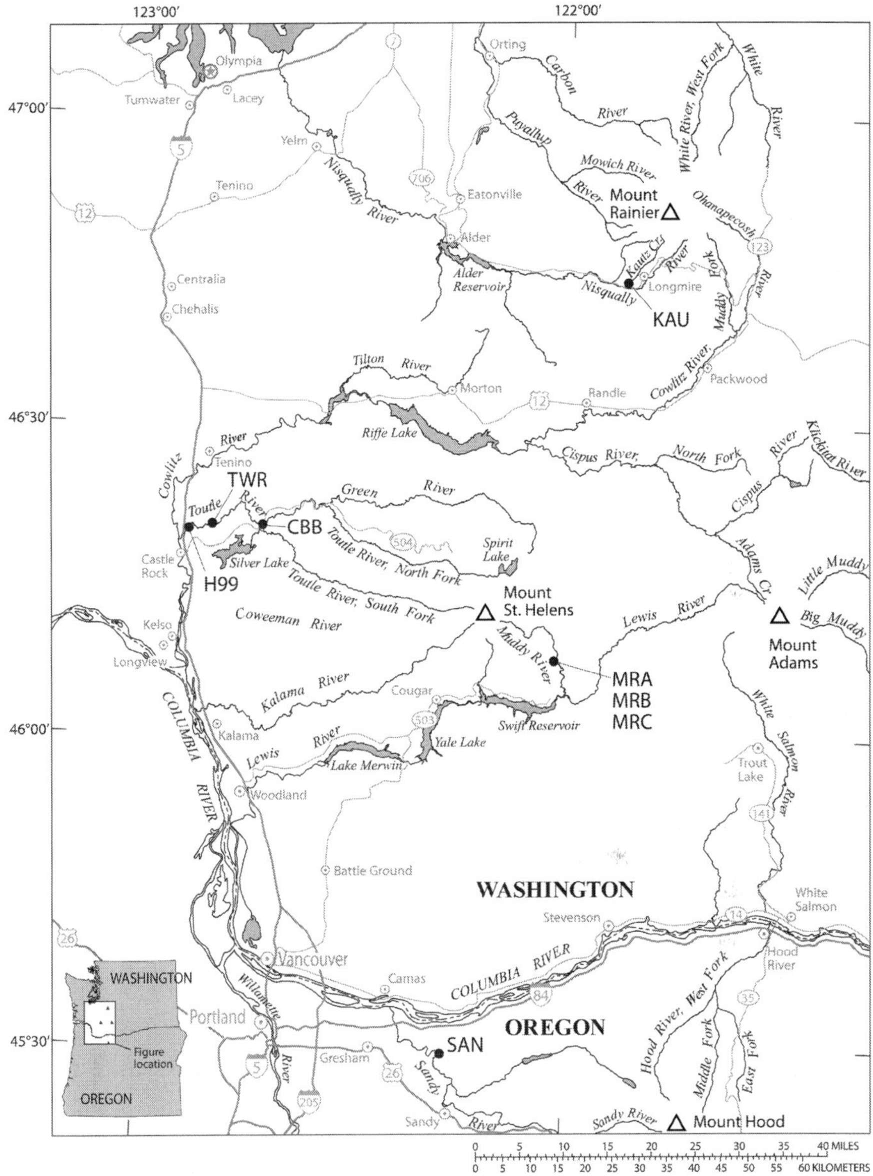

Fig. 2 Location map of study sites (*black dots*) downstream from volcanoes in the Cascades

gravel on the terrace. At Mount St. Helens three sites each on the lower Toutle and Muddy Rivers were examined where debris-flow lahars, triggered by the 1980 eruption, deforested and capped pre-existing low terraces with compact muddy gravelly sands. The Mount Hood site on the lower Sandy River is an alluvial point bar composed of loose gravelly sand, created by a large flood in February 1996. This site was sampled in 2004, but all seedlings were removed by another flood in November 2006—a reminder of the importance of survival, not just germination.

Study site selection criteria, in addition to knowing the time of surface formation, were ease of access to sites and close proximity to seed sources. All the study sites were wholly or partially barren surfaces immediately following deposition of the sediment that formed or covered the terraces or flood plains, and vegetation sprouting from propagules or buried roots beneath the fresh deposits was minimal or absent.

Conifer seedling populations at study sites were dominated by Douglas-fir, the predominant conifer species in the old- and second-growth forests that were adjacent to the study sites. Tree cores were taken with increment borers, generally at BH. At each site, the largest (and presumed oldest) trees were sampled, assuming that they would include at least some of the truly oldest trees growing on the site, as has been done in other studies (e.g., McCarthy and Luckman 1993). In areas where trees were too small to core at BH, they were cored or sectioned within 5 cm of ground level.

Apparent tree ages were determined by counting annual growth rings in the cores, using a template to estimate missing rings in off-centre cores. Under favourable growing conditions, Douglas-fir seedlings can reach 5–10 cm in height in the first growing season (Fowells 1965); for seedlings growing less than 5 cm in the first season, the first year's ring could be missed.

3 Dating of Young Surfaces

Germination lag times (GLT), growth times from germination to breast height (BHGT), and colonization time gaps (CTG) for all the study sites are fully listed in Pierson (2007) and summarized in Table 1. The results show that the largest trees are not always the oldest trees. For example, at one site the two oldest trees were the seventh and ninth largest out of a sample of 11 (22% and 28% smaller in circumference than the largest tree at the site).

GLT for the single oldest Douglas-fir at each of the 8 study sites varies from 1 year (germination in spring of growing season following land-surface formation) to 14 years. Mean GLT for the oldest trees at the study sites is 3 (3.0 ± 2.1) years, or 2 (2.3 ± 1.5) years if the corrected value from site KAU is used (see explanation in Pierson 2007), whereas the mean of the single largest trees from each of the study sites is 4 (3.8 ± 2.9) years. Mean GLT for the five oldest trees at each site varies from 3 (3.0 ± 1.2) years to 8 (7.6 ± 0.5) years among the different sites, with an overall mean of 5 (4.8 ± 2.5) years. The overall mean for the five largest trees at each site is also 5 years (5.3 ± 2.8).

BHGT values show that Douglas-fir seedlings took from 3 to 10 years to grow to BH among the different study sites. Mean BHGT for all the sites where rings could be counted both at ground level and at BH ($n = 32$) was 6 (5.9 ± 2.0) years. Averages among the different sites vary between 5 and 7 years.

Table 1 Summary of mean values for breast height growth time (*BHGT*), germination lag time (*GLT*) and colonization time gap (*CTG*) for all study sites

Site	BHGT (in years)	GLT (in years)	CTG (in years)
Kautz Creek, Mount Rainier: Terrace capped with lahar deposits			
Oldest tree	6 (est)	7 (1)[c]	13 (7)[c]
Largest tree	6 (est)	8	14
Mean of 5 oldest trees[a]	6.0 ± 0	7.6 ± 0.5	13.6 ± 0.5
Mean of 5 largest trees[a]	6.0 ± 0	8.6 ± 0.5	14.6 ± 0.5
Muddy River, Mount St. Helens: Terrace capped with lahar deposits			
Mean of single oldest trees at each site (n = 3)[a]	5.3 ± 0.6	2.3 ± 2.3	7.7 ± 2.1
Mean of single largest trees at each site (n = 3)[a]	5.0 ± 1.0	2.7 ± 2.9	7.7 ± 2.1
Mean of 5 oldest trees at each site (n = 14)[a,b]	5.5 ± 0.7	4.1 ± 2.4	9.7 ± 2.3
Mean of 5 largest trees at each site (n = 14)[a,b]	5.4 ± 0.8	4.4 ± 2.7	9.8 ± 2.4
Toutle River, Mount St. Helens: Terrace capped with lahar deposits			
Mean of single oldest trees at each site (n = 3)[a]	6.7 ± 0.6	3.0 ± 0	9.7 ± 0.6
Mean of single largest trees at each site (n = 3)[a]	5.7 ± 2.3	5.0 ± 2.6	10.7 ± 0.6
Mean of 5 oldest trees at each site (n = 12)[a,b]	5.8 ± 1.7	4.8 ± 2.2	10.6 ± 3.1
Mean of 5 largest trees at each site (n = 12)[a,b]	5.6 ± 2.2	5.3 ± 2.3	10.8 ± 3.4
Sandy River, Mount Hood: Alluvial flood bar			
Only tree per site	6	1	7
Summary of study area			
Mean of single oldest trees at each site (n = 8)[a]	6.0 ± 0.8	3.0 ± 2.1 (2.3 ± 1.5)[c]	9.0 ± 2.3 (8.3 ± 1.7)[c]
Mean of single largest trees at each site (n = 8)[a]	5.8 ± 1.3	3.8 ± 2.9	9.5 ± 2.7
Mean of 5 oldest trees at each site (n = 32)[a, b]	5.8 ± 1.1	4.8 ± 2.5	10.5 ± 2.8
Mean of 5 largest trees at each site (n = 32)[a, b]	5.5 ± 1.4	5.3 ± 2.8	10.8 ± 3.1

[a]Means of BHGT and GLT use some estimate values
[b]If less than 5 trees at a site, values from all trees used
[c]Using GLT = 1 year for site KAU (Sigafoos and Hendricks 1969)

CTG for the single oldest tree at each of the 8 study sites ranges from 6 to 13 years, with a mean of 9 (9.0 ± 2.3) years. Using the corrected GLT of 1 year for site KAU, mean single-tree CTG reduces to 8 (8.3 ± 1.7) years. CTG based on the largest single tree at each of the sites ranges from 7 to 14 years and has a mean of 10 (9.5 ± 2.7) years. For the study as a whole, CTG for the largest trees at each of the sites averages only 1 year more than that for the oldest trees at those sites. Using the means of the five oldest and five largest trees at each site (rather than the single oldest and largest) provides better age agreement. Although the means of the five oldest trees at each of the 8 study sites vary between 8 and 14 years, the five-tree mean of all the sites is 11 (10.5 ± 2.8) years. This is approximately the same as the five-tree mean CTG for the largest trees from all the sites, which is 10.8 ± 3.1 years. In other words, it takes, on average, 11 years from the time a new land surface emerges and stabilizes until the first five Douglas-fir seedlings become established on that surface and survive to grow to BH. Furthermore, use of five-tree mean ages of the largest-circumference trees approximates the CTG for a newly colonized land surface as well as the ages of the five oldest trees on that surface.

4 Limitations and Error Sources

Even though this study showed that the dating of first colonizers growing on a newly formed geomorphic surface can be achieved with reasonable precision when the age of the surface is already known, various sources of error are important when applying this approach to undated surfaces: (1) the oldest tree cannot be selected *a priori*, because the oldest tree is not necessarily the largest (see Larsen and Bliss 1998); (2) it may not be possible to ascertain that the oldest tree sampled belongs to the first cohort of colonizers; and (3) various errors are inherent in dendrochronologic methods.

These potential errors can be addressed in various ways. Tree-selection error can be greatly reduced by taking the mean age of the five largest trees growing on a geomorphic surface instead of the single largest. Identification of first-cohort trees requires independent verification that a stand-destroying disturbance such as wildfire, flooding, or logging has not removed the original oldest trees and "reset the clock". Clues to post-landform disturbance include finding charred wood on or in the soil, age structure analysis, the presence of contrasting surface deposits, or the presence of cut stumps (commonly widely spaced in old selectively logged areas). It may also be possible to verify that the landform is not older than the average life span of the tree species being sampled by using historical accounts or radiocarbon dating. Alternatively, trees may be older than the landform if survivors protrude up through a thin deposit from a buried soil on an older surface. Such partially buried survivors can be identified by their boles showing no stump flare or butt swell near the ground surface. Dendrochronologic errors (off-centre cores, false or light rings, missing rings, etc.) are considered in other references (e.g. Stokes and Smiley 1968; Phipps 1985; Yamaguchi 1989).

It is also important to note that natural variability of GLT and BHGT can be large, both within and between sites. Local variability in colonization rates may even exceed differences based on altitude or regional climate (Desloges and Ryder 1990). Douglas-fir seedlings became established at some sites within the first year following surface formation, but at others no seedlings survived in the first 5 years and only a relative handful of trees became established within the first decade following surface formation. During typically dry Northwest summers, high ground-surface temperatures and paucity of shade on barren ground are likely limiting factors in early seedling survival; shade from other pioneer tree species may be required.

Thus, the combined error from tree selection, site variability, and dendrochronologic sources for trees cored at ground level is about −5 to +18 years when the single largest tree on a landform is sampled, and approximately ±5 years when the five largest trees are sampled. If trees are cored at breast height, variable seedling growth rates increase combined error to −7 and +20 years for sampling the single largest tree and ±7 years when the five largest trees are sampled.

Study results (Pierson 2007) indicate that two approaches are available to researchers seeking to use Douglas-fir ages to date a geomorphic surface of relatively young but unknown age in lowland southwestern Washington and northwestern Oregon. The simpler approach is to find the largest tree on the surface and add the largest-tree GLT (4 years) or CTG interval (10 years) to the ring-count age of the sampled tree, depending on the level at which the sampled tree is cored, in order to estimate the age of the surface. In this study the largest single trees average a year younger than the actual oldest trees, and the largest tree at one site was 8 years younger than the oldest tree. McCarthy and Luckman (1993) suggest various sampling strategies to improve accuracy. Alternatively, one can determine the mean ring-count age of the five largest trees growing on a surface, and then add the mean GLT for the five largest trees (5 years) or CTG (11 years) to estimate surface age. The oldest-tree and largest-tree means are approximately the same, thus this latter approach should give a better estimate of the age of the geomorphic surface than sampling single trees.

5 Summary

Ages of young (<500 year) geomorphic surfaces inferred from ring-count ages of Douglas-fir growing on those surfaces (summarized here from Pierson 2007) will have plus/minus error bars on the order of 10 years—not as precise as some applications of dendrochronology. However, tree-ring dating of surfaces is an improvement in precision over standard radiocarbon dating, which for some calibrated young ages can have 2-sigma calibrated age ranges of as much as 300 years, although specialized techniques can improve this error. The lowland river valley GLT (or "ecesis interval") values of 4 or 5 years from this study (depending on number of trees sampled) contrast with (1) similar values of 1–5 years for slightly higher elevation sites on Cascade volcanoes (Hupp et al. 1987; Larsen and Bliss 1998) and more northerly lowland areas in British Columbia (Desloges and Ryder 1990), and (2) values of 5–60 years for Cordilleran montane sites near glacial limits, summarized in McCarthy and Luckman (1993).

References

Desloges JR, Ryder JM (1990) Neoglacial history of the Coast Mountains near Bella Coola, British Columbia. Can J Earth Sci 27(2):281–290

Fowells HA (1965) Silvics of forest trees of the United States, Agricultural handbook No. 271. US Department of Agriculture, Forest Service, Washington, DC

Frenzen PM, Krasny ME, Rigney LP (1988) Thirty-three years of plant succession on the Kautz Creek mudflow, Mount Rainier National Park, Washington. Can J Bot 66(1):130–137

Frenzen PM, Hadley KS, Major JJ, Weber MH, Franklin JF, Hardison JH, Stanton SM (2005) Geomorphic change and vegetation development on the Muddy River mudflow deposit. In: Dale VH, Swanson FJ, Crisafulli CM (eds) Ecological responses to the 1980 eruption of Mount St. Helens. Springer, Berlin, pp 75–91

Hupp CR, Osterkamp WR, Thornton JL (1987) Dendrogeomorphic evidence and dating of recent debris flows on Mount Shasta, northern California, US Geological Survey professional paper 1396B. US Government Printing Office, Washington, DC, pp 1–39

Larsen DR, Bliss LC (1998) An analysis of structure of tree seedling populations on a lahar. Landsc Ecol 13(5):307–322

McCarthy DP, Luckman BH (1993) Estimating ecesis for tree-ring dating of moraines—a comparative study from the Canadian Cordillera. Arct Alp Res 25(1):63–68

Phipps RL (1985) Collecting, preparing, crossdating, and measuring tree increment cores, US Geological Survey water-resources investigations report 85–4148. US Department of Interior, Geological Survey, Reston

Pierson TC (2007) Dating young geomorphic surfaces using age of colonizing Douglas fir in southwestern Washington and northwestern Oregon, USA. Earth Surf Process Landf 32(6):811–831

Savage M, Brown PM, Feddema J (1996) The role of climate in a pine forest regeneration pulse in the southwestern United States. Ecoscience 3(3):310–318

Sigafoos RS, Hendricks EL (1969) The time interval between stabilization of alpine glacial deposits and establishment of tree seedlings, US Geological Survey professional paper 650B. US Geological Survey, Boulder, pp B89–B93

Stokes MA, Smiley TL (1968) An introduction to tree-ring dating. University of Chicago Press, Chicago

Winter LE, Brubaker LB, Franklin JF, Miller EA, DeWitt DQ (2002) Initiation of an old-growth Douglas-fir stand in the Pacific Northwest: a reconstruction from tree-ring records. Can J For Res 32(6):1039–1056

Yamaguchi DK (1989) Using dendrochronology to date late Holocene geologic events. In: Forman SL (ed) Dating methods applicable to Quaternary geologic studies in the western United States, Utah Geologic and Mineral Survey, Miscellaneous Publications 89–7. Utah Geologic and Mineral Survey, Salt Lake City, pp 10–24

Lichenometric Dating of Debris Avalanche Deposits with an Example from the French Alps

Vincent Jomelli

1 Introduction

Debris flow is a dominant mass-movement process in mountain areas all over the world and represents a significant natural hazard. Consequently, knowledge of the frequency and magnitude of debris flows is important for hazard and risk assessments (Rickenmann 1999).

As debris flows and debris avalanches are often triggered by intense summer rainfalls (Caine 1980; Blijenberg 1998), future climatic changes could have an impact on the frequency of these processes (Jomelli et al. 2007a, 2009). However, before establishing a cause-and-effect relationship between global warming and the incidence of geomorphic processes, not only must the natural variability of extreme weather events be examined but detailed information obtained on past process dynamics on debris-flow cones, alluvial fans or floodplains is also required (Jomelli et al. 2004; Stoffel et al. 2008). Records of past debris-flow activity may be particularly useful in the recognition of past process dynamics and precipitation events.

Unfortunately there are only very few records available on past debris-flow activity. Most detailed records on past debris flows are either based on chronostratigraphic studies made in northern Europe or the United States (e.g., Brazier et al. 1988; Jonasson et al. 1993; Matthews et al. 1997, 2009; Blikra and Nemec 1998; Blair 1999), or on tree rings in the Alps (Strunk 1991, 1992; Stoffel et al. 2005, 2008). Very few studies are, in contrast, based on lichen measurements. This chapter focuses on this approach by firstly presenting an overview of the different techniques used to document past changes in debris-flow activity (including debris floods and

V. Jomelli (✉)
CNRS-UMR 8591, Laboratoire de géographie Physique, 1 Place A. Briand
92195 Meudon, France
e-mail: jomelli@cnrs-bellevue.fr

M. Schneuwly-Bollschweiler et al. (eds.), *Dating Torrential Processes on Fans and Cones*, Advances in Global Change Research 47, DOI 10.1007/978-94-007-4336-6_14,

debris avalanches) in mountain areas, and secondly by analyzing changes in debris avalanche activity in one area of the French Alps over the past two centuries.

2 A Short Review of Lichen Methods

Lichenometry was first used in the 1950s by botanist Beschel (1961) to date paleo-glacial extents in the Alps. It has been used in a wide range of mountain environments primarily to date glacial deposits. This relative dating technique has many advantages. It is inexpensive (except for the time required for field sampling) and does not require specific laboratory equipment as long as the identification of the genus is made. It can be used in various environments and especially at high altitudes that are generally poorly colonized by vegetation, making classical methods such as ^{14}C or dendrochronology difficult to apply. In addition, lichenometry may complement to or jointly with ^{14}C and or cosmogenic dating techniques (Jomelli et al. 2011). Generally the shortest or the longest axis of concentric lichens growing on different surfaces is measured because these can be rapidly and easily measured in the field. The method involves measuring rock lichens (symbiotic association between an algae and a fungus) on control surfaces of known age to derive a size-age relationship commonly named a growth curve. The growth curve is then used to determine the age of lichens growing on other surfaces whose ages are unknown. This link between lichen diameters and ages can be obtained by either recording a growth rate over a period of several years (Haworth et al. 1986; Karlen and Black 2002) or by measuring lichens on dated surfaces (Innes 1982, 1985a, b).

Specific local environmental conditions that vary over time and space influence the ecology of lichens and consequently their growth rate, which can depend on intraspecific competition and climatic conditions, as well as on other parameters like lithology. In addition, the identification of lichen species is complex thus introducing further uncertainty (Innes 1985b). Generally, periglacial researches involve measurement of crustose lichens such as *Rhizocarpon*. *Rhizocarpon geographicum* with thallus of different shades of yellow, and areoles round to angular is the most frequent lichen species selected in research studies. However, other yellow *Rhizocarpon* such as *R. macrosporum* or *R. lecanorinum* is also relevant. Consequently, as far as possible, the relationship between the size of lichens and their age must respect homogeneous characteristics. The positive relationship between the size of the lichens and the size of the sample area (Innes 1984) must also be taken into consideration when lichens are measured in the field.

Since the very early stages of development of lichenometry, two distinct approaches have been used to collect and implement lichen measurements. One approach uses all lichen sizes in a population while the other takes only the largest lichens into account. Both approaches are valid but their use implicitly refers to specific statistical theories that must be taken into account in the derivation of the growth curve and in the uncertainties associated with calculating the age of the lichens.

2.1 *The Approach Using the Whole Lichen Sizes*

This approach was developed in the 1960s by Benedict (1967) and consists in measuring a large number of lichens (not only the biggest) on a dated surface and in repeating the experiment on different dated and undated surfaces. After some methodological improvements, the approach has since been used to document debris-flow activity in different environments (Winchester 1984, 1988; Winchester and Harrison 1994, 2000; Menounos 2000; Winchester and Chaujar 2002). Typically between 100 and 400 lichen measurements are used to document a population. Measurements are displayed as size–frequency histograms. Then the best-fitted slope of the size–frequency (log 10) distribution versus the age of each dated surface provides the age of different features. Innes (1984) argued that the form of the size–frequency distribution would range from truncated log–normal distribution to Poisson, depending primarily on the age of the population being sampled. The size–frequency distribution of a young population describes an exponential curve with colonization increasing in the smallest size ranges. As a population ages, colonization declines due to less rock space available for newcomers, and the distribution displays a positive skew.

The size–frequency distribution method has the advantage of being based on a large number of measurements. Hence it is implicitly assumed that measurements are statistically representative of natural variability of lichen sizes. In other words, the variability of lichen growth – for example due to climatic conditions that are favourable for some and not for others – is taken into account. Moreover, there is a delay before a rock is colonized after exposure with widely different delays between lichens due to microclimate conditions and rock rugosity. This delay induces changes in the lichen size and is also taken into account. The approach is also good for distinguishing surfaces that are diachronous or that do not have a similar history.

2.2 *The Approach Using Extremes*

The second approach is based on analyzing measurements of the diameter of several large lichens. The underlying assumption is that the lichens with the largest diameters were among the first to colonize a surface. These specimens are assumed to be those growing under optimal conditions (Beschel 1950, 1961; Locke et al. 1979; Innes 1985b). The field sampling method consists of measuring exclusively the largest lichen colonizing a boulder, and repeating the experiment on many different boulders and on each geomorphic feature, for example on a debris-flow deposit. This procedure is carried out on dated and undated surfaces.

Different methods have been proposed to derive the growth curve and calculate the age from these extreme measurements. The first technique developed by Beschel (1961), consisted in ordering the measurements for each dated and undated surface and keeping only the largest recorded diameter. To minimize errors, the largest lichen was ignored if its size exceeded that of the second largest by ≥20%. However other researchers argued that the largest thallus is not necessarily contemporaneous with the deposit, and preferred to compute the average of several maxima per surface, the 5, 10, or 30 largest diameters for instance (McCarroll 1993). The approach has been used successfully to date debris flows in Scotland (Innes 1982), Norway (Innes 1985a), Svalbard (André 1995), Iceland (Decaulne et al. 2005), Tien Shan (Solomina et al. 1994), the French Alps (Van Steijn 1991, 1999; Jomelli et al. 2002; Helsen et al. 2002), and in the United States (Wilkerson and Schmid 2003). Another technique was recently proposed that consisted of describing the largest lichen diameters (Cooley et al. 2006; Naveau et al. 2007; Jomelli et al. 2007b) not only by computing averages as was previously the case, but by modelling the entire lichen measurements here referred to as the GEV (Generalized Extreme Value) technique. Because these lichens are extremes they do not follow a normal distribution. To conform with the nature of the data (extremes, i.e. the largest lichens), the generalized extreme value distribution is used and the lichen distribution computed for each debris avalanche. Maximum lichen measurements are modelled by varying the distribution of GEV parameters. These parameters are then used to derive a growth curve and to estimate ages in a single step.

Based on this procedure, a Bayesian model is built, i.e. the GEV and growth function parameters are treated as random variables with prior distributions. These prior distributions come from prior information about dated surfaces and biological knowledge (for example, the slope of the growth function is positive; the y-axis value is equal to 0). A Monte Carlo Markov Chain (MCMC) procedure is applied making it possible to estimate uncertainties associated with the ^{14}C technique (Reimer et al. 2009). MCMC methods are a class of algorithms for sampling from probability distributions (in our case we use the prior distribution) based on constructing a Markov Chain. The convergence proprieties of MCMC allow a good approximation of posterior distribution (Cooley et al. 2006) of parameters after a large number of iterations (more than 100,000). At each iteration, all the parameters (GEV parameters and the parameters of the growth curve), are updated one-at-a-time until the best combination of parameters is found. This allows computation of an empirical distribution for each parameter. In particular, the age distribution of each undated deposit is obtained. Confidence intervals for the age of the undated deposit are computed from the mean and from the variance of the previous age distribution. Finally, uncertainties are estimated taking into account measurement errors, especially variability of natural lichen growth. This approach has several advantages over other methods based on extremes or whole lichen sizes (Jomelli et al. 2007b, 2010), particularly due to its ability to provide robust confidence intervals.

3 A Chronology of Debris Avalanches in the Massif des Ecrins, French Alps

3.1 Study Area

Debris-avalanche deposits were selected in the La Selle catchment located in the northern part of the *Massif des Ecrins* (45°00′S, 6°30′E; Fig. 1a, b). The valley is about 6 km long, E-W oriented and consists of broad steep granite walls (100–600 m). At the base of these rock walls, 21 debris cones were investigated on north facing and 11 on south facing slopes. Eleven of the northern cones are dominated by glaciers whereas only one cone (Rama cone) on the southern slope is affected by a glacier. Most cones were probably mainly established by rockfalls and snow avalanches based on morphometric (apex and distal slope angles) and sedimentologic characteristics (class size; Jomelli and Francou 2000). However, debris-avalanche deposits with lobes and levees were observed on each of these debris-cone surfaces. The selected debris cones are relatively large with a mean surface area of about 2.5 ha (maximum 5.5 ha). The average slope angle of the cones is 25°, which is relatively steep and supports the polygenic hypothesis.

All debris avalanche deposits are located between 1,720 m asl and 2,450 m asl. The 0°C annual isotherm is close to 2,400 m on northern slopes and close to 2,600 m on southern slopes. The study area is dominated both by the northern alpine climate and (sub) Mediterranean influences from the south. In summer, heavy rainstorms can occur locally, and are responsible for debris-avalanche activity on the cones, e.g. the event that occurred on July 27, 2003 on the Rama debris cone. At the meteorologic station of St Christophe (1,570 m asl), the annual average temperature for the period 1961–2000 was 5.9°C and annual average precipitation 985.7 mm. A recent analysis of climate trends in this region revealed a significant increase in precipitation exceeding 20 mm/day in summer for the same period as well as a significant decrease in the number of freezing days (Jomelli et al. 2007c).

4 Method

The inventory and cartography of debris avalanches in the study area were carried out by analyzing aerial photographs taken in 1952, 1960, 1967, 1971, 1974, 1981, 1986, 1989, 1994, 1996, 1999, and 2000 and field observations performed at a yearly basis between 1995 and 2003. The scale of the aerial photographs was between 1:15,000 and 1:30,000. Debris avalanches were dated by combining three approaches. Aerial photographs were combined with historical documents combining engraving, terrestrial photographs, scientific reports (*Office National des Forêts*, *Restauration des Terrains en Montagnes*, *Archives départementales*) and lichenometry. Only well-defined debris-avalanche deposits showing clear lobes

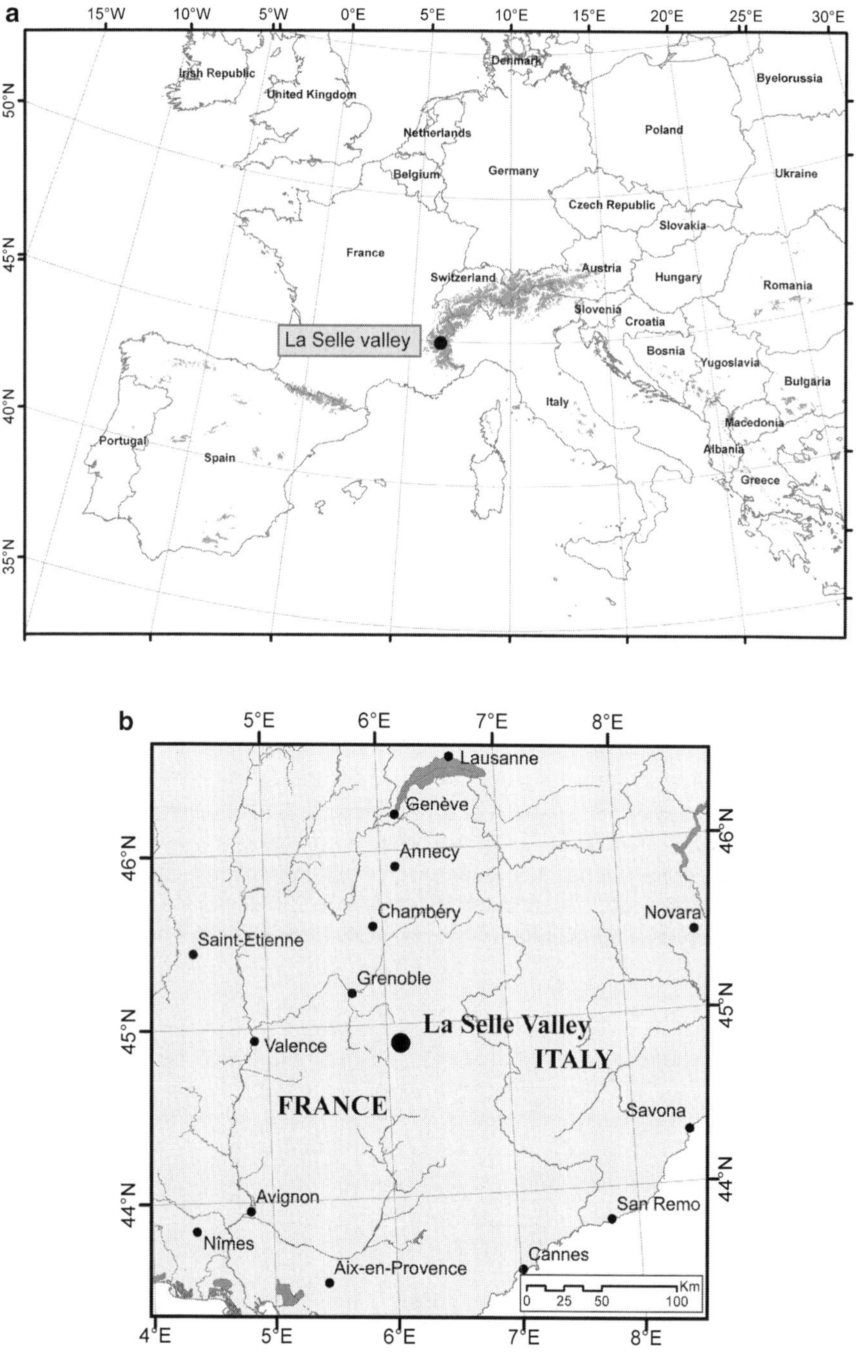

Fig. 1 (**a**, **b**) Location of the study area

and levees were used. Lichen measurements were performed as follows: 50 rock boulders with an a-axis between 20 and 60 cm were selected along the lobe and levees of each deposit. Deposits with coalescent levees were not sampled. On each block, we measured the smallest diameter of the largest of *Rhizocarpon* sp. thalli on the face exposed to the supposed trajectory of the debris avalanche. One measurement was made per block. Anomalous lichen shapes were rejected to reduce the risk of coalescence. Measurements were made with a precision of 1 mm. To make sure that lichens measured on two levees belonged to the same unit, lichen distributions from each levee were compared using non-parametric tests.

Dated surfaces described in Pech et al. (2003) and Jomelli and Pech (2004) were used to calibrate the relationship between the size of the largest lichens and their age. The oldest control point was located in a moraine dated to the beginning of the nineteenth century. Debris avalanches older than the beginning of the nineteenth century could not be dated and were disregarded. Ages of debris avalanches were obtained by applying the GEV method mentioned above and described in detail in previous papers (Naveau et al. 2007; Jomelli et al. 2007b).

Debris-avalanche frequencies were reconstructed for individual debris cones. Statistical analysis of long-term trends was based on the intervals between the occurrences of debris avalanches and minimum, maximum, mean and Weibull median. Debris-avalanche recurrence intervals were calculated for each debris cone. In the following, statistical analyses were performed for the entire time period (1815–2003) and for the sub-periods 1815–1850, 1851–1900, 1901–1950, and 1951–2003. Finally, a Student's t-test was used to check for significant changes in mean debris-avalanche recurrence intervals between the periods investigated.

5 Results

The analysis of variations in the frequency of debris avalanches was based on 118 events that occurred since 1850 on both sides of the La Selle valley (Fig. 1). Figure 2 shows clear spatial variations with about 75% of the debris avalanches occurring on northern slopes. Moreover the northern slope is currently much more active than the southern slope: In fact, 87% of the debris avalanches triggered since 1951 took place on northern slope. On both sides of the valley, the highest frequencies were observed on debris cones below rock faces with glaciers. For instance, 15 debris avalanches were triggered below the Burlan Glacier. In contrast, a maximum of four events was observed on debris cones without glaciers. The influence of glaciers on the frequency of debris avalanches has already been observed in other parts of the *Massif des Ecrins* (Jomelli et al. 2007c).

Figure 3a, b show the frequency of debris avalanches for individual cones on both sides of the valley. On most debris cones, more debris avalanches were triggered during the twentieth century than in the nineteenth century. However this trend should be considered with caution because the inventory of events cannot be considered as complete. Indeed, in a system with a high degree of activity,

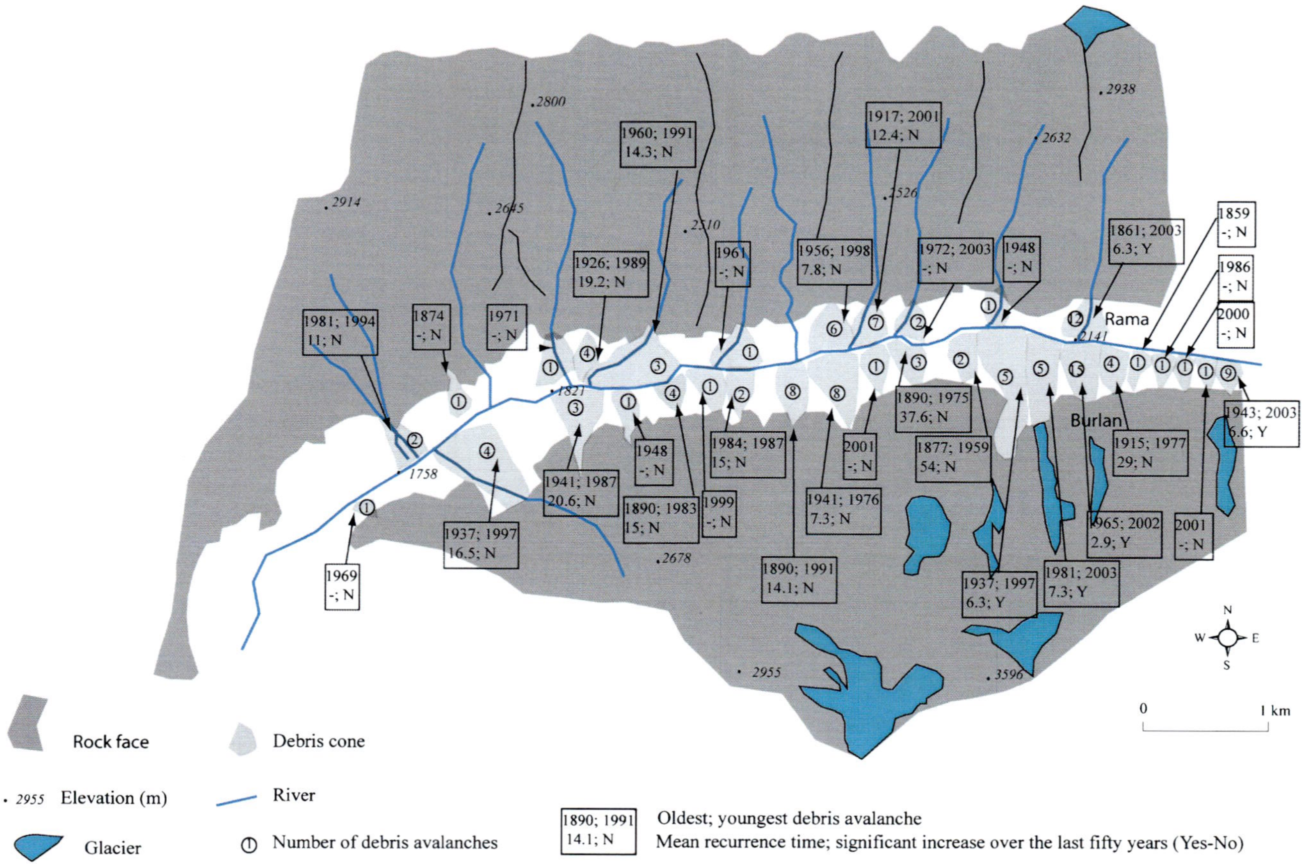

Fig. 2 La Selle valley and the 31 debris cones

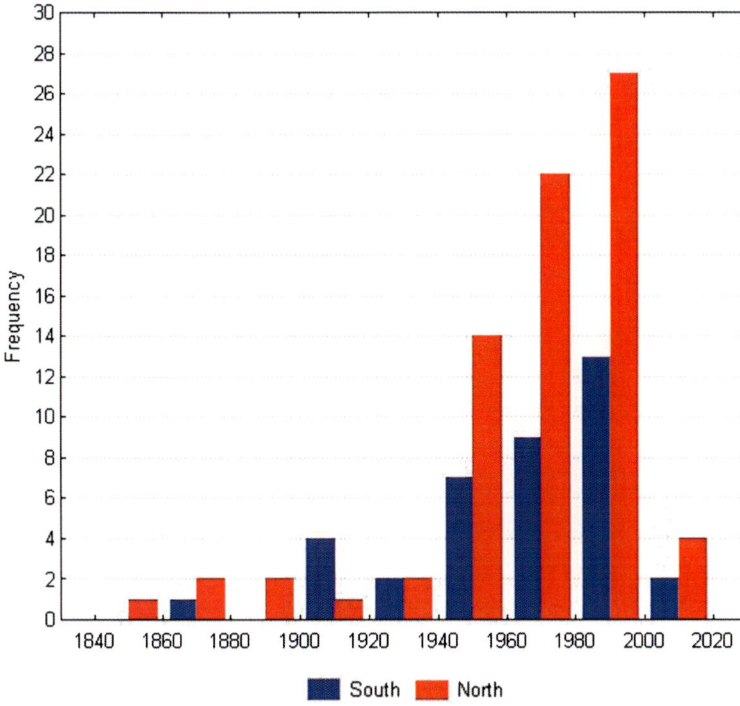

Fig. 3 Frequency of debris avalanches on the south and north sides of the La Selle valley

younger deposits can easily cover or sometimes erode older flows. The overall decadal frequency of debris avalanches in the La Selle valley is shown in Fig. 4. On average, two and three events in 10 years on the right and left side of the valley respectively, were identified during the time period considered. A slight trend towards more debris avalanches was observed for the most recent 30-year period. Differences in debris-avalanche activity are clearly visible and we identify decades with a significantly larger number of occurrences between 1980 and 1999 (i.e., 41 events on 17 debris cones). In contrast, low activity was observed before the 1960s and especially before the 1940s.

The minimum interval between two debris avalanche events was generally between 1 or 2 years. In the valley-wide chronology, debris avalanches represent a very common geomorphic process with a total of 144 intervals between 1815 and 2003 and a mean recurrence interval of 1.94, meaning that more than one event occurred in one of the valley torrents every second year. However, recurrence intervals varied considerably between the debris cones with a standard deviation for the whole valley of 15.9. In general, recurrence intervals were shorter on cones located below glaciers. The Burlan debris cone exhibited the greatest activity with a mean recurrence interval of 2.93 years for the period 1815–2009. We used a Student's t-test to check if changes in the temporal occurrence of debris avalanches

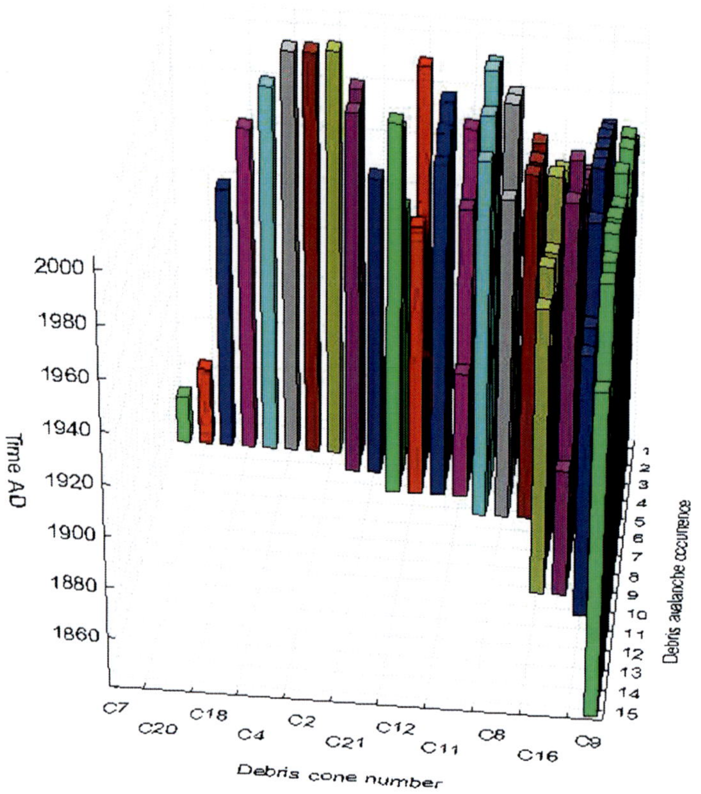

Fig. 4 Debris avalanche occurrence in the La Selle valley

were significant at a level of 0.125. In the case of the Rama debris cone, we did not observe any significant changes in the occurrence of events between the four time periods investigated. In contrast, some cones underwent a significant increase in the number of reconstructed debris-avalanche events in the last 50 years compared to previous periods (Fig. 2). A similar trend was also apparent when the nineteenth was compared with the twentieth century. However, on most debris cones, the number of events was not large enough to reveal a significant trend. The general increase in the number of events observed in the valley reconstruction may result from sample skew and should be considered with caution.

6 Discussion and Conclusion

Lichen measurements were used to assess the frequencies and changes in the temporal occurrence of events on 32 debris cones of the La Selle valley (French Alps). We were able to reconstruct a total of 118 events for the period AD

1815–2003. It should be noted that the frequencies of debris avalanches reported here represent minimum frequencies. Although lichenometry is a cost-effective technique, easy to use and provides reliable results, only a limited number of projects have been realized so far. One major concern was probably related to the resolution of the dating technique based on lichens, which is not annual, and the lack of robust confidence intervals associated with ages. Such difficulties can be compensated for by analyzing longer periods of observations (several decades for instance) and applying recent improvements in lichenometry (Jomelli et al. 2007b, 2010).

The analysis conducted in La Selle valley revealed that the overall frequency of debris avalanches has increased since AD 1815 in general and during the last 50 years in particular. This increase in debris-avalanche activity is, however, to a certain degree biased due to the fact that in a system with a high degree of activity, younger deposits can easily cover or sometimes erode older flows. However, we believe that the higher debris-avalanche activity in the last 50 years on some debris cones may not only be the result of sample skew but representing a phenomenon observed in other valleys with sufficiently large data (Jomelli et al. 2003) as well. In La Selle Valley, such a trend depends largely on the geomorphic context, which is mainly characterized by large rock walls covered by glaciers. Analyses conducted in the same region but in other valleys without glaciers revealed an opposite trend (Jomelli et al. 2003, 2004, 2007c) characterized by a decrease in the number of debris avalanches especially in low-altitude valleys (<2,400 m asl) in the twentieth century. A decrease in debris avalanches (without glaciers) in the twentieth century has also been observed in other parts of the French Alps (Van Steijn 1991; Helsen et al. 2002) and in the Swiss Alps (Bollschweiler and Stoffel 2010), demonstrating that climatic changes will not necessarily and unequivocally lead to an increase in event frequency but that the occurrence of future events would rather depend on the local geomorphic context (Jomelli et al. 2007c).

The enhanced activity in La Selle valley is also probably due to climatic conditions and may be explained by a significant increase in extreme summer precipitation events (Jomelli et al. 2007c) as well as an increase in summer temperatures causing regional glacial recession (Vincent et al. 2005). Indeed, it is well known that two conditions are needed for a debris avalanche to be triggered (Caine 1980; Van Steijn 1999; Pech and Jomelli 2001): long duration or high-intensity rainfalls and a large volume of rock debris. In the case of La Selle valley, the retreat of glaciers would produce morainic accumulations that could easily be transported by heavy rainfall. To improve our understanding of relationships between debris avalanches and climate conditions in this region, an analysis of a larger area of the French Alps is currently underway.

Acknowledgements The authors gratefully acknowledge P. Pech for his help in the field. This study was supported by the ANR SCAMPEI and the EU-FP7 ACQWA, and Circle Arnica projects.

References

André MFA (1995) Holocene climate fluctuations and geomorphic impact of extreme events in Svalbard. Geogr Ann 77:241–250

Benedict JB (1967) Recent glacial history of an alpine area in the Colorado Front Range, USA. I. Establishing a lichen growth curve. J Glaciol 6:817–832

Beschel RE (1950) Flechten als Altersmasstab rezenter Moränen. Zeitschrift fur Gletscherkunde und Glazialgeologie 1:152–161

Beschel RE (1961) Dating of rock surfaces by lichen growth and its application to glaciology and physiography (lichenometry). In: Raasch GO (ed) Geology of the Arctic II. 1st International Proceedings, Arctic Geology, 1960. University of Toronto Press, Toronto, pp 1044–1062

Blair TC (1999) Sedimentology of the debris-flow dominated Warm Spring Canyon alluvial fan, Death Valley, California. Sedimentology 46:941–957

Blijenberg HM (1998) Rolling stones? Triggering and frequency of hillslope debris flows in the Bachelard Valley, southern French Alps. PhD thesis, University of Utrecht, Utrecht, 257 pp

Blikra LH, Nemec W (1998) Postglacial colluvium in western Norway: depositional processes, facies and palaeoclimatic record. Sedimentology 45:909–959

Bollschweiler M, Stoffel M (2010) Changes and trends in debris-flow frequency since AD 1850 – results from the Swiss Alps. The Holocene 20:907–916

Brazier V, Whittington GW, Ballantyne CK (1988) Holocene debris cone evolution in Glen Etive, Western Grampian Highlands, Scotland. Earth Surf Process Landform 13:525–531

Caine N (1980) The rainfall intensity-duration control of shallow landslides and debris flows. Geogr Ann 62:23–27

Cooley D, Naveau P, Jomelli V, Rabatel A, Grancher D (2006) A bayesian hierarchical extreme value model for lichenometry. Environmetrics 17:555–574

Decaulne A, Saemundson O, Petursson O (2005) Debris flow triggered by rapid snowmelt: a case study in the Gleiarhjalli area Northwestern Iceland. Geogr Ann 87:487–500

Haworth L, Calkin PE, Ellis J (1986) Direct measurement of lichen growth in the central Brooks range, Alaska, USA, and its application to lichenometric dating. Arct Alp Res 18:289–296

Helsen MM, Koop PJM, Van Steijn H (2002) Magnitude-frequency relationship for debris flows on the fan of the Chalance torrent, Valgaudemar (French Alps). Earth Surf Process Landform 27:1299–1307

Innes JL (1982) Lichenometric dating of debris-flow deposits in the Scottish highlands. Earth Surf Process Landform 8:579–588

Innes JL (1984) The optimal sample size in lichenometric studies. Arct Alp Res 16:233–244

Innes JL (1985a) Lichenometric dating of debris-flow deposits on Alpine colluvial fans in Southern Norway. Earth Surf Process Landform 10:519–524

Innes JL (1985b) Lichenometry. Prog Phys Geogr 9:187–254

Jomelli V, Francou B (2000) Comparing characteristics of rockfall talus and snow avalanche landforms in an alpine environment using a new methodological approach. Geomorphology 35:181–192

Jomelli V, Pech P (2004) Little Ice Age impacts on avalanche boulder tongues in the French Alps (massif des Ecrins). Earth Surf Process Landform 29:553–564

Jomelli V, Bertran P, Kunesch S (2002) Le cône de la Momie: un nouveau type de dépôt proglaciaire mis en place depuis la fin du Petit Age Glaciaire. Quaternaire 13:257–263

Jomelli V, Chochillon C, Brunstein D, Pech P (2003) Hillslope debris flows frequency since the beginning of the 20th century in the French Alps. In: Rickenmann D, Chen Ch (eds) Debris flow hazards mitigation. Millpress, Rotterdam, pp 127–137

Jomelli V, Pech P, Chochillon C, Brunstein D (2004) Geomorphic variations of debris flows and recent climatic change in the French Alps. Clim Chang 64:77–102

Jomelli V, Déqué M, Brunstein D, Grancher D (2007a) Probabilités d'occurrence des coulées de débris de versant dans les Alpes françaises au 21em siècles estimées à partir des sorties du modèle ARPEGE. Géomorphologie 93:2001–2212

Jomelli V, Grancher D, Naveau P, Cooley D (2007b) Assessment study of lichenometric methods for dating surface. Geomorphology 86:131–143
Jomelli V, Brunstein D, Grancher D, Pech P (2007c) Is the response of hill slope debris flows to recent climate change univocal? A case study in the Massif des Ecrins (French Alps). Clim Chang 85:119–137
Jomelli V, Déqué M, Brunstein D, Grancher D (2009) Impacts of future climate change (2070–2100) on debris flow occurrence: a case study in the Massif des Ecrins (French Alps). Clim Chang 97:171–191
Jomelli V, Naveau P, Cooley D, Grancher D, Brunstein D, Rabatel A (2010) A response to Bradwell's commentary on recent statistical studies in lichenometry. Geogr Ann 92(4):485–487
Jomelli V, Khodri M, Favier V, Brunstein D, Ledru MP, Wagnon P, Blard PH, Sicart JE, Braucher R, Grancher D, Bourlès D, Braconnot P, Vuille M (2011) Irregular tropical glacier retreat over the Holocene epoch driven by progressive warming. Nature 474(7350):196–199
Jonasson C et al (1993) Holocene debris-flow activity in northern Sweden. In: Frenzel B, Matthews JA, Gläser B (eds) Solifluction and climate variations in the Holocene, vol 11, Paläoklimaforschung. Gustav Fischer Verlag, Stuttgart, pp 179–195
Karlen W, Black J (2002) Estimates of lichen growth-rate in northern Sweden. Geogr Ann 84:225–232
Locke WW, Andrews JT, Webber PJ (1979) A manual for lichenometry. Br Geomorphol Res Group Tech Bull 26:1–45
Matthews JA, Dahl SO, Berrisford MS, Nesje A, Dresser PQ, Dumayne-Peaty L (1997) A preliminary history of Holocene colluvial (debris flow) activity, Leirdalen, Jotunheimen, Norway. J Q Sci 12:117–129
Matthews JA, Dahl SO, Dresser PQ, Berrisford MS, Lie O, Nesje A, Owen G (2009) Radiocarbon chronology of Holocene colluvial (debris-flow) events at Sletthamn, Jotunheimen, southern Norway: a window on the changing frequency of extreme climatic events and their landscape impact. The Holocene 19:1107–1129
McCarroll D (1993) Modelling late-Holocene snow-avalanche activity; incorporating a new approach to lichenometry. Earth Surf Process Landform 18:527–539
Menounos B (2000) A Holocene debris flow chronology for an Alpine Catchement, Colorado Front Range. In: Slaymaker A (ed) Geomorphology, human activity and global environmental change. Wiley, New York, pp 118–149
Naveau P, Jomelli V, Cooley D, Grancher D, Rabatel A (2007) Modeling uncertainties in lichenometry studies with an application: the tropical Andes (Charquini Glacier in Bolivia). Arct Antarct Alp Res 39:277–288
Pech P, Jomelli V (2001) Rôle du cône apical dans le déclenchement des coulées de débris alpines du Massif du Dévoluy, Hautes-Alpes, France. Geographie Physique et Quaternaire 1:47–61
Pech P, Jomelli V, Baumgart-Kotarba M, Bravard J-P, Chardon M, Jacob N, Kedzia S, Kotarba A, Raczkowska Z, Tsao C (2003) A lichenometric growth curve in the French Alps: Ailefroide and Veneon valleys; Massifs des Ecrins. Geodinamica Acta 16:187–193
Reimer PJ, Baillie MGL, Bard E, Bayliss A, Beck JW, Blackwell PG, Bronk Ramsey C, Buck CE, Burr GS, Edward RL, Friedrich M, Grootes PM, Guilderson TP, Hajdas I, Heaton TJ, Hogg AG, Hughen KA, Kaiser KF, Kromer B, McCormac FG, Manning SW, Reimer RW, Richards DA, Southon JR, Talamo S, Turney CSM, van der Plicht J, Weyhenmeyer CE (2009) Intcal09 and Marine09 radiocarbon age calibration curves, 0–50,000 years cal BP. Radiocarbon 4:1111–1150
Rickenmann D (1999) Empirical relationships for debris flows. Nat Hazard 19:47–77
Solomina O, Savoskul O, Cherkinsky AE (1994) Glacier variations, mudflow activity and landscape development in the Aksan Valley (Tian Shan) during the late Holocene. The Holocene 4:25–31
Stoffel M, Lievre I, Conus D, Grichting MA, Raetzo H, Gartner HW, Monbaron M (2005) 400 years of debris-flow activity and triggering weather conditions: Ritigraben, Valais, Switzerland. Arct Antarct Alp Res 37:387–395

Stoffel M, Conus D, Grichting MA, Lievre I, Maitre G (2008) Unraveling the patterns of late Holocene debris-flow activity on a cone in the Swiss Alps: chronology, environment and implications for the future. Glob Planet Change 60:222–234

Strunk H (1991) Frequency distribution of debris flows in the Alps since the Little Ice Age. Zeitschrift für Geomorphologie 83:71–81

Strunk H (1992) Reconstructing debris flow frequency in the southern Alps back AD 1500 using dendrogeomorphological analysis. In: IAHS (ed) Erosion, Debris flow and Environment in Mountain Regions, Proceedings of the Chengdu Symposium, IAHS Publication, 209, Chengdu, pp 299–306

Van Steijn H (1991) Frequency of hillslope debris flows in part of the French Alps. Bull Geomorphol 19:83–90

Van Steijn H (1999) Frequency of hillslope debris flows in the Bachelard valley (French Alps). In: Panizza M, Soldati M, Bertacchini M, Van Asch W, Malmusi S (eds) Programme in geomorphology. Arnold, London, pp 15–24

Vincent C, Le Meur E, Six D (2005) Solving the paradox of the end of the Little Ice Age in the Alps. Geophys Res Lett 32:L09706. doi:10.1029/2005GL022552

Wilkerson FD, Schmid GL (2003) Debris flows in Glacier National Park, Montana: geomorphology and hazards. Geomorphology 55:317–328

Winchester V (1984) A proposal for new approach to lichenometry. Br Geomorphol Res Group Tech Bull 33:3–20

Winchester V (1988) An assessment of lichenometry as a method for dating recent stone movements in two stone circles in Cumbria and Oxfordshire. Bot J Linn Soc 96:57–68

Winchester V, Chaujar RK (2002) Lichenometric dating of slope movements, Nant Francon, North Wales. Geomorphology 47:61–74

Winchester V, Harrison S (1994) A development of the lichenometric method applied to the dating of glacially influenced debris flows in Southern Chile. Earth Surf Process Landform 19:137–151

Winchester V, Harrison S (2000) Dendrochronology and lichenometry: an investigation into colonization, growth rates and dating on the east side of the North Patagonian Icefield, Chile. Geomorphology 34:181–194

Lake Sediments as Archives of Recurrence Rates and Intensities of Past Flood Events

Adrian Gilli, Flavio S. Anselmetti, Lukas Glur, and Stefanie B. Wirth

1 Introduction

Palaeoflood hydrology is an expanding field as the damage potential of floods and flood-related processes is increasing with the population density and the value of the infrastructure. Assessing the risk of these hazards in mountainous terrain requires knowledge about the frequency and severness of such events in the past. A wide range of methods is employed using diverse biologic, geomorphic or geologic evidences to track past flood events. Impact of floods are studied and dated on alluvial fans and cones using for example the growth disturbance of trees (Stoffel and Bollschweiler 2008; Schneuwly-Bollschweiler and Stoffel 2012: this volume) or stratigraphic layers deposited by debris flows, allowing to reconstruct past flood frequencies (Bardou et al. 2003). Further downstream, the classical approach of palaeoflood hydrology (Kochel and Baker 1982) utilizes geomorphic indicators such as overbank sediments, silt lines and erosion features of floods along a river (e.g. Benito and Thorndycraft 2005). Fine-grained sediment settles out of the river suspension in eddies or backwater areas, where the flow velocity of the river is reduced. Records of these deposits at different elevations across a river's

A. Gilli (✉) • S.B. Wirth
Geological Institute, ETH Zurich, CH-8092 Zurich, Switzerland
e-mail: adrian.gilli@erdw.ethz.ch

F.S. Anselmetti
Eawag, Department of Surface Waters, CH-8600 Dübendorf, Switzerland

Institute of Geological Sciences, University of Bern, Baltzerstrasse 1+3, CH-3012 Bern, Switzerland
e-mail: flavio.anselmetti@geo.unibe.ch

L. Glur
Department of Surface Waters, Eawag, CH-8600 Dübendorf, Switzerland
e-mail: lukas.glur@eawag.ch

M. Schneuwly-Bollschweiler et al. (eds.), *Dating Torrential Processes on Fans and Cones*, Advances in Global Change Research 47, DOI 10.1007/978-94-007-4336-6_15,

profile can be used to assess the discharge of the past floods. This approach of palaeoflood hydrology studies was successfully applied in several river catchments (e.g. Ely et al. 1993; Macklin and Lewin 2003; O'Connor et al. 1994; Sheffer et al. 2003; Thorndycraft et al. 2005; Thorndycraft and Benito 2006). All these different reconstruction methods have their own advantages and disadvantages, but often these studies have a limited time coverage and the records are potentially incomplete due to lateral limits of depositional areas and due to the erosional power of fluvial processes that remove previously deposited flood witnesses. Here, we present a method that follows the sediment particle transported by a flood event to its final sink: the lacustrine basin.

Lakes offer a great possibility to record flood events as their suspension load is deposited on the lake floor as a distinct detrital layer. This detrital layer contrasts in the ideal case strongly with the 'normal' background sedimentation, which mainly consists of authigenically-produced minerals, organic remains and clastic particles. Detrital flood layers are detected visually or by a suite of different sedimentological and geochemical methods. Flood events that have been documented either by historical sources or by instrumentation were detected in lacustrine sediment cores (Bussmann and Anselmetti 2010; Chapron et al. 2002; Gilli et al. 2003; Lamoureux 2000; Siegenthaler and Sturm 1991) confirming the concept of using lake sediments as flood archive. Major advantage of the lacustrine archive is the continuous mode of the background sedimentation in the deep lake areas, which offers the potential for the reconstruction of a temporally complete flood record. The resolution of a lake-based flood reconstruction is highly dependent on the overall sedimentation rate, but in the case of annually laminated sediments the flood record could be resolved down to the season. Past flood events can be scaled by thickness and grain-size distribution of the detrital layer as they reflect flood dynamics.

Although modern limnogeology started to understand the involved processes between high river-runoff and the deposition of detrital layers in lakes through underflows in the 1970s (e.g. Lambert and Hsü 1979; Sturm and Matter 1978), the limnogeologic community did not immediately realize the value of the lake sediments for the reconstruction of flood history. For Switzerland, the severe damage of several flood events in the Swiss Alps and its foreland in summer 1987 spurred different disciplines to reconstruct flood frequencies and intensities. Siegenthaler and Sturm (1991) systematically sampled the sedimentary archive of Lake Uri, which drained an area of severe flood damage. They documented in detail the spatial characteristics of the 1987 flood layer in Lake Uri and were able to identify further strong flood events that occurred in the last 700 years. This intensified flood-risk evaluation after the 1987-flood event was amplified by the fact that Switzerland experienced only few incidences of natural disasters between 1882 and 1976 (Pfister 2009) and, therefore, was not aware of the natural variability of such devastating events. Especially in Europe and North America, the number of flood reconstructions using lake sediments (e.g. Arnaud et al. 2005; Bussmann and Anselmetti 2010; Chapron et al. 2002; Dapples et al. 2002; Eden and Page 1998; Gilli et al. 2003; Mazzucchi et al. 2003; Nesje et al. 2001; Noren et al. 2002; Osleger et al. 2009; Parris et al. 2010) increased exponentially in the last 15 years underlining the great potential of this method.

2 Depositional Processes in Lakes

Extreme flood events mobilize and entrain large amounts of sediments that are consequently fed into the river drainage. The amount and the grain size of the sediment particles entrained by a flow are strongly dependent on the hydraulic energy. The coarser fraction of these mobilized particles may remain in the catchment, forming alluvial fans and cones. The finer particles, which are kept in suspension, are transported from the catchment area to the next downstream lake, which is the final sink for nearly the entire riverine sediment load (Fig. 1).

When a sediment-laden river enters a lake, the sediment plume is redistributed depending on the density stratification of the lake and the density of the entering water, which is a function of temperature, suspension load and ion content (salinity). The density contrast between sediment-laden river water and the ambient lake water determines if the sediment plume forms an over-, inter- or underflow in the lake

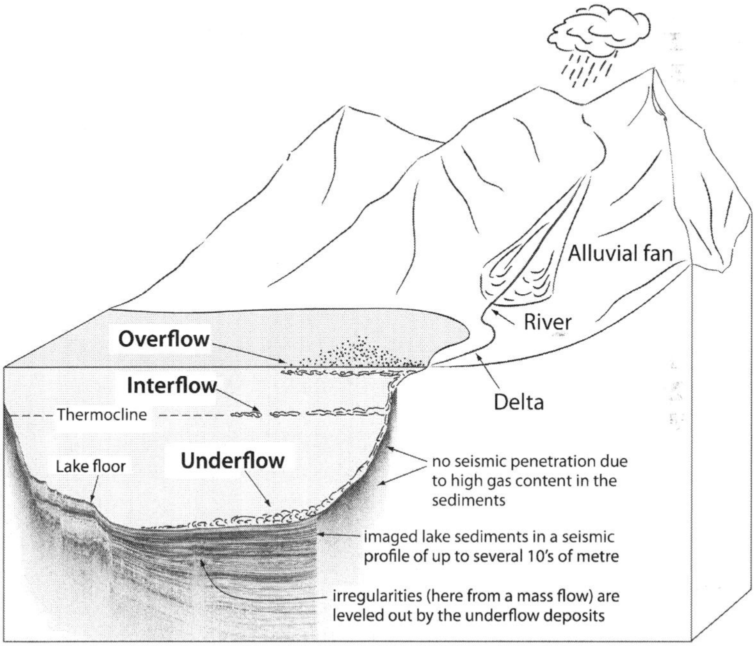

Fig. 1 Schematic drawing of a lake and its catchment area illustrating in particular the different distribution paths of a river-borne sediment plume in a lake as *over-*, *inter-*, or *underflow* (Giovanoli 1990; Sturm and Matter 1978). Coarser sediment particles mobilized through a flood are mainly deposited on alluvial fans or in the delta, whereas the finer particles are entering the lake as sediment plume. High-density flows, generated by heavy rainfall events, form underflows depositing their particles as specific turbidite layers, which focus in the deepest lake area. Lakes dominated by underflows have a flat and near horizontal lake bottom. Seismic investigations image the subsurface providing a seismic stratigraphy that allows prediction of sediment type and selection of ideal coring sites

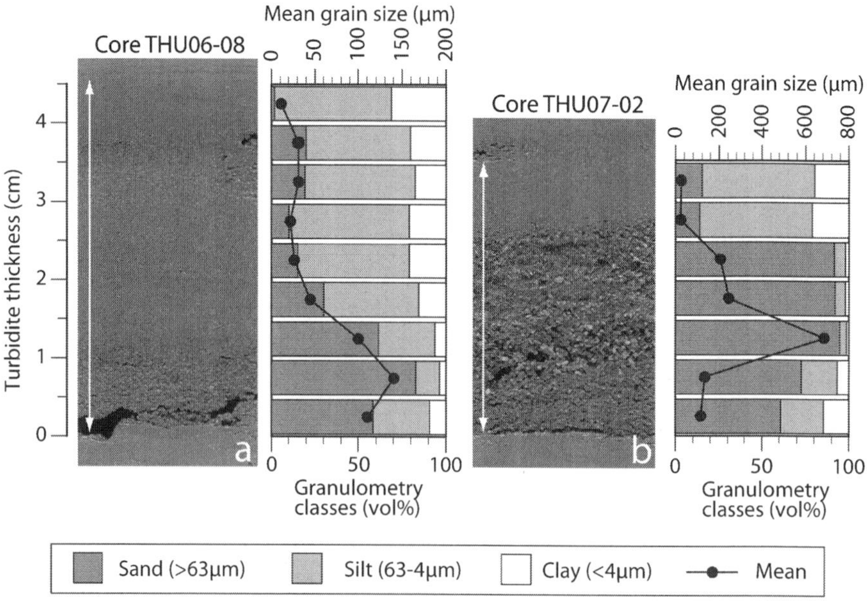

Fig. 2 Grain-size measurements across two flood layers in Lake Thun, Switzerland (Wirth 2008; Wirth et al. 2011). *Vertical white double-arrow* indicates extent of flood layer. (**a**) This turbidite shows an upcore increase in grain size between 3 and 4 cm indicating, in addition to the coarse base, a second peak in river runoff. (**b**) Turbidite layer with clearly inversed grading at the base as a result of the waxing river runoff

(Giovanoli 1990). Regular runoff waters are usually less dense than the lake water and thus rather form over- and interflows, from where the suspended fine particles slowly settle out and eventually become deposited on the lake floor. Extreme events, on the other hand, are so suspension-laden that the high density of the flow allows the water to break through the lake stratification to proceed as turbiditic underflow (hyperpycnal flow) to the deepest area of the lake (Fig. 1). Once the deepest lake area is reached, the underflow spreads out and flow velocities decrease. Consequently, the sediment particles are deposited forming characteristic layers termed 'turbidites'. The depositional characteristics of a turbidite are dependent on temporal runoff evolution (i.e. the shape of the hydrograph) of a flood event as the grain size is a direct function of the current velocity (Mulder et al. 2001, 2003). Turbidites comprise a fining upward sequence representing the waning of the discharge (Fig. 2). Sometimes, turbidites also record the increasing underflow velocity during the initiation of a flood by a coarsening upward basal unit (Fig. 2b). This concept of directly linking the rising and falling river discharge with the reverse and subsequently normal graded turbidite deposition was recently questioned for the marine realm (Lamb and Mohrig 2009), although for the smaller lacustrine setting this concept probably still is valid.

2.1 Selecting Lakes with Promising Flood Records

The selection of appropriate lakes is crucial for this approach of flood reconstruction. The understanding of the transport and depositional processes of flood-transported particles allows us to set criteria to select lakes holding promising flood records. The following two criteria are highly relevant:

1. *Well-defined morphobathymetric depocentre*
   Underflows transport the suspension load of a flood along the delta slope towards the deepest part of the lake, where flow velocity decreases and larger particles start to settle down on the lake floor. Lakes with underflow-dominated sedimentation are characterized morphologically by a flat and horizontal basin floor, where the flood turbidites accumulate and consequently level out inherited topography (Fig. 1). This flat floor is the prime target area to recover cores for flood reconstruction studies. A detailed knowledge about the lake's bathymetry gained from map studies, detailed depth soundings or reflection seismic surveying is therefore a prerequisite for a successful flood reconstruction. Although in most of the cases, the deepest lake part is the depocentre for the underflows, possible sills between sub-basins may prevent the underflows reaching the lake's centre. This needs to be taken into account with complex lake basin morphologies. Furthermore, reflection seismic information from the subsurface is needed to show the stratigraphic architecture of the sedimentary basin fill (see Fig. 1). Potential unconformities related to large mass movements, bottom currents, or gas escape features may seriously hamper the continuous nature of the sediment succession. Only a detailed seismic stratigraphy allows to select ideal coring locations, define sediment thickness and predict sediment types.
2. *Geomorphic indications around the lake*
   As a consequence of the high detrital influx to a lake during flood events, the following geomorphic indications around the lake should be present.

   (a) A certain relief in the catchment area is necessary in order to erode material that can be transported into the lake.
   (b) The lake should have evident inflows, however, they should ideally only be activated in the case of an extreme event. That provides a certain threshold with an 'on/off signal' so that only the larger 'extreme' events are recorded in the lacustrine archive.
   (c) A delta structure should be present at the major inflows of a lake assuring a persistent detrital influx in the past.

The following criterion deals with the recognition of the flood layers in the sedimentary record. This can often only be evaluated after preliminary investigations using short sediment cores.

3. *Contrast between flood deposits and regular background sedimentation*
   The lithology of the flood deposits should contrast to the regular background sediments, so that these events can be lithologically and geochemically recognized. This implies that the regular background sedimentation should not

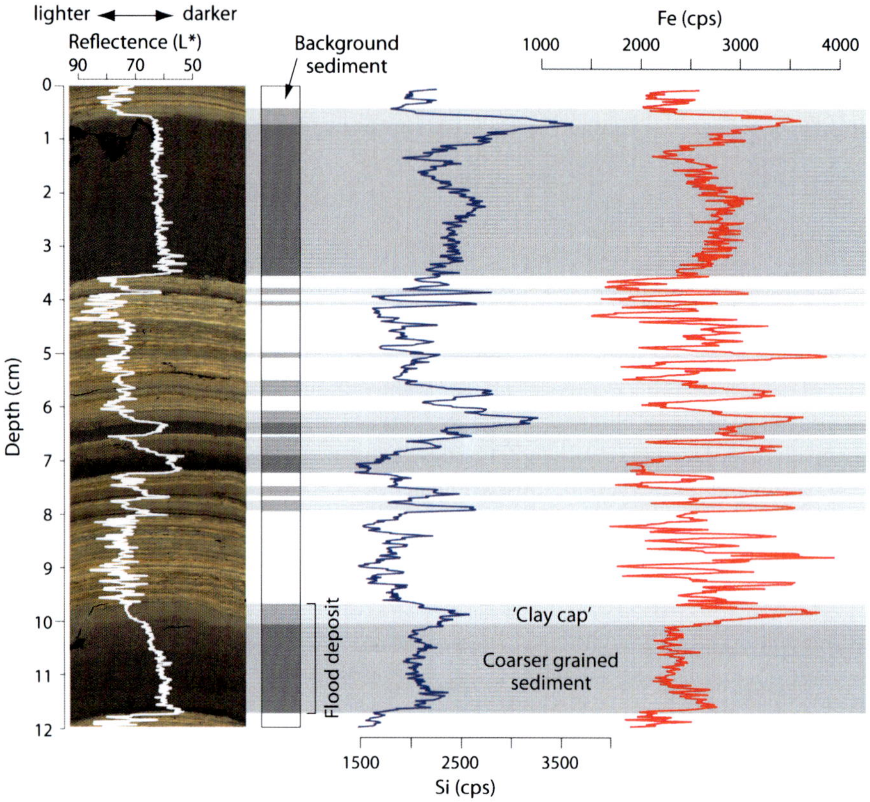

Fig. 3 Flood layers in a core section (core LL081-A3a) of Lake Ledro, Italy, with relative concentration of silicium (Si) and iron (Fe) measured continuously by an XRF core scanner (measuring interval: 0.2 mm). *Dark-coloured* flood layers stand out prominently in the finely-laminated calcareous background sediment (note the lightness difference reported in reflectance L^*). Flood layers (*grey bars*) are characterized by increased Si and Fe content, especially pronounced in the fine-grained *top part* of the flood layer ('clay cap') marked in *light grey bars*

consist solely of detrital material. If organic materials dominate the background sediments, the detrital layers are in general lighter than the otherwise very dark background sediments (e.g. Lake Schwendi, Gilli et al. 2003). But if the debris background sediments manly consist of authigenic carbonates, the detrital layers are considerably darker than the normal sediment (e.g. Lake Ledro, see Fig. 3). In either setting, detrital layers stand out prominently when compared to the background sediments and can easily be identified.

Beside the described criteria for selecting a lake system to record recognizable flood layers, more technical aspects to recover the expected flood record are as well of importance for a successful study.

4. *Coring ability to recover desired time interval*
   The sedimentation rate of a lacustrine archive can vary strongly between the different lake systems. In particular, the contribution of the general detrital influx to the total sedimentation can make a large difference to total sediment thickness. Standard coring techniques with various piston-coring methods are capable of reaching ~20 m sub lake-floor depth, so that sediment thickness for the desired time interval should not exceed this range. Deeper core recovery can otherwise only be reached by complex and expensive commercial drilling operations. If a complete Holocene section is the target, rather small lakes with little inflow are ideal (Gilli et al. 2003; Theiler 2003). Larger perialpine lakes (e.g. Lake Uri, Lake Brienz, both Switzerland) have substantially larger sedimentation rates reaching values of up to 10 m per 1,000 years (Girardclos et al. 2007; Siegenthaler and Sturm 1991).

3 The Signature of Flood Events in the Lacustrine Archive

A key step in palaeoflood studies using lake sediments is the secure identification of detrital layers as the result of a flood event. As the sedimentation of each lake system is influenced by a large variety of different factors, we can only discuss general rules here, which clearly need to be adapted to each individual lacustrine system (Mulder and Chapron 2011). Ideally, the detrital layers should sharply contrast in terms of sedimentology and geochemical composition with the regular background sediment (Sletten et al. 2003). In these cases, the detrital layer can be often identified by visual observation even at the millimetre scale, a process, which can be supported by image analysis (e.g. Rodbell et al. 1999) (Fig. 3).

The sedimentology of the flood layers is characterized by a near pure terrestrial mineralogical composition and consequently a low organic content. The underflow-related depositional processes (see part "Depositional processes in lakes") cause a fining-upward grain-size pattern with sometimes an inverse graded base from temporal waxing of the flood-related runoff. The top part of the detrital layer consists of the finest particles and has often a distinctive light colour (Fig. 3), commonly referred as 'clay cap'. Sometimes, several stacked graded intervals within the same flood layer point towards multiple water discharge peaks (Fig. 2). In rather rare cases, the base of flood layers consists of aquatic macrophytes (Gilli et al. 2003), which were entrained by the underflow. On the basis of the characteristics of these detrital layers, previous studies applied, next to visual observations, one or a combination of several proxies to identify such layers. The three most common applied proxies are (1) grain size, (2) total organic carbon (often expressed as loss of ignition, i.e. LOI), and (3) magnetic susceptibility. These proxies are interpreted as follows:

1. *Grain size*: Grain-size measurements are almost routinely used to confirm a fining-upward gradation overlying a coarser base (Fig. 2). Recently, Parris

et al. (2010) proposed end-member modelling of the particle-size distribution and demonstrated that the coarsest end-member is a sensitive indicator for the base of detrital flood layers.
2. *Total organic carbon* (*TOC*): Caused by the high minerogenic content of the flood-related material, the organic-carbon content is usually low in the flood layers. This proxy is successfully applicable only in rather eutrophic lakes, where the background sedimentation is high in organic material offering sufficient contrast to the detrital signature.
3. *Magnetic susceptibility*: Depending on the catchment's geology, the detrital layers are often enriched in para- and ferromagnetic minerals reflecting high magnetic susceptibility values (Chapron et al. 2005; Wolfe et al. 2006). More elaborated proxies like isothermal remanent magnetization (IRM) have also been applied to identify surface-soil-derived material in lacustrine sediments (Boe et al. 2006; Thorndycraft et al. 1998).

Beside these widely applied proxies, various additional methods are proposed to better identify detrital layers in the lacustrine archive.

- *Elemental composition*: Elements like Si, Al, Fe, K, and Ti are typical for detrital sediments and thus their concentration is elevated in the detrital layers (Revel-Rolland et al. 2005). Elemental concentration is determined either on discrete samples or by continuous non-destructive XRF core scanning techniques (Fig. 3) with resolutions down to the sub-millimetre scale.
- *Bulk density*: Especially the coarser base of detrital layers is characterized by elevated bulk densities, which can be determined by continuous gamma-ray attenuation measurements performed on a standard multi-sensor core logging device (Gilli et al. 2003).
- *Image analysis*: Detrital layers can be detected and mapped with image analysis (Fig. 3) if there is a clear colour contrast between detrital layer and background sediments.
- *Carbon/nitrogen ratios*: Organic matter of lacustrine origin reveals a distinct lower C/N ratio (i.e. 4–10) compared to vascular land plants (>20) (Meyers and Teranes 2001). Several studies showed an increase of the C/N ratio in the detrital flood layers (Brown et al. 2000; Osleger et al. 2009; Wolfe et al. 2006)
- *Stable carbon isotopes on organic material*: As with the C/N ratio, carbon isotopes can be used to differentiate between aquatic or terrestrial origin of the organic matter (Bierman et al. 1997). In general, rather higher carbon isotopic ratios ($\delta^{13}C$) are typical for terrestrial plants and found in the detrital layers (Osleger et al. 2009; Wolfe et al. 2006)
- *Pollen assemblages*: Pollen spectra of detrital layers show a clear shift towards more robust pollen types, like Liguliflorae and Filicales pollens, as they remain in the catchment and are subsequently washed-in, when more vulnerable pollen types are already corroded and decayed (Thorndycraft et al. 1998).
- *Sr and Nd isotopes*: If the lake's catchment consists of different geologic terrains, strontium and neodymium isotopes of the detrital particles stand out prominently when compared to the values from background sediment (Revel-Rolland et al. 2005).

The clear identification and attribution of a detrital layer as a past flood event can sometimes be challenging as similar layers could also be the result of mass movements (i.e. slope failures) triggered by earthquakes (Chapron et al. 1999; Schnellmann et al. 2002; Strasser et al. 2006), delta collapse (Girardclos et al. 2007) or lake level fluctuations (Anselmetti et al. 2009). But these mass-movement related turbidite layers have different structural and compositional characteristics and they are – in general – rather thicker than flood layers (Siegenthaler and Sturm 1991; Sturm et al. 1995). Mass-movement related turbidite layers show only a thin fining-upward sequence followed by a thick homogenous interval with no inverse grading at the base. With the exception of turbidites triggered by delta collapses, the mass-movement layers contain reworked lacustrine background sediment with a geochemical composition different from the catchment-derived detrital signal.

Detailed sedimentological analysis of detrital layers is further of importance to scale intensities of past flood events. Assessing past flood events intensity is always based on a constant availability of sediment during the observed period. The amount of sediment, however, that is mobilized by intense precipitation and runoff may be also dependent on anthropogenic activity, as various forms of land-use expose the soils and increase their sensitivity to become eroded. Nevertheless, a positive relationship between suspended particle concentration and river discharge (Mulder et al. 2003) enables a scaling of the flood event by the detrital layer thickness. This relationship is specific for each river basin, but follows a power-law function making the sedimentary record very sensitive to variations in the river discharge of extreme floods (Mulder et al. 2003). So far, only a few studies used this relationship to assess the flood magnitude of past events (e.g. Brown et al. 2000; Bussmann and Anselmetti 2010; Irmler et al. 2006). These studies always use the imprint of a documented historic flood event in the sedimentary archive to relatively scale prehistoric flood events. A proxy even less applied in evaluating past flood events in the sedimentary archive is grain-size distribution. Grain-size measurements are often carried out to identify detrital layers, but as the grain-size distribution also reflects hydrologic conditions, in particular transport capacity, during a flood event, it could be applied to assess the intensity of past floods.

4 Dating Methods for Lake Sediments

Once the flood layers have been identified, age dating of the sedimentary succession is required to establish a flood chronology. This can be achieved applying various dating techniques, which eventually result in a 'best-fit' age model for the recovered stratigraphic section. For this purpose, one has to consider that a sedimentary sequence consists of background sediments, characterized by low sedimentation rate, which is punctuated or 'interrupted' by individual event-layers that were deposited in short time (i.e. hours to days) thus resulting in a high sedimentation rate. Consequently, in order to interpolate between age markers, those rapidly deposited layers need to be removed from the sedimentary succession before the age model is defined. The resulting interpolation assumes a stable background sedimentation rate between age markers, which is often the best-possible approach.

The uppermost part of the sediments that were deposited in the last century can be dated measuring the gamma-activity of radionuclides, in particular ^{210}Pb and ^{137}Cs (Appleby 2001, and references therein). ^{137}Cs emissions into the environment started with the onset of atmospheric nuclear bomb testing in the mid-1950s culminating in 1963. A second maximum was produced upon the accident of Chernobyl in 1986, offering thus two marker horizons that can be detected by gamma spectroscopy. ^{210}Pb-activities, in contrast, are a product of in situ produced 'supported' as well of 'unsupported' sources from a decay chain, the latter of which allows calculation of sedimentation rates for the last $\sim$150 years assuming various sedimentation models (Appleby 2001; von Gunten et al. 2009). Furthermore, the occurrence of combustion products, such as spherical carbonaceous particles (SPE) or various other combustion-derived aerosols reflect local to regional use of burning agents and can be compared to anthropogenic and industrial history (Thevenon and Anselmetti 2007; von Gunten et al. 2009), providing additional dating possibilities for the most recent section.

Radiocarbon dating of terrestrial organic matter is probably the most common and powerful dating tool that is used with lacustrine sediments. The success of this method relies on identification of organic macro-remains of terrestrial origin, as aquatically produced biomass often incorporates 'old' carbon, which is not in equilibrium with the atmosphere (Deevey et al. 1954). This radiocarbon reservoir effect causes an offset towards too old radiocarbon ages, which for this reason prevents building a robust age model.

Further possibilities for providing age markers are the occurrences of volcanic tephra layers of known age, which can be tracked over large distances. In Swiss lakes, the Laacher See Tephra (originating from the Eifel area in Germany) and the Vasset-Killian Tephra from an eruption in the Massif Central in France (Hajdas et al. 1993) can often be identified and may refine a radiocarbon-based age model (Schnellmann et al. 2006).

Annually laminated background sediments (i.e. varves, see Zolitschka 2006 for an overview) provide probably the most accurate dating by layer counting, as annual resolution cannot be obtained by the other methods. However, as varves rarely occur throughout the entire section, varve chronology may be 'floating' and therefore be used together with other independently determined age markers providing thus a relative rather than an absolute chronology of annual resolution.

Eventually, the age model may be adjusted by making reference to prominent historic event markers as earthquakes, rock falls or large floods leave their unique fingerprint behind. Several studies (e.g. Bussmann and Anselmetti 2010; Gilli et al. 2003; Siegenthaler and Sturm 1991) linked detrital layers within the ^{137}Cs and ^{210}Pb-dated interval to documented extreme flood events producing higher accuracies of the independently dated succession. Overall, defining an age model is usually a combined application of various methods described above, which together maximize range, resolution, and accuracy of the ages that are then assigned to each individual flood layer.

5 Determining the Seasonality of Flood Events

Flood reconstructions mainly focus on the frequency and intensity of past heavy-precipitation events. This allows the identification of past time periods of increased flood activity with implications for general atmospheric circulation patterns. However, knowledge of past atmospheric processes leading to severe floods would substantially be improved if the individual flood layers could be seasonally resolved. With such high-resolution records, conclusions about – for example – a change in frequency of floods caused by summer thunderstorms could be drawn.

Varved sediments allow a precise chronology of the lacustrine record down to the season based on the alternating deposition of distinct laminae during the annual cycle (Zolitschka 2006). The position of detrital flood layers within the seasonal alternating laminae of a varve may be used to date the season they were deposited. Depending on the thickness of the varves, the position of the flood layer within a varve is visually determined either on thin sections, on high-resolution core images, or by the naked eye. To apply the concept of seasonal flood layer dating, a possible erosion of the underlying laminae needs to be considered. Thin section analysis is used to identify an erosional base of the detrital layer, but a safe seasonal assignment of the flood layer can be achieved by determining the season of the next overlying varve lamina.

Unfortunately, continuously varved lacustrine sections over longer time intervals are quite rare and if present, the lake setting is often not sensitive to flood events. For this reason only very few seasonal resolved flood records are documented in the literature (Lamoureux 2000; Mangili et al. 2005). With the above outlined large gain of knowledge about flood-triggering atmospheric processes, the search for lacustrine flood records in varved lacustrine sections and its detailed, but time-consuming analysis, is highly desirable.

6 Occurrence of Flood Events in the Past – Case Studies

The following two studies illustrate the application of lake sediments as archive for flood events.

6.1 Flood History of Lake Lauerz, Central Switzerland

Lake Lauerz is a small perialpine lake (surface area: 3 km^2, catchment area: 69 km^2) located in Central Switzerland that records heavy precipitation events with typical flood layers. Piston cores were retrieved in the deepest part of the main lake basin allowing a composite record of nearly 10 m length (Bussmann

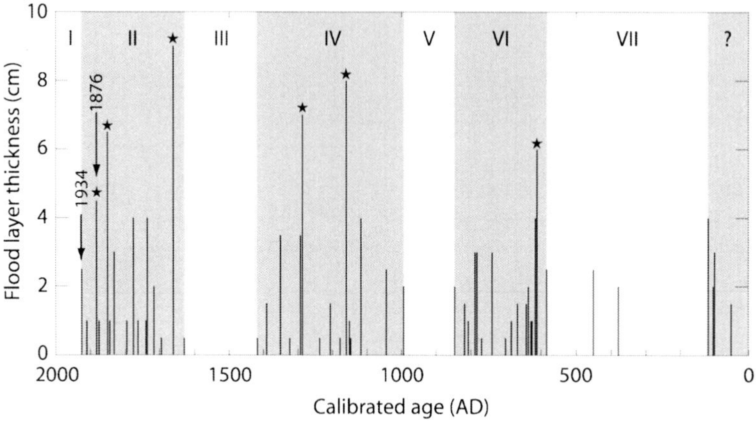

Fig. 4 Chronology of flood layers determined from a sediment core from Lake Lauerz (Switzerland) covering the last 2,000 years (Bussmann and Anselmetti 2010). Intervals of frequent flood events are *shaded* and labelled with *even roman numbers*. The layer thickness of the flood event from 1934 AD allows a scaling of the storm magnitude of past events. Six flood events (*marked with stars*) were as severe as the historically documented flood event from 1876, which caused large damages in Central Switzerland (reprinted by permission from the Swiss Geological Society, © 2010)

and Anselmetti 2010). The sediments were dated with four radiocarbon ages to establish a core chronology comprising the last 2,000 years (Fig. 4). A high bulk sedimentation-rate of up to 1 cm year^{-1} provides a unique high temporal resolution of palaeofloods in Lake Lauerz. 54 single flood events with a layer thickness between 0.5 and 9 cm were identified. Interestingly, the flood events are not evenly distributed in the past 2,000 years, but rather cluster in three periods (Fig. 4). An increased occurrence of heavy precipitation events is clearly documented between 1940–1630 AD, 1420–990 AD and 850–580 AD in the palaeoflood record of Lake Lauerz (Bussmann and Anselmetti 2010).

Under the assumption that layer thickness is a proxy for storm magnitude (Brown et al. 2000), the flood layers detected during the instrumental period in the twentieth century were used to scale past flood events. The largest flood during the period of precipitation measurements occurred in 1934 AD. This event caused severe devastation around Lake Lauerz and was recorded with a 2.5 cm thick detrital layer in the lake basin. On the basis of the established 2,000-year palaeoflood record, 17 of the 54 detected events were recorded with an even larger layer thickness and are thus interpreted to reflect a larger flood intensity than the 1934 AD event. One of the most severe flood events in Central Switzerland during historic times, which is recorded with layer thickness of 4.5 cm in Lake Lauerz, occurred in 1876 AD. Considering the event layer thicknesses, five earlier flood events had a similar or even larger storm magnitude pointing towards the occurrence of severe runoff events in the past 2,000 years (Bussmann and Anselmetti 2010).

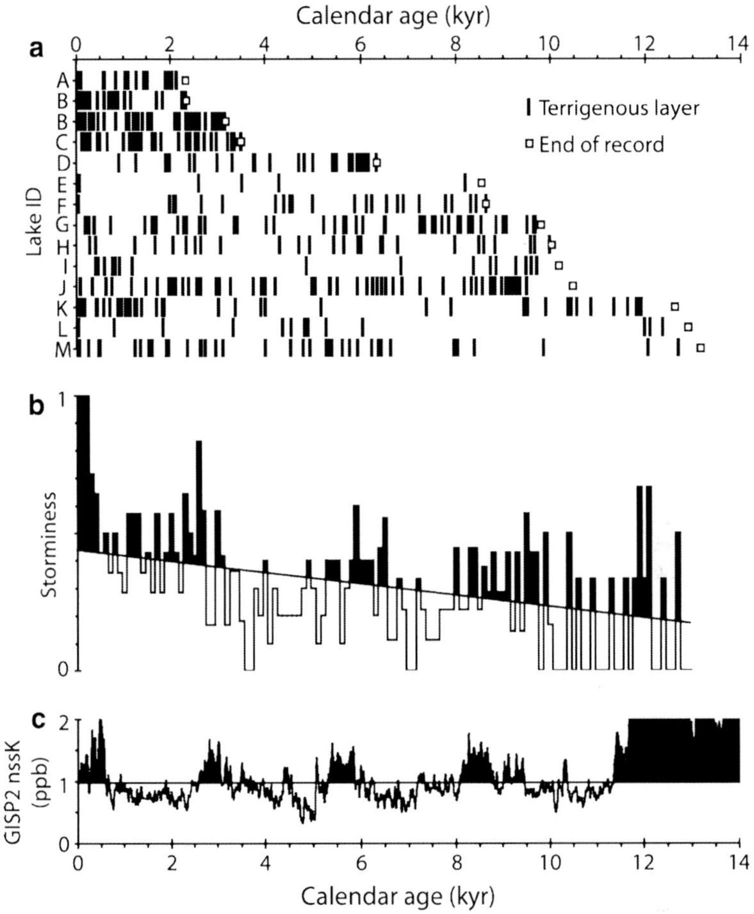

Fig. 5 Holocene palaeoflood records from northeastern United States (Noren et al. 2002). (**a**) Palaeoflood records from 13 individual lakes. (**b**) Weighted compilation of all flood records arranged in 100-year bins. Linear increasing trend of the storminess toward modern is explained by the authors by delta progradation leading to an increased sensitivity of the coring site to record flood events. (**c**) GISP2 non-sea-salt (n.s.s.) K concentration with values above the superimposed linear regression *shaded* (reprinted by permission from Macmillan Publishers Ltd: Noren et al. 2002)

6.2 Flood History from Lakes in Northeastern United States

Noren et al. (2002) conducted a comprehensive study about millennial-scale periodicity of exceptional rainfall events in the northeastern United States covering the entire Holocene. Unique in the design of this study is the stacking of the chronological occurrence of detrital layers from 13 individual lakes (Fig. 5a, b). This multi-archive approach is less susceptible to the influence of only locally occurring events. The results have therefore more paleoclimatic significance when compared to studies dealing solely with a single lake basin.

Lakes investigated within this study fulfil clearly the criteria for promising lacustrine flood reconstruction (cf. section "Selecting lakes with promising flood records") with well-defined morphobathymetric depocentres, a steep surrounding relief, evident inflows and delta structures, as well as predominantly organic background sedimentation contrasting with the detrital flood layers. The identification of flood layers was achieved by visual logging and high-resolution X-radiography (providing information on internal sediment structures), as well as magnetic susceptibility, loss-on-ignition and grain-size measurements.

Spectral analysis and correlations with proxies from the Greenland Ice Sheet Project 2 (GISP2) ice-core record could connect the established flood record with large-scale atmospheric circulation patterns (Fig. 5b, c). Periods with high flood activity show a recurrence interval of about 3,000 years and correlate in the GISP2 record with periods characterized by higher concentrations of non-sea-salt potassium (n.s.s. K). Increased concentrations of these aerosols are attributed to enhanced storminess in the high-latitude North-Atlantic region. Hence, the results indicate that precipitation intensity in the northeastern United States is controlled by large-scale atmospheric circulation modes such as the Arctic Oscillation (AO) (Noren et al. 2002).

7 Conclusions

Lakes have a great potential to record flood events by the deposition of detrital layers through turbiditic underflows. The major advantages of the lacustrine approach in palaeoflood reconstructions compared to analysis of alluvial fans and cones are the continuous mode of sedimentation, the high temporal resolution and the long time period covered by lake sediments. These advantages led to an increasing number of studies exploring the lacustrine archives for palaeoflood reconstructions. The results of these studies are very promising and clearly indicate periods of different flood frequencies for the Holocene. However, these studies cover almost exclusively a single catchment and therefore may include local events. Compilations of different lacustrine palaeoflood records, as for example applied by Noren et al. (2002), overcome this limitation and are able to build regionally robust flood reconstructions. With the now available limnogeologic data sets of past floods, linkages with other scientific communities like geomorphologists and dendrochronologists, need to be established to compile palaeoflood records from different archives. This combined approach (as outlined by Irmler et al. 2006) will be of great importance to understand the sensitivity of each flood archive and will ultimately lead to the most complete flood reconstructions in terms of frequency and intensity.

Acknowledgements The authors thank the Swiss National Science Foundation (SNF) for the financial support within the project 'FloodAlp' (Project No. 200021–121909) to initiate research on the Holocene flood history in Switzerland and northern Italy using lake sediments. Figures 1 and 3 contains scientific results form a joint project on Lake Ledro, Italy whose co-funding by the French ANR (project LAMA, directed by M. Magny and N. Combourieu-Nebout) is kindly acknowledged. The cores from Lake Thun shown in Fig. 2 were collected as part of a project in cooperation with Stephanie Girardclos, University of Geneva.

References

Anselmetti FS, Ariztegui D, De Batist M, Gebhardt AC, Haberzettl T, Niessen F, Ohlendorf C, Zolitschka B (2009) Environmental history of southern Patagonia unraveled by the seismic stratigraphy of Laguna Potrok Aike. Sedimentology 56:873–892

Appleby PG (2001) Chronostratigraphic techniques in recent sediments. In: Last WM, Smol JP (eds) Tracking environmental change using lake sediments, vol 1, Basin analysis, coring and chronological techniques. Kluwer Academic, Dordrecht, pp 171–201

Arnaud F, Revel M, Chapron E, Desmet M, Tribovillard N (2005) 7200 years of Rhone river flooding activity in Lake Le Bourget, France: a high-resolution sediment record of NW Alps hydrology. The Holocene 15:420–428

Bardou E, Fournier F, Sartori M (2003) Paleoflood reconstruction at Illgraben torrent (Switzerland): a current need for event frequency estimation. In: Thorndycraft VR, Benito G, Barriendos M, Llasat C (eds) Palaeofloods, historical floods and climate variability: applications in flood risk assessment. Proceedings of the PHEFRA international workshop, Barcelona, pp 53–59, 16–19 October 2002

Benito G, Thorndycraft VR (2005) Palaeoflood hydrology and its role in applied hydrological sciences. J Hydrol 313:3–15

Bierman P, Lini A, Zehfuss P, Church A, Davis PT, Southon J, Baldwin L (1997) Postglacial ponds and alluvial fans: recorders of Holocene landscape history. GSA Today 7:1–8

Boe AG, Dahl SO, Lie O, Nesje A (2006) Holocene river floods in the upper Glomma catchment, southern Norway: a high-resolution multiproxy record from lacustrine sediments. The Holocene 16:445–455

Brown SL, Bierman PR, Lini A, Southon J (2000) 10,000 yr record of extreme hydrologic events. Geology 28:335–338

Bussmann F, Anselmetti FS (2010) Rossberg landslide history and flood chronology as recorded in Lake Lauerz sediments (Central Switzerland). Swiss J Geosci 103:43–59

Chapron E, Beck C, Pourchet M, Deconinck J-F (1999) 1822 earthquake-triggered homogenite in Lake Le Bourget (NW Alps). Terra Nova 11:86–92

Chapron E, Desmet M, De Putter T, Loutre MF, Beck C, Deconinck JF (2002) Climatic variability in the northwestern Alps, France, as evidenced by 600 years of terrigenous sedimentation in Lake Le Bourget. The Holocene 12:177–185

Chapron E, Arnaud F, Noel H, Revel M, Desmet M, Perdereau L (2005) Rhone River flood deposits in Lake Le Bourget: a proxy for Holocene environmental changes in the NW Alps, France. Boreas 34:404–416

Dapples F, Lotter AF, van Leeuwen JFN, van der Knaap WO, Dimitriadis S, Oswald D (2002) Paleolimnological evidence for increased landslide activity due to forest clearing and land-use since 3600 cal BP in the western Swiss Alps. J Paleolimnol 27:239–248

Deevey ES, Gross MS, Hutchinson GE, Kraybill HL (1954) The natural C^{14} content of materials from hard-water lakes. Proc Natl Acad Sci USA 40:285–288

Eden DN, Page MJ (1998) Palaeoclimatic implications of a storm erosion record from late Holocene lake sediments, North Island, New Zealand. Palaeogeogr Palaeoclimatol Palaeoecol 139:37–58

Ely LL, Enzel Y, Baker VR, Cayan DR (1993) A 5000-year record of extreme flood and climate-change in the southwestern United States. Science 262:410–412

Gilli A, Anselmetti FS, Ariztegui D, McKenzie JA (2003) A 600-year sedimentary record of flood events from two sub-alpine lakes (Schwendiseen, Northeastern Switzerland). Eclogae Geol Helv 96(Supplement 1):S49–S58

Giovanoli F (1990) Horizontal transport and sedimentation by interflows and turbidity currents in Lake Geneva. In: Tilzer MM, Serruya C (eds) Large lakes: ecological structure and function. Springer, Berlin/Heidelberg, pp 175–195

Girardclos S, Schmidt OT, Sturm M, Ariztegui D, Pugin A, Anselmetti FS (2007) The 1996 AD delta collapse and large turbidite in Lake Brienz. Mar Geol 241:137–154

Hajdas I, Ivy SD, Beer J, Bonani G, Imboden D, Lotter AF, Sturm M, Suter M (1993) AMS radiocarbon dating and varve chronology of Lake Soppensee – 6000 to 12000 ^{14}C years BP. Clim Dyn 9:107–116

Irmler R, Daut G, Mäusbacher R (2006) A debris flow calendar derived from sediments of lake Lago di Braies (N. Italy). Geomorphology 77:69–78

Kochel RC, Baker VR (1982) Paleoflood hydrology. Science 215:353–361

Lamb MP, Mohrig D (2009) Do hyperpycnal-flow deposits record river-flood dynamics? Geology 37:1067–1070

Lambert A, Hsü KJ (1979) Non-annual cycles of varve-like sedimentation in Walensee, Switzerland. Sedimentology 26:453–461

Lamoureux S (2000) Five centuries of interannual sediment yield and rainfall-induced erosion in the Canadian High Arctic recorded in lacustrine varves. Water Resour Res 36:309–318

Macklin MG, Lewin J (2003) River sediments, great floods and centennial-scale Holocene climate change. J Quaternary Sci 18:101–105

Mangili C, Brauer A, Moscariello A, Naumann R (2005) Microfacies of detrital event layers deposited in Quaternary varved lake sediments of the Piànico-Sèllere Basin (northern Italy). Sedimentology 52:927–943

Mazzucchi D, Spooner IS, Gilbert R, Osborn G (2003) Reconstruction of Holocene climate change using multiproxy analysis of sediments from Pyramid Lake, British Columbia, Canada. Arct Antarct Alp Res 35:520–529

Meyers PA, Teranes JL (2001) Sediment organic matter. In: Last WM, Smol JP (eds) Tracking environmental change using lake sediments, vol 2, Physical and geochemical methods. Kluwer Academic, Dordrecht, pp 239–269

Mulder T, Chapron E (2011) Flood deposits in continental and marine environments: character and significance. In: Slatt RM, Zavala C (eds) Sediment transfer from shelf to deep water - Revisiting the delivery system: AAPG Studies in Geology 61:1–30

Mulder T, Migeon S, Savoye B, Faugères J-C (2001) Inversely graded turbidite sequences in the deep Mediterranean: a record of deposits from flood-generated turbidity currents? Geo-Mar Lett 21:86–93

Mulder T, Syvitski JPM, Migeon S, Faugères J-C, Savoye B (2003) Marine hyperpycnal flows: initiation, behavior and related deposits. A review. Mar Pet Geol 20:861–882

Nesje A, Dahl SO, Matthews JA, Berrisford MS (2001) A 4500-yr record of river floods obtained from a sediment core in Lake Atnsjoen, eastern Norway. J Paleolimnol 25:329–342

Noren AJ, Bierman PR, Steig EJ, Lini A, Southon J (2002) Millennial-scale storminess variability in the northeastern United States during the Holocene epoch. Nature 419:821–824

O'Connor JE, Ely LL, Wohl EE, Stevens LE, Melis TS, Kale VS, Baker VR (1994) A 4500-year record of large floods on the Colorado River in the Grand-Canyon, Arizona. J Geol 102:1–9

Osleger DA, Heyvaert AC, Stoner JS, Verosub KL (2009) Lacustrine turbidites as indicators of Holocene storminess and climate: Lake Tahoe, California and Nevada. J Paleolimnol 42:103–122

Parris AS, Bierman PR, Noren AJ, Prins MA, Lini A (2010) Holocene paleostorms identified by particle size signatures in lake sediments from the northeastern United States. J Paleolimnol 43:29–49

Pfister C (2009) The "Disaster Gap" of the 20th century and the loss of traditional disaster memory. Gaia 18:239–246

Revel-Rolland M, Arnaud F, Chapron E, Desmet M, Givelet N, Alibert C, McCulloch M (2005) Sr and Nd isotopes as tracers of clastic sources in Lake Le Bourget sediment (NW Alps, France) during the Little Ice Age: palaeohydrology implications. Chem Geol 224:183–200

Rodbell DT, Seltzer GO, Anderson DM, Abbott MB, Enfield DB, Newman JH (1999) An 15,000-year record of El Nino-driven alluviation in southwestern Ecuador. Science 283:516–520

Schnellmann M, Anselmetti FS, Giardini D, McKenzie JA, Ward SN (2002) Prehistoric earthquake history revealed by lacustrine slump deposits. Geology 30:1131–1134

Schnellmann M, Anselmetti FS, Giardini D, McKenzie JA (2006) 15,000 years of mass-movement history in Lake Lucerne: implications for seismic and tsunami hazards. Eclogae Geol Helv 99:409–428

Schneuwly-Bollschweiler M, Stoffel M (2012) Dendrogeomorphology – tracking past events with tree rings. In: Schneuwly-Bollschweiler M, Stoffel M, Rudolf-Miklau M (eds) Dating torrential processes on fans and cones – methods and their application for hazard and risk assessment, Advances in global change research. Springer, Dordrecht/Heidelberg/London/New York

Sheffer NA, Enzel Y, Benito G, Grodek T, Poart N, Lang M, Naulet R, Coeur D (2003) Paleofloods and historical floods of the Ardeche River, France. Water Resour Res 39(12):1376

Siegenthaler C, Sturm M (1991) Die Häufigkeit von Ablagerungen extremer Reuss-Hochwasser. Die Sedimentationsgeschichte im Urnersee seit dem Mittelalter. In: Ursachenanalyse der Hochwasser 1987 – Ergebnisse der Untersuchungen. Mitteilungen des Bundesamtes für Wasserwirtschaft, vol 4, Bern, pp 127–139

Sletten K, Blikra LH, Ballantyne CK, Nesje A, Dahl SO (2003) Holocene debris flows recognized in a lacustrine sedimentary succession: sedimentology, chronostratigraphy and cause of triggering. The Holocene 13:907–920

Stoffel M, Bollschweiler M (2008) Tree-ring analysis in natural hazards research – an overview. Nat Hazard Earth Syst Sci 8:187–202

Strasser M, Anselmetti FS, Fäh D, Giardini D, Schnellmann M (2006) Magnitudes and source areas of large prehistoric northern Alpine earthquakes revealed by slope failures in lakes. Geology 34:1005–1008

Sturm M, Matter A (1978) Turbidites and varves in Lake Brienz (Switzerland): deposition of clastic detritus by density currents. Spec Publ Int Assoc Sedimentol 2:147–168

Sturm M, Siegenthaler C, Pickrill RA (1995) Turbidites and 'homogenites' – a conceptual model of flood and slide deposits. In: Publication of IAS-16th regional meeting sedimentology, vol 22, Paris, p 140

Theiler A (2003) Die Sedimente des Seeli (Seelisberg, Uri) – Starkniederschläge im Holozän. Unpublished diploma, Geological Institute, ETH Zurich

Thevenon F, Anselmetti FS (2007) Charcoal and fly-ash particles from Lake Lucerne sediments (Central Switzerland) characterized by image analysis: anthropologic, stratigraphic and environmental implications. Quaternary Sci Rev 26:2631–2643

Thorndycraft VR, Benito G (2006) Late Holocene fluvial chronology of Spain: the role of climatic variability and human impact. Catena 66:34–41

Thorndycraft V, Hu Y, Oldfield F, Crooks PRJ, Appleby PG (1998) Individual flood events detected in the recent sediments of the Petit Lac d'Annecy, eastern France. The Holocene 8:741–746

Thorndycraft VR, Benito G, Rico M, Sopena A, Sánchez-Moya Y, Casas A (2005) A long-term flood discharge record derived from slackwater flood deposits of the Llobregat River, NE Spain. J Hydrol 313:16–31

von Gunten L, Grosjean M, Beer J, Grob P, Morales A, Urrutia R (2009) Age modeling of young non-varved lake sediments: methods and limits. Examples from two lakes in Central Chile. J Paleolimnol 42:401–412

Wirth S (2008) Lake Thun sediment record: 300 years of human impact, flood events and subaquatic slides. Unpublished MSc thesis, Department of Earth Sciences, ETH Zurich, doi:10.3929/ethz-a-005676930

Wirth SB, Girardclos S, Rellstab C, Anselmetti FS (2011) The sedimentary response to a pioneer geo-engineering project: tracking the Kander River deviation in the sediments of Lake Thun (Switzerland). Sedimentology 58:1737–1761

Wolfe BB, Hall RI, Last WM, Edwards TWD, English MC, Karst-Riddoch TL, Paterson A, Palmini R (2006) Reconstruction of multi-century flood histories from oxbow lake sediments, Peace-Athabasca Delta, Canada. Hydrol Process 20:4131–4153

Zolitschka B (2006) Varved lake sediments. In: Elias SA (ed) Encyclopedia of Quaternary science. Elsevier, Amsterdam, pp 3105–3114

Dating Fan Deposits with Cosmogenic Nuclides

Susan Ivy-Ochs, Miriam Dühnforth, Alexander L. Densmore, and Vasily Alfimov

1 Introduction

Cosmogenic nuclides build-up predictably with time within the lattices of minerals exposed to cosmic rays (Lal 1991). The cosmogenic nuclide concentration extracted from the suitable minerals and measured with either accelerator mass spectrometry (AMS) or static noble gas mass spectrometry is a direct measure of how long the rock or sediment has been exposed at or near the surface of the earth. The nuclides most widely used for dating are the radionuclides ^{10}Be, ^{14}C, ^{26}Al, and ^{36}Cl and the stable noble gases ^{3}He and ^{21}Ne (Table 1). Each nuclide has a set of minerals that are most suited for measurement (Ivy-Ochs and Kober 2009). With this palette of nuclides a broad spectrum of landforms composed of nearly all rock types can be dated and myriad geomorphic problems can be addressed. In 2001, Gosse and Phillips published a comprehensive overview of cosmogenic nuclide methods available to earth scientists. Further summaries especially with information on field considerations and possible pitfalls are given by Cerling and Craig (1994), Bierman

S. Ivy-Ochs (✉)
Laboratory of Ion Beam Physics, ETH Zurich, 8093 Zurich, Switzerland

Department of Geography, University of Zurich, 8057 Zurich, Switzerland
e-mail: ivy@phys.ethz.ch

M. Dühnforth
Institute of Arctic and Alpine Research (INSTAAR), University of Colorado, Boulder, CO 80303, USA
e-mail: miriam.duhnforth@colorado.edu

A.L. Densmore
Department of Geography, Durham University, Durham DH1 3HP, UK
e-mail: a.l.densmore@durham.ac.uk

V. Alfimov
Laboratory of Ion Beam Physics, ETH Zurich, 8093 Zurich, Switzerland
e-mail: alfimovv@phys.ethz.ch

M. Schneuwly-Bollschweiler et al. (eds.), *Dating Torrential Processes on Fans and Cones*, Advances in Global Change Research 47, DOI 10.1007/978-94-007-4336-6_16,

Table 1 Summary of nuclide characteristics

Nuclide	Half-life	Other isotopes	Suitable minerals	Target elements	Production rate[a] atoms g^{-1} $year^{-1}$	Applicable time range[b]
^{10}Be	1.4 Myr	^{9}Be	Quartz	O	5	Exposure ages to several million years
^{14}C	5.73 kyr	^{12}C, ^{13}C	Quartz	O	16	To 20,000 years
^{26}Al	0.7 Myr	^{27}Al	Quartz	Si	31	To several million years
^{36}Cl	0.3 Myr	^{35}Cl, ^{37}Cl	All rock types including carbonates	K, Ca, ^{35}Cl	Composition dependent ca. 10 granite; ca. 20 limestone	To one million years
^{3}He	stable	^{4}He	Olivine, pyroxene	Many	115	ca. 1,000 to millions of years
^{21}Ne	stable	^{20}Ne, ^{22}Ne	Quartz, olivine, pyroxene	Si, Mg	20	ca. 40,000 to millions of years

[a]Nuclide production rates at sea level and high latitude (Gosse and Phillips 2001)
[b]Applicable age range depends on rock surface or landform weathering rate

et al. (2002), Cockburn and Summerfield (2004), Ivy-Ochs and Kober (2009) and Ivy-Ochs and Schaller (2010).

Cosmogenic nuclide dating is a unique and powerful tool never before available to geomorphologists. Both depositional landforms, like moraines, landslide deposits, fluvial terraces, or alluvial fans, as well as erosional bedrock landforms, such as roches moutonnées, fault footwall faces, or landslide niche bedrock surfaces, can be dated directly. Exposure dating is particularly well suited for the study of alluvial fans; one can assess both the duration of intervals of sediment aggradation as well as the timing of lobe abandonment. In the past geomorphology has relied on ^{14}C for the (indirect) dating of landforms. Questions of timing of fan construction that previously were difficult to solve due to the limitation of finding organic material for radiocarbon dating or due to the 45 kyr upper limit of radiocarbon dating can now be addressed with cosmogenic nuclides. Moreover, the relationship of the organic material to the landform can never be unequivocally established, it is simply older than the enclosing fan sediment. Sites unsuitable for luminescence techniques, such as those that have not been exposed to light long enough for re-zeroing or those with sediments that are too coarse for luminescence methods (gravels), can be dated with cosmogenic nuclides. Depending on the fan's character and degradation rate, dating of fans can as well far exceed the upper limit of luminescence techniques (several hundred thousand years). Alluvial deposits on the order of Millions of years old have been dated with cosmogenic nuclides (Dühnforth 2007; Matmon et al. 2009). Luminescence dating gives the time since the sediment was buried while exposure dating yields the timing of fan abandonment.

Reflecting the variety of processes through which fans form and the relatively long timescales of their preservation (kyr to Myr) several different cosmogenic nuclide approaches can be applied. These may be subdivided into nuclide build-up methods like surface exposure dating and decay methods such as burial dating. The dating results may then serve as the basis for drawing inferences about catchment and fan processes, the contribution of different forcing mechanisms to changes in fan aggradation or abandonment and their variation with time. In this respect we present a case study by Dühnforth et al. (2007, 2008) where boulders on debris-flow fan lobes in Owens Valley, California, were exposure dated with ^{10}Be.

2 Cosmogenic Nuclide Systematics

Galactic and solar cosmic rays (primarily protons) are constantly bombarding the earth. Due to interactions with target nuclei in the atmosphere a cascade of secondary particles is produced which at the surface of the earth is dominated by neutrons and muons. Impingement of these secondary cosmic ray particles causes nuclear reactions within the lattices of exposed minerals whereby cosmogenic nuclides are produced. Reactions include spallation, muon-induced reactions and low-energy (epithermal and thermal) neutron capture reactions (Lal and Peters 1967; Lal 1991). In quartz, more than 95% of ^{10}Be and ^{26}Al production is through spallation reactions.

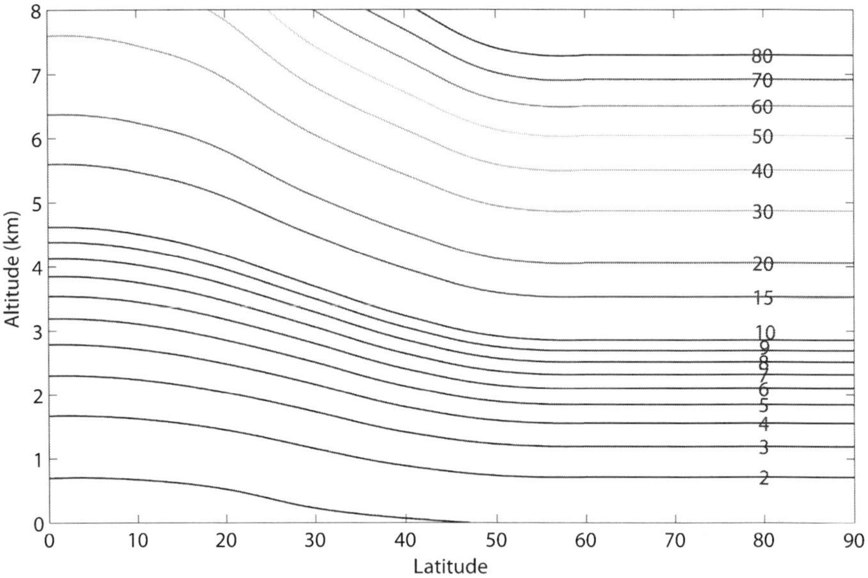

Fig. 1 Variation of nuclear disintegrations (approximation of variation of spallation production rate) as a function of geomagnetic latitude and altitude on the surface of the Earth (Based on Lal and Peters 1967; Lal 1991)

The cosmic ray flux decreases with decreasing altitude. Consequently, production of cosmogenic nuclides is lower at lower altitude than at higher altitude. The magnetic field shields the earth from cosmic rays but this effect is less at the poles. Thus nuclide production is higher there than at lower latitudes (Fig. 1). This means that in order to calculate an exposure age, the sea level high latitude production rate must be scaled to the latitude, longitude and altitude of the site (Gosse and Phillips 2001). Several schemes for this scaling have been presented (cf. Balco et al. 2008). For a given data set results can be calculated based on the different schemes by using the University of Washington/CRONUS-Earth website (http://hess.ess.washington.edu/math/; Balco et al. 2008 and update). In addition, published data can be recalculated based on different scaling schemes to afford consistent comparison of data from different sites. It is important that scaling formalities used to calculate exposure ages for samples are consistent with those that were used to determine the production rate. In the documentation for the CRONUS calculator and Gosse and Phillips (2001) the appropriate equations are given so that researchers can as well set-up their own calculation files.

In addition to the scaling of the production rate to the site latitude, longitude and altitude, several additional corrections to the production rate may be required (Gosse and Phillips 2001). Shielding by surrounding topography and the dip of the sampled rock surface may reduce the cosmic ray flux that the surface 'sees.' Production of cosmogenic nuclides decreases with depth into rock or sediment. Figure 2 shows

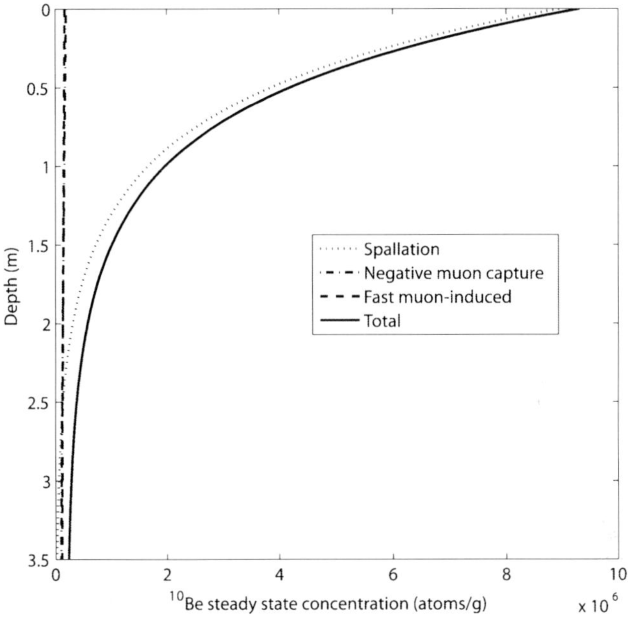

Fig. 2 Production rate of ^{10}Be in quartz as a function of depth. Total production is a composite of the production by neutron spallation, stopped muons, and fast muons

the decrease of the ^{10}Be production rate with depth. Note that production due to spallation decreases much more rapidly with depth than production due to muon interactions. In contrast to the other nuclides which are dominated by production due to spallation, ^{36}Cl is produced also through low-energy (thermal and epithermal) neutron capture on ^{35}Cl (Phillips et al. 2001). Accordingly, but depending on rock composition, the ^{36}Cl depth profile may differ significantly from that of ^{10}Be (Gosse and Phillips 2001; Ivy-Ochs and Schaller 2010). Alfimov and Ivy-Ochs (2009) and Schimmelpfennig et al. (2009) give a detailed treatment of ^{36}Cl production mechanisms and their behaviour with depth. In any case, the production rate must be corrected for the thickness of the sample processed, which amounts to only a few percent if the sample is no more than several centimetres thick.

Both geological and methodological factors affect the upper and lower age limits that can be determined with cosmogenic nuclides. The lower age limit is controlled by methodological concerns such as the background (carrier) correction and the sensitivity of the accelerator set-up for ^{10}Be (^{14}C, ^{26}Al, ^{36}Cl). ^{10}Be exposure ages in the range of several hundred years have been reported recently (Akçar et al. 2012). For the radionuclides, the upper age limit or oldest age it is possible to determine is controlled by the nuclide half-life (Table 1) and the weathering rate of the rock surface (Fig. 3). Radionuclides build up in exposed minerals but because they are unstable, they also decay. With time secular equilibrium or saturation is reached. At secular equilibrium the number of nuclides produced per unit time is equivalent to

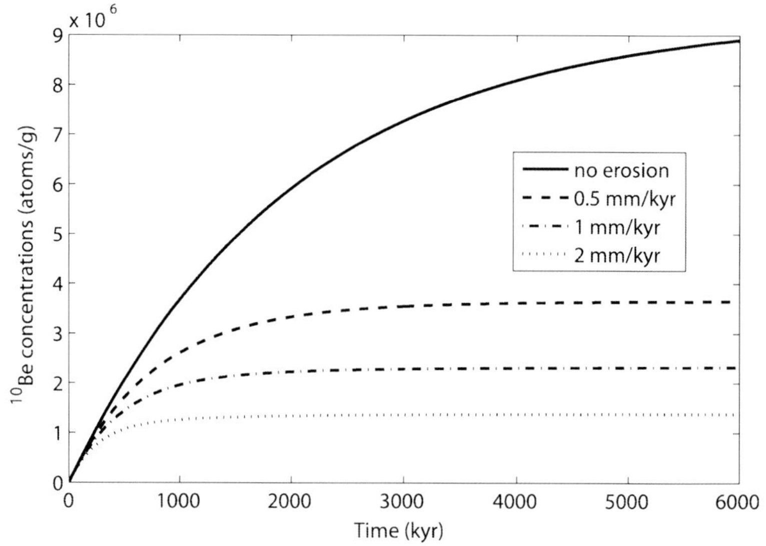

Fig. 3 Increase of the ^{10}Be concentration with time in minerals exposed to cosmic rays taking into account different steady-state erosion rates of the exposed surface (0.5 mm/kyr, 1 mm/kyr, 2 mm/kyr). Secular equilibrium, where radionuclide gain due to production equals loss due to radioactive decay (and erosion), is approached earlier for more rapid erosion rates

the number that decay. The concentration is at steady state and does not increase no matter how much longer the rock or sediment is exposed. As shown in Fig. 3 the upper nuclide concentration growth with time curve begins to flatten out (saturation is approached) after three to four half-lives. When the rock surface is eroding, the uppermost layers of rock, which contain the highest nuclide concentrations, are lost due to erosion. Nuclide saturation is reached earlier (after a shorter exposure period) (lower lines on Fig. 3 for several erosion rates). In regions where bedrock erosion rates are slow (for example in arid and hyper-arid deserts) exposure ages up to several million years have been determined. In contrast, in terrains where rocks are eroding somewhat more rapidly (a good estimate of crystalline bedrock erosion rate is about 5 mm/kyr or less, Cockburn and Summerfield 2004) the oldest exposure age it is possible to measure is controlled by the erosion rate and may be only a few hundred thousand years.

3 Cosmogenic Nuclide Dating of Fan Deposits

Alluvial fans build up through a variety of processes, including fluvial flow, overland flow, debris flow, rock avalanche, and rockfall (Bull 1977; Blair and McPherson 1994). Thus there exists a broad range of surface morphologies and sedimentary architectures to consider in dating fan deposits. The appropriate cosmogenic nuclide dating method is largely dependent on the character of the fan surface and sediment, which in turn is a function of the depositional processes and age of the fan. More information about timing of aggradation, avulsion, fan-head incision and lobe abandonment may be gained by combining several methods. For fans with very coarse deposits, the top surfaces of large boulders can be exposure dated. Older fans where boulders have collapsed or those that never had surface boulders are dated with individual or amalgamated clast samples either from the top surface of the fan or from a depth profile down into the sediment. Deeply buried fan deposits can be dated with nuclide decay methods such as burial and isochron dating, whereby measurement of two different nuclides is required.

To be exposure dated the fan lobe surface must no longer be active. Abandonment results as fans are moved laterally along a fault away from the locus of deposition or as a result of avulsion or fan head incision events. The latter may be climatically controlled, reflecting changes in sediment flux, or tectonically controlled (e.g. Bull 1991). Exposure dating of offset abandoned fan lobes has been widely used to estimate fault slip rates (Bierman et al. 1995; Siame et al. 1997; Zehfuss et al. 2001; Mériaux et al. 2005; Frankel et al. 2007; Lee et al. 2009; Machette et al. 2008). Slip distances are determined through offset stream piercing points or by reconstructing the pre-abandonment cross profile (curvature) of the fan surface. Many fan-dating studies were done in arid and hyper-arid environments. There the preservation of fan surface morphology is good and the record of fan construction can extend to millions of years. Especially where the goal is to determine tectonic slip rates the dated fans are located along mountain fronts. Studies in fans that are within mountain ranges are fewer and timescales are necessarily much shorter. There, fans form only after the main valleys became ice-free at the end of the most recent large-scale glaciation. Based on exposure dating results aggradation events in fans fed by glaciated catchments are linked to glacier expansions (Barnard et al. 2006). Aggradation of other intramontane fans is fed by rock avalanche or rock fall events (Seong et al. 2009). The first of such studies to be performed on fans in the Alps, and under more humid conditions than any of the previous studies, is now being carried out by Peter Schuerch and colleagues at Durham University, UK, as part of a wider study of debris-flow fan dynamics (Schürch et al. 2010). These workers examine the possibility of exposure dating boulders on debris-flow deposits at Illgraben in the Rhône Valley in the Swiss Alps (Schuerch, Densmore, Ivy-Ochs unpublished data). Whether the goal is to use the dating of a fan surface to determine fault slip rates or to study the timing of aggradation and avulsion events, the fan deposits and fan surface morphology should be mapped in detail before sampling.

3.1 *Exposure Dating of Boulders*

Fans formed by debris flows, rock avalanches or rockfalls often have large boulders on the surface which may be candidates for surface exposure dating. Good results have been obtained from dating boulders that are large (>1.5 m high), broad, flat-topped and in stable positions (away from channels) but also away from steep valley margins. When original surface morphology is present, boulders on top of debris-flow levees or snouts are preferred (Nishiizumi et al. 1993; Bierman et al. 1995; Siame et al. 1997; Zehfuss et al. 2001; Matmon et al. 2005; Benn et al. 2006; Ritz et al. 2006; Ryerson et al. 2006; Dühnforth et al. 2007; Frankel et al. 2007; Spelz et al. 2008; Lee et al. 2009). To obtain an exposure date that most closely represents the time since landform construction, sampled boulders must have (i) been continuously exposed in the same position (not shifted), (ii) never been covered by sediment vegetation or snow, (iii) undergone only minimal surface weathering or erosion (not spalled), and (iv) experienced single-stage exposure (no inheritance). The first three will lead to exposure ages that are younger than the depositional age of the landform while the last will lead to ages that are older.

Boulders are preferable to smaller clasts in that they generally contain less inheritance. Inheritance or pre-exposure refers to nuclide concentrations that are already in the rock or sediment before it was deposited. These excess nuclide concentrations lead to exposure ages that are older than the time since deposition on the fan. Large boulders are produced through rapid large-scale mass wasting events such as landslides or rock fall. They thus enter the system in the first cycle and are transported to the fan in debris flows or other flood events. Although inherited nuclide concentrations may be acquired in the bedrock setting, outer boulder surfaces (with the highest nuclide concentrations) are usually eroded off during fall and/or transport. Never-the-less there are several ways that boulders may have inheritance. Short transport distances in non-glaciated catchments may lead to more fan boulders having inherited nuclide concentrations. Windows into underlying older deposits may also lead to apparent older outliers on mapped fan lobes (Dühnforth et al. 2007). Whether boulders or clasts are analyzed the inheritance contribution will be proportionally higher in relatively young landforms.

3.2 Exposure Dating of Clasts

Fans formed by fluvial or overland flow often do not have large boulders. On older fan surfaces (> ca. 150–200 kyr) most of the surface boulders, if present at all, will have disintegrated (Matmon et al. 2006; Ritz et al. 2006). For dating such fans individual clasts (cobbles, pebbles) or amalgamated clast or sediment samples are analyzed (Brown et al. 1998; Mériaux et al. 2005; Machette et al. 2008). The cosmogenic nuclide concentration in any individual clast may be higher or lower with respect to the actual timing of fan deposition. Too low nuclide concentration suggests that the small clasts are derived from degradation of larger clasts or boulders; they were thus partially shielded during exposure on the inside of the boulder. Clasts with too low nuclide concentrations can also result from periodic burial reflecting (bio- or cryo-) turbation of the upper tens of centimetres of sediment, i.e. the clasts come from deeper in the deposit and were partially shielded. Clasts may contain excess nuclide concentrations (inheritance). Inheritance in the clasts can be acquired during hillslope weathering in the catchment especially where the denudation rate there is low. Clasts may also have been exposed during storage in terraces in the catchment before final deposition on the fan. Scavenging or reworking of sediment due to rilling or incision of older fan deposits at higher elevation to younger deposits downstream may be another explanation for inheritance in clasts. The presence of inherited nuclide concentrations would lead to an age that is older than the time of fan deposition. The technique of amalgamating >50 quartz-rich clasts of similar size from the fan surface into one sample is thought to cancel out these random effects (Anderson et al. 1996; Repka et al. 1997). Clasts should be collected from flat stable surfaces away from modifying channels, which where possible exhibit intact desert pavement and dark desert varnish. Analysis of clasts from the active channel at the catchment mouth may but does not unequivocally provide information about inheritance. The inheritance may result from intermittent storage of sediment within the catchment. As both reworking of fan material and catchment processes vary in time and space, a 'typical' inheritance contribution for a given fan may not necessarily exist (Matmon et al. 2005).

3.3 *Depth-Profile Dating*

One way to deal with inheritance is to analyze several samples along a depth profile (Hancock et al. 1999). Amalgamated clast samples (>50 clasts) of similar size or samples of coarse sand (too coarse to have been wind blown) are taken every 10–20 cm down to a depth of 2–3 m within a single depositional unit in the fan deposit (Fig. 4). Similar to exposure dating of boulders or clasts a relatively stable upper fan surface is best suited to depth profiling. Both surface erosion (indicated by convex, as opposed to flat, tread morphology) and aggradation (indicated by a wind-blown silt layer below the top surface pavement) should be minimal since fan lobe abandonment. In a sediment profile where there is no inheritance the nuclide concentrations with depth will mimic the production depth profile (Fig. 2). When inheritance is present in the analyzed clasts, the model curve will be shifted to the right (red line in Fig. 4). The amount of this shift (indicated by the arrow) shows the excess nuclide concentration (inheritance) contained in the sediment. Such a uniform shift also suggests that the sampled section was deposited relatively rapidly (likely no more than several thousand years) and that the incoming sediment contained roughly similar inherited nuclide contribution. The shape of the curve reveals the age and erosion rate of the fan (Vassallo et al. 2005; Ritz et al. 2006; Matmon et al. 2006; Machette et al. 2008). In Fig. 5 depth profiles reflecting continuous single-stage exposure with no inheritance with various exposure periods and different erosion rates of the fan upper surface are shown. Best fit chi-squared methods are used to seek unique solutions for the fit of the exposure age and erosion rate curve to the data. At some sites the depth profile data reveal breaks in deposition that were difficult to detect in the field. Two units separated by a hiatus are present. The curve in (Fig. 6) indicates deposition then abandonment of the lower unit

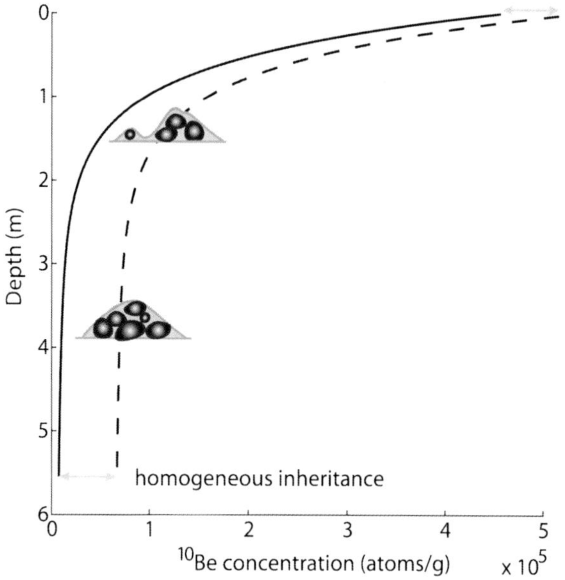

Fig. 4 Principles of age determination in sediment by analyzing several samples along a depth profile. The nuclide concentration in a sample from the surface is a composite of the inherited and the post-depositional nuclide concentration. The *solid line* depicts ^{10}Be concentration with depth in a deposit where there is no inheritance. The *dashed line* depicts nuclide concentration with depth in a deposit with inheritance. The amount of inheritance can be determined by the offset of the two lines at depth (*arrow*). Probability density distribution of nuclide concentrations of random, amalgamated clast samples taken along the depth profile is shown schematically

followed by a period of exposure then deposition of the overlying unit and a period of exposure. Note as well that the upper unit has a different inheritance signal which may point to changes in the drainage configuration in the catchment such as stream capture or scavenging of a terrace upstream. Often nuclide concentrations measured in samples from the top 20–50 cm are the same (Fig. 7). This suggests mixing of the top sediment (Phillips et al. 1998). A surface sample would in this case yield an age that is younger than the age determined with the depth profile. Depth profile dating is time consuming as it requires several samples from a single landform. Never-the-less it is a good way to learn about inheritance as well as the abandonment age and degradation rate of the fan surface.

3.4 Dating with Two Nuclides Using Differential Decay of ^{10}Be and ^{26}Al

The above described methods use the build-up of cosmogenic nuclides. One can also use the decay of cosmogenic nuclides to date fan deposits, which includes burial

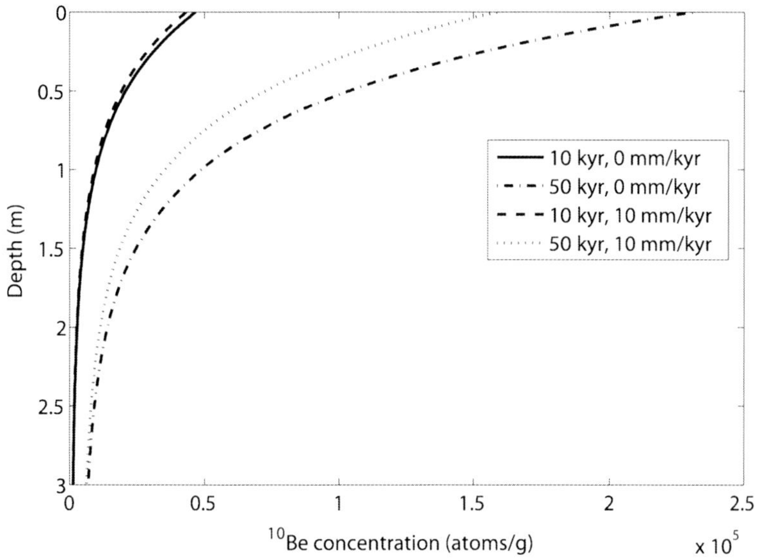

Fig. 5 Figure showing the evolution of the shape of the depth profile curve with time. ^{10}Be concentration produced at depth in sediment after continuous exposure (10, 50 kyr) with no erosion and (10, 50 kyr) when the rock surface is eroding at 10 mm/kyr

Fig. 6 ^{10}Be depth profile for two units. *Black line* depicts ^{10}Be concentration with depth in a deposit continuously exposed for 100 kyr where there is no inheritance (*lower part* of the curve). Following this period of exposure the *upper unit* was deposited on *top*. Note that the *upper unit* has different nuclide inheritance in the clasts. After deposition the sediment was exposed for another 50 kyr. Concentrations in the *lower unit* increase only to a small degree because they are buried under 2 m of the overlying sediment

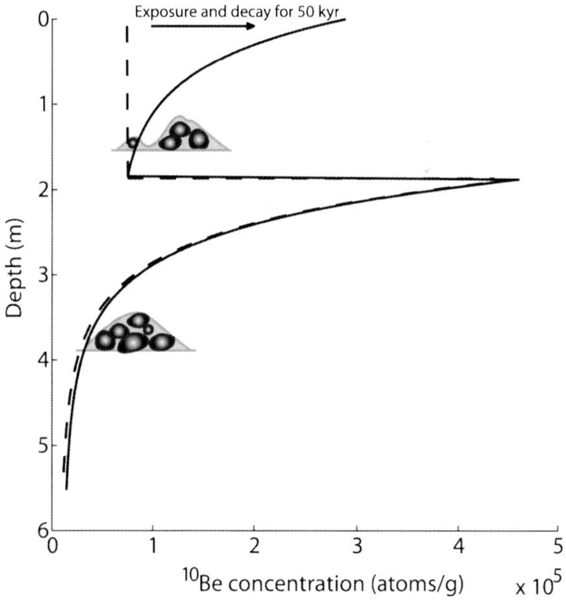

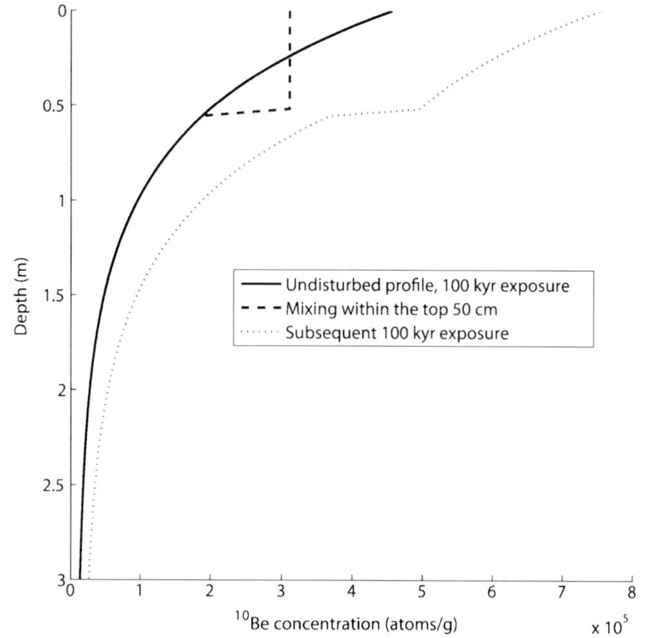

Fig. 7 The concentration of ^{10}Be with depth in sediment. *Black solid line* shows an undisturbed profile after 100 kyr of exposure. *Dashed line* indicates resulting concentrations reflecting mixing of the upper 50 cm of sediment due to bio- or cryo-turbation. *Dotted line* shows concentration increase if the mixing is halted all together and the sediment is exposed for another 100 kyr

(Granger 2006) and isochron dating (Balco and Rovey 2008). In burial dating it is the non-zero initial concentration (the same as inheritance described above) that is used to date how long the sediment has been buried. Cosmogenic nuclides build up in sediment exposed to cosmic rays in the catchment during hillslope weathering and during transport. After rapid burial, production stops and both ^{10}Be and ^{26}Al decay. As ^{26}Al decays twice as fast as ^{10}Be, the $^{26}Al/^{10}Be$ ratio decreases with time of burial. Burial dating differs from exposure dating in that with time the nuclide concentration decreases. Because of the resulting low concentrations and as a result of the typical errors of ^{10}Be and ^{26}Al AMS measurements, the lower limit of burial dating is 100–200 kyr. Flat-topped landforms, like abandoned fan lobes, where the upper surface is quasi-stable since cessation of deposition can be suitable for burial dating (Granger 2006 and references therein). Samples of amalgamated clasts or sand grains are taken from more than 30 m depth (where possible) so that production from the top of the surface down into the sediment is nearly negligible. Recently, isochron dating, a variation of burial dating, was proposed by Balco and Rovey (2008). The age determination is also based on differential decay of ^{26}Al and ^{10}Be, but in several sediment samples taken along a depth profile.

4 Case Study: Variations in Debris-Flow Deposition Patterns on Fans in Owens Valley, California

Dühnforth et al. (2007) used detailed mapping and ^{10}Be dating of boulders on fan surfaces to examine the spatial and temporal pattern of fan build-up on the west side of Owens Valley (California). The dated fans are constructed of sediment delivered by streams draining the east side of the Sierra Nevada. The catchments are presently not glacierized but underwent periods of minor to full glacial occupation in the past (Brocklehurst and Whipple 2002). Debris-flow deposits are ubiquitous on the fans along the eastern Sierra Nevada, which suggests that debris-flow deposition is the dominant process for fan aggradation in this area (Whipple and Dunne 1992). Dühnforth et al. (2007) investigated two adjacent fans (Fig. 8) to the west of Independence, the Shepherd and Symmes Creek fans. Even though the fans are adjacent, they show a striking difference in fan head morphology with minor incision at the Symmes Creek fan head (<5 m) and deep entrenchment of about 40 m at the Shepherd Creek fan head.

Prior to sampling for cosmogenic nuclide analysis a relative chronology of fan deposition and incision was established. Variations in the debris-flow morphology

Fig. 8 *Shaded* relief map showing location of the study area (Modified from Dühnforth et al. 2007). Symmes (*upper*) and Shepherd Creek catchments are shown in *white*

Fig. 9 Relative and absolute chronology of fan deposition on Symmes and Shepherd Creek fans (modified from Dühnforth et al. 2007). Patterned lobes (Symmes and Shepherd A-F) represent the relative chronology based on mapping of debris-flow deposits with similar geomorphic signature and cross-cutting relationships between channels. Points indicate sample locations for cosmogenic ^{10}Be surface exposure dating. *Inset* shows arithmetic mean of sample ages on each fan lobe. Shepherd Creek fan preserves a longer record of fan deposition due to fan head incision and lobe abandonment. In contrast, Symmes Creek fan has been completely resurfaced in the Holocene. Details of sample preparation and age calculations are given in Dühnforth et al. (2007)

across the entire fan surface could be used to distinguish different depositional periods on the fan. This was based on debris-flow channel cross-cutting relationships, abrupt topographic changes that form clear fan surface boundaries for tens to hundreds of meters, short-wavelength (1–10 m) surface relief, boulder weathering, and the characteristic geometric signature of the debris-flow deposits. The latter includes channel width, boulder size, levee height, and run-out length. Dühnforth et al. (2007) define a fan lobe as a distinct area on the fan where fan deposition is active during a limited period of time until sediment supply to the fan temporarily ceases or significantly decreases, and deposition subsequently shifts to another area on the fan.

On the Shepherd Creek fan Dühnforth et al. (2007) identified at least six different depositional lobes with distinct variations in both, the overall fan surface characteristics and the degree of boulder weathering. The lobes are named A to F in order of decreasing age (Fig. 9). Numerous granitic or granodioritic boulders were sampled for ^{10}Be exposure dating on each lobe (Fig. 10). Details of sample preparation and age calculations are given in Dühnforth et al. (2007).

Fig. 10 View of debris-flow channel and snout (ending in *centre* of photo) on Shephard Creek fan, Owen's Valley, California. Channel width is about 40 m and boulder height is about 1 m. The tops of the largest boulders were sampled for cosmogenic ^{10}Be dating to determine the timing of lobe formation (Dühnforth et al. 2007). Note person in *centre left* for scale

Lobe A is the oldest depositional lobe with a smooth surface topography that is mantled by grus. Debris-flow channels are difficult to identify as diffusional processes have largely modified the surface. But local concentrations of large boulders may represent former debris-flow levees and snouts. On average, the boulders are 1–3 m in diameter; a few are even larger with diameters up to 5 m. Some of these boulders have a well-developed desert varnish coating and show no clear signs of erosion. Others are fractured and clearly show that spalling removed areas from the boulder surface. The oldest relative age was confirmed by the absolute mean age of 83.5 ± 3.6 kyr of the dated boulders ($n = 3$).

Lobe B has clearly recognizable, radially oriented channel and ridges, which Dühnforth et al. (2007) interpret as former debris-flow channels. As the boundaries between the debris-flow deposits and the fan surface are fuzzy, they did not measure the channel geometries. Nonetheless, the character of the boulder surfaces is similar to lobe A but the boulder sizes are on average slightly smaller with 1–2 m. The absolute age of lobe B is slightly younger than lobe A with 78.4 ± 3.2 kyr (dated boulders $n = 6$).

Lobe C has a smooth surface topography without channel-levee morphology. Instead, clast-rich lobate deposits of less than 0.5 m thickness are commonly present. On average, the deposits are 15 m wide, and they terminate in subdued, coarse-grained snouts. The deposits are very clast- and boulder-rich with diameters generally less than 1 m. Both clasts and boulders are in comparison to lobes A and B unvarnished and relatively unweathered. The apparent difference in the preservation of the debris-flow boulders is reflected in the absolute age of lobe C; the average age of lobe C based on three dated boulders is, at 33.5 ± 4.1 kyr, significantly younger than the ages of lobes A and B.

The morphology of the debris-flow deposits changes significantly on lobe D. A prominent debris-flow channel–levee morphology is characteristic for this lobe. Well-developed channels vary between 25 and 35 m in width and have a mean channel depth of 1–2 m. The levees are boulder-rich, boulders are fresh and unweathered, and generally have sizes of 1–2 m in diameter. Three dated debris-flow boulders give an average exposure age of 27.5 ± 5.2 kyr.

Lobe D is crosscut by the younger deposits of lobe E. Lobe E has large debris-flow channels with mean widths of 50–60 m and 2–3 m depth. One channel stands out in particular with its maximum width of 100 m and minimum run-out length of 3–4 km. Boulders are scarce on this lobe and those that are present are small in diameter with less than 0.5 m. This lobe has an average age of 21.8 ± 5.5 kyr ($n = 7$ dated boulders). Initiation of fan-head incision post-dates deposition of lobe D and thus occurred after 21.8 ± 5.5 kyr.

The most recent deposits are present on lobe F. This lobe consists of inset terraces, the floor of the incised channel at the Shepherd fan head, and the equivalent depositional area further downstream. The debris-flow channels on this lobe are 15–25 m wide and less than 2 m deep. Boulder abundance is very high, and the boulder sizes are on average 1–2 m in diameter. The absolute dating confirms the hypothesis that lobe F represents the area of most recent deposition on the Shepherd Creek fan, and the lobe yields a mean age of 7.4 ± 3.1 kyr based on five boulders. The youngest age on lobe F (2.9 ± 0.2 kyr) furthermore suggests that inheritance on the Shepherd Creek fan appears to be negligible.

The dating results show that deposition on the Shepherd Creek fan surfaces occurred throughout both glacial and interglacial (interstadial) time periods. But based on the large extent of debris-flow deposits of glacial ages on the Shepherd Creek fan, debris-flow deposition is markedly enhanced during times of glacial occupation in the fan catchments. The Shepherd Creek fan ages point to fan deposition during marine isotope (MIS) stage 2 and stage 4. Evidence for the latter period of enhanced fan aggradation and subsequent fanhead incision adds to the growing body of evidence for a marked and widespread pulse of alluviation during MIS 4 around the North American southwest (Bull 1991; Gosse et al. 2004; Benn et al. 2006; Frankel et al. 2007; Nichols et al. 2007; Spelz et al. 2008; Lee et al. 2009).

In stark contrast to the morphology of the Shepherd Creek fan, the Symmes Creek fan shows nearly uniform surface characteristics. Therefore, Dühnforth et al. (2007) suggest that this fan comprises a single depositional lobe. On average, the debris-flow channels are 20–30 m wide and 2–3 m deep. The debris-flow channels are inset into the fan surface and well-defined levees with fresh, unweathered boulders are preserved along both sides of the channels. The areas between the channels are very smooth and mantled by grus. Based on the four measured samples the average exposure age of boulders on this lobe is 4.4 ± 2.2 kyr. This indicates that nearly the entire Symmes Creek fan was re-surfaced during the Holocene. The young ages (the youngest is 2.60 ± 0.2 kyr) point to the presence of very little nuclide inheritance in boulders from the Symmes Creek catchment.

Dühnforth et al. (2007) attribute the presence of solely Holocene ages on the Symmes Creek fan to the fact that the lack of fan head incision prevents the preservation of older fan lobes at the surface and instead leads to continuous re-surfacing of the entire fan surface. At Shepherd Creek the deep incision of the fan head led to abandonment of the older Pleistocene debris-flow deposits and a downfan shift of the younger depositional lobe. A longer depositional record is seen in the surface deposits at the Shepherd Creek fan. These results emphasize that the degree of fan head incision plays an important role for the length of the preserved depositional record and the implied incompleteness of the record through any single fan lobe.

At Symmes Creek fan there is no evidence for glacial modification in the drainage basin, and consequently the longitudinal valley profile is a classic fluvial profile with its smooth and concave-up shape. Sediment is stored throughout the drainage basin and readily available. The sediment transport-limiting factor is the transport capacity of the fluvial system. There are no overdeepenings or bedrock steps that could serve as barriers for sediment transport. The lack of sediment-retaining mechanisms leads to continuous sediment export from the basin over glacial-interglacial time scales. Flows that are exiting the Symmes catchment are generally sediment-saturated such that they have no capacity to entrain additional sediment and to incise the fan head (Dühnforth et al. 2008).

In contrast, in the Shepherd Creek drainage basin, the glacial modification of the valley reaches about halfway down from the drainage divide. About 90% of the entire sediment in this catchment is stored in this glacially modified part of the basin. Characteristic bedrock barriers and overdeepenings, which today are occupied by small alpine lakes, are ideal locations to store sediment, and furthermore likely reduce or even prevent sediment export from the drainage basin until the next glaciation. The sedimentary processes in the Shepherd Creek catchment vary significantly from glacial to interglacial (interstadial) conditions. During glacial times, the glacier acts as an efficient sediment transport agent and entrains sediment throughout the drainage basin. Sediment exits the catchment in saturated flows, which serve to aggrade the fan surface. During interglacial and interstadial times, sediment is trapped in the glacially modified part of the basin. Flows exiting the catchment will therefore be sediment-undersaturated and able to incise the fan head. Furthermore, the cosmogenic nuclide dating of the fan surfaces supports this conceptual model as the onset of the fan head incision overlaps with the timing of complete deglaciation dated at sites further north in the Sierra Nevada (Clark and Gillespie 1997; James et al. 2002). Dühnforth et al. (2008) could show that all along the eastern side of the Sierra Nevada, fans with deeply incised heads drain catchments that were moderately to extensively glaciated in the past. The potential for storage is higher the more area within a drainage basin has been modified by glaciers, especially over several glaciations.

5 Summary and Conclusions

Abandoned alluvial fans are especially well suited to dating with cosmogenic nuclide methods. The spectrum of cosmogenic nuclide methods available for dating of fan deposits is broad and in part dependent upon the size of the clasts found on the fan surface. Younger fans, especially those built up by debris flows may be dated by analyzing boulder surfaces as shown by the Owens Valley case study presented here. In comparison to other dating methods, it is only with cosmogenic nuclides that the timing of deposition of fan boulders can be determined directly. On older alluvial fans (more than several 100 kyr) boulders crumble and degradation of the fan surface leads to smoothing out of the original bar and swale morphology (Matmon et al. 2006). Such fans can be dated by determination of cosmogenic nuclide concentrations in samples of amalgamated or individual clasts or with bulk coarse sediment from the surface and in depth profiles. Again this is the only method where the fan upper surface is dated directly. Depth profiles are especially helpful for estimating inheritance, which can be a serious concern for the dating of alluvial fans. Inheritance can have been acquired in the catchment, in intermediate storage, during transport, or as clasts are moved from older to younger lobes. Reflecting the rezeroing effect of glaciers on cosmogenic nuclide concentrations in bedrock, inheritance may be greater in fan deposits originating in unglaciated catchments. For even older deposits burial or isochron dating methods using two nuclides are called upon. Burial and isochron methods show when the analyzed sediment was buried, nuclide production ceased and nuclide concentrations change due to decay only. In contrast, exposure dating of boulder tops, amalgamated clast samples or depth-profile dating date abandonment of the upper fan surface, recording either movement away from the stream due to faulting or incision at the fan head and shifting of the locus of deposition. Never-the-less, important questions remain about whether dates represent the interval of lobe construction or whether they pinpoint timing of fan abandonment. Although most fan studies have aimed at determining tectonic slip rates, such studies and those directed at studying the fans themselves have led to a marked increase in our understanding of fan evolution and related catchment processes through time.

Acknowledgements The authors thank B. Guralnik and M. Stoffel for several suggestions for improving this manuscript.

References

Akçar N, Deline P, Ivy-Ochs S, Alfimov V, Hajdas I, Kubik PW, Christl M, Schlüchter C (2012) The AD 1717 rock avalanche deposits in the upper Ferret Valley (italy): a dating approach with cosmogenic ^{10}Be. J Quat Sci 27:383–392

Alfimov V, Ivy-Ochs S (2009) How well do we understand production of ^{36}Cl in limestone and dolomite? Quat Geochronol 4:462–474

Anderson RS, Repka JL, Dick GS (1996) Explicit treatment of inheritance in dating depositional surfaces using ^{10}Be and ^{26}Al. Geology 24:47–51
Balco G, Rovey CW II (2008) An isochron method for cosmogenic-nuclide dating of buried soils. Am J Sci 308:1083–1114
Balco G, Stone JO, Lifton NA, Dunai TJ (2008) A complete and easily accessible means of calculating surface exposure ages or erosion rates from ^{10}Be and ^{26}Al measurements. Quat Geochronol 3:174–195
Barnard PL, Owen LA, Finkel RC (2006) Quaternary fans and terraces in the Khumbu Himal south of Mount Everest: their characteristics, age and formation. J Geol Soc 163:383–399
Benn DI, Owen LA, Finkel RC, Clemmens S (2006) Pleistocene lake outburst foods and fan formation along the eastern Sierra Nevada, California: implications for the interpretation of intermontane lacustrine records. Quat Sci Rev 25:2729–2748
Bierman PR, Gillespie AR, Caffee MW (1995) Cosmogenic ages for earthquake recurrence intervals and debris flow fan deposition, Owens Valley, California. Science 270:447–450
Bierman PR, Caffee MW, Davis PT, Marsella K, Pavich M, Colgan P, Mickelson D, Larsen J (2002) Rates and timing of earth surface processes from in-situ produced cosmogenic ^{10}Be. In: Grew E (ed) Beryllium: mineralogy, petrology, and geochemistry, Reviews in mineralogy and geochemistry. Mineralogical Society of America, Washington, DC, pp 147–204
Blair TC, McPherson JG (1994) Alluvial fan processes and forms. In: Abrahams AD, Parsons AJ (eds) Geomorphology of desert environments. Chapman & Hall, London, pp 354–402
Brocklehurst SH, Whipple KX (2002) Glacial erosion and relief production in the eastern Sierra Nevada, California. Geomorphology 42:1–24
Brown ET, Bourlès DL, Burchfiel BC, Qidong D, Jun L, Molnar P, Raisbeck GM, Yiou F (1998) Estimation of slip rates in the southern Tien Shan using cosmic ray exposure dates of abandoned alluvial fans. Geol Soc Am Bull 110:377–386
Bull WB (1977) The alluvial fan environment. Prog Phys Geogr 1:222–270
Bull WB (1991) Geomorphic responses to climate change. Oxford University Press, New York
Cerling TE, Craig H (1994) Geomorphology and in-situ cosmogenic isotopes. Annu Rev Earth Planet Sci 22:273–317
Clark DH, Gillespie AR (1997) Timing and significance of late-glacial and Holocene glaciation in the Sierra Nevada, California. Quat Int 38(39):21–38
Cockburn HAP, Summerfield MA (2004) Geomorphological applications of cosmogenic isotope analysis. Prog Phys Geogr 28:1–42
Dühnforth M (2007) Sediment flux and deposition on arid-region fans, eastern California, USA. ETH dissertation No. 17020, Swiss Federal Institute of Technology, Zurich
Dühnforth M, Densmore AL, Ivy-Ochs S, Allen PA, Kubik PW (2007) Timing and patterns of debris flow deposition on Shepard and Symmes creek fans, Owen Valley, California, deduced from cosmogenic ^{10}Be. J Geophys Res 112:F03S15. doi:10.1029/2006JF000562
Dühnforth M, Densmore AL, Ivy-Ochs S, Allen PA (2008) Controls on sediment evacuation from glacially modified and unmodified catchments in the Sierra Nevada, California. Earth Surf Process Landf 33:1602–1613
Frankel KL, Dolan JF, Owen LA, Finkel RC, Hoeft JS (2007) Spatial variations in slip rate along the Death Valley-Fish Lake Valley fault system from LiDAR topographic data and cosmogenic ^{10}Be geochronology. Geophys Res Lett 34. doi:10.1029/2007GL030549
Gosse JC, Phillips FM (2001) Terrestrial in situ cosmogenic nuclides: theory and application. Quat Sci Rev 20:1475–1560
Gosse J, McDonald E, Pederson J, Stockli D, Lee J, Stockli L, Yang G (2004) Oxygen isotope stage 4 sediment dominates the U.S. southwest alluvial record. Geol Soc Am Abstr Programs 36:307
Granger DE (2006) A review of burial dating methods using ^{10}Be and ^{26}Al. In: Siame LL, Bourles DL, Brown ET (eds) In situ cosmogenic nuclides and their applications in earth sciences, Geological Society of America special paper 415. Geological Society of America, Boulder, pp 1–16

Hancock GS, Anderson RS, Chadwick OA, Finkel RC (1999) Dating of fluvial terraces using ^{10}Be and ^{26}Al profiles, Wind River, Wyoming. Geomorphology 27:41–60
Ivy-Ochs S, Kober F (2009) Surface exposure dating with cosmogenic nuclides. Quat Sci J 57:157–189
Ivy-Ochs S, Schaller M (2010) Examining processes and rates of landscape change with cosmogenic radionuclides. In: Froehlich K (ed) Environmental radionuclides – tracers and timers of terrestrial processes. Elsevier, Amsterdam, pp 232–294
James LA, Harbor J, Fabel D, Dahms D, Elmore D (2002) Late Pleistocene glaciations in the northwestern Sierra Nevada, California. Quat Res 57:409–419
Lal D (1991) Cosmic ray labeling of erosion surfaces: in-situ nuclide production rates and erosion models. Earth Planet Sci Lett 104:424–439
Lal D, Peters B (1967) Cosmic ray produced radioactivity on the Earth. In: Handbuch der Physik. Springer, Berlin, pp 551–612
Lee J, Garwood J, Stockli DF, Gosse J (2009) Quaternary faulting in Queen Valley, California-Nevada: implications for kinematics of fault-slip transfer in the eastern California shear zone-Walker Lane belt. Geol Soc Am Bull 121:599–614
Machette MN, Slate JL, Phillips FM (2008) Terrestrial cosmogenic-nuclide dating of alluvial fans in Death Valley, California, Geological Survey professional paper 1755. U.S. Geological Survey, Reston
Matmon A, Schwartz DP, Finkel R, Clemmens S, Hanks T (2005) Dating offset fans along the Mojave section of the San Andreas fault using cosmogenic Al-26 and Be-10. Geol Soc Am Bull 117:795–807
Matmon A, Nichols K, Finkel RC (2006) Isotopic insights into smoothening of abandoned fan surfaces, Southern California. Quat Res 66:109–118
Matmon A, Simhai O, Amit R, Haviv I, Porat N, McDonald E, Benedetti L, Finkel R (2009) Desert pavement-coated surfaces in extreme deserts present the longest-lived landforms on Earth. Geol Soc Am Bull 121:688–697
Mériaux A-S, Tapponnier P, Ryerson FJ, Xu X, King G, Van der Woerd J, Finkel RC, Li H, Caffee MW, Xu Z, Chen W (2005) The Aksay segment of the northern Altyn Tagh fault: tectonic geomorphology, landscape evolution, and Holocene slip rate. J Geophys Res Solid Earth 110(B4):B04404
Nichols KK, Bierman PR, Eppes MC, Caffee M, Finkel R, Larsen J (2007) Timing of surficial process changes down a Mojave Desert piedmont. Quat Res 68:151–161
Nishiizumi K, Kohl CP, Arnold JR, Klein J, Dorn R, Fink D, Middleton R, Lal D (1993) Role of in situ cosmogenic nuclides ^{10}Be and ^{26}Al in the study of diverse geomorphic processes. Earth Surf Process 18:407–425
Phillips WM, McDonald EV, Reneau SL, Poths J (1998) Dating soils and alluvium with cosmogenic ^{21}Ne depth profiles: case studies from the Pajarito Plateau, New Mexico, USA. Earth Planet Sci Lett 160:209–223
Phillips FM, Stone WD, Fabryka-Martin JT (2001) An improved approach to calculating low-energy cosmic-ray neutron fluxes near the land/atmosphere interface. Chem Geol 175:689–701
Repka JL, Anderson RS, Finkel RC (1997) Cosmogenic dating of fluvial terraces, Fremont River, Utah. Earth Planet Sci Lett 152:59–73
Ritz J-F, Vassallo R, Braucher R, Brown ET, Carretier S, Bourlès DL (2006) Using in situ–produced ^{10}Be to quantify active tectonics in the Gurvan Bogd mountain range (Gobi-Altay, Mongolia). In: Siame LL, Bourles DL, Brown ET (eds) In situ–produced cosmogenic nuclides and quantification of geological processes, Geological Society of America special paper 415. Geological Society of America, Boulder, pp 61–86
Ryerson FJ, Tapponnier P, Finkel RC, Mériaux A-S, van der Woerd J, Lasserre C, Chevalier ML, Xu XW, Li HB, King GCP (2006) Application of morphochronology to the active tectonics of Tibet. In: Siame LL, Bourles DL, Brown ET (eds) In situ–produced cosmogenic nuclides and quantification of geological processes, Geological Society of America special paper 415. Geological Society of America, Boulder, pp 61–86

Schimmelpfennig I, Benedetti L, Finkel R, Pik R, Blard PH, Bourles D, Burnard P, Williams A (2009) Sources of in-situ Cl-36 in basaltic rocks: Implications for calibration of production rates. Quat Geochronol 4:441–461

Schürch P, Densmore AL, Rosser NJ, McArdell BW (2010) Controls on sediment transfer and storage on debris-flow fans. Geophys Res Abstr 12:EGU 2010–488

Seong YB, Bishop MP, Bush A, Clendon P, Copland L, Finkel RC, Kamp U, Owen LA, Shroder JF (2009) Landforms and landscape evolution in the Skardu, Shigar and Braldu Valleys, Central Karakoram. Geomorphology 103:251–267

Siame LL, Bourlès DL, Sebrier M, Bellier O, Castano JC, Araujo M, Perez M, Raisbeck GM, Yiou F (1997) Cosmogenic dating ranging from 20 to 700 ka of a series of alluvial fan surfaces affected by the El Tigre fault, Argentina. Geology 25:975–978

Spelz RM, Fletcher JM, Owen LA, Caffee MW (2008) Quaternary alluvial-fan development, climate and morphologic dating of fault scarps in Laguna Salada, Baja California, Mexico. Geomorphology 102:578–594

Vassallo R, Ritz J-F, Braucher R, Carretier S (2005) Dating faulted alluvial fans with cosmogenic ^{10}Be in the Gurvan Bogd mountain range (Gobi-Altay, Mongolia): climatic and tectonic implications. Terra Nova 17:278–285

Whipple KX, Dunne T (1992) The influence of debris-flow rheology on fan morphology, Owens Valley California. Geol Soc Am Bull 104:887–900

Zehfuss PH, Bierman PR, Gillespie AR, Burke RM, Caffee MW (2001) Slip rates on the Fish Springs Fault, Owens Valley, California, deduced from cosmogenic ^{10}Be and ^{26}Al and soil development on fan surfaces. Geol Soc Am Bull 113:241–255

Radiocarbon Dating: Alluvial Fan/Debris Cone Evolution and Hazards

Richard Chiverrell and Matthias Jakob

1 Introduction

Radiocarbon dating has contributed considerably to the understanding of alluvial fan and debris cone evolution, as well as their hazard and risk potential. This development mirrors the contributions made in investigations of other depositional environments since invention of the technique in the 1950s (Cook and van der Plicht 2007; Jull 2007). Radiocarbon dating involves measuring the amount of the radioisotope ^{14}C preserved in fossil organic materials and using the rate of radioactive decay to calculate the age for a given sample. The application of radiocarbon dating to secure the development and evolution of alluvial fans requires the deposition and preservation of organic materials within the sedimentary stratigraphy. The dating technique has been used widely to reconstruct the development of alluvial fans in temperate to sub-arctic regions where environmental conditions are conducive to the presence and preservation of organic materials. The age limit to the application of radiocarbon dating ranges from approximately 45,000 years to typically several hundred years. Thus the technique is relevant for constraining the age of alluvial fans since the middle of the last cold stage, Marine Isotope Stage (MIS) 2 (45–11.5 ka) through the current interglacial (11.5–0 ka), the Holocene.

The evolution of alluvial fans can be attributed to tectonic activity, catchment landuse practices, climate fluctuations and extreme events such as storms, stand-replacing wildfires and large landslides, modulated by the internal dynamics of the

R. Chiverrell (✉)
School of Environmental Sciences, University of Liverpool, Roxby Building, Liverpool L69 7ZT, UK
e-mail: rchiv@liv.ac.uk

M. Jakob
BGC Engineering Inc, Suite 500-1045, Howe Street, Vancouver, BC, Canada, V6Z 2A9
e-mail: mjakob@bgcengineering.ca

M. Schneuwly-Bollschweiler et al. (eds.), *Dating Torrential Processes on Fans and Cones*, Advances in Global Change Research 47, DOI 10.1007/978-94-007-4336-6_17,

depositional environment. As a consequence, uncovering the temporal evolution of alluvial fan systems can sometimes be linked to such key drivers affecting the feeder catchment (e.g. Chiverrell et al. 2007, 2008, 2009b).

Basin entry and tributary junction landforms comprise a process continuum that extends from steep angle debris cones dominated by rockfall and debris flows to lower-angle fans whose sediments can be attributed to mostly alluvial processes (Miall 1996). Considerable variation in the nature of the environmental history can be discerned from both the geomorphology and stratigraphy of alluvial fans. The deposits that accumulate across this continuum from colluvial to alluvial processes-dominated systems range from sediment gravity-flow deposits, poorly sorted and often matrix-supported angular boulders and gravels, high energy often distributary fluvial deposits, poorly sorted clast-supported boulders and gravels, to lower energy well sorted sand and silt alluvial deposits. In larger alluvial fan systems this range of deposits can occur as a spatially variable sequence from coarser gravels in fan-head to finer grained deposits in more distal settings. As a consequence there is an inherent complexity to the sedimentary and geomorphic record, which challenges our understanding of the long-term evolution of alluvial fans, and the development of robust chronological control for the sediments is often the key.

2 Methodological Considerations

2.1 Scatter and Calibration of Radiocarbon Ages

Radiocarbon dating has provided important temporal control over the evolution of alluvial fan systems; however there are a number of methodological considerations that limit what can be achieved by radiocarbon dating. First, radiocarbon measurements need to be corrected for the way in which the Earth's geomagnetic field intensity, solar variability (de Vriess wiggles) and changes of the global carbon cycle have affected the relative concentrations of carbon isotopes in the atmosphere through time (Burr 2007; van der Plicht 2007; Fig. 1). This explains the discrepancy between radiocarbon and calendar years before present (BP), where present is defined as AD 1950. These relationships are well understood from research investigating the radiocarbon ages of known age samples, typically tree rings. Series of individual tree rings, whose ages are known to the year through dendrochronological methods, have been radiocarbon-dated to produce dendrocalibration curves to correct radiocarbon ages into calendar years (Fig. 1) (e.g. the INTCAL04 calibration curve: Reimer et al. 2004). Software packages like Oxcal (Ramsey 2008b) and CALIB (Stuiver et al. 2005) calibrate radiocarbon measurements and their associated measurement errors into calendar years. For individual age determinations and the reporting of radiocarbon results, the Trondheim convention (Stuiver and Kra 1986), proposes a listing of the raw measurement, the measurement error, the materials dated, the method of measurement and the unique

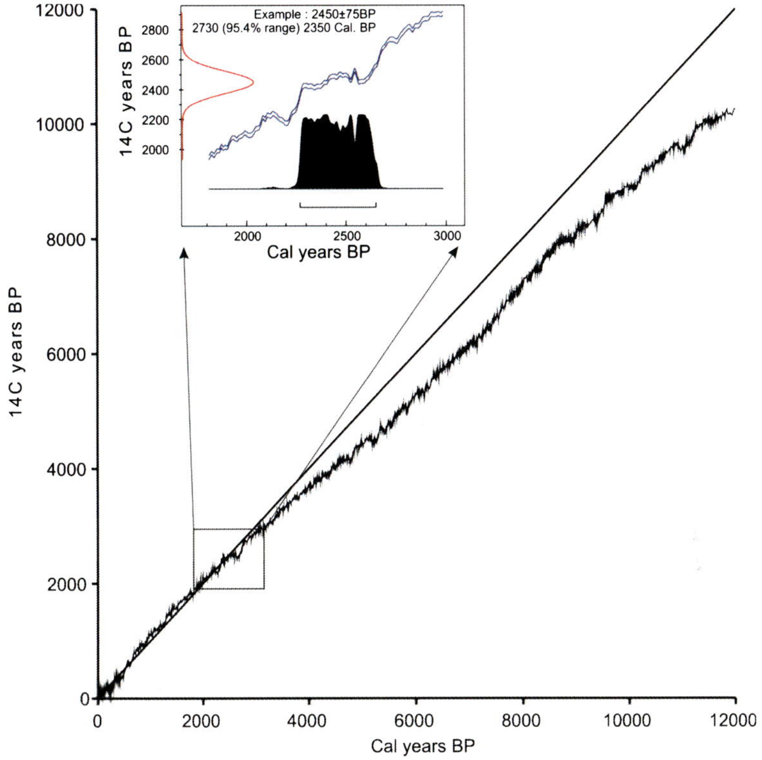

Fig. 1 The INTCAL04 dendrocalibration curve (Reimer et al. 2004) and a one-to-one line showing the relationship between radiocarbon and calendar years before present (*BP*), where present is defined as AD1950. *Inset* shows the calibration of an exemplar radiocarbon measurement of 2,450 ± 75BP undertaken using Oxcal (Ramsey 2008b) and the INTCAL04 curve (Reimer et al. 2004)

laboratory code for that sample. The calibrated ages should be quoted in calibrated years (either BP or AD/BC) giving the two standard deviation age range (95% confidence interval), along with calibration software and curve used to undertake the calibration (Fig. 1).

Another important constraint on radiocarbon dating is that because radiocarbon dating is a probabilistic process each measured age and associated measurement error, for example the age 2,475 ± 75 radiocarbon years BP (Fig. 1). In fact each radiocarbon measurement forms a scatter around the true age of the sample and repeated measurements of the same sample materials will produce different results. To demonstrate this, we have simulated 20 ages which could be obtained for a single stem of a hypothetical plant incorporated and preserved in a debris-flow levee on an alluvial fan during 300 BC. The ages were simulated using Oxcal and reflect the natural scatter and errors that would be generated by repeat dating of the same sample. Figure 2 shows the individual and the cumulative probability function

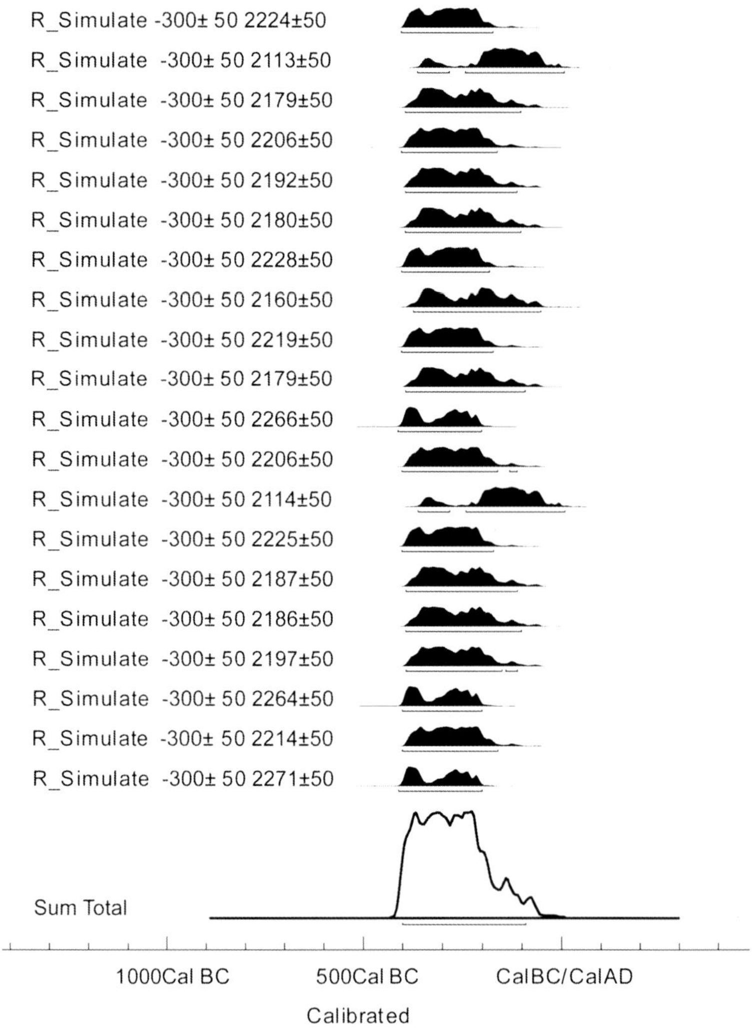

Fig. 2 Series of 20 simulated (R_simulate) radiocarbon ages for 300 BC, and each with a uniform error of ±50 years. The R_Simulate function generates radiocarbon dates with random but realistic errors for a known age event. The radiocarbon ages were calibrated and the cumulative probability of all the ages (Sum Total) calculated using the Sum command in Oxcal 4.1 (Ramsey 2008b) and the INTCAL04 curve (Reimer et al. 2004)

(CPF) produced by radiocarbon calibration of these 20 hypothetical radiocarbon ages with errors of ±50 BP, which shows the considerable range of ages in dating the debris flow. Awareness of the impact of scatter on the imprecision of individual radiocarbon measurements is important, because it is an addition to the measurement errors and dendrocalibration ranges (2 sigma) that are listed routinely for individual age determinations. Allowing for measurement error, the likely scatter and calibration range of an individual radiocarbon measurement provides an age range at best of 200–300 years. Obtaining a series of ages for a stratigraphic or geomorphic sequence and statistical treatment of the age-stratigraphy relationships can decrease some of the sample bias introduced by scatter on the overall age control. Refining the chronology further is only possible by obtaining greater numbers of radiocarbon ages of the same geophysical event, using other dating techniques (e.g. optically-stimulated luminescence dating) or information (archaeological horizons and artefacts), and by modelling the geochronologic control (see Sect. 3).

2.2 Depositional Contexts and Directional Lags

Alluvial fans and debris cones provide a wide range of depositional contexts (e.g. peat bogs, former channels, debris lobes and overbank alluvium) in which organic materials can be preserved, and materials dated control the information that can be gathered from each radiocarbon measurement. These contexts are similar in character to the ensembles described for riverine environments by Lewin et al. (2005). The nature of the depositional environment is important for two reasons. Firstly, not all contexts provide information about geomorphic changes, for example detrital wood incorporated in the middle of a unit of alluvial fan gravel provides no geochronologic information about the onset and cessation of aggradation of that gravel. Any age estimate is a time marker floating in the total duration of unit deposition. Secondly, there is often a directional age relationship between the dated material and the geomorphic change, for example a valley floor peat deposit that gets buried by prograding alluvial fan deposits can be radiocarbon-dated but that radiocarbon measurement has to pre-date the overlying fan deposits. Thus there is a directional lag, often of unknown duration, between the radiocarbon dated materials and the geomorphic change that is perhaps the subject of the investigation. This lack of contemporaneity is typical of radiocarbon ages obtained in geomorphic research.

It is critical to assess the nature of directional lags between radiocarbon ages and changes in alluvial fan systems. Radiocarbon measurements provide ages that can give minimum or older than constraint, *terminus ante quem* (TAQ), for example the relationship between the basal layers of peat deposits that overlie an older alluvial fan unit. Other contexts, for example radiocarbon measurements obtained for organic soil or peat deposits buried beneath alluvial fan gravels have a younger than or *terminus post quem* (TPQ) relationship with the overlying geomorphology. Short-lived plant detritus (e.g. leaves or seeds) incorporated within a debris-flow unit would have a contemporaneous relationship with the geomorphic event. This issue of directional time-lags has an impact on, and should be incorporated in, the strategies employed to interrogate series of radiocarbon ages because unless both (i) the nature of the environment or environmental change at the radiocarbon measurement and (ii) the nature of the contemporaneity relationship between the change and the radiocarbon measurement are known and acknowledged, the information derived can be meaningless.

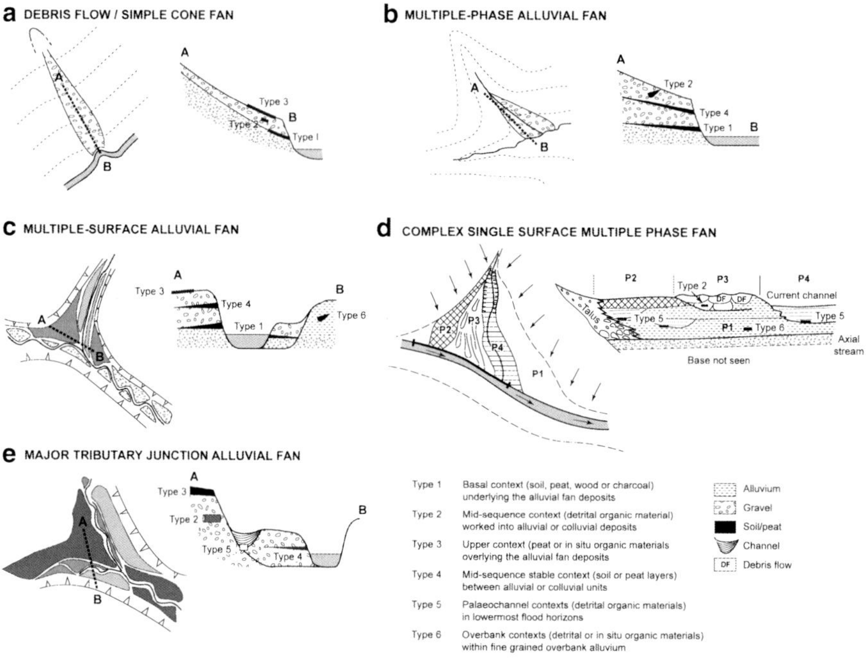

Fig. 3 Geomorphic and sedimentary contexts for radiocarbon dating of geomorphic changes in alluvial fans and debris cones

There are many depositional contexts on alluvial fans and debris cones that provide materials for radiocarbon dating, and they are reviewed focusing on their contribution in securing the timing of geomorphic development (Fig. 3). The types of debris cone and alluvial fan range across simple gully-fed single-phase debris cones or fans (i.e. those laying down a structureless single unit of gravel perhaps pertaining to a single event), multiple-phase gully-fed single surface fans (i.e. those with evidence for repeating phases of activity and stability), complex larger single surface multiple phase fans, multiple surface or terraced alluvial fans, and major tributary alluvial fans. *Type I* contexts are those that underlie and predate a debris flow or alluvial fan unit, and typically the materials dated are from stable types of environments where soil, peat or other organic-rich deposits have developed (Fig. 3a–c). Here the age would provide maximum age control over later stages of fan development, but any erosional hiatus or delay between the organic material and overlying unit would render the duration of the lag difficult to gauge. In the case of small hillslope gully or debris-flow systems (e.g. Harvey et al. 1981; Chiverrell et al. 2007, 2008, 2009b) dating Type 1 contexts can secure gully or debris-flow inception (Fig. 3a). Type 2 contexts are organic materials from within an alluvial fan unit (Fig. 3a, b) and typically have been reworked by alluvial or colluvial processes, which have an unknown temporal relationship with the cessation of the unit and could predate the onset in the case of older reworked materials. Type 3 contexts are further stability horizons, for example a peat, where the organic materials overlie and postdate an alluvial fan unit (Fig. 3a–c) but could do so by an unknown lag encompassing the period between fan aggradation and peat or soil inception. Type 4 contexts are horizons reflecting stability within a sediment sequence, for example an organic-rich unit between two alluvial fan units (Fig. 3b, c). Here radiocarbon assays on the organic layer can secure the duration of the period of stability, and provide a maximum age for the underlying fan sediments and a minimum age for the overlying phase of alluvial fan development. Some caution is needed with interpretation of radiocarbon dates from these contexts because of the potential for a lag between stabilization and the onset of organic sedimentation. The upper surface can also be truncated by erosion preceding deposition of the overlying alluvial fan deposits.

All these types of setting are typically associated with the active channel and/or debris lobe and tend to repeat, but become more complicated in larger and longer duration alluvial fans with either multiple phase terraced morphologies (e.g. Harvey and Renwick 1987) or multiple phased/lobed alluvial fans with a more uniform surface topography (e.g. Jakob and Friele 2010; Jakob and Weatherly 2005). These environments introduce the complication of a range of former channel and overbank contexts that often also contain dateable materials (Lewin et al. 2005). The fills (Type 5) of former channels after they have been cut-off or avulsed can contain organic materials, but the accumulation of a back-channel fill can lag the abandonment event (Fig. 3d, e). Overbank settings, particularly on lower angle alluvial fans introduce other problems, including understanding the relationship to changes in the channel zone. Care is needed with regard to interpretation of the stratigraphy when radiocarbon dating of organic materials within an overbank alluvial sequences on fans (Type 6), because these types of context often provide an age within the accumulation of the sequence, but are difficult to relate to phases of fan aggradation (Fig. 3d, e).

2.3 *Problematic Materials*

The application of radiocarbon dating has long been constrained by materials that are difficult to date or where the nature of the materials constrains what can be achieved by radiocarbon dating. For example, the "hard water effect" associated with dating materials in areas that have substrates rich with carbonate, because the abundance of old carbon in bedrock and other parent materials (sediments and soils) can produce ages that are anomalously old. Reworking of organic materials is a further problem on alluvial fans and debris cones where geomorphic processes are dynamic and reworking of sediments is widespread often including the organic materials used for dating. As a result the materials selected for dating are often not contemporaneous with the time of deposition. These problems can be reduced to some extent by targeting *in situ* organic materials, for example peat beds, or through the careful selection of short-lived soft tissue plant macrofossils, for example seeds and deciduous leaves that date to a single year (Chiverrell et al. 2009a). Radiocarbon measurements on the whole sample or organic fractions (humic acids and the humin fraction) should perhaps also be avoided if possible owing to the recycling of organic materials to the alluvial system. Certain fossil types have their own problems. For example detrital wood stemming from trees can have been reworked or the heartwood of mature trees can have a considerable age, hundreds of years, whereas targeting the outer ring-wood reduces any lag between the dateable materials and the time of the depositional process. Charcoal can also be problematic, with the age of the materials related to the time of burning rather than the time of deposition on the fan although debris flows associated with wildfires often occur within a few years of the wildfire (i.e. Cannon and Gartner 2005) and are therefore well within the dating error.

At some sites chronologies for alluvial fans have been secured through radiocarbon dating of buried soils or paleosols. However, radiocarbon analyses on contemporary soils highlight a significant problem with residence time of the soil carbon (Scharpenseel and Becker-Heidmann 1993): Replenishment with new carbon and transformations of carbon into organic fractions each result in differing ages. Thus though dating of buried soils is possible assuming the soil carbon has been isolated or locked-in, the targeting different fractions at various depths through the former soil for radiocarbon may produce a range of ages. Harvey (1996, 1997) tested this for soils buried beneath alluvial fan gravels in the Howgill Fells (NW England), by dating the uppermost and lowest horizons of a buried soil and analyzing the humin and humic acid fractions within the soil organic matter. This followed the approached outlined by Matthews (1985, 1993) when radiocarbon dating soils buried beneath glacial sediments associated with the late Holocene re-advance of glaciers in Norway. In these cases humin appeared the least mobile fraction in the buried soils and generally provide older radiocarbon ages, whereas humic acids were mobile and often provided younger dates (Matthews 1993). Between them humin fraction towards the base and the humic acid fraction from the top of a buried soil produced radiocarbon measurements that appear to secure the duration of soil accumulation. Individually the ages also gave older than constrained for underlying and younger than constrained for overlying alluvial fan sediments respectively (Harvey 1996, 1997).

3 Geomorphic Sequences and Modelling Approaches

Geomorphologists are increasingly modelling the geochronologic control produced to constrain landform evolution using Bayesian modelling (Chiverrell et al. 2009a, 2010). Bayesian approaches can be used to interrogate the age control for sediment sequences, generating age–depth models and assisting the comprehension of events (Ramsey 1995, 2008a, 2009). They are applied more widely in lakes and peat bogs where the depositional processes are sequential and largely uninterrupted. However, these approaches are increasingly being used in geomorphic studies where the depositional processes comprise spatially variable phases of erosion, non-deposition and deposition (Chiverrell et al. 2009a, b). In many cases geomorphologists can build theoretical models for the order of events, the *prior* model in Bayesian nomenclature, and the series of age measurements represent the *likelihood* information for example for an individual alluvial fan. The *prior* model and *likelihood* information (e.g. radiocarbon ages) can then be explicitly combined and tested by Bayesian modelling producing *posterior* probability densities (modelled age ranges) to better constrain the geomorphic evolution (Ramsey 2008a; Chiverrell et al. 2009a). These techniques allow non-contemporaneous relationships and directional lags to be defined explicitly and tested in a sequence of age determinations. In geomorphic settings the 'Sequence' model is probably the most applicable, and requires only that the relative order of events can be discerned or hypothesized rather than the rate or changes in rate of deposition (Ramsey 2008a, 2009). These approaches assist in also identifying outlier ages and problems with conceptual models, though the problems still require explanation and dealing with, for example excluding an age affected by reworked organic materials.

The approach advocated and demonstrated below is to interpret the landscape and sediment record to build one or more theoretical models for the relative order of events and the chronological control. This approach allows greater numbers of age determination, directional lags and differing contemporaneity relationships to be incorporated in a single or several relative order models. Bayesian modelling (Buck et al. 1996; Blockley et al. 2004) is then used to test these hypothetical 'relative order' models. The modelling tests the expected chronological order of the dating control and identifies outlier ages and/or contexts that are out of sequence. These models can breakdown and produce low agreement indices, when the modelled *posterior* probability densities do not overlap with the *likelihood* probability distributions (radiocarbon ages). In these low agreement cases there are either problems with the relative order model or individual ages. In either case the problems need explaining, for example identifying erroneously too-old ages because of the reworking of organic materials. In addition, whilst a simple calibrated radiocarbon age range is broadly an accurate estimate of the age of a sample, we may wish for more information, for example the timing of geomorphic changes between or within a series of dated samples. The (pseudo-) stratigraphical relationships between dated samples can be used to refine the dating control implicit in a series of ages and to generate probabilities for events that have not been explicitly dated. It must be emphasized that the modelled probability functions or age estimates produced by these Bayesian approaches are interpretative, and will change as further data are added or if modelled from different perspectives.

To demonstrate the application of Bayesian modelling of a series of radiocarbon measurements for an individual alluvial fan/debris cone, the Blakethwaite fan in the Howgill Fells, northwest England, is presented here as an example (Fig. 4) (Harvey 1996; Chiverrell et al. 2007). At Blakethwaite debris-slide deposits provide evidence of small slope failure. The debris slide overlies charcoal dated to $2{,}975 \pm 85$ years BP, and is buried by soil, the formation of which is secured by dating the humin fraction of the base of the lower buried soil to $1{,}900 \pm 60$ years BP. Overlying this soil, alluvial fan gravels associated with the development of an extensive feeder

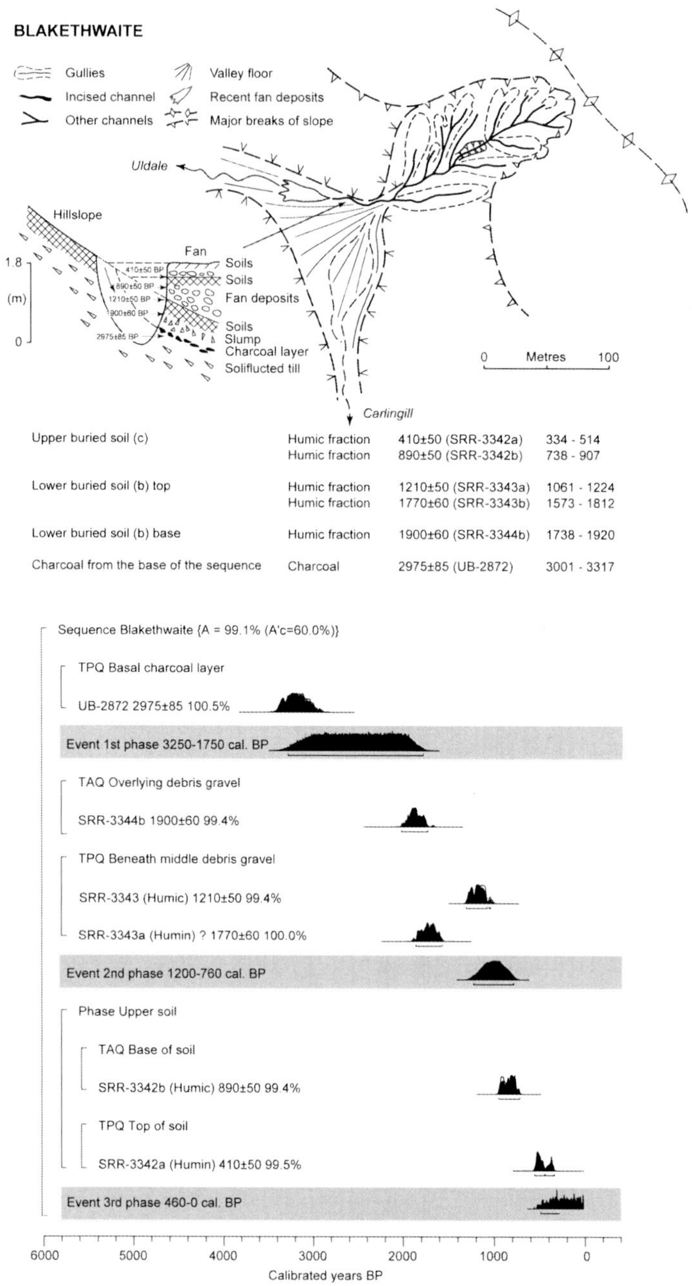

Fig. 4 Geomorphology, stratigraphic sections and geochronology of the Blakethwaite alluvial fan, Howgill Fells (after Harvey 1997; Chiverrell et al. 2007). Bayesian analysis of the relative order model and the probability distributions of dating control for the Blakethwaite alluvial fan, with the *large square brackets down the left hand side* along with the OxCal keywords define the relative order model, with the event A, B and C the estimated ages for the phases of gully instability and alluvial fan progradation

gully network are constrained by radiocarbon dating of the humic acid fraction of the top of the lower soil to after 1,210 ± 50 years BP and to before 890 ± 50 years BP using the humin fraction from base of an overlying upper buried soil. Younger alluvial fan gravels burying the upper soil reflect a later phase of alluvial fan activity and associated gully re-activation that has been constrained to after 410 ± 50 years BP by dating the humic acid fraction of the top of the upper soil (Harvey 1996). The sections at Blakethwaite form a continuous sequence getting younger from bottom to top and contain three phases of hillslope instability and debris cone/alluvial fan development. It is possible to arrange the stratigraphy and dating control into a relative order or Sequence model. Bayesian modelling (Buck et al. 1996; Ramsey 2008a, 2009) was employed to test this hypothetical model (Fig. 4) of the relative chronological order of events using the Oxcal software (Ramsey 2001, 2008b). These relative order models cannot rule out individual dates, but they can highlight potential problems.

The model shows good agreement between the radiocarbon measurements and the proposed relative order (agreement index = 99.1%: threshold 60% for acceptable agreement is a little arbitrary but broadly equivalent to a X^2 test); but to achieve this, SRR-3343-a from the middle soil has been excluded owing to the low agreement index of this age (denoted by '?' on Fig. 4). The 1st phase is constrained by TPQ constraint from basal charcoal and TAQ constraint from the lower soil and occurred in the period 3,250–1,750 cal. years BP. The two later phases of hillslope instability and alluvial fan progradation, 2nd and 3rd phases, are also constrained by bracket TPQ and TAQ dating to 1,200–760 and 460–0 cal. years BP, respectively. The Bayesian analysis of the dating control (Fig. 4) has rigorously tested the sequence at Blakethwaite, incorporated TAQ and TPQ constraining ages, and assessed the relative merits of each age estimate in relation to the geomorphic question of the timing of gully incision and alluvial fan aggradation at Blakethwaite in the central Howgill Fells.

4 Case Studies

4.1 Alluvial Fan Development and Hillslope Gullying in NW England

The drivers of the environmental changes recorded in the accumulation of alluvial fan deposits vary considerably, in the uplands of the northern British Isles much of the debate has focused on the interplay between climate, extreme events (storm flows) and human alteration of the landscape (Harvey et al. 1981). The past decade has seen concerted efforts to refine the chronology for alluvial fan sedimentation and associated development of hillslope gully systems in the uplands of northwest England and southwest Scotland. There, over 60 radiocarbon measurements have been obtained for organic-rich units associated with alluvial fan deposits (see Chiverrell et al. 2007, 2008, 2009b; Foster et al. 2008). Most of these alluvial fans and debris cones are small (0.01 km^2) fed by gully systems cut into hillslopes mantled with glacial and soliflucted diamictons left after the last ice age (Marine Isotope Stage 2). At each site organic soils or peat either overlie, underlie or lie within alluvial fan gravels, and the phases of alluvial fan aggradation either occur after (TPQ) or before (TAQ) (Type 1–3 contexts: Fig. 3). This regional alluvial fan database (Fig. 5) highlights distinct phases of aggradation and associated incision in the feeder catchment, the identification of these phases requires interpretation of both TPQ and TAQ radiocarbon dating control. Bayesian analysis of the entire database (Fig. 5; Chiverrell et al. 2011) tests the conformability of the dating control at individual sites and builds the TAQ and TPQ control into the overall analysis. The objective was to constrain the timing of episodes of incision in the catchment and alluvial fan aggradation. Figure 5 (Chiverrell et al. 2011) shows episodes of alluvial fan aggradation during three phases during the late Holocene: 200 cal. BC – cal. AD 300, cal. AD 900–1,300, and after cal. AD 1500. These episodes overlap with substantial increases in human usage of these uplands removing forest cover and increasing grazing pressure (Harvey et al. 1981; Chiverrell et al. 2007). Though the lowering of erosional thresholds on hillslopes appears critical in conditioning the flux of sediment to alluvial fans, individual erosion events or high-magnitude floods are also important and often responsible for the inception and rejuvenation erosion in the feeder gully systems (Wells and Harvey 1987).

4.2 Radiocarbon Dating to Aid in the Development of a Chronology of Debris Flows on Cheekye Fan, British Columbia

The previous case study dealt with an interesting scientific and historical issue of explaining fan activity in association with human landscape interference. This case study deals with a single fan complex that has been particularly well studied (i.e. Mathews 1952, 1958; Friele et al. 1999; Ekes and Hickin 2001; Ekes and Friele 2003; Friele and Clague 2002a, b, 2005, 2009; Clague et al. 2003), but which can be a surrogate, in principle, for thousands of developed fans worldwide.

Debris-flow fans are preferred locations for human settlement in many parts of the world. In many such cases, flood hazard of the main valley bottom river has been perceived as posing the higher risk. This can likely be attributed to the significantly higher frequency of floods on the adjacent floodplains compared to rare catastrophic debris flows or hyperconcentrated flows on fans. Alluvial and colluvial fan complexes characterized by rare events are often perceived as "stable" landforms by non-experts, and apparently suited for residential development because of their comparatively low gradients, well-drained conditions, easy water access and often scenic environments, for example along steep-sided lakes, overlooking valleys or ocean shores. This applies particularly to cases where there have either not been

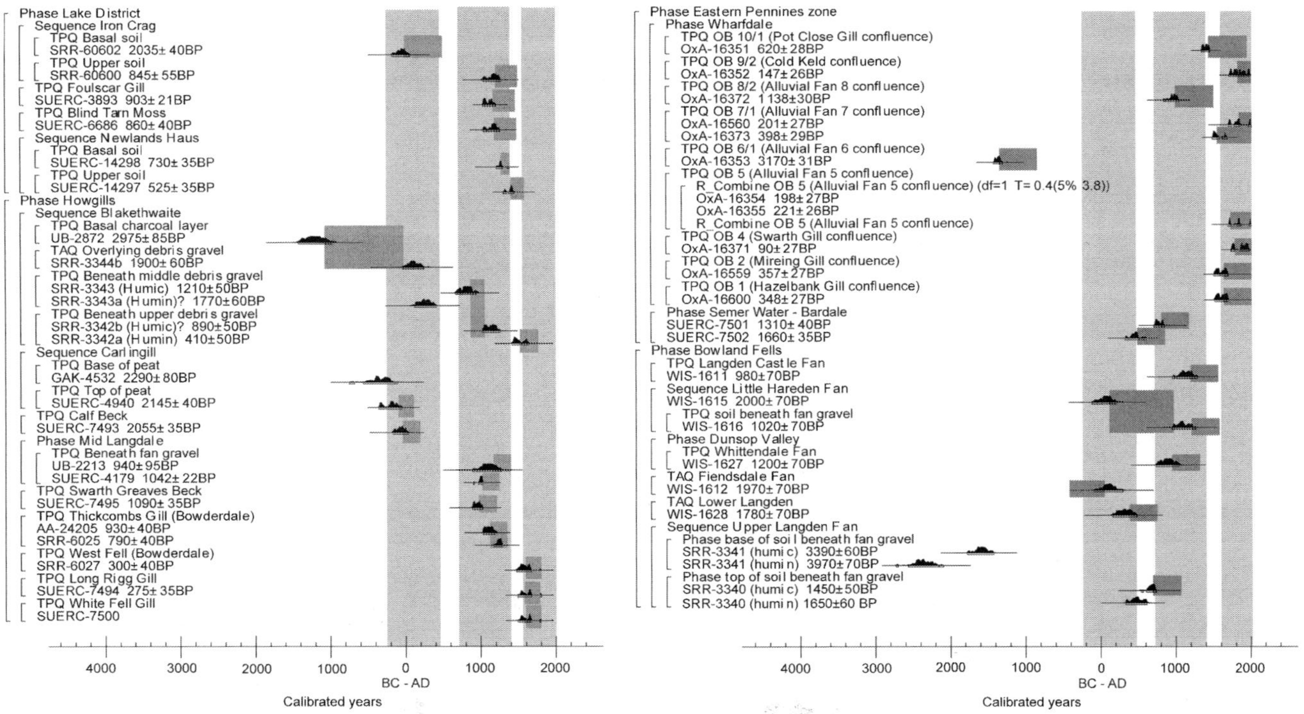

Fig. 5 Individual probability functions for alluvial fans/debris cones in NW England displayed as a Bayesian model (after Chiverrell et al. 2007, 2008, 2009b, 2011), with the dating for individual sites modelled as sequences where relevant. Each age estimate is identified a providing TPQ or TAQ control on episodes of gully or fluvial incision. Timing of incision episodes in relation to each age estimate is identified by *dark shading*, with regionally coherent phases in *lighter shade*

debris flows for several decades or high-frequency debris flows remain in the channel leading to the misguided perception that the observed recent past is a key to all future events. These are the same characteristics that have led to the desire of developing on the Cheekye River fan some 70 km north of the City of Vancouver.

The push for development on this fan that encompasses some 10 km^2 and whose headwaters drains a Quaternary volcano, has led to a large series of studies that have made this fan the best studied one in Canada (see Jakob and Friele 2010 for a listing of all prior studies). One of the central issues to determine if the fan is suited for development is that of the frequency of fan-forming events as well as their respective magnitudes. Development of reliable long-term debris flow frequency-magnitude relationships for large fan complexes is only possible through a detailed investigation program that entails test trenching, identification of individual debris-flow events and radiocarbon dating. Other absolute dating methods such as dendrochronology only reach back hundreds of years at best, are discontinuous with increasing age due to the lack of datable material and very rare high magnitude events often destroy the evidence of previous smaller events. Dendrochronology is therefore not suited to provide a continuous chronology of events throughout the Holocene. Neither is dendrochronology suited to reconstruct magnitude of flows into the deeper past because this requires a three-dimensional view of the fan stratigraphy.

For Cheekye fan, prior work by Friele and Clague (2005) has provided information on the evolution of the fan which was largely based on radiocarbon dating of buried organics (mostly wood fragments) and ground-penetrating radar that allowed differentiation of variable fan materials and underlying sediments (Ekes and Hickin 2001). Previous radiocarbon dates were supplemented with an additional 16 samples in the latest study to refine frequency-magnitude relationships (Jakob and Friele 2010). At each sample location, the thickness of the unit was measured in which the sample was obtained or which was overlying the sample. With almost 100 excavator-dug trenches up to 6 m deep this allowed the spatial tracing of individual deposits and, ultimately, a measurement of debris-flow deposit volumes, at least for the largest events, which dominate the hazard and risk for the low-frequency-high magnitude spectrum of debris-flow occurrence. Trenching was complemented with natural exposures and a large prior excavation for a garbage dump.

The dating of individual debris flows on the fan was further supplemented by radiocarbon dates of flows that had overrun a bedrock sill upstream of the fan apex to deposit debris in nearby Stump Lake (Clague et al. 2003). Flows that had not avulsed from the modern-day channel of Cheekye River and were thus not part of the radiocarbon record could be dated via dendrogeomorphic methods. While this method did not allow a reconstruction of flow volume, measurement of cross-sections between impact-scarred trees allowed an estimate of flow peak flow once debris-flow velocities had been back-calculated.

These combined efforts allowed a compilation of frequency-magnitude relationship for the entire Holocene (see Jakob 2012, *this volume*) which, even though it cannot ever be complete, provided a reasonable input for subsequent numerical debris-flow modelling and quantitative risk assessment for urban development on the fan. This study has demonstrated that through rigorous application of radiocarbon dating methods paired with other absolute dating methods, the stratigraphy, event chronology and, by inference, debris-flow hazards can be quantified with a reasonable degree of accuracy. Lamentably, such studies are rare because of their high price tag. These inevitable costs will, in most cases, deny application of the same rigor to other fans, where monetary resources are not available, or where the appetite of land developers for a substantial upfront investment in hazard assessments is low. Experience has shown that in-depth studies of debris-flow hazards are requested in most cases after an extreme event has led to substantial damage and/or loss of life.

For the practitioner, a few remarks can be helpful as general guidance:

- In many cases, radiocarbon dating is the only absolute dating method that provides an accurate enough dataset on debris-flow occurrence at time scales exceeding that of dendrochronology. This is especially important if existing legislation requests that frequency-magnitude relationships be established for return periods of several hundred or thousands of years.
- Radiocarbon dating will need to be performed in conjunction with a thorough understanding of the fan evolution from a geomorphic and stratigraphic point of view.
- Single ages for individual debris-flow deposits will often not provide enough certainty to accept a given radiocarbon measurement as the true age of occurrence as discussed in the text above.
- Many, if not most debris-flow events will not cover the entire fan which necessitates a well thought-out test trenching program that allows a representative sample. It is not required to identify each debris flow that may have occurred in the past as the highest frequency events may have little influence on the outcome of the hazard or risk assessment.
- Since most radiocarbon measurements are not exact, it is sometimes difficult to determine if two close age estimates indeed represent the same event. Other techniques as for example, grain size analysis, and detailed description of the sequence of depositional facies must underpin and complement the age control.

5 Concluding Remarks

Radiocarbon dating, as one of the few absolute dating methods suitable for Holocene or late Pleistocene applications, is an invaluable tool to decipher the formation chronology of fans. However careful selection of the materials for radiocarbon care and rigor in the strategies employed to recover chronological control for fans and debris cones are required. The testing of geomorphic and chronological models, as with other environments, offers considerable scope for generating well constrained time series and is an essential precursor to understanding the drivers of change in alluvial fan and debris cone environments. It is also essential for detailed hazard and risk fan studies that require a multi-method approach in reconstructing event frequencies and magnitudes over multi-century or even millennial time windows. Fan formation and fan hazard studies cannot be viewed in isolation but are closely linked. An understanding of fan formation and internal sedimentary architecture forms the basis for a hazard investigation.

Acknowledgements The authors would like to thank Pete Marshall (Chronologies) for comments and discussions on the Bayesian modelling of radiocarbon measurements. Thanks to Sandra Mather of the University of Liverpool Cartographics Unit for her work on the diagrams.

References

Blockley SPE, Lowe JJ, Walker MJC, Asioli A, Trincardi F, Coope GR, Donahue RE, Pollard AM (2004) Bayesian analysis of radiocarbon chronologies: examples from the European Late-glacial. J Quat Sci 19(2):159–176

Buck CE, Cavanagh WG, Litton CD (1996) The Bayesian approach to interpreting archaeological data. Wiley, Chichester

Burr GS (2007) Radiocarbon dating: causes of temporal variations. In: Elias SE (ed) Encyclopedia of quaternary science. Elsevier, Amsterdam, pp 2924–2941

Cannon SH, Gartner JE (2005) Wildfire-related debris flow from a hazards perspective. In: Jakob M, Hungr O (eds) Debris flows and related phenomena. Springer/Praxis, Berlin, pp 363–381

Chiverrell RC, Harvey AM, Foster GC (2007) Hillslope gullying in the Solway Firth – Morecambe Bay region: responses to human impact and/or climatic deterioration? Geomorphology 84:317–343

Chiverrell RC, Harvey AM, Hunter SY, Millington J, Richardson NJ (2008) Late Holocene environmental change in the Howgill Fells, Northwest England. Geomorphology 100:41–69

Chiverrell RC, Foster GC, Thomas GSP, Marshall P, Hamilton D (2009a) Robust chronologies for landform development in fluvial environments. Earth Surf Process Landf 34:319–328

Chiverrell RC, Foster GC, Marshall P, Harvey AM, Thomas GSP (2009b) Coupling relationships: hillslope-fluvial linkages in the Hodder catchment, NW England. Geomorphology 109:222–235

Chiverrell RC, Foster GC, Marshall P, Thomas GSP (2010) Sediment transmission and storage: the implications for reconstructing landform development. Earth Surf Process Landf 35:4–15

Chiverrell RC, Thorndycraft VR, Hoffmann TO (2011) Cumulative probability functions and their role in evaluating the chronology of geomorphological events during the Holocene. J Quat Sci 26(1):76–85

Clague JJ, Friele PA, Hutchinson I (2003) Chronology and hazards of large debris flows in the Cheekye River basin, British Columbia, Canada. Environ Eng Geosci 8:75–91

Cook GT, van der Plicht J (2007) Radiocarbon dating: conventional method. In: Elias SE (ed) Encyclopedia of quaternary science. Elsevier, Amsterdam, pp 2899–2911

Ekes C, Friele PA (2003) Sedimentary architecture and post-glacial evolution of Cheekye fan, southwestern British Columbia, Canada. In: Bristow CS, Jol HM (eds) Ground penetrating radar in sediments, Geological Society special publication 211. Geological Society, London, pp 87–98

Ekes C, Hickin EJ (2001) Ground penetrating radar facies of the paraglacial Cheekye fan, southwestern British Columbia, Canada. Sediment Geol 143:199–217

Foster GC, Chiverrell RC, Harvey AM, Dearing JA, Dunsford H (2008) Catchment hydrogeomorphological responses to environmental change in the Southern Uplands of Scotland. Holocene 18:935–950

Friele PA, Clague JJ (2002a) Readvance of glaciers in the British Columbia Coast Mountains at the end of the last glaciation. Quat Int 87:45–58
Friele PA, Clague JJ (2002b) Younger Dryas readvance in Squamish River valley, southern Coast Mountains, British Columbia. Quat Sci Rev 21:1925–1933
Friele PA, Clague JJ (2005) Multifaceted hazard assessment of Cheekye fan, a large debris flow fan in southwestern British Columbia. In: Jakob M, Hungr O (eds) Debris-flow hazards and related phenomena. Springer/Praxis, Heidelberg, pp 659–683
Friele PA, Clague JJ (2009) Paraglacial geomorphology of quaternary volcanic landscapes in the southern Coast Mountains, British Columbia. In: Knight J, Harrison S (eds) Periglacial and paraglacial processes and environments, Geological Society special publications, 320. Geological Society, London, pp 219–233
Friele PA, Ekes C, Hickin EJ (1999) Evolution of Cheekye fan, Squamish, British Columbia: Holocene sedimentation and implications for hazard assessment. Can J Earth Sci 36:2023–2031
Harvey AM (1996) Holocene hillslope gully systems in the Howgill Fells, Cumbria. In: Brooks SM, Anderson MG (eds) Advances in hillslope processes, vol 2. Wiley, Chichester, pp 731–752
Harvey AM (1997) Coupling between hillslope gully systems and stream channels in the Howgill Fells, northwest England: temporal implications. Géomorphol Relief Process Environ 1:3–20
Harvey AM, Renwick WH (1987) Holocene alluvial fan and terrace formation in the Bowland Fells, northwest England. Earth Surf Process Landf 12:249–257
Harvey AM, Oldfield F, Baron AF, Pearson GW (1981) Dating of post-glacial landforms in the central Howgills. Earth Surf Process Landf 6:401–412
Jakob M (2012) Events on fans and cones: recurrence interval and magnitude. In: Schneuwly-Bollschweiler M, Stoffel M, Rudolf-Miklau F (eds) Dating torrential processes on fans and cones – methods and their application for hazard and risk assessment. Springer, Dordrecht/Heidelberg/London/New York
Jakob M, Friele P (2010) Frequency and magnitude of debris flows on Cheekye River, British Columbia. Geomorphology 114:382–395
Jakob M, Weatherly H (2005) Debris flow hazard and risk assessment, Jones Creek, Washington. In: Hungr O, Fell R, Couture R, Eberhardt O (eds) Landslide risk management. Taylor & Francis Group, London, pp 533–541
Jull AJT (2007) Radiocarbon dating: AMS method. In: Elias SE (ed) Encyclopedia of quaternary science. Elsevier, Amsterdam, pp 2911–2918
Lewin J, Macklin MG, Johnstone E (2005) Interpreting alluvial archives: sedimentological factors in the British Holocene fluvial record. Quat Sci Rev 24:1873–1889
Mathews WH (1952) Mount Garibaldi, a supraglacial Pleistocene volcano in southwestern British Columbia. Am J Sci 250:553–565
Mathews WH (1958) Geology of the Mount Garibaldi map area, southwestern British Columbia, Canada. Part II: geomorphology and Quaternary volcanic rocks. Geol Soc Am Bull 69:179–198
Matthews JA (1985) Radiocarbon dating of surface and buried soils: principals, problems and prospects. In: Richards KS, Ellis S, Arnett RR (eds) Geomorphology and soils. Allen and Unwin, London, pp 269–288
Matthews JA (1993) Radiocarbon dating of arctic-alpine palaeosols and the reconstruction of Holocene palaeoenvironmental change. In: Chambers FM (ed) Climate change and human impact on the landscape. Chapman & Hall, London, pp 83–100
Miall AD (1996) The geology of fluvial deposits: sedimentary facies, basin analysis, and petroleum geology. Springer, Berlin, 582 pp
Ramsey CB (1995) Radiocarbon calibration and analysis of stratigraphy: the OxCal program. Radiocarbon 37:425–430
Ramsey CB (2001) Development of the radiocarbon calibration program OxCal. Radiocarbon 43:355–363
Ramsey CB (2008a) Deposition models for chronological records. Quat Sci Rev 27:42–60
Ramsey CB (2008b) OxCal 4.0 manual, online web resource. Oxford. http://c14.arch.ox.ac.uk
Ramsey CB (2009) Bayesian analysis of radiocarbon dates. Radiocarbon 51:337–360

Reimer PJ, Baillie MGL, Bard E, Bayliss A, Beck JW, Bertrand CJH, Blackwell PG, Buck CE, Burr GS, Cutler KB, Damon PE, Edwards RL, Fairbanks RG, Friedrich M, Guilderson TP, Hogg AG, Hughen KA, Kromer B, McCormac FG, Manning SW, Ramsey CB, Reimer RW, Remmele S, Southon JR, Stuiver M, Talamo S, Taylor FW, van der Plicht J, Weyhenmeyer CE (2004) IntCal04 terrestrial radiocarbon age calibration, 26–0 ka BP. Radiocarbon 46:1029–1058

Scharpenseel HW, Becker-Heidmann P (1993) Twenty five years radiocarbon dating of soils; paradigm of erring and learning. Radiocarbon 34:541–549

Stuiver M, Kra RS (eds) (1986) Calibration issue, proceedings of the 12th international ^{14}C conference. Radiocarbon 28(2B):805–1030

Stuiver M, Reimer PJ, Reimer R (2005) CALIB radiocarbon calibration. Execute version 6.0html. http://intcal.qub.ac.uk/calib/. Belfast

van der Plicht J (2007) Radiocarbon dating: variations in atmospheric ^{14}C. In: Elias SE (ed) Encyclopedia of quaternary science. Elsevier, Amsterdam, pp 2924–2931

Wells SG, Harvey AM (1987) Sedimentologic and geomorphic variations in storm generated alluvial fans, Howgill Fells, northwest England. Geol Soc Am Bull 98:182–194

Luminescence Dating of Alluvial Fans and Cones

Andreas Lang

1 Introduction

Alluvial fans and debris cones are important buffers in the sediment cascade from mountains to the sea. Interactions between tectonic, climate, and base-level controls often render fans and cones dynamic geomorphic features – even on timescales of millennia and beyond (Harvey 2010). Unraveling such dynamics has proven difficult mainly because establishing chronologies for fans and cones is challenging: Buried organic remains are usually scare (Chiverrell and Jakob 2012, *this volume*), dendrochronology and lichenometry (Schneuwly-Bollschweiler and Stoffel 2012, *this volume*; Jomelli 2012, *this volume*) will often be limited to cone and fan surfaces and help accessing the recent past only (Lang et al. 1999), and rarely are tephra layers preserved or synsedimentary carbonates developed that would permit radiometric dating. Luminescence dating can offer an alternative as it allows determining the time when sediments were laid down. But limitations exist that are related to (i) limited exposure of sediment grains to daylight during the short and rapid transport of many fan deposits and (ii) low luminescence sensitivity of quartz from many high mountain environments. After a brief introduction on luminescence dating techniques, an overview of relevant dating applications is given and the possibilities and limitations of applying luminescence dating to alluvial fan and debris cones are discussed.

A. Lang (✉)
School of Environmental Sciences, University of Liverpool, Liverpool L69 7ZT, UK
e-mail: lang@liv.ac.uk

M. Schneuwly-Bollschweiler et al. (eds.), *Dating Torrential Processes on Fans and Cones*, Advances in Global Change Research 47, DOI 10.1007/978-94-007-4336-6_18,

2 Luminescence Phenomena

Luminescence is the emission of light from non-conducting solids in addition to their blackbody radiation. It is an electro-optical phenomenon by which trapped electrons recombine under photon emission. Luminescence dating exploits the fact that ionizing radiation from natural radioactivity and cosmic rays produces free electrons, some of which are trapped at the metastable energy levels of defect sites common in many natural crystal lattices. Since these charges accumulate with time, their amount and thus the intensity of the luminescence signal can be used for dating. Many minerals have lattice characteristics that allow for luminescence to occur. In practice, however, mainly quartz and feldspar are used for dating applications. Luminescence dating is one of several techniques grouped as dosimetric dating where the age information is carried by the size of the radiation damage. This is in contrast to radiometric techniques where the radioactive decay products carry the age information. Within the group of dosimetric techniques and together with electron spin resonance dating, luminescence forms the subgroup of trapped charge dating techniques.

Luminescence dating techniques can be subdivided according to the kind of energy supplied for stimulation into thermoluminescence (TL, stimulated by heat) or photon-stimulated luminescence (stimulated by light). The latter is further subdivided according to the photon energy used in optical stimulated luminescence (OSL, stimulated by visible light) and infrared stimulated luminescence (IRSL or IR-OSL, stimulated by infrared), sometimes even further to more specific wavebands like 'green-light stimulated luminescence'. Often OSL and IRSL dating are referred to as 'optical dating'. The development of luminescence-based dating techniques goes back to 1960, when Grögler et al. (1960) used TL for dating burnt archaeological materials. The observation of green-stimulated luminescence in quartz by Huntley et al. (1985) was the beginning of dating sediments with OSL. A further step forward took place when Hütt et al. (1988) discovered the IRSL emission from feldspars. Since that time the application of luminescence dating has seen a multitude of applications and dramatic technological developments. While the target in the early years was mainly on archaeological materials and windblown sediments, the development of single aliquot (Murray and Wintle 2000) and single grain dating (Lamothe et al. 1994) opened up applications to numerous sedimentary environments. The major advantage that OSL and IRSL have over TL is the restriction to highly light-sensitive charge traps. This results in an improved dating precision and allows applications to sediments which were shortly and/or weakly exposed to light before burial. The latter aspect is of paramount importance when dealing with all water-lain sediments.

3 Luminescence Dating of Sediments

3.1 Dating Principles

In luminescence dating the age information is carried by the size of the radiation damage of a crystal lattice (shown in a schematic overview in Fig. 1). Starting with mineral formation environmental radioactivity will lead to increasingly more trapped electrons which in turn will result in a higher luminescence signal. This growth can go on until all defect sites are occupied by trapped charges and the corresponding luminescence signal is saturated. The saturation effect defines the maximum age range for the technique – usually about 100 ka but can vary dramatically depending on the environmental dose-rate (the rate at which ionizing radiation is impacting on the mineral grains). Light exposure will reduce the luminescence signal as trapped electrons can absorb photons and escape from their traps. This may happen during erosion and transport of a mineral grain and so reset the 'luminescence clock'. After the mineral grain is shielded from light during burial the impact of natural radioactivity will again increase the number of trapped charges and build up the luminescence signal. In the lab the luminescence signal of a sample will be measured. The intensity of the luminescence released during stimulation is a function of the energy dose accumulated in the mineral and thus of the exposure time ('age') to natural radiation. Using artificial radiation the sensitivity of the sample to radiation is tested and the equivalent dose (D_e) determined that induces a signal equivalent to the signal acquired during burial.

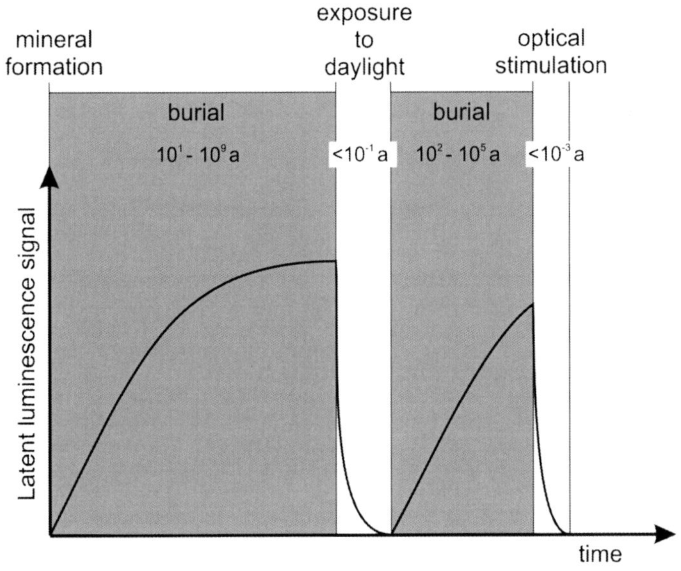

Fig. 1 Scheme of luminescence build up and depletion (see text for details)

The speed at which signal builds up during burial depends on (i) the environmental dose-rate, i.e. the energy dose delivered to the mineral grain per unit time which is directly dependent on the amount of radionuclides in the sediment and its immediate surroundings (for details see Aitken 1994) and (ii) the luminescence sensitivity of a mineral grain. The second is defined as the amount of luminescence emitted per unit dose absorbed. In many mountain regions the luminescence sensitivity of quartz is low – often even too low to allow for dating applications. Most probably this is linked to the rather fresh character of many sediment particles in geomorphically dynamic high mountain environments: Most sediment particles have only recently been eroded from unweathered rock and are undergoing the very first cycles of erosion, transport, deposition, and reworking (i.e. cycles of luminescence depletion and build up). Only after many such cycles will the luminescence sensitivity increase – as nicely shown for fluvial deposits by Pietsch et al. (2008) – a phenomenon common for electronic processes in crystal lattices. This can also explain the high luminescence sensitivity of quartz in old sedimentary basins like central Australia (Fitzsimmons et al. 2010), where sediments have been recycled for millions of years. In many mountain environments feldspar has the advantage of higher luminescence sensitivity. It has additional advantages of (i) carrying radioisotopes internally (most significant: ^{40}K) that deliver an internal dose-rate and make feldspars less dependent on changes in external components of environmental dose-rate and (ii) usually displaying a far higher saturation dose than quartz and thus offering the possibilities of dating older sediments. Feldspars do however show quantum-mechanical tunnelling resulting in signal loss (anomalous fading; Wintle 1973; Poolton et al. 2009) which greatly complicates reliable dosimetric measurements.

The luminescence age of a sample is calculated as the quotient between D_e and environmental dose-rate. Determining the present day environmental dose-rate is usually straight forward. Determining the environmental dose-rate representative of the whole burial period can be rather tricky. In cases where any post depositional changes have taken place (weathering, pedogenesis, hydromorphic alterations, compaction, etc.) additional geochemical measurements are needed to carefully quantify and model the time-dependent variability (e.g. Nathan and Mauz 2008).

3.2 Luminescence Bleaching of Sediments

The process resetting the luminescence ‘clock’ is the last exposure of quartz or feldspar mineral grains to daylight (Fig. 1), which is likely to occur while sediments are eroded, transported and deposited. The phenomenon that the signal is erased during daylight exposure (bleached) gives luminescence dating the advantage over other dating methods to directly date the deposition of clastic sediments (e.g., Lian and Roberts 2006). The speed of the resetting will depend on the type of mineral and the type of luminescence stimulated. OSL resetting is orders of magnitude faster than TL resetting (Godfrey-Smith et al. 1988) and in favourable light conditions will only require seconds of light exposure to fully erase all signal. Still, for many sedimentary environments daylight exposure during erosion transport and deposition will be limited. Daylight intensity and spectral composition will be reduced during cloudy conditions (Lang 1994) and depending on the type of sediment transport process the chance for daylight exposure may be rather small. For aeolian sediments, sufficient exposure to daylight to reset the luminescence signal is commonplace. For alluvial and slope deposits, complete signal resetting is hampered by various factors.

1. Several processes are involved in eroding and transporting sediment along slopes to form fans and cones. The important processes are sheet, rill and gully erosion, mass movements (e.g. debris flows, sliding, falling) and soil creep (e.g., solifluction). While these processes act on hillslope material to move it downslope, the sediment itself may move as a rigid block as in the case of sliding. Thus, in respect of exposure to daylight, only a small fraction of sediment grains involved in the process (e.g. only the outside layer of sediment grains on a rigid block) will be exposed to daylight. If material is transported in suspension, the intensity and spectral composition of the daylight is subdued by the suspended load (Ditlefsen 1992). Also, high-magnitude events will provide limited daylight exposure of sediments, and may even be erosive at their base and incorporate sub-surface deposits with high residual luminescence.
2. A large variety of sediment source materials exists, leading to a large variety in sediment characteristics, e.g., grain size distribution, mineral composition or organic content. Thus, certain sediments tend to be transported as single mineral grains, like well-sorted dry sand and bedload, while others are prone to coagulation, like clay-dominated sediments. The latter often hampers sufficient zeroing of the luminescence signal, because the inner grains of an aggregate are shielded from daylight exposure (Lang 1994). The same may result from mineral coatings, where e.g., iron, manganese or carbonate coatings hamper the daylight to penetrate into the mineral grains (Singhvi et al. 1986).
3. The catchment morphometry and the erosion process define the sediment path, sediment travel distance and sediment sink. Due to the usually rather small catchments of alluvial fans and cones the travel distance tends to be relatively short. This decreases the probability of efficient bleaching, because the duration of light exposure is limited and events may occur fully during unfavourable daylight conditions (e.g., transport entirely at night).

There are other factors that can increase the chance of daylight exposure: (i) hillslope sediments are often eroded, transported and accumulated not in a single event, but temporary storage on the hillslope occurs, before a further mobilization takes place (Lang and Hönscheidt 1999). An incremental transport increases the probability of exposure to daylight and thus bleaching. (ii) Bioturbation and mechanical processes in the soil and regolith ensure that mineral grains are frequently exposed to daylight before the sediment grains are eroded and in many cases also after deposition before the sediment grains are covered through further sedimentation.

If the bleaching was insufficient a residual signal will remain in a sediment grain and the luminescence age derived from the sediment will overestimate the true age. As in other sedimentary environments, excluding insufficient bleaching is a necessary requirement for successful luminescence dating of fan and cone deposits. This is demonstrated by investigations of modern deposits, where residual luminescence signals were identified, which would result in age overestimations of hundreds or even thousands of years (Porat et al. 2001).

3.3 Detection of Insufficiently Bleached Sediments

Recent years have seen big improvements for detecting and handling insufficient bleaching. In the past the techniques available to detect insufficient bleaching were based on comparing different luminescence signals or the degree of bleaching of different mineral grains. Earlier studies explored the lower light sensitivity of TL compared to OSL. For example Wintle et al. (1993) compare TL and IR-OSL of colluvial samples from Natal (South Africa), and find that TL ages are up to 10 ka older than IR-OSL ages. However, IR-OSL ages are also overestimating the expected age, and Wintle et al. (1993) conclude that in this setting IR-OSL is only suitable for dating older colluvial deposits where the residual age is insignificant.

Various grain size fractions may also show different degrees of bleaching. In luminescence dating, generally two different fractions are used: fine grains (4–11 μm) and coarse grains (90–300 μm). For alluvial sediments, several case studies have shown that coarser grains seem to be better bleached than finer fractions (e.g., Olley et al. 1998; Wallinga 2002).

Differing bleaching properties of different OSL components offer another way of detecting insufficient bleaching. Initially differentially bleached signal components within an OSL shine-down curve were compared (Aitken and Xie 1992; Lang and Wagner 1997). Later it was shown that the OSL signal of quartz consists of fast, medium, and slow components (e.g., Bulur 1996; Bailey 2000) of which the fast component bleaches most rapidly. Singarayer et al. (2005) show how this can be utilized to determine the extent of bleaching.

The introduction of single aliquot techniques (e.g., Duller 1991; Roberts et al. 1998; Murray and Wintle 2000) made it possible to develop alternative approaches to detect insufficient bleaching. Here, the differences between aliquots or single grains are used. The earlier attempts were based on comparing an aliquot's (or single grain's) natural luminescence intensity and its D_e (e.g., Li 1994; Colls et al. 2001). More recently, the analyses of D_e scatter has succeeded most other techniques. If D_e estimates are based on single grains or small aliquots insufficient bleaching can be unraveled from the scatter (e.g., Clarke 1996). When plotting the D_e values as histograms (Fig. 2), well bleached samples show narrow and almost symmetrical distributions, whereas broad distributions can be characteristic for insufficiently bleached samples. Often strongly skewed distributions are obtained

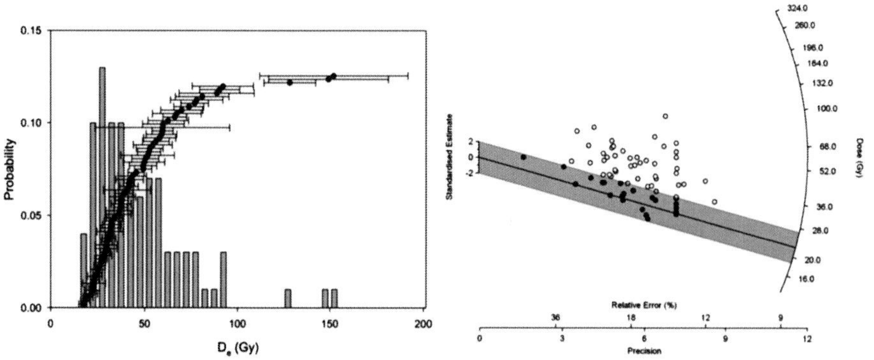

Fig. 2 Equivalent dose distribution of a poorly bleached glacio-fluvial sample (data from Thrasher 2008). *Left graph*: Histogram of D_e values. The highly skewed distribution is a clear indicator of poor bleaching. *Right graph*: Radial plot of D_e values. The same data as for the *left graph* displayed as radial plot (Galbraith et al. 1999). In this modified scatter plot all D_e values are plotted according to their precision: values known more precisely are plotted further to the *right*. The *shaded area* is *centred* on the equivalent dose calculated with the minimum age model

when the majority of the grains were well bleached but also a number of poorly bleached grains are present (e.g., Olley et al. 1999; Thomas et al. 2005).

Unfortunately, all scatter based techniques can only be used for coarse grain (sand-sized) sediment separates. For fine grains the number of grains per measurement is too large ($>10^6$) and any scatter will be averaged out.

3.4 Equivalent Dose Determination of Insufficiently Bleached Sediments

Several techniques have been developed to derive a true equivalent dose from a distribution of D_e values in an insufficiently bleached sample (for an overview see Bailey and Arnold 2006). The most commonly used are MAM: minimum age model, CAM: central age model, FMM: finite mixture model, L5%: mean of the lowest 5% (for details see Galbraith and Laslett 1993). All these techniques are based on statistical analyses of D_e distributions and aim at isolating the component of sufficiently bleached aliquots within a distribution. As prerequisite a large number of D_e estimates (>50; Rodnight et al. 2006) from single grains or small aliquots are necessary. Bailey and Arnold (2006) evaluate the different statistical approaches and provide decision support criteria to choose which model is best suited for analysing different types of dose distribution. Following their decision tree allows using observable characteristics of a D_e distribution (like skewness and kurtosis) to choose which age model best to be used for calculating the burial dose.

Using such models is not without problems as uncertainty exists about the causes of a broad distribution. Besides the bleaching history, there are also other causes for D_e scatter, such as microdosimetry (different environmental dose rates for different grains, e.g., Kalchgruber et al. 2003) or luminescence characteristics (Rodnight et al. 2006) and different saturation levels of grains. Using the lowermost D_es may be erroneous as low-dose outliers can exist that are measurement artefacts due to the luminescence characteristics of the low-dose aliquots (Rodnight et al. 2006). Thus, when insufficient bleaching is detected, a conservative approach is to use the median of a D_e distribution to determine a maximum age estimate for the deposition.

4 Luminescence Dating Applications

Luminescence dating of alluvial fans has been applied in many environmental settings. Recent applications cover most continents and involve small cones as well as megafans. Many studies are aiming to discriminate between climate and tectonic forcing mechanisms on longer-term fan evolution. Other studies utilise luminescence dated alluvial fan deposit to establish rates of tectonic movement or for unraveling human impact on the environment.

Sinha et al. (2010) investigate Late Pleistocene-Holocene deposits of the Ganga fan on the Himalayan mountain front. Quartz OSL dating was used to constrain terrace abandonment and to link fluvial evolution to monsoon activity. As the authors show, terrace formation on the alluvial fan was dominantly controlled by climate change which is of special importance in the highly tectonically active region of the Himalayan mountain front. Further West on the Brahmaputra River Srivastava et al. (2009) place fan terrace deposition between 15 and 10 ka, also based on a quartz OSL chronology. They attribute the aggradation to semi-arid climatic conditions and link the subsequent incision to increased discharge and tectonic activity along the Ranaghat Fault.

Constraining fault slip rates is the aim of a study on alluvial fans in the mountains of Iran (Le Dortz et al. 2009). Based on quartz OSL dating of alluvial fan deposits the authors constrain the Holocene rate of slip along the Anar fault to at least 0.8 mm ka^{-1}. A similar approach is used by Nissen et al. (2009a) on the Har-Us-Nuur fault in the Mongolian Altai. They establish a maximum dextral slip-rate of 2.4 ± 0.4 mm a^{-1}. OSL age estimates of alluvial fan deposits, along with measurements of the lateral offset are used by Berger et al. (2010) to estimate the Late Quaternary slip-rate of the northern Greenville fault zone in the Eastern San Francisco Bay Area California to about 2 mm a^{-1}.

Using quartz OSL Nissen et al. (2009b) establish ages of ~15 ka, ~20 ka and ~85 ka for fault cut alluvial deposits on the Har-Us-Nuur fault, a major, right-lateral strike-slip fault bounding the eastern margin of the Altai. The authors derive Late Quaternary shortening rate across the area of active thrust of 0.7–2.4 mm a^{-1}, representing a significant part of the ~7 mm a^{-1} convergence across the whole Altai range.

OSL dating has also been used in intramontane basin settings to unravel the interaction between basin bounding fault activity, and alluvial fan and fluvial activity. Zembo et al. (2009) use OSL dated fan deposits in the Apennines to understand the opening of an intramontane basin. The authors show how climatic variations and alterations of tectonic movement produce shifts in basin fill depositional style. In the eastern Cordilleras of South America Sancho et al. (2008) and Spencer and Robinson (2008) study complex segmented alluvial fans and based on OSL dating of quartz unravel the morpho-tectonic evolution of the basin. Further North and on the western slopes of the Andes, Schlunegger et al. (2012, *this volume*) use feldspar IRSL to understand sediment aggradation and dissection. They show that enhanced precipitation increased hillslope delivery to the channel network by debris flows in upper part of the catchment and alluviation of the trunk stream in the lower catchment. The study also shows that over time the locus of sediment deposition propagates upstream starting from the valley outlet.

To constrain the time of alluvial fan abandonment in the arid Negev Porat et al. (2010) determine quartz OSL ages that range from $\sim$90 to $\sim$5 ka. The chronology is used to inform studies on tectonics, landscape evolution and paleoclimate.

The studies showcased above clearly demonstrate the importance that luminescence dating has for establishing chronologies for alluvial fans on the 10^3 to 10^5 year time scale. They also highlight the current limitations which are related to poor bleaching, as well as insufficient luminescence sensitivity of some quartz and signal instability of some feldspar signals. OSL chronologies have proven essential for understanding sediment cycling through time, identification of periods of alluvial storage and erosion, quantification of rates of sediment accumulation and storage, and constraining the timing of fault activity.

5 Conclusion and Perspective

Recent developments in luminescence dating enable dating of non-aeolian sediments with often limited pre-depositional daylight exposure. Luminescence dating studies have shown that in many environments OSL dating of alluvial fan and cone deposits is possible despite the usually restricted transport distances. OSL dating in such environments still represents a challenge and sufficient bleaching must be checked for. This can be achieved by several techniques of which recently the study of the D_e distribution of small aliquots and single grains has proven most valuable (e.g. Bailey and Arnold 2006). Statistical techniques exist to isolate a well bleached population within the D_e distribution and base an age estimate on this population only.

In any case the validity of a chronology needs to be thoroughly tested using both, internal and external controls. For luminescence dating multiple samples from one sedimentary unit can provide an internal control, as can samples taken in stratigraphic order. Even more important are external controls based on independent age estimates. To establish a robust chronology several techniques need to be applied to unravel any potential systematic error. This has to be done taking the 'quality' of age information into account (direct age, 'older than' age or 'younger than' age, Lang et al. 1999) together with the uncertainty of the chronometric data (i.e. precision and accuracy, see Scott et al. (2007) for an overview). This is still not common practice even so newly developed age models offer big improvements for sequences of alluvial sediments. Statistical analysis of age information based on the probability distribution of an age estimate and Bayesian reasoning have originally been developed for ^{14}C data in continuous sedimentary sequences. These analyses allow to overcome the inherent uncertainties of isolated ^{14}C ages, and to identify the leads, lags, or synchrony between different events (Blaauw and Christen 2005). Even more, these techniques allow the identification of outliers and their removal (for an overview see: Ramsey 2008). The recent developments of these techniques go beyond ^{14}C-dating and now allow inclusion of other chronometric data and age information (Millard 2006). Also, spatially and temporally discontinuous data sets can now be analysed based on their stratigraphic relationship, a technique that is especially useful for alluvial fan and cone deposits (Chiverrell et al. 2008).

To reduce possible difficulties with poor bleaching, besides a thorough sedimentologic analysis, a careful geomorphic evaluation of a sampling site and its catchment is necessary before samples should be taken for luminescence dating. The occurrence of insufficient bleaching can significantly be reduced when taking into account that:

1. The transport distance of the sediments should be as large as possible, as this enhances the possibility of intermittent sediment storage on slopes and in channels, and thus increases the potential exposure time of the sediment grains to daylight.
2. Locations where sediments can originate from high magnitude events should be avoided as sub-surface sediments with high residual luminescence may be incorporated.

In alluvial fan environments complete resetting of the OSL signal is the exception rather than the rule. In consequence, detecting and eliminating insufficient bleaching is a necessary requirement.

References

Aitken MJ (1994) Optical dating – a nonspecialist review. Quat Sci Rev 13(5–7):503–508

Aitken MJ, Xie J (1992) Optical dating using infrared diodes: young samples. Quat Sci Rev 11:147–152

Bailey RM (2000) The slow component of quartz optically stimulated luminescence. Radiat Meas 32(3):233–246

Bailey RM, Arnold LJ (2006) Statistical modelling of single grain quartz De distributions and an assessment of procedures for estimating burial dose. Quat Sci Rev 25(19–20):2475

Berger GW, Sawyer TL, Unruh JR (2010) Single- and multigrain luminescence dating of sediments related to the Greenville fault, eastern San Francisco Bay area, California. Bull Seismol Soc Am 100:1051–1072

Blaauw M, Christen JA (2005) The problems of radiocarbon dating. Science 308(5728):1551–1553

Bulur E (1996) An alternative technique for optically stimulated luminescence (OSL) experiment. Radiat Meas 26:701–709

Chiverrell R, Jakob M (2012) Radiocarbon dating: alluvial fan/debris cone evolution and hazards. In: Schneuwly-Bollschweiler M, Stoffel M, Rudolf-Miklau M (eds) Dating torrential processes on fans and cones – methods and their application for hazard and risk assessment, Advances in global change research. Springer, Dordrecht/Heidelberg/London/New York

Chiverrell RC, Foster GC, Thomas GSP, Marshall P, Hamilton D (2008) Robust chronologies for landform development in fluvial environments. Earth Surf Process Landf. doi:10.1002/esp.1720

Clarke ML (1996) IRSL dating of sands: bleaching characteristics at deposition inferred from the use of single aliquots. Radiat Meas 26(4):611–620

Colls AE, Stokes S, Blum MD, Straffin E (2001) Age limits on the late quaternary evolution of the upper Loire River. Quat Sci Rev 20(5–9):743–750

Ditlefsen C (1992) Bleaching of K-Feldspars in turbid water suspensions – a comparison of photoluminescence and thermoluminescence signals. Quat Sci Rev 11(1–2):33–38

Duller GAT (1991) Equivalent dose determination using single aliquots. Nucl Tracks Radiat Meas 18(4):371–378

Fitzsimmons KE, Rhodes EJ, Barrows TT (2010) OSL dating of southeast Australian quartz: a preliminary assessment of luminescence characteristics and behaviour. Quat Geochronol 5: 91–95

Galbraith RF, Laslett GM (1993) Statistical models for mixed fission track ages. Nucl Tracks Radiat Meas 21:459–470

Galbraith RF, Roberts RG, Laslett GM, Yoshida H, Olley JM (1999) Optical dating of single and multiple grains of quartz from jinmium rock shelter, northern Australia, part 1, experimental design and statistical models. Archaeometry 41:339–364

Godfrey-Smith DI, Huntley DJ, Chen WH (1988) Optical dating studies of quartz and feldspar sediment extracts. Quat Sci Rev 7:373–380

Grögler N, Houtermans FG, Stauffer H (1960) Ueber die Datierung von Keramik und Ziegel durch Thermolumineszenz. Helv Phys Acta 33:595–596

Harvey AM (2010) Local buffers to the sediment cascade: debris cones and alluvial fans. In: Burt TP, Allison RJ (eds) Sediment cascades – an integrated approach. Wiley, Chichester, pp 153–180

Huntley DJ, Godfreysmith DI, Thewalt MLW (1985) Optical dating of sediments. Nature 313(5998):105–107

Hütt G, Jaek I, Tchonka J (1988) Optical dating – K-Feldspars optical-response stimulation spectra. Quat Sci Rev 7(3–4):381–385

Jomelli V (2012) Lichenometric dating of debris avalanche deposits with an example from the French Alps. In: Schneuwly-Bollschweiler M, Stoffel M, Rudolf-Miklau M (eds) Dating torrential processes on fans and cones – methods and their application for hazard and risk assessment, Advances in global change research. Springer, Dordrecht/Heidelberg/London/New York

Kalchgruber R, Fuchs M, Wagner GA (2003) Evaluating dose-rate distributions in natural sediments using a-Al2O3:C grains. Radiat Meas 37:293–297

Lamothe M, Balescu S, Auclair M (1994) Natural Irsl intensities and apparent luminescence ages of single feldspar grains extracted from partially bleached sediments. Radiat Meas 23(2–3): 555–561

Lang A (1994) Infrared stimulated luminescence dating of Holocene reworked silty sediments. Quat Sci Rev 13(5–7):525–528

Lang A, Hönscheidt S (1999) Age and source of colluvial sediments at Vaihingen-Enz, Germany. Catena 38(2):89–107

Lang A, Wagner GA (1997) Infrared stimulated luminescence dating of Holocene colluvial sediments using the 410 nm emission. Quat Sci Rev 16(3–5):393–396

Lang A, Moya J, Corominas J, Schrott L, Dikau R (1999) Classic and new dating methods for assessing the temporal occurrence of mass movements. Geomorphology 30(1–2):33–52

Le Dortz K, Meyer B, Sébrier M, Nazari H, Braucher R, Fattahi M, Benedetti L, Foroutan M, Siame L, Bourlès D, Talebian M, Bateman MD, Ghoraishi M (2009) Holocene right-slip rate determined by cosmogenic and OSL dating on the Anar fault, Central Iran. Geophys J Int 179(2):700–710

Li SH (1994) Optical dating: insufficient bleached sediments. Radiat Meas 23:563–567

Lian OB, Roberts RG (2006) Dating the quaternary: progress in luminescence dating of sediments. Quat Sci Rev 25(19–20):2449

Millard AR (2006) Bayesian analysis of Pleistocene chronometric methods. Archaeometry 48:359–375

Murray AS, Wintle AG (2000) Luminescence dating of quartz using an improved single-aliquot regenerative-dose protocol. Radiat Meas 32(1):57–73

Nathan RP, Mauz B (2008) On the dose-rate estimate of carbonate-rich sediments for trapped charge dating. Radiat Meas 43:14–25

Nissen E, Walker RT, Molor E, Fattahi M, Bayasgalan A (2009a) Late quaternary rates of uplift and shortening at Baatar Hyarhan (Mongolian Altai) with optically stimulated luminescence. Geophys J Int 177(1):259–278

Nissen E, Walker RT, Bayasgalan A, Carter A, Fattahi M, Molor E, Schnabel C, West AJ, Xu S (2009b) The late Quaternary slip-rate of the Har-Us-Nuur fault (Mongolian Altai) from cosmogenic Be-10 and luminescence dating. Earth Planet Sci Lett 286(3–4):467–478

Olley J, Caitcheon G, Murray A (1998) The distribution of apparent dose as determined by optically stimulated luminescence in small aliquots of fluvial quartz: implications for dating young sediments. Quat Sci Rev 17(11):1033–1040

Olley JM, Caitcheon GG, Roberts RG (1999) The origin of dose distributions in fluvial sediments, and the prospect of dating single grains from fluvial deposits using optically stimulated luminescence. Radiat Meas 30(2):207–217

Pietsch TJ, Olley JM, Nanson GC (2008) Fluvial transport as a natural luminescence sensitiser of quartz. Quat Geochronol 3(4):365–376

Poolton NRJ, Kars RH, Wallinga J, Bos JAA (2009) Direct evidence for the participation of band-tails and excited-state tunnelling in the luminescence of irradiated feldspars. J Phys Condens Matter 21:485505

Porat N, Zilberman E, Amit R, Enzel Y (2001) Residual ages of modern sediments in an hyperarid region, Israel. Quat Sci Rev 20(5–9):795–798

Porat N, Amit R, Enzel Y, Zilberman E, Avni Y, Ginat H, Gluck D (2010) Abandonment ages of alluvial landforms in the hyperarid Negev determined by luminescence dating. J Arid Environ 74:861–869

Ramsey CB (2008) Deposition models for chronological records. Quat Sci Rev 27:42–60

Roberts R, Bird M, Olley J, Galbraith R, Lawson E, Laslett G, Yoshida H, Jones R, Fullagar R, Jacobsen G, Hua Q (1998) Optical and radiocarbon dating at Jinmium rock shelter in northern Australia. Nature 393(6683):358–362

Rodnight H, Duller GAT, Wintle AG, Tooth S (2006) Assessing the reproducibility and accuracy of optical dating of fluvial deposits. Quat Geochronol 1:109–120

Sancho C, Pena JL, Rivelli F, Rhodes E, Munoz A (2008) Geomorphological evolution of the Tilcara alluvial fan (Jujuy Province, NW Argentina): Tectonic implications and palaeoenvironmental considerations. J S Am Earth Sci 26(1):68–77

Schlunegger F, Norton K, Steffen D (2012) Timing the interface between mass wasting and fluvial processes with OSL. In: Schneuwly-Bollschweiler M, Stoffel M, Rudolf-Miklau M (eds) Dating torrential processes on fans and cones – methods and their application for hazard and risk assessment, Advances in global change research. Springer, Dordrecht/Heidelberg/London/New York

Schneuwly-Bollschweiler M, Stoffel M (2012) Dendrogeomorphology – tracking past events with tree rings. In: Schneuwly-Bollschweiler M, Stoffel M, Rudolf-Miklau M (eds) Dating torrential processes on fans and cones – methods and their application for hazard and risk assessment, Advances in global change research. Springer, Dordrecht/Heidelberg/London/New York

Scott MA, Cook GT, Naysmith P (2007) Error and uncertainty in radiocarbon measurements. Radiocarbon 49:427–440

Singarayer JS, Bailey RM, Ward S, Stokes S (2005) Assessing the completeness of optical resetting of quartz OSL in the natural environment. Radiat Meas 40(1):13–25

Singhvi AK, Deraniyagala SU, Singupta D (1986) Thermoluminescence dating of Quaternary red-sand beds: a case study of coastal dunes in Sri Lanka. Earth Planet Sci Lett 80:139–144

Sinha S, Suresh N, Kumar R, Dutta S, Arora BR (2010) Sedimentologic and geomorphic studies on the Quaternary alluvial fan and terrace deposits along the Ganga exit. Quat Int 227:87–103

Spencer JQG, Robinson RAJ (2008) Dating intramontane alluvial deposits from NW Argentina using luminescence techniques: problems and potential. Geomorphology 93(1–2):144–155

Srivastava P, Bhakuni SS, Luirei K, Misra DK (2009) Morpho-sedimentary records at the Brahmaputra River exit, NE Himalaya: climate-tectonic interplay during the Late Pleistocene-Holocene. J Quat Sci 24(2):175–188

Thomas PJ, Jain M, Juyal N, Singhvi AK (2005) Comparison of single-grain and small-aliquot OSL dose estimates in <3000 years old river sediments from South India. Radiat Meas 39: 457–469

Thrasher IM (2008) Optically stimulated luminescence dating of ice-marginal palaeosandar from the last Irish Sea Ice-Stream. Unpublished PhD thesis, Department of Geography, University of Liverpool, UK

Wallinga J (2002) On the detection of OSL age overestimation using single-aliquot techniques. Geochronometria 21:17–26

Wintle A (1973) Anomalous fading of thermoluminescence in mineral samples. Nature 245: 143–144

Wintle AG, Li SH, Botha GA (1993) Luminescence dating of colluvial deposits from Natal, South-Africa. S Afr J Sci 89(2):77–82

Zembo I, Panzeri L, Galli A, Bersezio R, Martini M, Sibilia E (2009) Quaternary evolution of the intermontane Val d'Agri Basin, Southern Apennines. Quat Res 72(3):431–442

Timing the Interface Between Mass Wasting and Fluvial Processes with OSL

Fritz Schlunegger, Kevin Norton, and Damian Steffen

1 Introduction

Debris flows are episodic, gravity-induced movements of granular solids mixed with soil, debris and water. Their occurrence is interpreted based on the sedimentary fabric of the deposits. In particular, the episodic, in-phase transport and deposition of granular solids, soil debris and water results in the formation of a matrix-supported fabric where clasts are embedded in a framework of fine-grained material (Blair 1999) (Fig. 1). The large flow velocities and bulk densities of debris flows result in locally large impact forces. Their typically large volumes, several thousands of km^3 per event (e.g. Illgraben debris-flow catchment, Central Alps; McArdell et al. 2007) combined with a poor temporal predictability cause them to be very hazardous. Although debris-flow processes have been extensively studied for their initiation (e.g. Wieczorek 1987; Iverson 1997) and their transformation from initially small flows to large hazardous events (e.g. Gallino and Pierson 1984; Fannin and Rollerson 1993; Jakob et al. 2005; Jakob and Hungr 2005; Santi et al. 2008), little is known about how changes in debris-flow activity have modified the development of mountainous landscapes particularly on orbital scales. This deficit in research has been mainly due to a lack of adequate chronological tools to precisely calibrate the ages of

F. Schlunegger (✉)
Institute of Geological Sciences, University of Bern, Baltzerstrasse 1+3, 3012 Bern, Switzerland
e-mail: schlunegger@geo.unibe.ch

K. Norton
School of Geography, Environment and Earth Sciences, Victoria University of Wellington, New Zealand
e-mail: kevin.norton@vuw.ac.nz

D. Steffen
Burchard GmbH, Büro für Geologie, Geotechnik und Naturgefahren, Sebastianplatz 1, 3900 Brig, Switzerland
e-mail: d.steffen@wasserschneelawinen.ch

M. Schneuwly-Bollschweiler et al. (eds.), *Dating Torrential Processes on Fans and Cones*, Advances in Global Change Research 47, DOI 10.1007/978-94-007-4336-6_19,

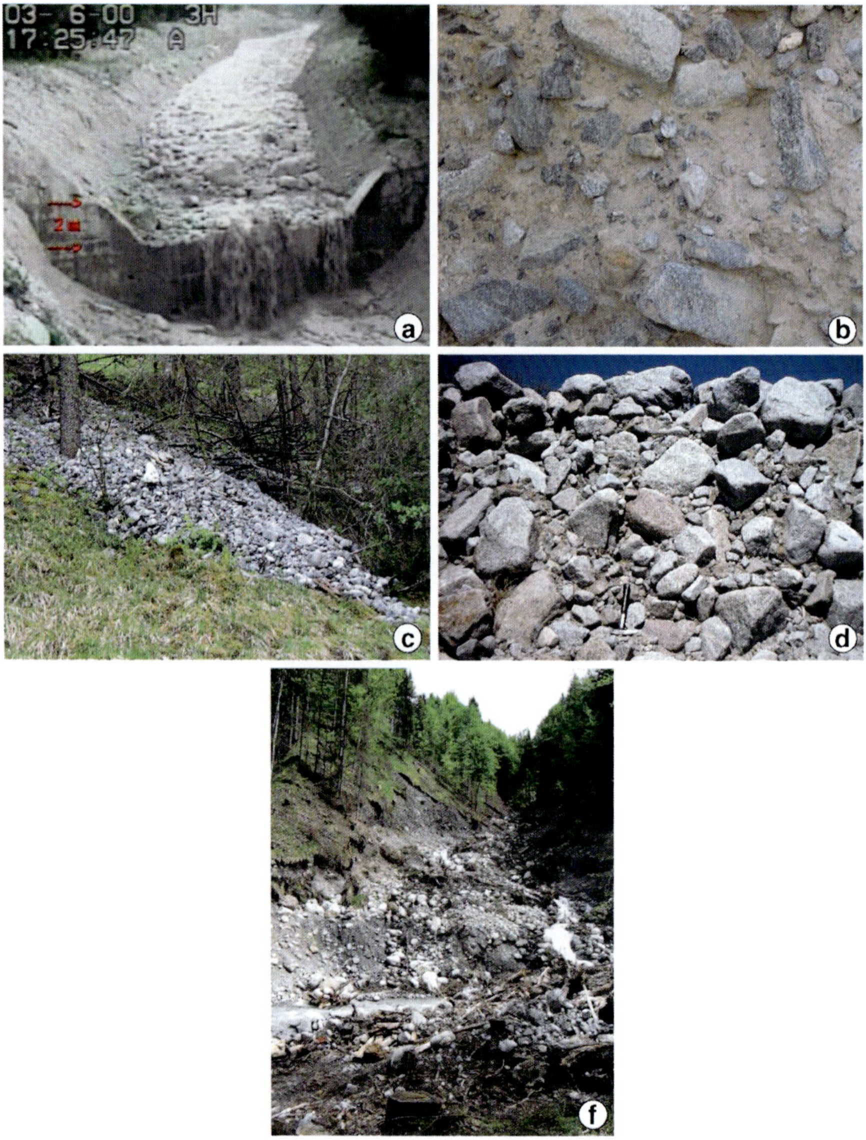

Fig. 1 Examples of debris-flow processes and debris-flow deposits. (**a**) Debris flow in the Illgraben catchment, Swiss Alps: Photo taken from the archive of the Swiss Federal Institute of Forest, Snow and Landscape Research WSL; (**b**) matrix-supported breccia deposited by a cohesive debris flow (Pisco valley, Peru); (**c**) and (**d**) granular flow deposits (Peru and Alps); sediment transfer was controlled by the turbulence and buoyancy of water; (**f**) matrix-supported debris-flow deposits in a tributary channel (Alps)

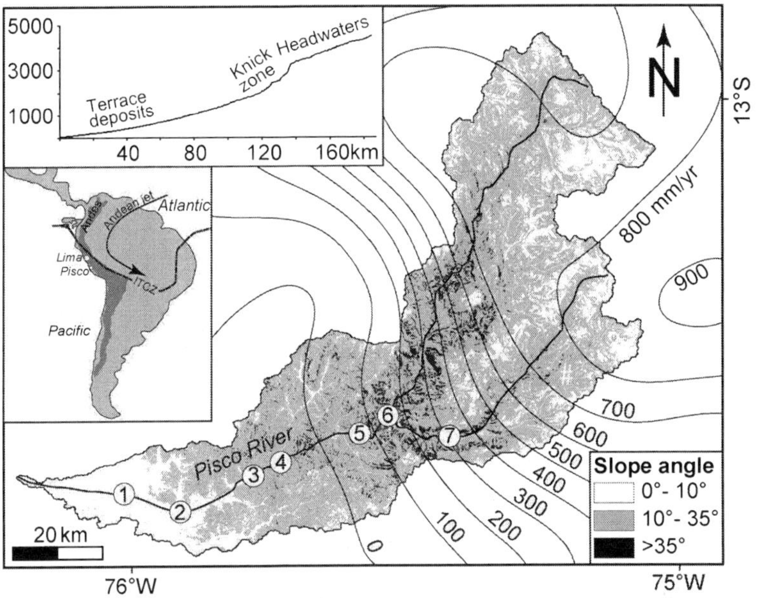

Fig. 2 Wind pattern in Southern America, longitudinal stream profile of Pisco River, and precipitation rate pattern in the Pisco drainage basin together with OSL sampling sites (modified after Steffen et al. 2009)

stratigraphic records (Blum and Törnqvist 2000). Steffen et al. (2009) presents for the first time optically stimulated luminescence (OSL) ages from Quaternary alluvial and colluvial terrace sequences in the Pisco valley of western Peru to reconstruct climate-driven changes in debris-flow activity for orbital time scale.

This chapter reports from the Steffen et al. (2009) study and illustrates how a change in debris-flow activity leads to a modification in the valley morphology. We will emphasize the importance of OSL dating for assessing the timing of changes, which in turn is a prerequisite for temporal correlations to well-studied palaeoclimate archives. In particular, we focus on the sedimentology and ages of cut-and-fill alluvial and colluvial terrace deposits in the Pisco valley located in western Peru at 13.5°S latitude (Fig. 2). These sedimentary archives were used to relate phases of sediment accumulation and erosion to climate-driven changes in debris-flow activity (Steffen et al. 2009). The ultimate scope is to present a well-studied example where OSL ages of deposits resulted in an improved understanding of how climate and climate changes have imprinted on the development of that particular catchment over orbital time scales.

2 Geologic and Geomorphic Setting

The western side of the Peruvian Andes is characterized by up to 14,000 km^2-large west-directed drainage systems that originate in the high Andes between 3,600 and 5,000 m a.s.l., and that discharge to the Pacific Ocean (Fig. 2). The bedrock underlying these basins comprises Tertiary volcanoclastic deposits in the uppermost portions of the drainage basins, Late Jurassic and Early Cretaceous sediments (e.g. sandstone, limestone and greywacke) and Cretaceous to Early Tertiary batholiths in the middle and lowermost reaches. The valley morphologies are characterized by two segments separated by distinct knickzones in the longitudinal stream profiles. These geomorphic features are interpreted to reflect the occurrence of headward erosion in response to a phase of an enhanced surface uplift in the Late Miocene (e.g. Schlunegger et al. 2006; Schildgen et al. 2007). Near the knickzones, the hillslopes are up to 60° steep, covered by regolith and dissected by debris-flow channels and shallow-seated landslides. Downstream the knickzones, the valleys widen from several tens to several hundreds of meters. The hillslopes are less steep compared to the situation at the knickzones (mostly less than 35°) and dissected by a dense network of bedrock debris-flow channels. Also downstream the knickzones, the valleys are filled by Quaternary deposits several tens of meters thick that form multiple terraces at various elevations. Preliminary mapping revealed that the thickest Quaternary valley-fill deposits are present near Lima, from where they thin to several meters towards the north and the south.

The Rio Pisco drainage basin, which is the focus of this chapter, is located on the western side of the Andes at 13.5°S latitude. It covers an area of 4,300 km^2 and extends ~200 km in an east-west direction. The Rio Pisco, the trunk stream in this valley, originates in the high Andes at 5,200 m above sea level and discharges to the Pacific Ocean.

3 Climate and Palaeoclimate Change

The precipitation rate pattern in western Peru shows a hyperarid situation at the coast, and semi-arid to semi-humid conditions in the high Andes. Along the Pacific coast, the climate is controlled by the Easter Island High and the Humboldt Current, which are responsible for the establishment of a persistent inversion thereby stabilizing the atmosphere and inhibiting convection. In the high Andes and in the headwaters of the studied catchment, the pattern of precipitation is closely related to the position of the subtropical jet stream and the ITCZ (Garreaud et al. 2003). When the position of the jet stream is on its southernmost position (during austral summer, from December to February), humidity from the tropical Amazon basin and the Atlantic is carried by easterlies to the Altiplano (Garreaud et al. 2003). In contrast, during austral winter between March and November, the subtropical jet stream and the ITCZ are located farther north, close to the equator. The results are persistently dry westerly winds with almost no precipitation on the Altiplano particularly during austral winter. In the high Andes, this results in a decrease in precipitation rates from ca. 1,000 mm/year near the equator at ca. 5°S (Mettier et al. 2009) to 700–800 mm/year at ca. 13°S near Pisco (Fig. 2), and to 200 mm/year at the border between Peru and Chile (18°S, Kober et al. 2007). Towards the Pacific, precipitation rates decrease to 50 mm/year in the region immediately downstream the knickzones 50 km farther west, and finally to zero at the coast.

On orbital time scales, shifts in the Earth's precession increased the heating contrasts between northern and southern hemisphere with the effect that upper-air easterlies became stronger leading to more precipitation on the Altiplano. This mechanism has been used to explain pluvial intervals on the Altiplano (e.g. Minchin or Tauca; see Baker et al. 2001a, b; Fritz et al. 2004). It is suggested here that stronger easterly winds also result in higher precipitation rates farther west, i.e. in the lower reaches of the transverse drainages.

4 Stratigraphy and Sedimentology

It has been shown, based on theoretical grounds, that an increase in precipitation rates and storm frequencies can result in alluvial and colluvial cut-and-fill sequences (Tucker and Slingerland 1997), but field-based observations are rather sparse. The Pisco valley serves as an example where climate change can result in variations in both fluvial and debris-flow processes (Steffen et al. 2009). The morphology of this valley is characterized by at least three terrace levels, two of which are readily identified in the field (Fig. 3). The highest (T1) terrace level is by far the most prominent and can be correlated over lateral distances of tens of kilometers downstream the knickzone. Each terrace level is underlain by a distinct stratigraphic sequence that is characterized by well-sorted, clast-supported conglomerates at the base and matrix- to clast-supported breccias at the top. The petrographic composition of the conglomerates displays the whole spectrum of bedrock geology of the Pisco River watershed (i.e., granites, metasediments and volcanoclastics), suggesting that the material was sourced from the entire catchment. As interpreted by Steffen et al. (2009), the sedimentologic fabric indicates deposition by the perennial Pisco River. The matrix-supported fabric and the monomict petrofacies of the breccias, however, are indicative of episodic supply of sediment by debris flows (Blair 1999) sourced from the bordering hillslopes and/or local tributaries. These flows form the top of the fans with well preserved convex-up upper surfaces.

The distinct succession in each section that is characterized by alluvial conglomerates at the base and colluvial debris-flow breccias at the top indicates that the locally-sourced fans prograded into the valley. It is likely that the supply of sediment from the tributaries and hillslopes resulted in sediment loads that exceeded the transport capacity of the Pisco River, causing sediment to accumulate and terrace levels

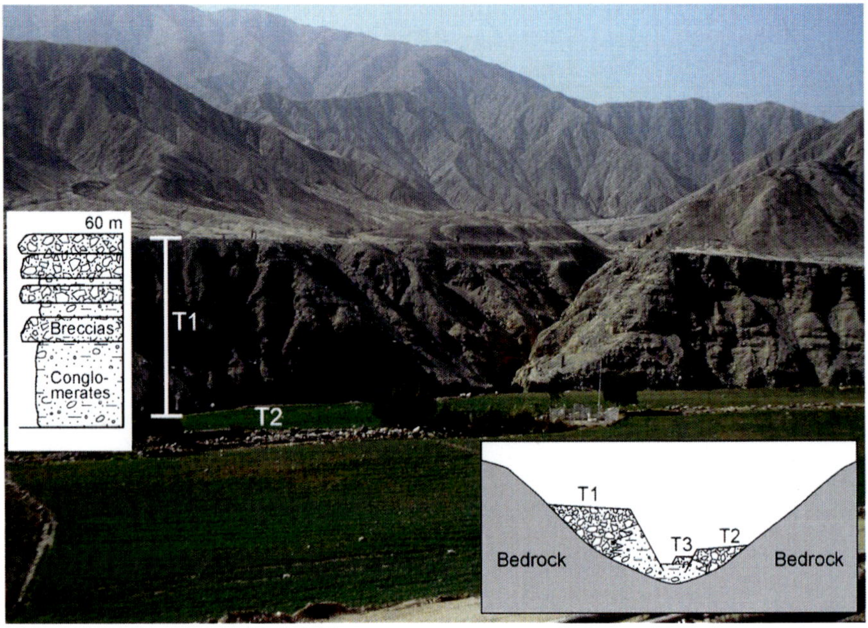

Fig. 3 Sedimentologic and stratigraphic architecture of cut-and-fill deposits in the Pisco valley

to form. During the erosion phase, however, the opposite situation was presumably the case. Hanging tributary valleys (Fig. 3) indicate that at present, erosion rates of the trunk stream (i.e. the Pisco River) are greater than those of the local sources.

5 Chronology of Terrace Deposits

The chronological framework of the Quaternary terrace deposits are based on a total of 28 samples collected from 7 sites (Steffen et al. 2009). For details regarding the methodological treatments of OSL samples in general the reader is referred to the previous chapter. Details about the dating of the Pisco samples are presented in Steffen et al. (2009, 2010).

The optically stimulated luminescence ages reveal that the two dominant terrace levels display distinctively different ages of sediment accumulation. The sequences of the highest terrace levels (T1) span the time interval between 54 and 38 ka ago (Fig. 4). The downstream terraces tend to yield older ages, indicating that sediment aggradation started at the valley outlet, with the locus of sediment deposition then propagating farther upstream. The destructive phases appear to follow the same pattern, with abandonment starting at the valley outlet and then propagating upstream. The lower terrace levels (T2) are made up of deposits that are dated between 27 and 16 ka old. The ages of the lowermost terrace sequences (T3) range between 11 and 4 ka.

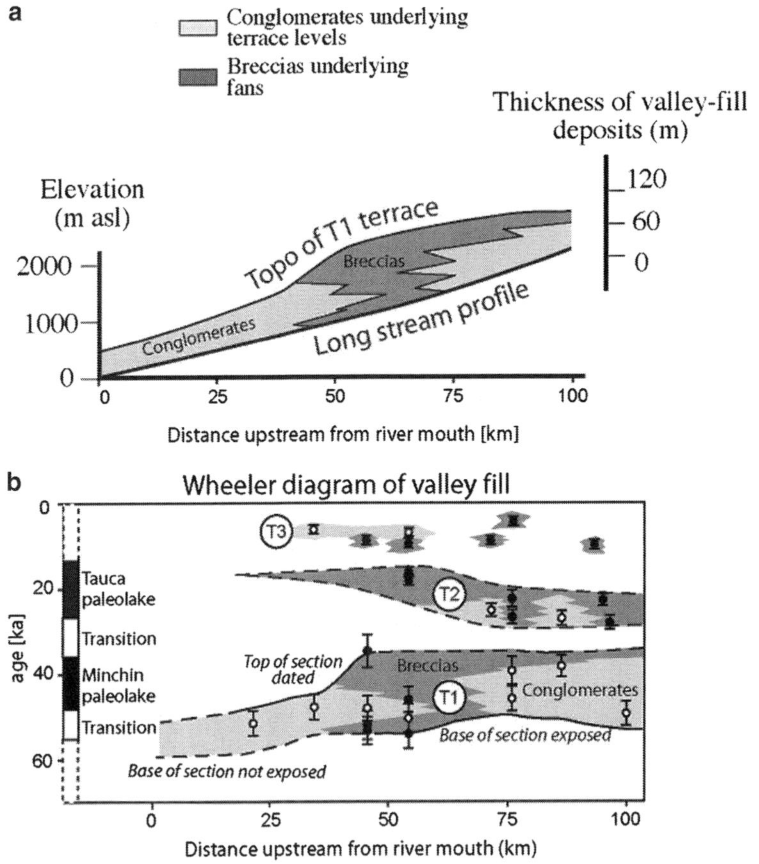

Fig. 4 (**a**) Pattern of sediment thickness of *T1* terrace for the reach between the Pacific ocean and the lower termination of the knickzone, and (**b**) chronology of *T1*, *T2* and *T3* terrace deposits (modified after Steffen et al. 2009)

6 Climate Change, Sediment Transfer by Debris Flows, and Stratigraphic Architecture

OSL dating of the valley fill deposits reveals that phases of sediment aggradation were contemporaneous with humid periods on the Altiplano. Aggradational phases occurred particularly during the Minchin (∼55–40 kyr) and Tauca (26–15 kyr) palaeolake cycles (Fig. 4). The sedimentologic data suggest that enhanced precipitation in the headwaters and the middle part of the catchment resulted in increased erosion and transport of sediment from the hillslopes into the channel network by debris flows, and towards the lower reaches by fluvial processes. An imbalance between sediment supply from hillslopes, and sediment transport capacity of the receiving trunk stream, in combination with a decreasing river gradient and an increasing valley width most probably resulted in deposition of sediment (Steffen et al. 2009). Not only were the headwaters affected by this climate change, but also the tributary catchments and hillslopes in the lower reaches that are currently experiencing an arid climate. The controls on sediment discharge in the tributary systems and mechanisms driving fan construction were considered to be local. Due to the absence of a vegetation cover, sediment discharge in the tributary drainage basins is very sensitive to changes in the rainfall pattern. A change towards a climate with a higher frequency of high-magnitude precipitation events most likely initiated the construction of fans. OSL ages additionally suggest that sediment aggradation started at the valley outlet, with the locus of sediment deposition then propagating farther upstream. Subsequent erosion and downcutting of the valley fill commenced during the same pluvial periods and continued throughout the following drier climates. Interestingly, the destructive phases also started at the valley outlet and then propagated upstream, presumably as the hillslope sediment reservoirs became depleted. This resulted in a reduction of sediment transfer by debris flows, and hence in sediment supplies below the sediment transport capacity of the trunk stream. We thus consider that the sedimentary fill of the Pisco valley, and presumably that in other mountain belts, records a transient, non-steady response of debris-flows and fluvial processes to climate change, where non-steady sediment flux by hillslope erosion is buffered for a limited time span. Accordingly, the terrace systems record the combined effect of allo- and autogenic mechanisms initiated by a shift towards a more humid climate.

Supply of sediment and formation of terraces and fans not only depend on climate and water discharge, but also to a large extent on the availability of unconsolidated material on hillslopes. Bedrock weathering, therefore, has presumably been the most important limiting process for terrace and fan construction, particularly in arid regions. This implies that the number of periods during which terraces and fans were formed only provides a minimum estimate of climate cycles. It is thus possible that further wet periods did occur, but they may not be recorded in the sedimentary deposits. This may be the case where the time scale of climate change is shorter than the time required for weathering and the production of new sediment.

7 Conclusions

Debris flows are hazardous processes resulting in large damages each year. Hence, there is a public interest in the understanding of the circumstances under which debris flows are triggered. This chapter synthesizes the results of a successful case study in which OSL dating facilitated the reconstruction of changes in debris-flow frequencies and magnitudes (Steffen et al. 2009). We emphasize here that the temporal resolution of this dating method does not allow us to assess the ages of each debris-flow event recorded in stratigraphic archives. However, it is useful for determining the circumstances leading to sediment accumulation and subsequent erosion, driven by changes in debris-flow activities. In particular, a phase of sediment accumulation is considered to have occurred when the paleoclimate shifted from hyperarid to more humid conditions. Sediment that was previously stored on hillslopes became removed and transferred to the main valley by debris flow. Most likely, the effect was an increase in the magnitude and frequency of debris flows, leading to a situation where sediment concentration presumably exceeded the transport capacity of the receiving trunk stream. The valley then experienced higher frequencies and magnitudes of floods, resulting in sediment accumulation on the valley floor. The resolution of OSL dating also allows determination of the time when valley erosion and recycling of the valley fill deposits started. In the case of the Pisco valley, erosion commenced during the pluvial period and presumably occurred when the hillslopes were liberated from the sedimentary cover. In this case, precipitation results in water discharge from the hillslopes with small sediment loads. The effect was an undercapacity of the trunk stream and erosion of the previously deposited valley fill. The OSL ages show that sediment accumulation occurred in response to a climate change rather than a particular climate, and that subsequent erosion is the autocylic consequence of the modification of the sediment discharge pattern.

References

Baker PA, Seltzer GO, Fritz SC, Dunbar RB, Grove MJ, Cross SL, Tapia P, Rowe HD, Broda JP (2001a) The history of South American tropical precipitation for the past 25,000 years. Science 291:640–643

Baker PA, Rigsby CA, Seltzer GO, Fritz SC, Lowenstein T, Bacher N, Veliz C (2001b) Tropical climate changes at millennial and orbital timescales on the Bolivian Altiplano. Nature 409: 698–701

Blair TC (1999) Sedimentology of the debris-flow-dominated Warm Spring Canyon alluvial fan, Death Valley, California. Sedimentology 46:941–965

Blum MD, Törnqvist TE (2000) Fluvial responses to climate and sea-level change: a review and look forward. Sedimentology 47:2–48

Fannin RJ, Rollerson TP (1993) Debris flows: some physical characteristics and behaviours. Can Geotech J 30:71–81

Fritz SC, Baker PA, Lowenstein TK, Seltzer GO, Rigsby CA, Dwyer GS, Tapia PM, Arnold KK, Ku TL, Luo S (2004) Hydrologic variation during the last 170,000 years in the southern hemisphere tropics of South America. Quat Res 61:95–104

Gallino GL, Pierson TC (1984) Polallie Creek debris flow and subsequent dma-break flood of 1980, East Fork Hood River Basin, Oregon. Open file-report US Geological Survey 84–578. US Department of Interior, Geological Survery, Portland, 37 pp

Garreaud R, Vuille M, Clement AC (2003) The climate of the Altiplano: observed current conditions and mechanisms of past changes. Palaeogeogr Palaeoclimatol Palaeoecol 194:5–22

Iverson RS (1997) The physics of debris flows. Rev Geophys 35:245–296

Jakob M, Hungr O (2005) Introduction. In: Jakob M, Hungr O (eds) Debris-flow hazards and related phenomena. Springer, Berlin, pp 1–8

Jakob M, Bovis M, Oden M (2005) The significance of channel recharge rates for estimating debris-flow magnitude and frequency. Earth Surf Process Landf 30:755–766

Kober F, Ivy-Ochs S, Schlunegger F, Baur H, Kubik PW, Wieler R (2007) Denudation rates and topography-driven rainfall threshold in northern Chile: multiple cosmogenic nuclide data and sediment budgets. Geomorphology 83:97–120

McArdell B, Bartelt P, Kowalski J (2007) Field observations of basal forces and fluid pore pressure in a debris flow. Geophys Res Lett 34. doi:10.1029/2006GL029183

Mettier R, Schlunegger F, Schneider H, Schwab M, Rieke-Zapp D (2009) Relationships between landscape morphology, climate and surface erosion in northern Peru at 5°S latitude. Int J Earth Sci 98:2009–2022

Santi PM, deWolfe VG, Higgins JD, Cannon SH, Gartner JE (2008) Sources of debris flow material in burned areas. Geomorphology 96:310–321

Schildgen TF, Hodges KV, Whipple KW, Reiners PW, Pringle MS (2007) Uplift of the western margin of the Andean plateau revealed from canyon incision history, southern Peru. Geology 35:523–526

Schlunegger F, Zelinger G, Kounov A, Kober F, Hüsser B (2006) Scale of relief growth in the Andes of northern Chile. Terra Nova 18:217–223

Steffen D, Schlunegger F, Preusser F (2009) Drainage basin response to climate change in the Pisco valley, Peru. Geology 37:491–494

Steffen D, Schlunegger F, Preusser F (2010) Late Pleistocene fans and terraces in the Majes valley, southern Peru, and their relation to climatic variations. Int J Earth Sci 99(8):1975–1989

Tucker G, Slingerland R (1997) Drainage basin responses to climate change. Water Resour Res 33:2031–2047

Wieczorek GF (1987) Effect of rainfall intensity and duration on debris flows in central Santa Cruz Mountains, California. In: Costa JE, Wieczorek GF (eds) Debris flows/avalanches: process, recognition and mitigation, Geological Society of America, reviews in engineering geology 7. Geological Society of America, Boulder, pp 93–104

Part III
Documentation and Monitoring

Debris-Flow Monitoring in Italy

Lorenzo Marchi and Pia Rosella Tecca

1 Introduction

Monitoring of debris flows, aimed at assessing their physical parameters, is important both for research purposes, such as the determination of the rheological behaviour and the calibration of mathematical models, as well as for planning countermeasures and designing warning systems.

Automated monitoring systems make an effective collection of debris-flow data possible. Different types of sensor systems, designed for both debris-flow detection and warning, and for research have been installed in several countries. Reviews of sensor types and monitoring systems have been presented by Itakura et al. (2005) and Arattano and Marchi (2008). This paper presents research experiences in two instrumented sites of the eastern Italian Alps: Moscardo Torrent and Acquabona (Fig. 1), where debris-flow monitoring systems were set up since 1990 and 1997, respectively.

2 Study Areas

The Moscardo Torrent basin covers an area of 4.1 km^2 and ranges in elevation from 890 m (fan apex) to 2,043 m. The rocky substratum of the basin is made of Carboniferous flysch, represented by highly fractured and weathered shale, slate, siltstone, sandstone and breccia. Quaternary deposits, mostly consisting of scree and landslide accumulations, are common in the basin. The steep slopes of the basin are prone to rockfalls and shallow slope failures, which supply large amounts of

L. Marchi (✉) • P.R. Tecca
CNR IRPI, Padova, Italy
e-mail: lorenzo.marchi@irpi.cnr.it; pia.tecca@irpi.cnr.it

M. Schneuwly-Bollschweiler et al. (eds.), *Dating Torrential Processes on Fans and Cones*, Advances in Global Change Research 47, DOI 10.1007/978-94-007-4336-6_20,

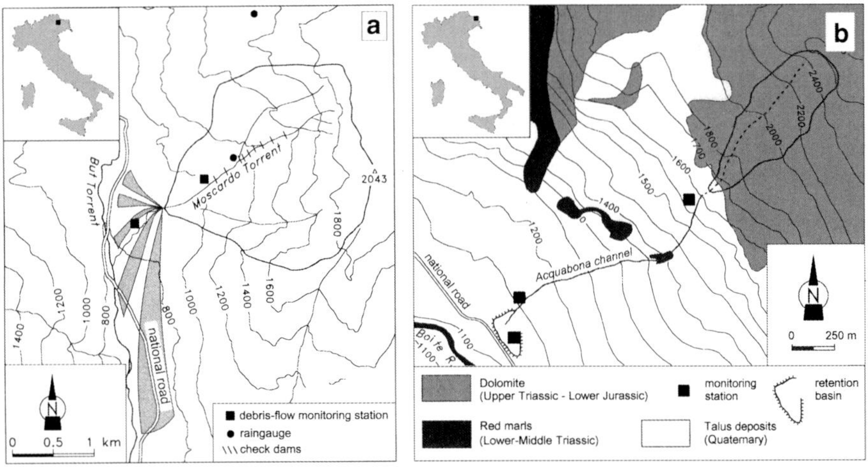

Fig. 1 Instrumented Italian catchments: (**a**) Moscardo Torrent, (**b**) Acquabona channel

debris to the channel. The channel has an average slope of 20° and a length of 2,760 m. Some concrete check dams, which are intended to prevent bed erosion and to stabilise channel banks in the middle and lower stretches of the main channel, have been built since the 1970s.

The Acquabona channel entrenches a scree slope developed at the outlet of a small, steep rock basin in the Dolomites. The rock basin is formed of Upper Triassic to Lower Jurassic massive dolomite and limestone cliffs and has a drainage area of 0.3 km^2. The maximum rock basin elevation is 2,667 m, the outlet elevation is 1,650 m, and its average slope is 43°. A thick talus covers the slope from the base of the rock cliffs to the valley floor; it consists of poorly sorted debris containing boulders up to 3–4 m in diameter and includes heterogeneous scree, alluvium and debris-flow deposits. The channel is 1,630 m long and has an average slope of 18°, ranging from 30° in the initiation area to 7° in the lower part. In order to contain the debris discharged by the flows and protect the national road at the downstream end of the channel, a retention basin was built at the channel outlet.

3 Monitoring Systems

The debris flows of the Moscardo Torrent have been recorded at two monitoring sites located, respectively, in the middle part of the alluvial fan, and in the lower part of the drainage basin (Fig. 1a). The downstream monitoring site (on the alluvial fan) has been equipped with ultrasonic sensors since 1990, and with seismic sensors and a video camera since 1995. The upstream monitoring site was instrumented with seismic sensors in 1995. Rainfall is recorded at two rain gauges (Fig. 1a). More details on the characteristics of the sensors, their installation and the evolution of the monitoring system can be found in Arattano (1999) and Marchi et al. (2002).

Debris flows were recorded in the Moscardo Torrent from 1990 to 1998 and from 2002 to 2006; the gap from 1999 to 2002 was due to the construction of check dams and dikes in the instrumented channel.

The Acquabona monitoring system consists of three on-site stations instrumented for debris-flows monitoring in the initiation area, in the middle and in the lower channel reach (Fig. 1b). The monitoring system, installed in summer 1997, is equipped with sensors for measuring rainfall, pore pressures in the mobile channel bottom, ground vibrations induced by the flows, debris-flow stage, total normal stress and fluid pore-pressure at the base of the flow. Two video cameras record debris flows in the upper initiation area and in the lower reach of the channel. Data from the on-site stations are radio-transmitted to an off-site station and stored in a host PC and are remotely retrieved via modem. A complete description of the technical characteristics of the monitoring system can be found in Genevois et al. (2000) and Tecca et al. (2003). Debris flows were recorded in the Acquabona channel from 1997 to 2001; from 2001 to 2005 only rainfall was measured.

4 Results

4.1 *Debris-Flow Frequency*

Debris-flow occurrence in the Moscardo and Acquabona catchments was documented by means of data from archives for the periods before the start of monitoring activities. Combining archive data with monitoring data results in a time series of debris-flow occurrences from 1898 in the Moscardo Torrent and from 1920 in the Acquabona channel (Marchi and Tecca 2006) (Fig. 2).

Since no major changes in stability conditions of channels and slopes have occurred in the studied catchments during the considered period, the increase of documented debris flows in the most recent period (Fig. 2) could be due to a change in the rainfall regime, with an increased frequency of intense storms, as well as

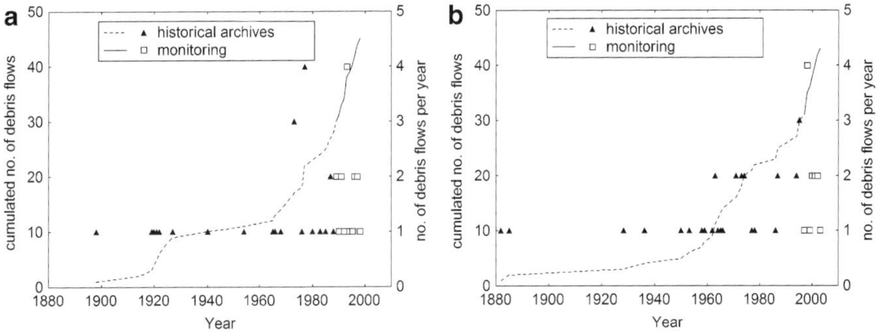

Fig. 2 Debris flows documented by historical documents and monitoring data. (**a**) Moscardo Torrent, (**b**) Acquabona channel

to a larger availability of information. Although it is difficult to assess the relative importance of these two factors, we argue that the better documentation of debris flows and the increased attention to the observation of hazardous natural phenomena are probably the most influencing reason for the increase of observed debris-flow frequency. Debris flows produced in catchments with active erosion and sediment transport phenomena, like Moscardo and Acquabona, are viewed by local people and by the experts involved in watershed management as 'normal' phenomena, unless they attain large magnitude. As a consequence, historical records often include only the largest events, whereas the small ones can remain unperceived.

4.2 Debris-Flow Hydrographs

Debris flows recorded at the downstream monitoring site of the Moscardo Torrent show relevant differences from event to event, especially regarding the recession limb. In some debris flows, the main surge is followed by a regular decrease in flow depth, and only a few, well-defined secondary waves occur after the main surge. In other cases, the recession limb is very irregular with abrupt fluctuations of the stage, indicating the occurrence of a number of smaller waves after the main surge (Marchi et al. 2002). However, debris flows on the alluvial fan of the Moscardo Torrent share an important common feature: the hydrograph is dominated by one major surge, characterised by a sharp rising limb, which corresponds to the passage of the debris-flow front at the monitored cross section. The main surge is often preceded by a precursory surge, which can be recognised on debris-flow hydrographs (Fig. 3). Video recordings of two events that occurred in 1996 show that the sediment-water mixture of the precursory surge is characterised by strong turbulence, indicating a density decidedly lower that the core of the debris flow (Marchi et al. 2002).

Further data on debris-flow waves is provided by the recordings of seismic sensors, which can be presented as plots of amplitude (absolute value of ground vibration) versus time. Differently from what is observed on the alluvial fan, debris-

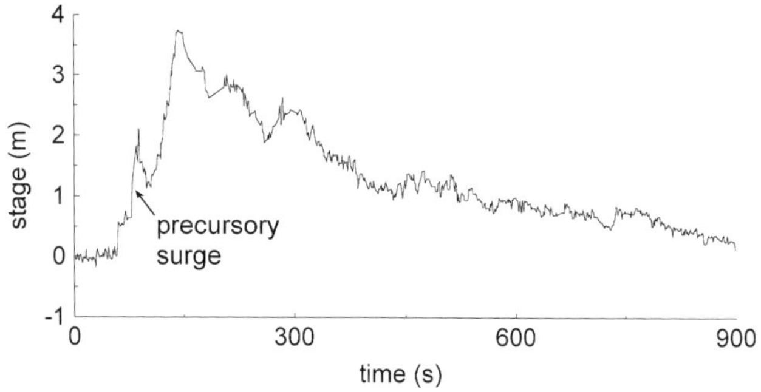

Fig. 3 Moscardo Torrent: debris-flow wave recorded by an ultrasonic sensor on the alluvial fan (debris flow of 8 July 1996)

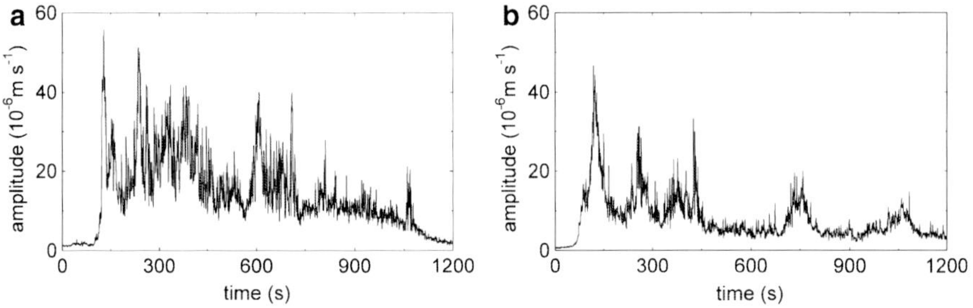

Fig. 4 Moscardo Torrent: debris-flow recording using ground vibration sensors at the two monitoring sites (debris flow of 24 July 1997). (**a**) upstream, (**b**) downstream

flows recordings at the upstream measurement site (Fig. 1a), equipped with seismic sensors, usually consist of an irregular sequence of several pulses, with the first only slightly more intense than the following ones (Fig. 4a), or with the highest intensity in the central part of the event: as discussed by Arattano (2000), in most events a well-defined front, corresponding to the main surge of the debris flow, seems to develop only on the alluvial fan (Fig. 4b).

Debris-flows hydrographs recorded at Acquabona display, as in the Moscardo Torrent, different features from event to event. This is exemplified by the three debris flows recorded in 1998 (Fig. 5). Features common to these three events are a sharp rising limb of the surges, followed by a regular decrease of flow stage (Genevois et al. 2001). Videos of debris flows occurred in 1998 evidence that the first solid surge was preceded by a thin water flow, which exhibited a high surface turbulence and was not sufficiently competent to suspend gravel-sized clasts. Due to its very small flow depth, the precursory surge, although visible in the videos, was not recorded by the ultrasonic sensor.

Peak flow depth was similar for the three debris flows, the event of 17 August attaining only slightly higher values. The main difference between the three debris flows is their different duration: the debris flows recorded in July (Fig. 5a, b) consisted of a unique surge, which lasted approximately 300 s (July 25) and 150 s (July 27), whereas the debris flow of 17 August (Fig. 5c) consisted of several surges of comparable intensity and similar shape, and lasted approximately 25 min. The longer duration of the 17 August debris flow, which resulted also in higher magnitude, is due to longer duration and higher intensity of the triggering rainstorm (Tecca and Genevois 2009).

4.3 *Flow Velocity*

Mean front velocity, which corresponds to the celerity of propagation of the debris-flow wave, can be computed as the ratio of the distance between two cross-sections, equipped with ultrasonic gauges or seismic sensors, to the time interval between the peaks of the debris-flow surge in the recorded hydrographs.

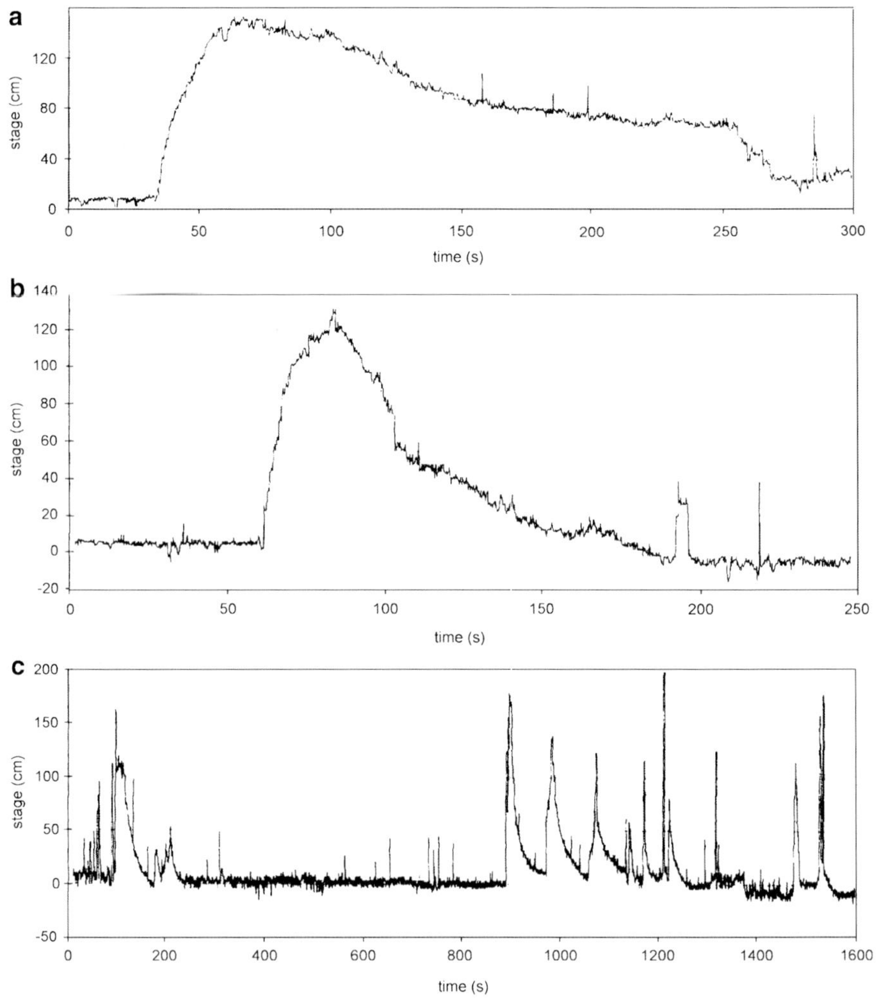

Fig. 5 Debris-flow hydrographs recorded at Acquabona in 1998: (**a**) 25 July, (**b**) 27 July, (**c**) 17 August

When a clearly defined front is not present and no common features can be recognised on the wave recorded at different cross-sections, this approach to the measurement of velocity cannot be applied. In these cases, the cross-correlation technique helps objectively defining the mean velocity of the flow. Arattano and Marchi (2005) have demonstrated that the assessment of debris-flow velocity by means of cross-correlation gives results consistent with those obtained calculating the ratio of the distance between the sensors to the time lag between the appearance of a signal peak at two consecutive detectors.

Video recordings have been used both in the Moscardo Torrent and in Acquabona for assessing the surface velocity of debris flows. The use of videos for surface

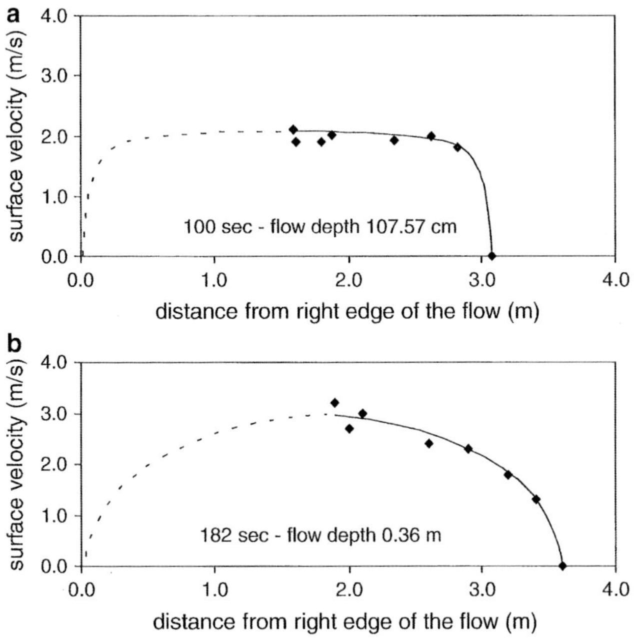

Fig. 6 Examples of horizontal velocity profiles at Acquabona. (**a**) Bingham fluid, (**b**) Newtonian fluid (debris flow of 17 August 1998)

velocity assessment is based on tracking on the video objects recognised on the surface of the flow, such as boulders, trees or logs, or particular features of the flow (e.g. secondary waves). The calibration of the scene with topographic surveys permits assessing the distance travelled by selected objects. The ratio of the distance travelled to the time elapsed corresponds to the mean surface velocity of the debris flow.

In the Moscardo Torrent, a calibration method, which associates points in 2D video images to points in the 3D space (Arattano and Grattoni 2000), was devised for the assessment of the distance travelled by objects on debris-flow surface. At Acquabona, video images were processed in order to track the motion of single particles, scaled to the actual distance, among successive frames and flow surface velocities were estimated. The distribution of velocities at the debris-flow surface shows that Bingham and Newtonian flows occur at different times of the same surge (Tecca et al. 2003) (Fig. 6).

4.4 Peak Discharge and Volume

Debris-flow peak discharge on the alluvial fan of the Moscardo Torrent was computed by multiplying the area of the monitored cross-sections by the flow velocity. The debris-flow volume was estimated by integrating discharge over the hydrograph. Approximations and assumptions of this approach to the estimation

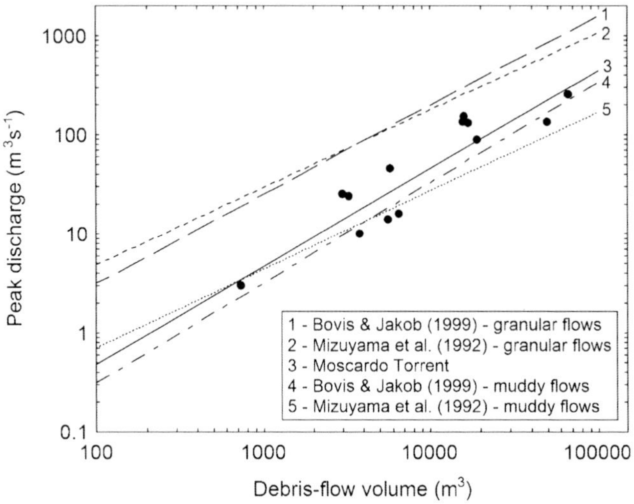

Fig. 7 Relation between debris-flow volume and peak discharge in the Moscardo Torrent and comparison with the regression lines by Mizuyama et al. (1992) and Bovis and Jakob (1999) for both muddy and granular flows

of debris-flow discharge and volume are discussed in Marchi et al. (2002). The same approach was used in Acquabona, where topographic surveys of debris-flow deposits in the retention basin made it possible to cross-check debris-flow volume.

Equation 1 expresses the relation between debris-flow volumes V (m^3) and peak discharge Q_p ($m^3\ s^{-1}$) in the Moscardo Torrent. Equation 1 was computed with least square regression and has a coefficient of determination, R^2 of 0.82. The comparison with the relations developed by Mizuyama et al. (1992) and Bovis and Jakob (1999) (Fig. 7) shows that the Moscardo data lie in the range of muddy debris flows. This agrees with the results of sedimentologic observations (Moscariello et al. 2002) and rheological tests (Coussot et al. 1998) on the characteristics of the Moscardo Torrent debris flows.

$$Q_p = 0.005V^{0.99} \tag{1}$$

The smaller size of the upslope basin causes the debris flow of Acquabona to have lower magnitude than those recorded in the Moscardo Torrent (Tecca and Genevois 2009). At Acquabona, the volume of the largest debris flows is transported by a number of surges of similar intensity occurring over the whole duration of the event (Fig. 5c), so that debris-flow volume does not display a clear relation with peak discharge.

5 Concluding Remarks

Data recorded in the instrumented catchments of Moscardo Torrent and Acquabona have found several applications in the domain of debris-flow studies. The monitoring systems made it possible to record the time of occurrence of debris flows with accuracy higher than normally achievable by means of post-event investigations in ungauged areas. This led to define relations between rainstorms and debris-flow occurrence (Deganutti et al. 2000; Tecca and Genevois 2009). Quantitative data recorded by ultrasonic and seismic sensors allowed identifying variables relevant for assessing debris-flow hazard and designing control measures, such as flow depth, peak discharge and volumes. Moreover, recorded debris-flow hydrographs have provided the experimental basis for numerical modelling of debris flows (Arattano and Savage 1994; Fraccarollo and Papa 2000; Arattano and Franzi 2003; Zanuttigh and Lamberti 2004).

Debris-flow monitoring makes it possible to evaluate advantages and limitations of various types of sensors for their possible use in warning systems. For instance, Arattano (1999, 2003) stresses the advantages of ground vibration sensors (geophones and seismometers) as detectors of debris flows for warning purposes. Valuable characteristics of ground vibration sensors are their safe installation on channel banks and the possibility of detecting debris flows before the arrival of the wave at the instrumented cross-section, which permit extending the warning time.

The valuable results achieved underscore that continuing debris-flow monitoring in the Italian Alps is of the utmost importance for improving the monitoring technologies, extending the time series and recognise possible variations in debris-flow response due to climatic changes. The experience so far gained has made it possible to move beyond the pioneering phase, in which debris-flow monitoring was carried out by research institutes and universities. Close collaborations with agencies of local governments are a promising way to carry on future monitoring activities. This ensures a valued support in the management of the monitoring instrumentation and permits an efficient transfer of experimental data and know-how to the agencies in charge of the protection against mountain hazards.

References

Arattano M (1999) On the use of seismic detectors as monitoring and warning systems for debris flows. Nat Hazard 20:197–213

Arattano M (2000) On debris flow front evolution along a torrent. Phys Chem Earth Part B 25(9):733–740

Arattano M (2003) Monitoring the presence of the debris-flow front and its velocity through ground vibration detectors. In: Wieczorek GF, Naeser ND (eds) Debris-flow hazards mitigation: mechanics, prediction, and assessment. Balkema, Rotterdam, pp 719–730

Arattano M, Franzi L (2003) On the evaluation of debris flows dynamics by means of mathematical models. Nat Hazard Earth Syst Sci 3(6):539–544

Arattano M, Grattoni P (2000) Using a fixed video camera to measure debris-flow surface velocity. In: Rickenmann D, Chen C (eds) Debris-flow hazards mitigation: mechanics, prediction, and assessment. Millpress, Rotterdam, pp 273–281

Arattano M, Marchi L (2005) Measurements of debris flow velocity through cross-correlation of instrumentation data. Nat Hazard Earth Syst Sci 5(1):137–142

Arattano M, Marchi L (2008) Systems and sensors for debris-flow monitoring and warning. Sensors 8(4):2436–2452

Arattano M, Savage WZ (1994) Modelling debris flows as kinematic waves. Bull IAEG 49:95–105

Bovis MJ, Jakob M (1999) The role of debris supply conditions in predicting debris flow activity. Earth Surf Landf Process 24(11):1039–1054

Coussot Ph, Laigle D, Arattano M, Deganutti A, Marchi L (1998) Direct determination of rheological characteristics of debris flow. J Hydraul Eng ASCE 124(8):865–868

Deganutti AM, Marchi L, Arattano M (2000) Rainfall and debris-flow occurrence in the Moscardo basin (Italian Alps). In: Wieczorek GF, Naeser ND (eds) Debris-flow hazards mitigation: mechanics, prediction, and assessment. Balkema, Rotterdam, pp 67–72

Fraccarollo L, Papa M (2000) Numerical simulation of real debris-flow events. Phys Chem Earth Part B 25(9):757–763

Genevois R, Tecca PR, Berti M, Simoni A (2000) Debris flow in Dolomites: experimental data from a monitoring system. In: Wieczorek GF, Naeser ND (eds) Debris-flow hazards mitigation: mechanics, prediction, and assessment. Balkema, Rotterdam, pp 283–291

Genevois R, Galgaro A, Tecca PR (2001) Image analysis for debris flow properties estimation. Phys Chem Earth 26(9):623–631

Itakura Y, Inaba H, Sawada T (2005) A debris-flow monitoring devices and methods bibliography. Nat Hazard Earth Syst Sci 5(6):971–977

Marchi L, Tecca PR (2006) Some observations on the use of data from historical documents in debris-flow studies. Nat Hazard 38(3):301–320

Marchi L, Arattano M, Deganutti AM (2002) Ten years of debris-flow monitoring in the Moscardo Torrent (Italian Alps). Geomorphology 46(1–2):1–17

Mizuyama T, Kobashi S, Ou G (1992) Prediction of debris flow peak discharge. In: International symposium Interpraevent 1992, Bern, Switzerland, vol 4, pp 99–108

Moscariello A, Marchi L, Maraga F, Mortara G (2002) Alluvial fans in the Italian Alps: sedimentary facies and processes. In: Martini P, Baker VR, Garzon G (eds) Flood and megaflood processes and deposits – recent and ancient examples. Blackwell Science, Oxford, pp 141–166

Tecca PR, Genevois R (2009) Field observations of the June 30, 2001 debris flow at Acquabona (Dolomites, Italy). Landslides 6(1):39–45

Tecca PR, Galgaro A, Genevois R, Deganutti AM (2003) Development of a remotely controlled debris flow monitoring system in the Dolomites (Acquabona, Italy). Hydrol Process 17(9):1771–1784

Zanuttigh B, Lamberti A (2004) Analysis of debris wave development with one-dimensional shallow-water equations. J Hydraul Eng ASCE 130(4):293–304

Documentation of Torrential Events

Hans Kienholz, Johannes Hübl, and Anton Loipersberger

1 Introduction

Accurate and comprehensive hazard assessment, as one part of integrated risk management, demands the application of a wide-range of methods (Fig. 1). Such methods include:

- the prediction of future events (i.e. foresighted indication, such as detailed analysis and assessment of the situation in the field as well as application of physical and numerical models describing the processes); and
- the evaluation of past events (retrospective indication, e.g. "silent witnesses", which are documents about historic events in the terrain as well as the evaluation of written documents/chronicles).

The predictive methods also depend on the experience gained through the assessment of past events. It is impossible to create accurate models without observations (measurements), monitoring and experience from field observations. Thus knowledge about former events is indispensable.

Many hazardous events are momentary (lasting only minutes to a few hours), while there may be a very long time span (years, decades or even centuries) between two recurring events (see example in Fig. 2). Hazard assessment usually has to

H. Kienholz (✉)
KiNaRis – Naturgefahren/Risikomanagement, Bern, Switzerland
e-mail: kienholz@kinaris.ch

J. Hübl
Institute of Mountain Risk Engineering, University of Natural Resources and Life Sciences, Vienna, Austria
e-mail: johannes.huebl@boku.ac.at

A. Loipersberger
Bavarian Environment Agency (LFU), Munich, Bavaria, Germany
e-mail: Anton.Loipersberger@lfu.bayern.de

M. Schneuwly-Bollschweiler et al. (eds.), *Dating Torrential Processes on Fans and Cones*, Advances in Global Change Research 47, DOI 10.1007/978-94-007-4336-6_21,

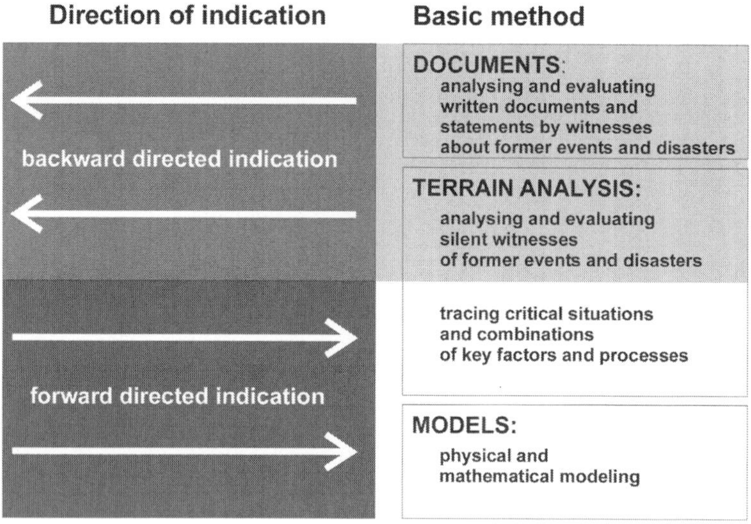

Fig. 1 Basic methods of hazard assessment (according to Kienholz in Hübl et al. 2002: p. I/4)

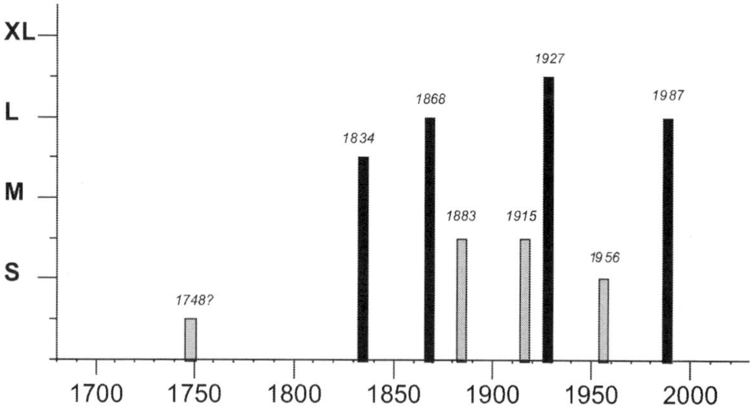

Fig. 2 Catastrophic torrential activity and debris flows affecting the debris fan of the Zavragia River in the Grisons (Switzerland) (According to Kienholz in Hübl et al. 2002: p. I/4) Magnitude of event (transported bedload) – **S**mall, **M**edium, **L**arge, e**X**tra **L**arge. Events larger than medium (*M*) size are indicated as *dark bars*, smaller ones as *light bars*

take place during the calm phase between the scarce catastrophic events. Thus, the documenting expert must have the ability to reconstruct the real processes in the course of a historic event. In addition to this, it is necessary to identify realistic scenarios for the damage event. It is needless to say that these results have to be based on all kinds of reliable and significant data ("hard facts") gathered from former events. This requires a comprehensive monitoring of the events themselves. However, in reality it is quite unusual that experts are present in place when such events occur. Therefore, it is desirable that those people who live and work close to the event monitor the processes and collect data. Experts would be alerted immediately in the case of an event, and thus would be able to collect data during, or at least immediately after, the event. Immediate measures, such as removal of debris from settlements and roads, are usually undertaken within a few hours. Hence, important silent witnesses are removed in the run out and sedimentation zones of the disastrous processes.

This perfection, however, is not realistic. Inhabitants of the disaster area are instead fully engaged in protecting and rescuing life and possessions. Also the experts and officers of the local governmental authorities are involved with rescue operations and immediate measures. People who incidentally try to document some aspects of the event (such as local eyewitnesses, tourists or journalists) usually focus on the damage caused, rather than the geomorphic process itself.

In order to improve the situation, an international group of experts intensively discussed this issue during three field workshops, and published the general concept of DOMODIS (= DOcumentation of MOuntain DISasters; Hübl et al. 2002). This publication includes organizational considerations, as well as general information on the natural processes, their features and indicators.

2 Different Contributors, Various Interests

Different contributors and customers are interested in a wide range of data about triggers and conditions of hazardous events and the relevant processes. Those people involved in the event management require actual data and first survey information. On the other hand, specialized scientists would like to gather very specific data about those aspects of the processes they are especially interested in. And in between, there are also the hazard experts and practitioners (e.g. civil engineers, forest engineers, etc.), from governmental agencies or private companies who are involved in mountain risk prevention.

The profound and specific data required by specialized scientists must be gathered by themselves, even if this is only possible some time after the event. It is essential that they are informed about the event as soon as possible, as well as given access to the data already gathered by the other contributors.

For those contributors who are involved in the event management, the time factor is crucial. They require quick and accurate information, yet they do not need detailed information about the process. In this case, the information should be gathered by those people who are on site.

DOMODIS focuses mainly on the lower, and partly on the medium, categories of contributors (Fig. 3) who require sound data that should be gathered during, or very soon after, the event. The data should be profound and reliable enough for hazard zoning, the conception and design of future preventive measures, and for the preparation for possible future events.

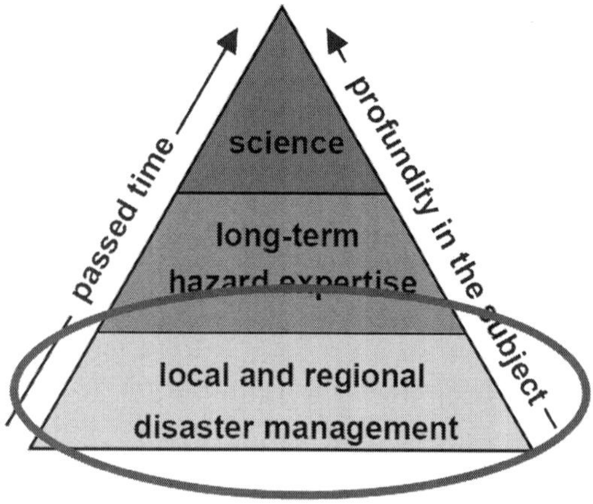

Fig. 3 Actuality and profundity of event documentation: Interests of different contributors (Hübl et al. 2002: p. I/7)

3 Organization and Implementation

In order to implement DOMODIS it is necessary to have a comprehensive administrative (and legal) framework at a national, state or provincial level.

The organization of DOMODIS in each country and province depends on various conditions including the divisions and duties of the various governmental agencies, availability of (own) experts in the case of an event, availability of private experts, practicable financing procedures, and financial restrictions.

Event management at national, provincial or municipal levels include many different activities that should be based on well prepared organizational structures. Many of the considered events, depending on their type (Fig. 4), require the triggering of very well-prepared as well as of ad-hoc activities. For example, activities include:

- communication between all involved contributors;
- rescue of human victims;
- reconnaissance trips (flights);
- removal of debris;
- restoration of supply lines (roads, railways, energy supply etc.); and
- warning systems.

Additionally, the event documentation must start as soon as possible after event occurrence.

The monitoring and documentation of the event must be carried out by experts who are not affected themselves and who are not in charge of rescue measures. To facilitate such documentation two major demands must be covered:

- Experts who can be called upon in case of an event must be instructed in a way to be able to provide such documentation in a standardized way and with the necessary grasp of the subject. This instruction is a part of the preparation of the event management.

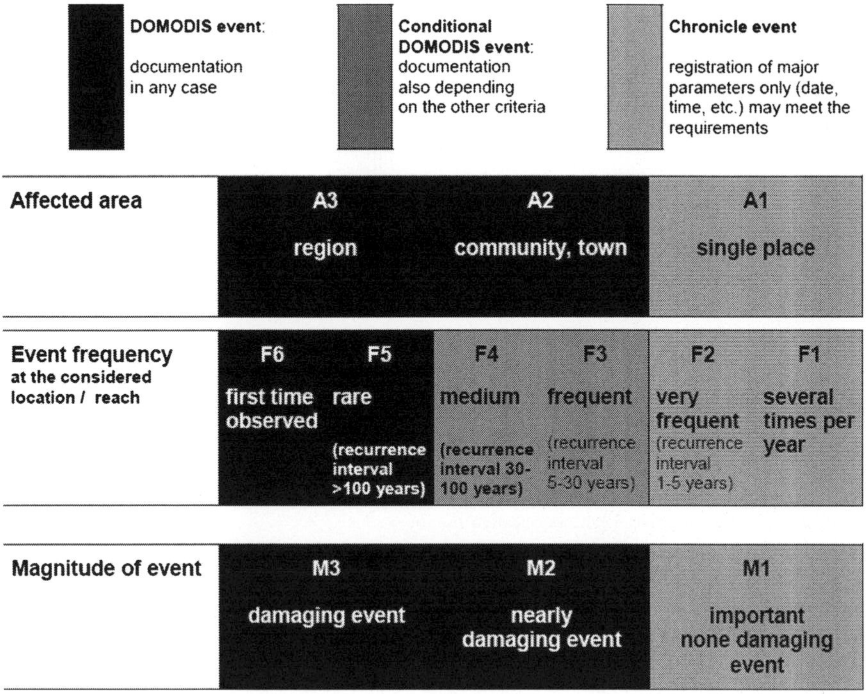

Fig. 4 Proposed classification of events: What are DOMODIS events (Kienholz in Hübl et al. 2002: p. I/7)? Example: A1–F3–M3 = single place event, frequent and damaging

- An organizational structure must be provided,
  - that allows such experts to be called upon and to coordinate their actions;
  - that supports the documentation by other appropriate means as to guarantee free access to the sites (e.g. by an official permit), to offer transportation, and to arrange the taking of air photographs;
  - that guarantees the compilation, archiving of, and free access to the collected data; and
  - that guarantees the basic funding of these actions.

Those events that affect man, goods and infrastructure require an optimized event management. Within sustainable event management it is essential to include all available information of past events, with or without respective damages, as well as that of current processes. How this documentation of the event can be integrated into the event management is outlined in the following.

In this context, the evaluation of historical data in community archives, authorities, monasteries etc. may also be helpful tools for a better assessment of hazards in a given situation.

Since the late 1990s, the event documentation has been implemented as part of the risk management in several European countries. Additionally, after severe and important hazardous events detailed event analyses were organized.

The Interreg-III-b-Project "DIS-ALP – Disaster and Information Systems of Alpine Regions" aimed for a standardization of the documentation procedures, and provided a handbook for instruction and training of the documentation work in the terrain (Kienholz et al. 2006).

4 What Events Should Be Documented?

Geomorphic processes occur everywhere at everytime: Water is flowing, and weathering, erosion at small scales, transportation and deposition of soil materials, etc. continue. However, one issue of documentation is those events that are of an important magnitude that may cause either:

- damage to man and/or valuable goods;
- damage to vegetation and ecology;
- changes to landscape and ecosystems;
- reduction of performance of technical construction works.

Most of such events last only a short time (minutes, hours, few days); while some other processes characterized by large masses but slow velocities (e.g. deep seated landslides or rock creeping) may occur continuously, periodically or episodically (years, decennials, centennials). However, the documentation of the latter is less critical. Thus the documentation should mainly concentrate on the events which are momentary in nature.

Besides the processes mentioned above, the documentation may also include all different event types, including those which are small in extent and are non-damaging. The reason for this is that they may provide information about processes, and about how well protective measures (e.g. defence structures) worked.

The time and cost expenditures which are really necessary and possible may depend on the category of the event, and it is up to the responsible governmental administrations to decide upon this. However, it should be considered that often the costs for good event documentation are less than one percent of the costs for rescue, clearance, restoration and the eventual mitigation measures. Furthermore, in many cases, the expenditures for mitigation measures may be better staked if the events are carefully analysed.

5 Tools for Recording

For accurate and concentrated recording in a disaster area under stressful conditions it is helpful, or even necessary, to rely on accurate tools. Thus, in a long-term preparatory stage it is necessary to provide such tools, to test already existing tools, to adopt them to local/regional circumstances, and to instruct the relevant persons, etc.

What tools are necessary and helpful for the event documentation may depend on the organizational situation. In the field these may include (see Hübl et al. 2002; Kienholz et al. 2006):

- checklists;
- forms for basic information;
- map legends; and
- illustrated examples.

In the field it is sometimes more practical to use simple checklists rather than to apply sophisticated forms. The first goal must be to gather all indispensable information, and only in a second step, the forms are completely filled in.

6 Data Management – Storage, Maintenance and Dissemination

Data collected by documenting and mapping damaging events have to be stored appropriately in order to be able to access them quickly for future planning and work. Therefore, it is very important to decide how the data are to be stored, who is maintaining the data-base, and how the data access can be organized. First of all, independent of the applied technical means, a capable data-base structure must be selected or created. It should be taken into account that the data will be used for several decades, and that their life-span corresponds to several generations of hard- and soft-ware. Hence, an emphasis must be placed on data organization (Fig. 5).

Certain minimum requirements should be determined at a state level. Also, the basic structure of the data-bank should be provided by the state. This structure should allow adaptations and completions at regional or municipal levels.

The goal of the data-base is to provide information on historical, mostly damaging, events. Emphasis must be given to the type and conditions of triggering processes, the controlling factors of the occurring process (e.g. vegetation, geology, meteorology, terrain conditions such as slope angle, aspect, etc.), and on the process itself including all specific characteristics (e.g. velocity of movement, volume, frequency, etc.), the effect (inclusive affected area), as well as possible damage. Based on that data-base, the following minimum requirement may be met.

Data about hazardous events typically refers to defined places or areas. Therefore, the data base has to include some geographical information. This may be done (also in the future) by well established mapping methods (e.g. hand-written numbers on a paper map). It may also be done by applying any Geographical Information System (GIS). Referring to this, the most important criterion to be considered is to provide an open system that can be adapted to future needs and possibilities.

After data collection and storage in a data-base, the information must be legally and technically accessible. Therefore, rules about disposal and use of the data must be defined.

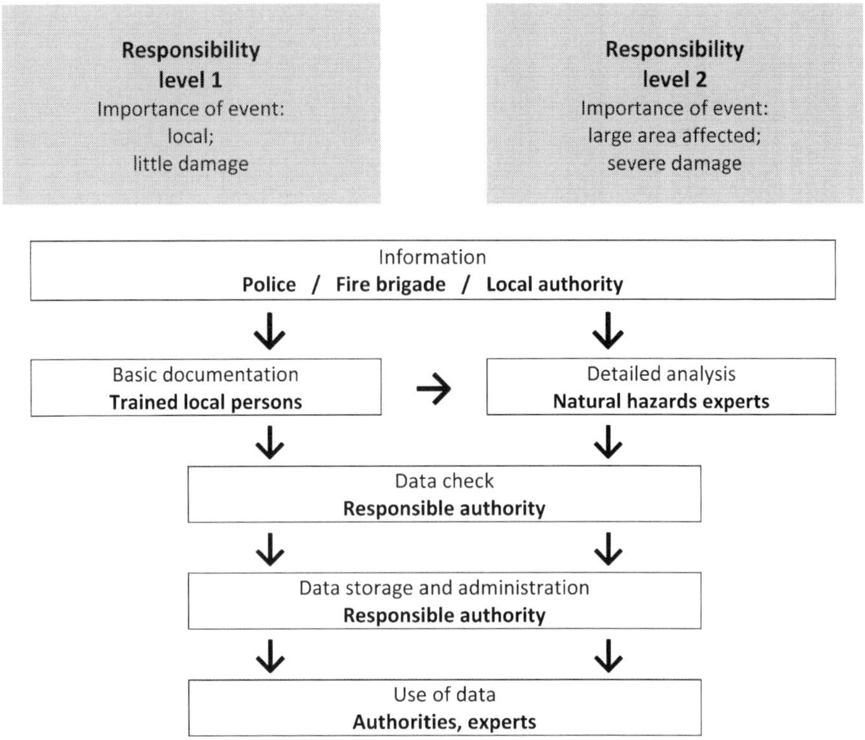

Fig. 5 General organization chart of the event documentation (modified from Kienholz et al. 2006: p. 13)

References

Hübl J, Kienholz H, Loipersberger A (2002) DOMODIS: documentation of mountain disasters – state of discussion in the European mountain areas. International Research Society INTERPRAEVENT. Klagenfurt (Austria). 60 pp. ISBN 3-901164-06-5

Kienholz H, Perret S, Schmid S (2006) Dokumentation von Naturereignissen. Feldanleitung. Alpensignale 4. Plattform Naturgefahren der Alpenkonvention (PLANALP), Innsbruck, Bern

Rainfall Thresholds for Possible Occurrence of Shallow Landslides and Debris Flows in Italy

Maria Teresa Brunetti, Fabio Luino, Carmela Vennari, Silvia Peruccacci, Marcella Biddoccu, Daniela Valigi, Silvia Luciani, Chiara Giorgia Cirio, Mauro Rossi, Guido Nigrelli, Francesca Ardizzone, Mara Di Palma, and Fausto Guzzetti

1 Introduction

In mountain regions worldwide, rainfall-induced landslides and associated debris flows erode slopes, scour channels, and contribute to the formation of alluvial fans that may harm humans and destroy buildings. Rainfall-induced slope failures are frequent and widespread in Italy, where individual rainfall events can result in single or multiple slope failures in small areas or in very large regions. Most of the harmful failures were rainfall-induced, and several were shallow slides or debris flows. In the 60-year period 1950–2009, casualties due to landslides were at least 6,349, an average of 16 harmful events per annum. The large number of harmful events indicates the considerable risk posed by rainfall-induced shallow landslides and debris flows to the population of Italy (Guzzetti et al. 2005a; Salvati et al. 2010).

In this work, we exploit information on rainfall events that have resulted in shallow failures and debris flows to define regional thresholds for the possible occurrence of rainfall-induced shallow landslides and debris flows in three regions in northern, central, and southern Italy. Following a review of methods for establishing the dependence of landslide occurrence on rainfall measurements, we illustrate three catalogues of rainfall events that have resulted in slope failures in the Abruzzo, Calabria and Lombardy regions. Next, we present the method used for the definition

M.T. Brunetti (✉) • S. Peruccacci • S. Luciani • M. Rossi • F. Ardizzone • F. Guzzetti
CNR IRPI, via Madonna Alta 126, 06128 Perugia, Italy
e-mail: MariaTeresa.Brunetti@irpi.cnr.it

F. Luino • M. Biddoccu • C.G. Cirio • G. Nigrelli • M.D. Palma
CNR IRPI, Strada delle Cacce 73, 10135 Torino, Italy
e-mail: fabio.luino@irpi.cnr.it; guido.nigrelli@irpi.cnr.it

C. Vennari • D. Valigi
Dipartimento di Scienze della Terra, Università degli Studi di Perugia, piazza dell'Università 1, 06123 Perugia, Italy

M. Schneuwly-Bollschweiler et al. (eds.), *Dating Torrential Processes on Fans and Cones*, Advances in Global Change Research 47, DOI 10.1007/978-94-007-4336-6_22,

of objective rainfall thresholds, and we apply the method to the available catalogues to determine new intensity-duration (ID) thresholds for possible occurrence of shallow landslides and debris flows in the three regions. We conclude by comparing the new thresholds to similar empirical thresholds proposed for possible landslide initiation in Italy.

2 Modelling Approaches

Two approaches have been proposed to evaluate the dependence of landslide occurrence (or lack of occurrence) on rainfall measurements (Reichenbach et al. 1998; Guzzetti et al. 2007). The first approach relies on process-based models (Montgomery and Dietrich 1994; Wilson and Wieczorek 1995; Wu and Sidle 1995; Iverson 2000; Crosta and Frattini 2003), and the second approach is based on empirical thresholds (Caine 1980; Reichenbach et al. 1998; Corominas 2000; Aleotti 2004; Wieczorek and Glade 2005; Guzzetti et al. 2007).

Process-based models rely upon the understanding – and the schematization – of the physical laws controlling slope instability, and attempt to extend spatially the simplified stability models widely adopted in geotechnical engineering. Stability conditions are evaluated chiefly by means of a static stability model where the local equilibrium of forces along a pre-defined slip surface is considered. Most commonly, the slip surface is assumed planar, at a fixed depth, and parallel to the topographic surface. Values for the pore fluid pressure are assumed, or obtained adopting rainfall infiltration models, coupled to surface and sub-surface runoff models.

The main problem of process-based models consists in the difficulty of obtaining reliable and accurate information necessary to construct and calibrate the distributed models. The problem is particularly severe where the models are applied to large areas. The simplest model requires eight parameters (i.e., terrain gradient, soil density, friction angle, cohesion, hydraulic conductivity, failure depth, water table depth, and rainfall history), of which only terrain gradient is known reasonably accurately over large areas through the use of digital terrain models. Although a stochastic approach may partly solve the problem of insufficient geotechnical and hydrological data, lack of distributed information constrains the applicability of process-based models to areas of limited extent, and mostly to the back analysis of individual rainfall events. This reduces the use of process-based models in operational landslide forecasting systems (Aleotti 2004). A second limitation of process-based models is that, for a given rainfall event, the models predict as unstable an area significantly larger than the area actually affected by the event landslides. The overestimation of the extent of the unstable areas has multiple reasons that must be considered when the model outputs are used in landslide forecasting systems. A different method consists in using process-based models to determine the amount of rainfall (i.e., the critical rainfall) that, when and where it infiltrates in the ground, can result in slope instability, and use this information to forecast possible landslide occurrence during a rainfall event, or using quantitative rainfall forecasts or estimates (Chang et al. 2007; Chiang and Chang 2008).

The second approach to relate landslide occurrence on rainfall measurements consists in the use of empirical thresholds. An empirical threshold defines the rainfall, soil moisture, or hydrological conditions that, when reached or exceeded, are likely to result in landslides (Crozier 1996; Reichenbach et al. 1998). Rainfall thresholds are the most common type of empirical thresholds for possible landslide occurrence, and are defined through the analysis of past rainfall events that have (or have not) resulted in slope failures. Rainfall thresholds are commonly classified on the geographical extent for which they are determined, and the type of rainfall information used to establish the threshold (Guzzetti et al. 2007).

Based on the geographical extent, rainfall thresholds can be loosely subdivided as global, national, regional, or local thresholds. A global threshold attempts to establish a general ("worldwide") minimum level below which landslides do not occur, independently of local morphological, lithological and land-use conditions and of local or regional rainfall pattern and history. Global thresholds were proposed by Caine (1980), Innes (1983), Clarizia et al. (1998), Crosta and Frattini (2001), Cannon and Gartner (2005), Hong et al. (2006) and Guzzetti et al. (2008). National thresholds are similar to global thresholds, and are established using information for a single nation (e.g., Brunetti et al. 2010). Regional thresholds are defined for areas extending from a few to several thousand square kilometres of similar meteorological, climatic, physiographic, and soil characteristics. Local thresholds consider the local climatic regime and geomorphic setting, and are applicable to single landslides or to groups of landslides in areas extending from a few to some hundreds of square kilometres.

Based on the rainfall measurements, empirical rainfall thresholds can be grouped in three broad categories: (i) thresholds that combine precipitation measurements obtained for specific rainfall events, (ii) thresholds that include the antecedent rainfall and soil moisture conditions (e.g., Terlien 1998; Crozier 1999; Glade 2000; Chleborad 2003; Aleotti 2004), and (iii) other thresholds (e.g., Jakob and Weatherly 2003). Thresholds established using precipitation measurements obtained from individual or multiple rainfall events can be further subdivided into: intensity–duration thresholds, thresholds based on the total event rainfall, rainfall event–duration thresholds, and rainfall event–intensity thresholds (for a review see Guzzetti et al. 2007, 2008). To account for meteorological and climate influence on slope stability, rainfall measurements (chiefly rainfall intensity) can be normalized to measures of climate, including the mean annual precipitation, the average number of rainy-days, and the rainy-days normal (Wilson and Jayko 1997).

3 Landslide and Rainfall Information

Information on rainfall events that have resulted in landslides exists in Italy (see e.g. Crosta 1998; Crosta and Frattini 2001; Aleotti 2004; Guzzetti et al. 2007, 2008; Brunetti et al. 2010, and references therein). For this work, we considered the existing information and added new information to obtain three regional catalogues

Table 1 Characteristics of the regional catalogues used to establish empirical rainfall thresholds for possible occurrence of shallow landslides and debris flows

	Lombardy	Abruzzo	Calabria
Number of rainfall events	173	85	113
Period (year)	9/1929–7/2008	8/2002–10/2009	1/1996–2/2010
Rainfall duration (h)	$1 < D < 91$	$1 < D < 600$	$2 < D < 1080$
Rainfall mean intensity (mm h^{-1})	$0.6 < I < 32$	$0.3 < I < 50$	$0.3 < I < 29$
Number of rain gauges	56	25	65
Area extent (km^2)	~8,000	~11,000	~15,000
Mean area per rain gauge (km^2)	~140	~430	~230
Elevation range (m)	150–3,769	0–2,914	0–2,267
Geographical location	Northern Italy	Central Italy	Southern Italy
Topography	Mountains	Hills, mountains	Hills, mountains
Climate	Mountain, cold	Mediterranean, mild	Mediterranean, warm

of rainfall events that have resulted in shallow landslides and debris flows, for the Lombardy region, northern Italy, for the Abruzzo region, central Italy, and for the Calabria region, southern Italy (Table 1).

Luino et al. (2008) searched historical archives, old chronicles, journals, and regional and local newspapers to obtain a comprehensive catalogue of rainfall induced soil slips and debris flows for the Sondrio and Brescia Provinces, Lombardy region. Systematic analysis of the historical information resulted in a catalogue of 661 individual landslides, in the 80-year period 1929–2008. The location and time of occurrence of the failures were obtained from the historical sources and the press. Newspapers were most useful to decide the timing of the failures, whereas the exact or approximate geographical coordinates of the failures were obtained in a GIS using large-scale digital base maps. Once the failures were located geographically, rainfall information for the individual landslides was obtained from a database of rainfall measurements made available by the Lombardy region. The database contained rainfall data for 56 rain gauges in the Brescia and Sondrio provinces, corresponding to an average density of one rain gauge every ~140 km^2. For each single landslide, selection of the most appropriate rain gauge was based on: (i) the location of the rain gauge, preferably placed in the same catchment as the landslide, (ii) the distance of the rain gauge to the landslide, not to exceed 5 km, and (iii) the availability of sufficient rainfall data for the considered landslide event.

The catalogue obtained by Luino et al. (2008) listed individual soil slips and debris flows caused by rainfall. For consistency with the information available for the Abruzzo (Brunetti et al. 2010) and the Calabria regions (see below), we transformed the list of 611 rainfall-induced landslides in a new catalogue of 160 rainfall events that have resulted in individual or multiple soil slips and debris flows in the Brescia and Sondrio provinces. During the process, landslides listed in separate records in the original catalogue that were triggered by an individual rainfall event, and for which the rainfall information was obtained from the same rain gauge, were merged in a single entry in the new catalogue of rainfall events with landslides. We completed the catalogue for the Lombardy region by adding 13 events, including 4 events in the Bergamo province and 9 events in the Sondrio province, obtained from a global catalogue of rainfall events with shallow landslides and debris flows prepared by Guzzetti et al. (2008). Thus, the final catalogue for the Lombardy region listed 173 rainfall events with shallow slope instabilities in an Alpine area of about 8,000 km^2.

For the Abruzzo region, central Italy, Brunetti et al. (2010) obtained a catalogue of 77 rainfall events with landslides in the 8-year period 2002–2009. The landslide information was obtained through the systematic analysis of regional and local newspapers. The exact or approximate location of the single landslides (or of groups of landslides triggered by the same rainfall event) was decided based on the historical information. The corresponding geographical coordinates for the landslides were obtained using Google Earth. Analysis of the historical sources revealed that most of the landslides in the catalogue were shallow, and small in size. For this work, we updated this regional catalogue searching new landslide information, and excluding landslides that could not be classified reasonably as shallow failures or debris flows. As a result, we obtained a new catalogue listing 85 rainfall events that have resulted in shallow landslides and debris flows in the Abruzzo region, in the period from August 2002 to October 2009.

Adopting a similar procedure, we searched more than 25 national, regional and local newspapers and internet resources for the Calabria region, southern Italy, and we obtained information on 191 individual rainfall events that have caused landslides, including 113 events that have resulted in shallow failures and debris flows, in the period from January 1996 to February 2010. Again, the timing and the location of the individual failures (or of groups of failures triggered by the same rainfall event) were decided through the analysis of the historical information, and the geographical coordinates for the landslides were obtained using Google Earth.

For the Abruzzo and the Calabria regions, rainfall information was obtained from a national database of rainfall measurements managed by the national Department for Civil Protection. The database contains rainfall measurements for 1950 rain gauges in Italy, including 25 rain gauges in the Abruzzo region and 65 rain gauges in the Calabria region. This corresponds to an average density of one rain gauge every $\sim$430 km^2 for Abruzzo region, and one rain gauge every $\sim$230 km^2 for Calabria. In the database, rainfall measurements are cumulated every 10, 15, 30 or 60 min, allowing for a detailed reconstruction of the individual rainfall events that have resulted in slope instabilities, in terms of rainfall duration D and rainfall mean intensity I. For Calabria, additional rainfall information was obtained from an on-line catalogue provided by the Regional Environmental Agency (http://www.arpacalabria.it/). For each individual landslide, up to four rain gauges were considered, and the most representative station was selected. Representativeness of a rain gauge was decided based on the geographical distance to the failure, the elevation of the rain gauge compared to the elevation of the landslide, and the location of a rain gauge with respect to the local topographical and morphological setting. In general, the selected rain gauge was the closest to the landslide, in the set of rain gauges located in the same valley. In other cases, for the selection of the representative rain gauge, elevation was considered more important than geographical distance to the landslide.

Table 1 summarizes the main characteristics of the three regional catalogues of rainfall events with shallow landslides and debris flows.

4 Definition of Rainfall Thresholds

Guzzetti et al. (2007) identified the need for objective methods to define reproducible thresholds of possible landslide occurrence from empirical rainfall data, and proposed a method based on Bayesian inference. Brunetti et al. (2010) adopted a Frequentist statistical approach, and proposed a different method for the definition of objective rainfall thresholds. In this work, we adopt the Frequentist approach to determine new regional rainfall thresholds for possible landslide occurrence in the Lombardy, Abruzzo, and Calabria regions.

To define a threshold, we start by plotting the empirical rainfall duration D and mean rainfall intensity I data in a single graph, and we find the best-fit line to the cloud of empirical data points (D, I). As an example, Fig. 1a shows the 173 (D, I) data points for the Lombardy region. To avoid problems associated with the fitting of data spanning multiple orders of magnitude (e.g., the least square minimization criteria may not work), we log-transform the empirical data, and we fit the distribution of the rainfall conditions log(I) vs. log (D) that have resulted in slope failures with a linear equation of the type $\log(I) = \log(\kappa) - \beta \cdot \log(D)$, which is entirely equivalent to the power law $I = \kappa \cdot D^{-\beta}$ in linear coordinates, where D is the duration of the rainfall event (in h), I is the rainfall mean intensity (in mm·h^{-1}), κ is the intercept, and β is the gradient of the power law curve. In Fig. 1a, the solid line is the best-fit line $\mathrm{T}_{\mathrm{L},50}$: $I = 10.93 \cdot D^{-0.42}$.

Next, for each rainfall event (D, I), we calculate the difference $\delta(D)$ between the logarithm of the event intensity log[$I(D)$] and the corresponding intensity value of the fit log[$I_f(D)$], $\delta(D) = \log[I(D)] - \log[I_f(D)]$. Then, we estimate the probability density of the distribution of δ, $p(\delta)$, using a kernel density estimation (KDE) technique (Venables and Ripley 2002), and we model the distribution through least square fitting using the Gaussian function

$$G(\delta) = \frac{1}{\sqrt{2\pi\sigma^2}} \cdot \exp\left[-\frac{(\delta - \mu)^2}{2\sigma^2}\right]$$

where μ is the mean and σ is the standard deviation. In Fig. 1d, the dashed line shows the empirical distribution (KDE) of the δ values for the Lombardy rainfall data, and the continuous line is the equivalent Gaussian model.

Based on the modelled distribution of δ, multiple thresholds can be defined, corresponding to different probabilities of exceeding thresholds. As an example, Fig. 1g portrays the 5% rainfall thresholds $\mathrm{T}_{\mathrm{L},5}$ for the Lombardy region. The distance

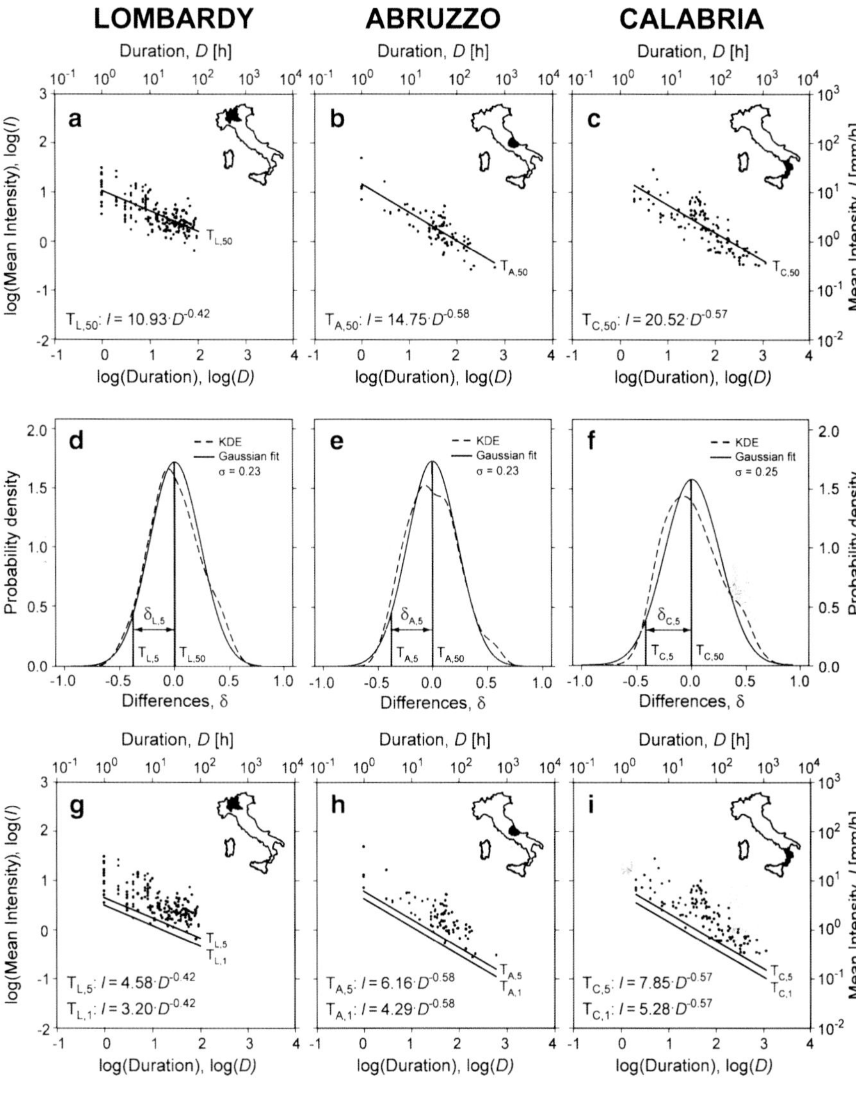

Fig. 1 Empirical rainfall (*D*, *I*) data, analysis, and new ID rainfall thresholds for the Lombardy (*left column*), Abruzzo (*central column*), and Calabria (*right column*) regions, Italy. Plots in the *upper row* show rainfall conditions that have resulted in shallow landslide and debris flows (*dots*), and the associated power law best-fit lines, $T_{L,50}$ for Lombardy, $T_{A,50}$ for Abruzzo, and $T_{C,50}$ for Calabria. *Inset* maps show geographical location of each region. *Graphs* in the central row show the kernel density estimation of the differences, δ, for 173 rainfall events in Lombardy, for 85 events in Abruzzo, and for 113 events in Calabria. Solid lines are the corresponding Gaussian model fits. *Vertical lines* for $T_{L,5}$, $T_{A,5}$, and $T_{C,5}$ are the thresholds that correspond to 5% probability of having landslides, for Lombardy, Abruzzo, and Calabria, respectively. Plots in lower row portray rainfall (*D*, *I*) conditions that have resulted in shallow landslide and debris flows (*dots*) in the three regions. *Solid lines* are the 1% ($T_{L,1}$, $T_{A,1}$, $T_{C,1}$) and the 5% ($T_{L,5}$, $T_{A,5}$, $T_{C,5}$) power law thresholds

$\delta_{L,5}$ between $T_{L,5}$ and $T_{L,50}$ is used to calculate the intercept of the 5% threshold curve. The 5% rainfall threshold $T_{L,5}$ is the curve parallel to the best-fit line $T_{L,50}$ (slope $= -\beta$), with intercept $\kappa_{L,5} = \kappa_{L,50} - \delta_{L,5}$. For the $T_{L,5}$ threshold, assuming the catalogue of rainfall events is sufficiently complete and representative for the Alpine portion of the Lombardy region, we can state that the probability of experiencing landslides triggered by rainfall below this threshold is less than 5%. The method can be used to determine thresholds for any probability of exceeding a threshold level (Brunetti et al. 2010).

5 New Regional Thresholds for Shallow Landslide Occurrence

We have exploited the catalogues of rainfall events that have resulted in shallow landslides and debris flows in the Lombardy, Abruzzo, and Calabria regions presented in Sect. 3, and the statistical method discussed in Sect. 4, to determine new rainfall thresholds for possible occurrence of shallow failures in the three regions. Results are summarized in Fig. 1.

In this figure, the three upper plots show the distributions of the rainfall conditions (D, I) that have resulted in shallow landslides and debris flows in the Lombardy region (Fig. 1a), in the Abruzzo region (Fig. 1b), and in the Calabria region (Fig. 1c). In the plots, the black lines are the best-fit lines to the empirical rainfall data points, obtained through linear fitting (least square method) of the log-transformed rainfall data [log(D), log(I)].

The three central plots in Fig. 1 show, for each region, the empirical distribution of δ (dashed lines) and the corresponding modelled Gaussian fit (solid lines). The spread of the modelled Gaussian distribution around the central value, $\mu = 0$, is measured by the standard deviation σ, which ranges between 0.23 (for Lombardy and Abruzzo) and 0.25 (for Calabria). The empirical distributions of the differences δ for the three regional data sets and the corresponding Gaussian models exhibit interesting similarities. First, the peaks of the empirical distributions are shifted to the left of the corresponding modelled distributions. We attribute the shift, which is largest for Calabria and reduced for Lombardy, to heterogeneities in the data (e.g., different lithological, morphological, and climatic conditions in each region). Second, the modelled distributions overestimate slightly the distribution of the empirical data, for $\delta \approx 0$. We attribute this chiefly to the reduced number of data points in the individual catalogues. Third, the empirical distributions exhibit a visible deviation from the Gaussian models for $\delta \approx 0.5$. Inspection of the catalogues revealed that this corresponds primarily to short duration – high intensity rainfall conditions, which are somewhat different from the majority of the other (D, I) conditions that have resulted in slope failures. Inspection of the catalogue revealed that the short duration – high intensity rainfall data pertain to a single regional meteorological event that has resulted in landslides in multiple sites in a large region, for which several rain gauges were available.

In Fig. 1, the three lower plots portray the empirical rainfall thresholds, for the three regions. The slopes of the power law curves representing the thresholds are similar, but not identical. The slope of the threshold defined for the Lombardy region (Fig. 1g), an area characterized by a mountain (cold) climate, is significantly lower ($\beta = 0.42$) than the thresholds defined for the Abruzzo ($\beta = 0.58$, Fig. 1h) and the Calabria ($\beta = 0.57$, Fig. 1i) regions, characterized by a Mediterranean climate. This is in agreement with previous findings by Guzzetti et al. (2007), and it indicates that in mountain areas mean rainfall intensity is more important than rainfall duration to decide when landslides occur, an important information to construct an operational landslide forecasting system (Brunetti et al. 2009). Further inspection of the threshold curves for the three regions reveals that, for the same short rainfall duration ($D < 7$ h), the mean rainfall intensity I required to initiate shallow landslides and debris flows, corresponding to 1% probability of having landslides, is lowest in the Lombardy region (Fig. 1g, $D = 2$ h, $I = 2.39$ mm·h^{-1}), which is characterized by a mountain climate, intermediate in the Abruzzo region (Fig. 1h, $D = 2$ h, $I = 2.87$ mm h^{-1}) that exhibits a mild Mediterranean climate, and largest in Calabria (Fig. 1i, $D = 2$ h, $I = 3.56$ mm·h^{-1}), where the climate is warm Mediterranean. Conversely, for long rainfall duration ($D > 30$ h), the mean rainfall intensity required to initiate shallow slope failures (at the 1% probability level) is lowest in Abruzzo, intermediate in Calabria, and largest in Lombardy. We conclude that in Italy, the rainfall conditions (D, I) that can result in shallow landslides and debris flows depend on the prevailing meteorological conditions and climate. The two are related, and conditioned on latitude and the physiographical setting.

6 Discussion and Conclusions

The new regional rainfall ID thresholds for possible occurrence of shallow landslides and debris flows established for the Lombardy, Abruzzo, and Calabria regions can be compared to similar empirical thresholds proposed in Italy for individual regions, single catchments, or local areas. Figure 2 portrays six regional (thick lines) and 11 local (thin lines) ID rainfall thresholds proposed in Italy, including our three new thresholds $T_{L,1}$ (for Lombardy), $T_{A,1}$ (for Abruzzo), and $T_{C,1}$ (for Calabria). Inspection of Fig. 2 indicates that the new regional thresholds are significantly lower than the regional thresholds previously proposed for Piedmont (Aleotti 2004), for Lombardy (Ceriani et al. 1994), and for the Campania region (Calcaterra et al. 2000). Our new Lombardy threshold, defined using rainfall data primarily for the Brescia and Sondrio provinces, is also significantly lower than the threshold proposed by Cancelli and Nova (1985) for Valtellina, a portion of the Sondrio province. Further, our new regional thresholds are significantly lower than the thresholds proposed for individual catchments in the Susa Valley, Piedmont (Bolley and Oliaro 1999), for the Moscardo torrent, Friuli (Marchi et al. 2002), for the Apuane Alps, Tuscany (Giannecchini 2005), and for the Valzangona area, Marche (Floris et al. 2004).

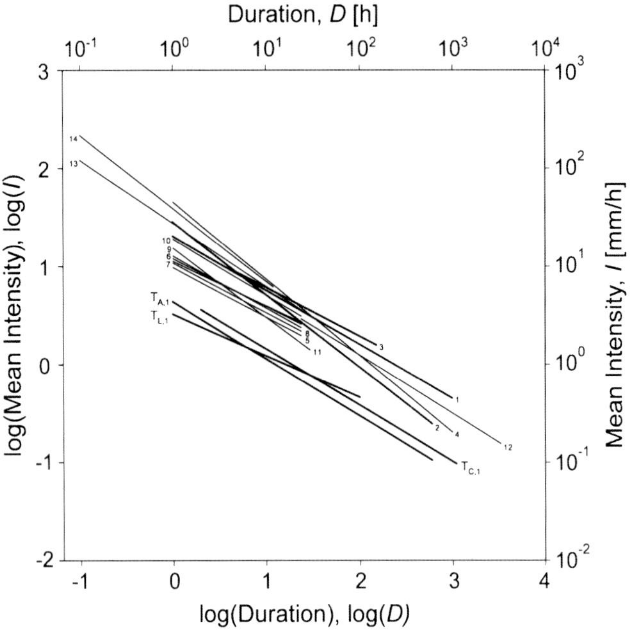

Fig. 2 Comparison of regional and local rainfall thresholds for possible landslide occurrence in Italy. Regional (*thick line*) and local (*thin line*) thresholds, and new 1% probability of exceeding the thresholds for the Abruzzo ($T_{A,1}$), Calabria ($T_{C,1}$), and Lombardy ($T_{L,1}$) regions (Source: 1, Ceriani et al. 1994; 2, Calcaterra et al. 2000; 3, Aleotti 2004; 4, Cancelli and Nova 1985; 5–10, Bolley and Olliaro 1999; 11, Marchi et al. 2002; 12, Floris et al. 2004; 13–14, Giannecchini 2005)

The fact that the new thresholds are systematically lower than previously established thresholds, even for the same general area (e.g., for Lombardy (Ceriani et al. 1994), or Valtellina (Cancelli and Nova 1985)) deserves an explanation. Even if we cannot exclude that the result is – at least partially – attributable to local climatic, meteorological, morphological, lithological, and land cover settings, we maintain that it depends primarily on the type and quality of the information used to define the thresholds. Most often, thresholds published in the literature are defined following major (i.e., extreme, catastrophic, rare) rainfall events that have resulted in abundant and widespread landsliding. With this respect, many of the published thresholds are – or are close to – "maximum" thresholds, i.e. thresholds that represent rainfall conditions above which there is a 100% chance of landslide occurrence (Crozier 1996; Reichenbach et al. 1998). Conversely, our thresholds are defined from catalogues that include low intensity (i.e., non catastrophic, more frequent) rainfall events that have resulted in single, or sparse landslides. Further, our catalogues are the result of a systematic and thorough analysis of the landslide and rainfall information available for the study areas. This has resulted in more complete lists of rainfall events with associated slope failures, which have provided better data sets for the definition of empirical thresholds based on statistical methods. We take this as an indication of the need for the systematic collection of reliable, accurate, and as complete as reasonably possible information on rainfall events that have resulted in individual or multiple slope failures, to be able to determine dependable rainfall thresholds for possible landslide initiation to be used in landslide forecasting systems (Brunetti et al. 2009).

The finding that in the areas investigated in this study, and possibly in other areas in Italy, shallow slope failures and associated debris flows can be triggered by less severe rainfall conditions than previously recognized has multiple implications. First, it is important for the design and the implementation of landslide forecasting systems (Brunetti et al. 2009, 2010). Second, it indicates that the temporal frequency of shallow failures is higher than previously considered, with consequences for hazard assessment (Guzzetti et al. 2005b, 2006) and risk evaluation. Third, it suggests that ongoing and foreseen climate and environmental changes, including human induced changes, can have a higher impact on landslide activity than expected.

Acknowledgements Part of the paper is reprinted from the work of Brunetti et al. (2010), and it is used with permission. Work supported by the Italian National Department for Civil Protection (DPC), which provided the national rainfall database for the period 2002–2009. MTB, SP and MR were supported by DPC grants. We thank G. Tonelli for the work on the rainfall database, and E. E. Brabb for reviewing the paper.

References

Aleotti P (2004) A warning system for rainfall-induced shallow failures. Eng Geol 73:247–265

Bolley S, Oliaro P (1999) Analisi dei debris flows in alcuni bacini campione dell'Alta Val Susa. Geoingegneria Ambientale e Mineraria, Marzo, 69–74

Brunetti MT, Peruccacci S, Rossi M, Guzzetti F, Reichenbach P, Ardizzone F, Cardinali M, Mondini A, Salvati P, Tonelli G, Valigi D, Luciani S (2009) A prototype system to forecast rainfall induced landslides in Italy. In: Picarelli L, Tommasi P, Urciuoli G, Versace P (eds) Rainfall-induced landslides: mechanisms, monitoring techniques and nowcasting models for early warning systems, vol 1. Naples, pp 157–161

Brunetti MT, Peruccacci S, Rossi M, Luciani S, Valigi D, Guzzetti F (2010) Rainfall thresholds for the possible occurrence of landslides in Italy. Nat Hazard Earth Syst Sci 10:447–458

Caine N (1980) The rainfall intensity-duration control of shallow landslides and debris flow. Geogr Ann Ser A 62:23–27

Calcaterra D, Parise M, Palma B, Pelella L (2000) The influence of meteoric events in triggering shallow landslides in pyroclastic deposits of campania, Italy. In: Bromhead E, Dixon N, Ibsen ML (eds) Proceedings of the 8th international symposium on landslides, vol 1. A.A. Balkema Publishers, Cardiff, pp 209–214

Cancelli A, Nova R (1985) Landslides in soil debris cover triggered by rainstorms in Valtellina (central Alps – Italy). In: Proceedings of the 4th international conference and field workshop on landslides. The Japan Geological Society, Tokyo, pp 267–272

Cannon SH, Gartner JE (2005) Wildfire-related debris flow from a hazards perspective. In: Jakob M, Hungr O (eds) Debris flow hazards and related phenomena. Springer, Berlin, pp 363–385

Ceriani M, Lauzi S, Padovan N (1994) Rainfall thresholds triggering debris-flows in the alpine area of Lombardia Region, central Alps – Italy. Man and Mountain '94 Proceedings, Ponte di Legno, pp 123–139

Chang K, Chiang SH, Lei F (2007) Analysing the relationship between typhoon-triggered landslides and critical rainfall conditions. Earth Surf Process Landforms 33(8):1261–1271

Chiang SH, Chang K (2008) Application of radar data to modelling rainfall-induced landslides. Geomorphology 103(3):299–309

Chleborad AF (2003) Preliminary evaluation of a precipitation threshold for anticipating the occurrence of landslides in the Seattle, Washington, DC. US geological survey open-file report: 03–463 F

Clarizia M, Gullà G, Sorbino G (1998) Sui meccanismi di innesco dei soil slip. In: Proceedings of International conference on prevention of hydro-geological hazards: The role of scientific research. Alba, Italy, 5–7 Nov 1996, vol 1, pp 585–597

Corominas J (2000) Landslides and climate. Keynote lecture. In: Bromhead E, Dixon N, Ibsen ML (eds) Proceedings of the 8th international symposium on landslides, vol 4. A.A. Balkema Publishers, Cardiff, pp 1–33

Crosta GB (1998) Regionalization of rainfall thresholds: an aid to landslide hazard evaluation. Environ Geol 35(2–3):131–145

Crosta GB, Frattini P (2001) Rainfall thresholds for triggering soil slips and debris flow. In: Mugnai A, Guzzetti F, Roth G (eds) Proceedings of the 2nd EGS Plinius conference on Mediterranean storms. Siena, Italy, 16–18 Oct 2001, pp 463–487

Crosta GB, Frattini P (2003) Distributed modelling of shallow landslides triggered by intense rainfall. Nat Hazard Earth Syst Sci 3(1–2):81–93

Crozier MJ (1996) The climate-landslide couple: a southern hemisphere perspective. Paleoclim Res 19 ESF(Special Issue 12):329–350

Crozier MJ (1999) Prediction of rainfall-triggered landslides: a test of the antecedent water status model. Earth Surf Process Landforms 24:825–833

Floris M, Mari M, Romeo RW, Gori U (2004) Modelling of landslide-triggering factors – a case study in the Northern Apennines, Italy. Lect Note Earth Sci 104:745–753

Giannecchini R (2005) Rainfall triggering soil slips in the southern Apuan Alps (Tuscany, Italy). Adv Geosci 2:21–24

Glade T (2000) Modelling landslide triggering rainfall thresholds at a range of complexities. In: Bromhead E, Dixon N, Ibsen ML (eds) Landslides in research, theory and practice. Thomas Telford, Cardiff, pp 633–640

Guzzetti F, Stark CP, Salvati P (2005a) Evaluation of flood and landslide risk to the population of Italy. Environ Manage 36(1):15–36

Guzzetti F, Reichenbach P, Cardinali M, Galli M, Ardizzone F (2005b) Probabilistic landslide hazard assessment at the basin scale. Geomorphology 72:272–299

Guzzetti F, Galli M, Reichenbach P, Ardizzone F, Cardinali M (2006) Landslide hazard assessment in the Collazzone area, Umbria, central Italy. Nat Hazard Earth Syst Sci 6:115–131

Guzzetti F, Peruccacci S, Rossi M, Stark CP (2007) Rainfall thresholds for the initiation of landslides in central and southern Europe. Meteorol Atmos Phys 98:239–267

Guzzetti F, Peruccacci S, Rossi M, Stark CP (2008) The rainfall intensity-duration control of shallow landslides and debris flows: an update. Landslides 5(1):3–17

Hong Y, Adler R, Huffman G (2006) Evaluation of the potential of NASA multi-satellite precipitation analysis in global landslide hazard assessment. Geophys Res Lett 33:L22402

Innes JL (1983) Debris flows. Prog Phys Geogr 7:469–501

Iverson RM (2000) Landslide triggering by rain infiltration. Water Resour Res 36(7):1897–1910

Jakob M, Weatherly H (2003) A hydroclimatic threshold for landslide initiation on the North Shore Mountains of Vancouver, British Columbia. Geomorphology 54:137–156

Luino F, Biddoccu M, Cirio CG, Nigrelli G, Di Palma M (2008) Definizione delle soglie pluviometriche d'innesco di frane superficiali e colate torrentizie: accorpamento per aree omogenee. Quaderni IReR, monografia, rapporto finale della convenzione fra CNR IRPI e Protezione Civile Regione Lombardia, 125 pp

Marchi L, Arattano M, Deganutti AM (2002) Ten years of debris-flow monitoring in the Moscardo Torrent (Italian Alps). Geomorphology 46:1–17

Montgomery DR, Dietrich WE (1994) A physically based model for the topographic control of shallow landsliding. Water Resour Res 30(4):1153–1171

Reichenbach P, Cardinali M, De Vita P, Guzzetti F (1998) Regional hydrological thresholds for landslides and floods in the Tiber River Basin (Central Italy). Environ Geol 35(2–3):146–159

Salvati P, Bianchi C, Rossi M, Guzzetti F (2010) Societal landslide and flood risk in Italy. Nat Hazard Earth Syst Sci 10:465–483

Terlien MTJ (1998) The determination of statistical and deterministic hydrological landslide-triggering thresholds. Environ Geol 35(2–3):124–130

Venables WN, Ripley BD (2002) Modern applied statistics with S. Springer, New York

Wieczorek GF, Glade T (2005) Climatic factors influencing occurrence of debris flows. In: Jakob M, Hungr O (eds) Debris flow hazards and related phenomena. Springer, Berlin/Heidelberg, pp 325–362

Wilson RC, Jayko AS (1997) Preliminary maps showing rainfall thresholds for debris-flow activity, San Francisco Bay Region, California. US Geological survey open-file report 97–745 F

Wilson RC, Wieczorek GF (1995) Rainfall thresholds for the initiation of debris flow at La Honda, California. Environ Eng Geosci 1(1):11–27

Wu W, Sidle RC (1995) A distributed slope stability model for steep forested basins. Water Resour Res 31:2097–2110

Part IV
Application of Event Dating in Practice

Hazard Assessment

Markus Zimmermann

1 Hazard Assessment: An Integral Part of Risk Management

The integrated management of risks from natural hazards (Bründl 2012, *this volume*) and the prevention or mitigation of disasters demand a thorough assessment of the prevailing hazards at the very start, as stipulated, e.g. in the Hyogo Framework for Action (HFA 2005). The reduction of risks on alluvial fans and debris cones is of particular interest as these land features serve as prime estate for villages and towns in the mountain environment, be it in industrialised or developing countries.

The demand for risk reduction measures is steadily growing with the fast development of fan and cone areas; in parallel, there is growing need for thorough assessments as well. In the past two to three decades much progress has been made to develop methods, approaches and tools for the documentation and analysis of torrential processes (Jakob 2005; Rickenmann 1999). The range of available instruments is rather wide and extends from empiric relations for the estimation of one individual parameter (e.g. Mizuyama et al. 1992) to the hydro-dynamic modelling of flow characteristics (e.g. Rickenmann and Scheidl 2012, *this volume*).

Although various methods and tools have been developed, uncertainties remain in the forecasting of individual debris flows (Zimmermann 2006), in particular when changing environmental factors (e.g. climate change) are considered (Wieczorek and Glade 2005). Nevertheless, practitioners require robust approaches and models for real-world solutions like hazard and risk maps which serve local planning processes or the development of risk reduction measures. Particularly the less developed world needs easy-to-apply methods which provide good results at an affordable price (ISDR 2007).

M. Zimmermann (✉)
NDR Consulting GmbH, Riedstrasse 5, 3600 Thun, Switzerland
e-mail: markus.zimmermann@ndr.ch

M. Schneuwly-Bollschweiler et al. (eds.), *Dating Torrential Processes on Fans and Cones*, Advances in Global Change Research 47, DOI 10.1007/978-94-007-4336-6_23,

This chapter provides an overview of some practical issues related to the assessment of torrent hazards in an Alpine environment. The focus is put on practical applications rather than on science.

2 Debris-Flow Parameters

Hazard assessments serve a wide range of purposes: Hazard maps and related instruments are used in local planning processes; the design of protective structures requires mainly quantitative information about the magnitude of possible events; and early warning, alert and evacuation systems depend heavily on the understanding of the hazardous process. The necessity for thorough assessments in highly hazardous environments is underlined in a number of publications, (e.g., UNDHA 1996; Petrascheck and Kienholz 2003; Zimmermann and Issa 2009). The basic questions to be addressed and answered remain the same:

- **What type of process**? What are the prevailing processes occurring in a particular torrential channel?
- **How big**? What is the total volume of the process? What is the peak discharge?
- **How far**? What is the travel distance? What areas are affected (lateral spreading)? How fast is the process (flow velocity) in a particular location along the channel?
- **How often**? What is the recurrence interval (magnitude-frequency relation)?

The clients for hazard and risk assessments demand accurate and transparent information about the prevailing conditions in a channel and on the fan/cone; i.e. they want clear and easy-to-understand answers to the questions listed above. On the other hand, the specialists are challenged to overcome the many uncertainties which are behind each of the relevant parameters. In general, ready made models are hardly working in a particular case. Therefore, it is of high importance to apply several approaches to estimate individual parameters. There are basically three independent approaches available to address hazards in a particular torrential system (see e.g., Eisbacher and Clague 1984). These are:

- **Historic approach**: this approach is based on written or orally communicated data and information about past activity of a torrent. In addition, data from various methods as outlined in part II of this book (Schneuwly-Bollschweiler et al. 2012) provide independent information. Generally, people's memory is limited to few years back for this type of problem. However, paramount situations may remain for decades in people's memory and they are able to clearly report conditions even 50 or more years back. In few places written documents enable the reconstruction of the event history over many centuries (e.g. Kienholz 1988).
- **Geomorphic approach** (outlined in Habersack and Kreisler 2012, *this volume*): this method is based on the interpretation of landforms and relicts which refer to past activity of a torrent (silent witnesses as introduced by Aulitzky 1972).

Of particular importance are features on the fan/cone, like levees, pronounced channels, or debris-flow lobes (cf. Costa 1988 or Pierson 2005). Dating methods support the chronological classification of the landforms. The approach also calls for a thorough assessment of the watershed to identify sediment sources and interpret the triggering factors.

- **Modelling approach**: In the past few decades a number of empiric statistical approaches (e.g. Rickenmann and Zimmermann 1993), empiric analytical models (e.g. Horton et al. 2008) as well as hydraulic simulation models (e.g. FLO2D or RAMMS; Cesca and D'Agostino 2008; Scheuner et al. 2009) for the practical application have been developed.

3 Relevant Parameters

Risk management, in particular the design of structural as well as the outline of non-structural measures, requires in-depth data and information about various parameters of debris or mud flows. These are (1) process occurrence; (2) magnitude; and (3) affected zones. The frequency of particular events and magnitude-frequency relations need to be considered as well.

3.1 What Type of Process? (Debris-Flow Occurrence)

Many channels in steep mountainous areas have the potential to produce large debris flows. A minimum of sediments must be destabilised in order to trigger a debris flow; the bed shear stress plays an important role. The general conditions of sediment transport, gradient and slope failures for the initiation of debris flows are summarised in Rickenmann and Scheidl (2012, *this volume*).

Debris flows occur in channels having a gradient of more than 25%, according to Takahashi (1991). However, debris flows can also be triggered on more gentle slopes if the watershed is larger than a few square kilometres. In general, a trend between the size of the watershed (availability of water) and the gradient of the starting zone can be observed (Fig. 1).

In most cases the field evidence (geomorphic approach) clearly indicates past activity: levees along the channels, the overall shape of the fan/cone (steepness, rouged surface) and superficial landforms on the fan/cone (levees and channels, lobes of unsorted debris etc.). These features are prominent in humid (Fig. 2a) as well as in the arid environment (Fig. 2b).

Features are less prominent on the fan/cone in places where mud flows are dominant or a mix of processes occurs. In those cases the geomorphic approach may concentrate on the conditions in the watershed and along the channel. The historic approach may provide additional relevant information.

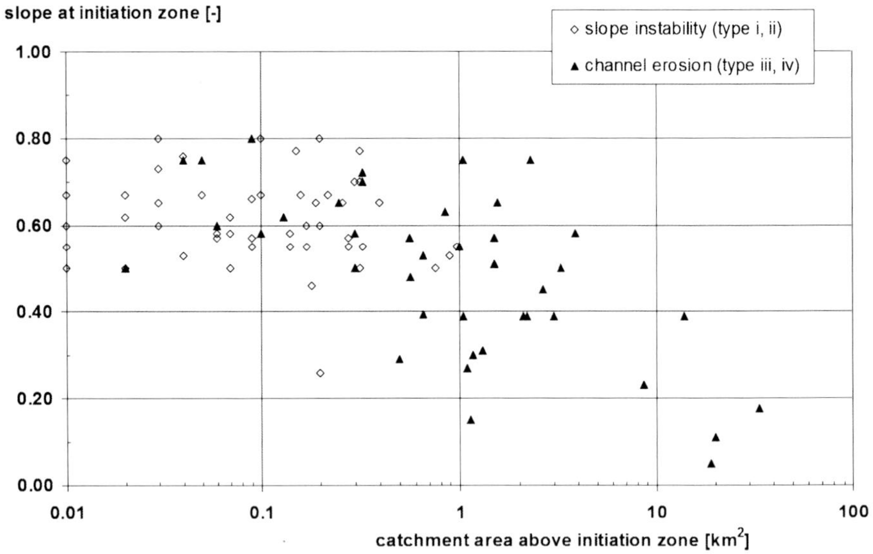

Fig. 1 Critical slope in the source area for the occurrence of debris flows (slope-type and channel-type) as a function of the size of the watershed. Examples from Switzerland (data from D. Rickenmann and M. Zimmermann)

The application of modelling approaches is useful for large areas as well as for individual cases. An exemplary study on debris-flow occurrence has been implemented already in the early 1990s in the San Mateo County, California (Mark 1992) using a spatial statistical model.

3.2 *How Big? (Magnitude)*

The magnitude of debris flows (total volume) is used for hazard mapping or for the design of structural measures, for instance sediment retention basins. Modelling approaches with single or multi-parameter empirical relations provide a rough figure about a so-called "maximum" event (Table 1); however, results may scatter in a wide range.

The values of these equations on a violent torrent in the Bernese Alps (Glyssibach; 2 km^2) vary from ~20,000 to ~135,000 m^3. Such results need to be carefully interpreted and cross-checked with other methods and approaches. A robust approach considers the erosional capacity (erosion section) of particular channel reaches [m^{3}m^{-1}]. Geologic, geomorphic and hydrologic characteristics of the channel (and of the whole watershed) need to be considered. Field investigations play an important role. A rather robust relation to estimate peak discharge has been developed by Mizuyama (1992) for coarse debris flows and mud flows.

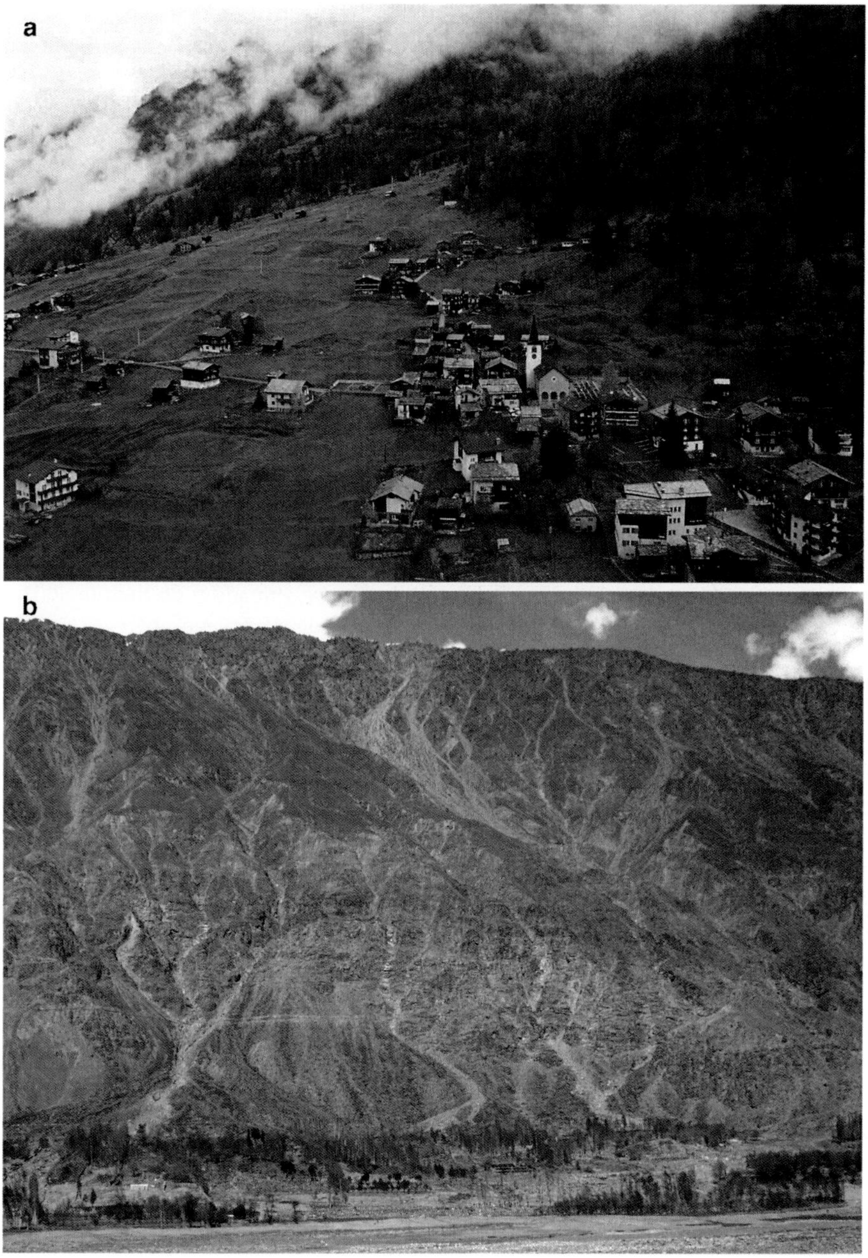

Fig. 2 (**a**) Cone of the Dorfbach, Randa (Switzerland): The typical debris-flow features are clearly visible, although the surface is well covered by vegetation (photo by the author, 1993). (**b**) Few of the many debris flows threatening Sherqila village (Northern Areas, Pakistan). In arid regions the typical landforms are well preserved (photo by the author, 2007)

Table 1 Estimate of debris-flow magnitude in the Glyssibach using different empirical relations

Relation	Source	Based on N observations	Applied for Glyssibach, m^3
$M = K\ Ac\ 100\ Jc$	Kronfellner-Kraus (1984)	1,420	135,000
$Ma = 13{,}600\ Ac^{0.61}$	Takei (1984)	551	20,000
$M = 27{,}000\ Ac^{0.78}$	Zeller (1985)	65	43,000
$M = Lc\ (110\text{–}250\ J_f)$	Rickenmann and Zimmermann (1993)	82	98,000
$Ma = 70\ Ac\ Jc^{1.28}\ IG$	D'Agostino and Marchi (2001)	84	38,000

The scatter of results is considerable M = "Maximum" event volume [m^3], Ma = Medium event volume [m^3], Ac = watershed area [km^2], Jc = average gradient of channel [−], J_f = average slope of fan [−], Lc = active channel lengths [m], K = torrential factor, IG = geological index

3.3 *How Far? (Affected Zones)*

Local planning processes demand accurate information and guidance about the affected areas on the fan or cone. The areas potentially affected depend on the travel distance of the process on the one side and the lateral spreading on the other. The three basic approaches provide valuable and independent information:

Historic documents often enable the determination of affected areas. Written reports, sometimes enriched with sketches, maps or photographs provide excellent information for particular locations (Fig. 3).

The interpretation of (depositional) features on the fan or cone as well as the fan or cone as a whole provides information about maximum extent (geomorphic approach). In the past 10 years the application of various types of models supports the hazard and risk mapping process. There are statistic empirical models, for instance, for the average friction slope (travel angle) for determination of the maximum travel distance (Fig. 4) or analytical empirical approaches (Horton et al. 2008, Fig. 5). Full hydro-dynamic simulation models (e.g. FLO2D) are new tools to analyse conditions on the fan or cone. An overview of such methods is given by Rickenmann (2005).

3.4 *How Often? (Magnitude-Frequency Relationship)*

Frequency analysis is common in meteorological and hydrological investigations to estimate the recurrence of particular events. However, for debris flows this approach fails in most cases because records are missing and the occurrence seems episodic. The historic approach may provide valuable information to estimate the occurrence (Zimmermann et al., 1997a). Written documents found for the Glyssibach revealed that this creek was rather active in the eighteenth and nineteenth century but showed a long period of inactivity in the twentieth century (Fig. 6). A major debris flow

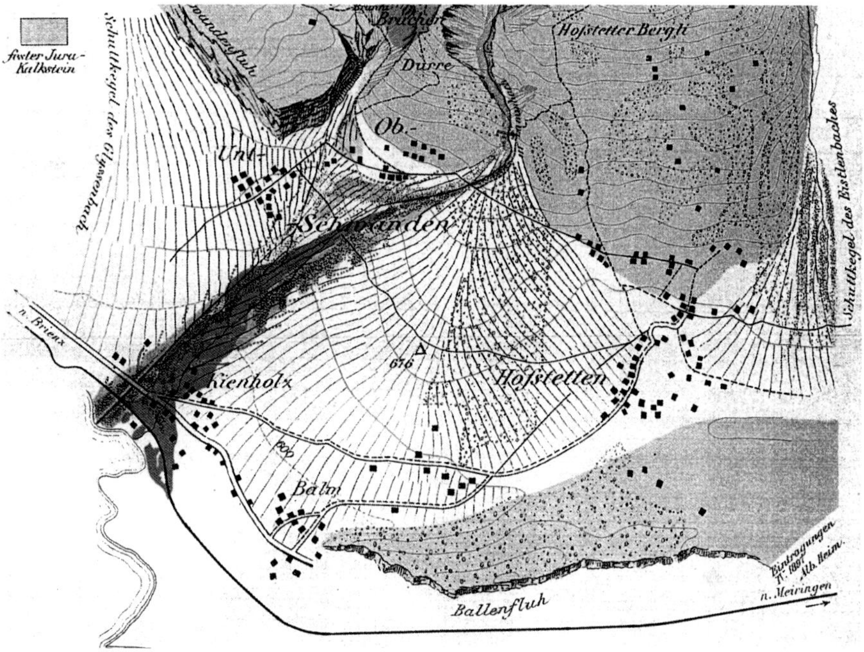

Fig. 3 Map by Albert Heim (1897), showing conditions after the major event 1896 in the Lammbach (Bernese Oberland). The dark surfaces on the cone indicate zones of deposition

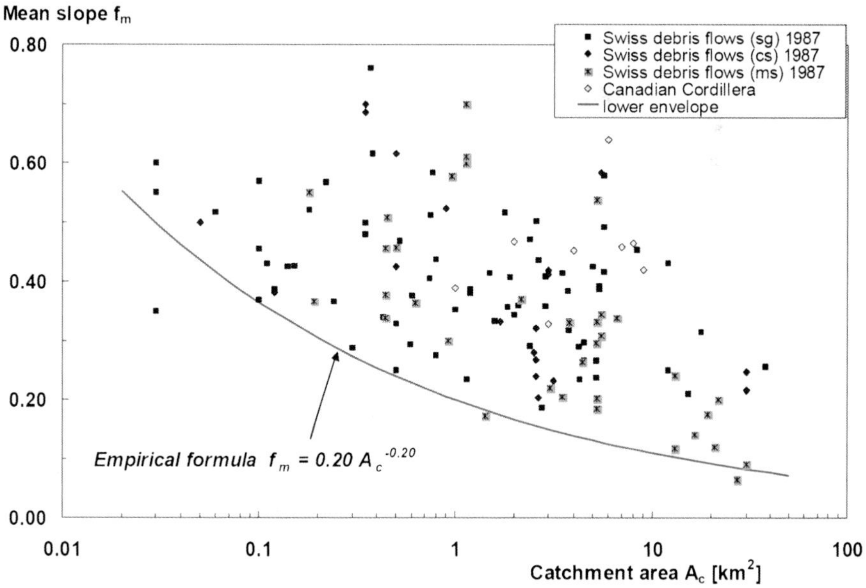

Fig. 4 Debris flows in the Swiss Alps (mainly 1987 events) and data from Canada. The friction slope (fm) as a function of the watershed area (Ac) (adapted from Zimmermann et al. 1997b)

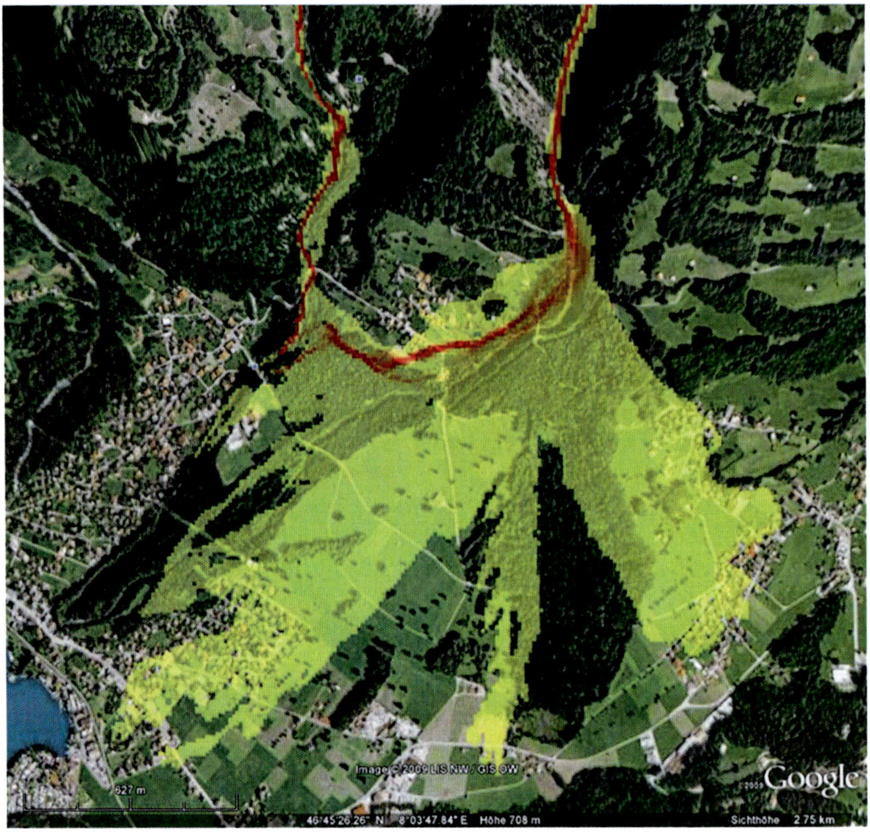

Fig. 5 Simulation of debris flows on the Lammbach fan (Bernese Oberland) using an approach by Horton (2008)

occurred in 2005 (Zimmermann 2006). Considering only the past few decades, this event has to be classified as "extraordinary". However, looking centuries back, the 2005 debris flow is no longer that exceptional. In addition, when considering past activity one has to consider climate change: is the past still relevant for the future?

4 The Practitioner's Challenge: Uncertainties

Hazard assessments play an important role in local planning processes. Practitioners, politicians and finally the local residents request clear and easy-to-understand answers to the basic questions. The approaches, methods and tools developed in the last 20 years provide the necessary technical information. However, a considerable amount of uncertainties is remaining. Conditions for the main parameters may be summarised as follows:

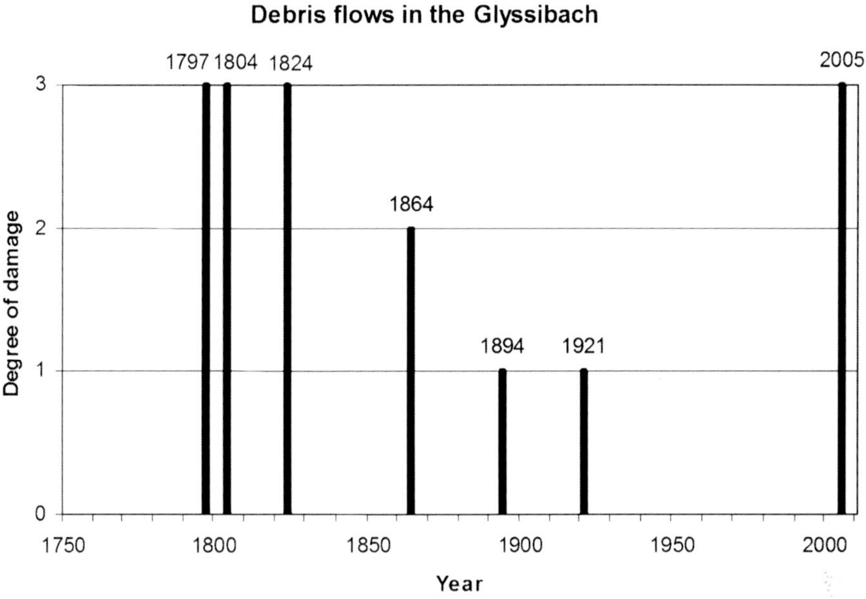

Fig. 6 Frequency analysis of the Glyssibach events: Results from investigations in the course of hazard mapping process (adapted from Zimmermann 2006)

- **Occurrence**: the certainty is rather high to address and identify the hazardous processes in a particular channel properly.
- **Magnitude**: the uncertainties are considerable for the total volume of a debris or mud flow. A factor of two to three has to be assumed.
- **Travel distance and affected area**: the information is reliable; however, flows may show behaviour which can hardly be forecasted.
- **Frequency**: relatively firm statements for the frequent small to mid-scale events. Major uncertainties remain for the recurrence interval of large-scale events. Climate change may even add to uncertainty.

Uncertainties have to be carefully addressed. They can be reduced to a certain extent by applying several methods on the same case. The communication of uncertainties to decision makers or the local population is as important as the result itself (e.g. Zimmermann 2005). The reliability of the assessment is perceived to be much better if uncertainties are properly explained.

References

Aulitzky H (1972) Möglichkeiten und Grenzen der Vorbeugung von Unwetterkatastrophen. Österreichische Wasserwirtschaft 24:8–17

Bründl M (2012) Dealing with natural hazard risks in Switzerland – the influence of hazard mapping on risk-based decision making. In: Schneuwly-Bollschweiler M, Stoffel M,

Rudolf-Miklau F (eds) Dating torrential processes on fans and cones – methods and their application for hazard and risk assessment. Springer, Dordrecht
Cesca M, D'Agostino V (2008) Comparison between FLO-2D and RAMMS in debris-flow modelling: a case study in the Dolomites. In: DeWrachien D, Brebbia CA, Lenzi MA (eds) Monitoring, simulation, prevention and remediation of dense debris flows II. Wit Press, Southampton, pp 197–206
Costa JE (1988) Rheologic, geomorphic and sedimentologic differentiation of water floods, hyperconcentrated flows, and debris flows. In: Baker VR, Kochel RC, Patten PC (eds) Flood geomorphology. Wiley, New York, pp 113–122
D'Agostino V, Marchi L (2001) Debris flow magnitude in the Eastern Italian Alps: data collection and analysis. Phys Chem Earth C 26(9):657–663
Eisbacher GH, Clague JJ (1984) Destructive mass movements in high mountains: hazard and management, Geological survey of Canada paper 84–16. Geological Survey of Canada, Ottawa
Habersack H, Kreisler A (2012) Sediment transport processes. In: Schneuwly-Bollschweiler M, Stoffel M, Rudolf-Miklau M (eds) Dating torrential processes on fans and cones – methods and their application for hazard and risk assessment., Advances in global change research. Springer, Dordrecht/Heidelberg/London/New York
Heim A (1897) Gutachten über Lammbach und Schwandenbach und deren Verbauung. Bericht an den Bundesrat
HFA (2005) Hyogo Framework for action 2005–2015: building the resilience of nations and communities to disasters. www.unisdr.org/hfa
Horton P, Jaboyedoff M, Bardou E (2008) Debris flow susceptibility mapping at a regional scale. 4e Canadian Conference on Geohazard. Université Laval, Québec, Canada
ISDR (2007) Words into action: a guide for implementing the Hyogo Framework. www.preventionweb.net/english/professional/publications/v.php?id=594
Jakob M (2005) Debris-flow hazard analysis. In: Jakob M, Hungr O (eds) Debris-flow hazards and related phenomena. Praxis-Springer, Heidelberg
Kronfellner-Krauss G (1984) Extreme Feststofffrachten und Grabenbildungen von Wildbächen. In: Proceedings Internationales Symposium Interpraevent, Villach, Bd. 2, pp 109–118
Mark RK (1992) Map of debris-flow probability. USGS, Miscellaneous Investigations Series, San Mateo County, California
Mizuyama T, Kobashi S, Ou G (1992) Prediction of debris flow peak discharge. Int Cong Interpraevent 4:99–108
Petrascheck A, Kienholz H (2003) Hazard assessment and mapping of mountain hazards in Switzerland. In: Rickenmann D, Chen CL (eds) Debris flow hazards mitigation: mechanics, prediction, and assessment. Millpress, Rotterdam, pp 25–38
Pierson CP (2005) Distinguishing between debris flows and floods from field evidence in small watersheds. USGS Fact Sheet 2004–3142
Rickenmann D (1999) Empirical relationships for debris flows. Nat Hazard 19:47–77
Rickenmann D (2005) Runout prediction methods. In: Jakob M, Hungr O (eds) Debris-flow hazards and related phenomena. Praxis-Springer, Heidelberg
Rickenmann D, Scheidl C (2012) Debris-flow runout and deposition on the fan. In: Schneuwly-Bollschweiler M, Stoffel M, Rudolf-Miklau F (eds) Dating torrential processes on fans and cones – methods and their application for hazard and risk assessment. Springer, Dordrecht
Rickenmann R, Zimmermann M (1993) The 1987 debris flows in Switzerland: documentation and analysis. Geomorphology 8:175–189
Scheuner T, Keusen HR, McArdell BW, Huggel C (2009) Murgangmodellierung mit dynamisch-physikalischem und GIS-basiertem Fliessmodell. Wasser Energie Luft 101(1):15–21
Schneuwly-Bollschweiler M, Stoffel M, Rudolf-Miklau M (2012) Dating torrential processes on fans and cones – methods and their application for hazard and risk assessment, Advances in global change research. Springer, Dordrecht/Heidelberg/London/New York
Takahashi T (1991) Debris flow, IAHR monograph series. Balkema, Rotterdam, 165 p
Takei A (1984) Interdependence of sediment budget between individual torrents and a river-system. In: Proceedings Internationales Symposium Interpraevent, Villach, Austria, Bd. 2, pp 35–48

UNDHA (1996) Experience and lessons learned from the management of major disasters: mudflows. United Nations Department of Humanitarian Affairs, Geneva
Wieczorek GF, Glade T (2005) Climatic factors influencing occurrence of debris flow. In: Jakob M, Hungr O (eds) Debris-flow hazards and related phenomena. Praxis-Springer, Heidelberg
Zeller J (1985) Feststoffmessung in kleinen Gebirgseinzugsgebieten. Wasser Energie Luft 77(7/8):246–251
Zimmermann M (2005) Analysis and management of debris flow risks at Sörenberg (Switzerland). In: Jakob M, Hungr O (eds) Debris-flow hazards and related phenomena. Springer, Berlin
Zimmermann M (2006) Prognose von Murganggefahren: Wie sicher ist sie? Bulletin für angewandte Geologie 11(2):65–73
Zimmermann M, Issa S (2009) Risk-conscious Reconstruction in Pakistan-administered Kashmir. A case study of the Chakhama Valley. Mountain research and development 29(3):202–210. http://dx.doi.org/doi:10.1659/mrd.1074
Zimmermann M, Mani P, Romang H (1997a) Magnitude-frequency aspects of Alpine debris flows. Eclogae Geol Helv 90:415–420
Zimmermann M, Mani P, Gamma P, Gsteiger P, Heiniger O, Hunziker G (1997b) Murganggefahr und Klimaänderung – ein GIS-basierter Ansatz. NFP31 Schlussbericht. Vdf Verlag, Zürich

Dealing with Natural Hazard Risks in Switzerland – The Influence of Hazard Mapping on Risk-Based Decision Making

Michael Bründl

1 Introduction

Fans and cones are the results of recurring geomorphic processes and they mark preferred areas for settlements and other forms of land use. Due to their position below larger slopes and valleys, processes like landslides, rock falls, debris flows and snow avalanches often endanger people, buildings and infrastructure. In the past centuries and decades, damage causing natural hazard events initiated the construction of structural mitigation measures like snow supporting structures against avalanches or check dams to reduce erosion in torrents. The goal of these efforts was to reduce the probability of damage causing events – decisions were primarily *hazard-based*. In the last 10 years the damage potential was increasingly taken into account, which led to *risk-based* decisions. The fundamental basis for both approaches is a solid knowledge of the event history allowing to derive realistic scenarios of potential future events (Jakob 2012, Chap. 6). Since the frequency and magnitude of these scenarios have a large impact on the calculated risk, dating of past events are of pivotal interest for risk analyses.

The risk concept first introduced and applied for dealing with complex technical systems in the 1970s and 1980s of the last century has also become a general framework for assessing the potential effects of natural hazards. Risk assessments are one part of an integral risk management, which has become a synonym for dealing with natural hazard risks in many countries. In this chapter we will first present the general concept of integral risk management; second, we present the risk concept in which we highlight the importance of hazard analysis; third, we show the influence of different spatial extent of intensity maps on the benefit-cost-ratio of mitigation measures. Although we mainly refer to the situation in Switzerland, the general conclusions hold for other examples in different countries.

M. Bründl (✉)
WSL Institute for Snow and Avalanche Research SLF, CH-7260 Davos Dorf, Switzerland
e-mail: bruendl@slf.ch

M. Schneuwly-Bollschweiler et al. (eds.), *Dating Torrential Processes on Fans and Cones*, Advances in Global Change Research 47, DOI 10.1007/978-94-007-4336-6_24,

2 The Risk Cycle

A widely acknowledged illustration for the various fields of natural hazard risk management is the risk cycle (Fig. 1), which categorizes different types of measures and actions related to natural disasters (Carson 1991; Ammann 2006). The cycle is a symbol for the recurring tasks in disaster management, which aim to learn from past events and to better prepare the society for potential events in future. All measures related to risk management are grouped into three categories: prevention, intervention and recovery. Although we refer in this chapter to natural hazard management, the general principle of the risk cycle can be transferred to other fields where one has to cope with undesired negative effects. In the following sections these categories are briefly described.

2.1 *Before an Event – Preparing the Intervention Phase*

In this phase, various indications for a natural hazard event, which potentially can cause damage to assets and/or persons become obvious. Persons and organizations take precautionary measures and prepare themselves for coping with the upcoming event. Warning messages and data reflecting the possible course of the expected process(es) have to be interpreted and safety measures have to be put in place. The affected population and other stakeholders have to be informed and continuously to be updated about the situation. The lead-time and the state of organization of intervention forces are of crucial importance for the success of measures at this stage.

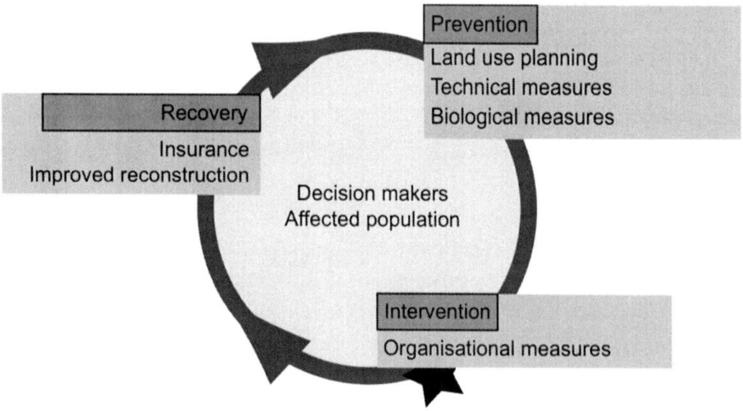

Fig. 1 Risk cycle illustrating the prevention, the intervention, and the recovery phase related to the management of natural disasters

2.2 During an Event – Intervention Phase

The expected event is occurring and intervention forces like e.g. fire brigades are engaged in reducing damage to buildings and infrastructure, evacuating persons and livestock from endangered places or saving lives. The course of the event and temporary protection measures are continuously observed and evaluated. Information of the affected population and communication among intervention and rescue teams are of primary concern. All efforts are concentrated on preventing and reducing further damages.

2.3 After an Event – The Recovery Phase

In the first phase after an event, recovering the most essential basic services has the highest priority. Depending on the severity of the event, the first actions are: fixing the drinking water supply, supplying food and medical care, repairing electricity, communication and traffic infrastructure.

In the second phase restoring the normal state is the major goal. This includes eliminating damages but also to learn from the event. Event documentation, event analyses and learning the lessons from coping with the event are a key for an improved adaptation in future (Kienholz et al. 2012, *this volume*). Systematic recording of typical characteristics of the process(es) in an inventory helps e.g., to estimate the frequency of scenarios for hazard and risk assessment.

2.4 Preparing for the Next Event – The Prevention Phase

Natural hazard disasters are a proof of existing mitigation measures as far as they exist in the affected area. The analysis of the event can give evidence, which measures worked well, and which did not. Therefore, in most cases, responsible organizations start to adapt and to extend existing mitigation measures after an event.

Dealing with natural hazards covers different, complementary types of measures (Fig. 2). One can distinguish between technical (structural measures), biological (protection by plants, e.g. trees), organizational measures (warning, closure or evacuation) and land use planning (land use restrictions, hazard maps).

An analysis of coping with natural hazards or disasters should always include a critical reflection of organizational issues. Questions to be answered are: Did a warning exist and did it reach the right persons at the right time? Did the organization meet the requirements of a successful management of the disaster? Improving the efficiency of an organization is a low cost non-structural measure that should always be taken into consideration while analyzing mitigation strategies.

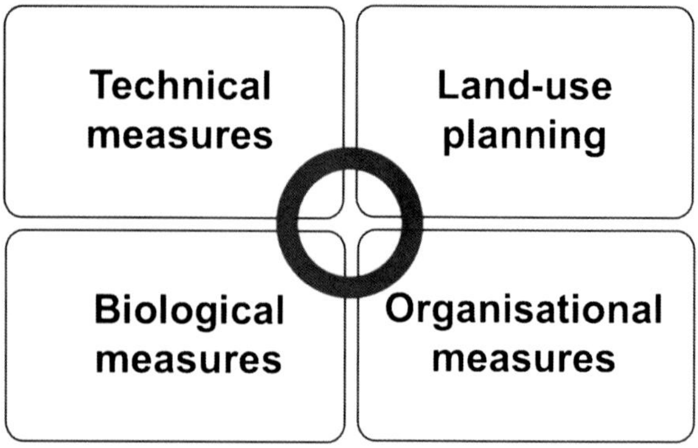

Fig. 2 Mitigation strategies for natural hazards should include different types of measures complementing each other

The goal for improving or building measures is to find the combination, which provides a maximum reduction of risk at a proportional investment of financial, personal and environmental resources. Proportional investment means that decision makers as representatives of the society have to decide, how much efforts should be taken to reduce future damage and to prevent future fatalities or injured people.

A widely used framework for identifying and analyzing mitigation strategies regarding their efficiency and effectiveness is the risk concept that will be presented in the following sections.

3 Posing Questions in Advance – The Risk Concept

The development of the risk concept started in the 1960s when organizations were looking for a framework to estimate and to quantify the potential negative consequences in case of an accident with complex technical systems (e.g., Rasmussen et al. 1975). The risk concept aims to answer the following guiding questions: "What can happen?", "What may happen?" and "What needs to be done?" (Kaplan and Garrick 1981). The risk concept (Fig. 3) has become also the guiding framework of the National Strategy for dealing with natural hazards in Switzerland (Ammann 2006; PLANAT 2004), which has been put into practice with guidelines and software tools (Borter 1999; Bründl 2009; Bründl et al. 2009).

3.1 Risk Analysis

The risk analysis consists of four components: hazard analysis, exposure analysis, consequence analysis and risk calculation.

The goal of a *hazard analysis* is to derive scenarios on which a risk analysis is based on and to characterize them by their frequency, their spatial extent and their physical impact (intensity), often denominated as magnitude-frequency-relationship. The most important basis for scenarios is data from past events.

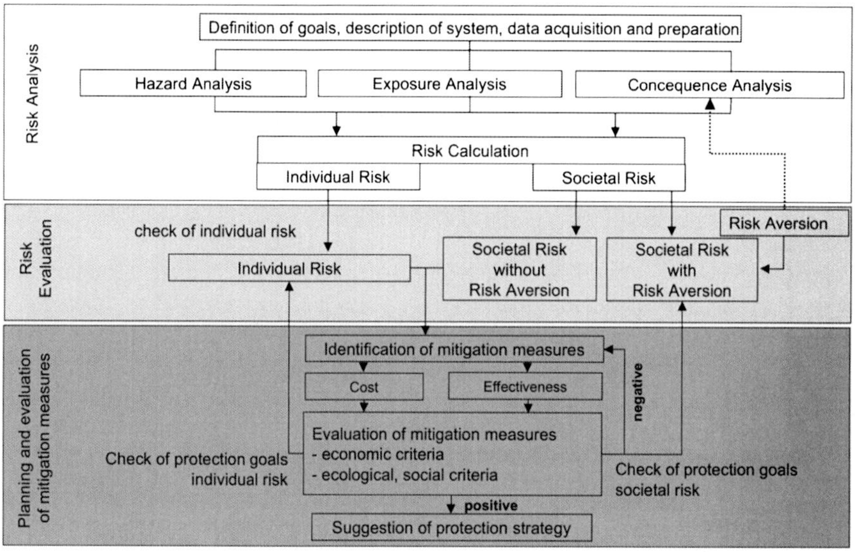

Fig. 3 Schematic illustration of the risk concept

Numerous methods exist for deriving the relation of frequency and magnitude of past events as outlined in several chapters of this book (see part II; Schneuwly-Bollschweiler et al. 2012, *this volume*). Documents and dating results from historic event are used as input for deducing potential future events. Although sound experiences from the past is the best information we can get, it is important to critically judge their validity towards prevailing environmental conditions, like e.g. existing vegetation cover and forest cover, land use, buildings, infrastructure and existing protection measures. In the second part of a hazard analysis, in the effect analysis, the potential physical impact of scenarios is determined. Supported by physical modeling delivering the spatial extent of scenarios and their impacts expressed in physical units (e.g. pressure of snow avalanches in kPa or height and velocity of debris for debris flows), the resulting intensity maps for each scenario are deduced. Realistic intensity maps with the best achievable quality are therefore of pivotal importance for a sound risk analysis. Wrong assumptions persist for the whole risk assessment and can significantly influence the calculated risk and the decision on mitigation measures. The meaning of uncertainties in the hazard analysis and their consequences for the calculated risk is shown by an example in Sect. 4 of this chapter.

In the *exposure analysis* persons and assets at risk are identified and their number, type, value and probability of exposure is assessed. Exposure varies with different types of objects, which can be either permanent (e.g. buildings) or mobile (e.g. persons on traffic routes). The probability of exposure for mobile objects (e.g. persons) is defined by the fraction of time the object or person is present in the endangered area in relation to the time period in which the process can occur.

Since the presence of persons can strongly vary within the considered time period, it can be appropriate to regard different exposure situations in which a varying number of persons are exposed during different time intervals. The consideration of different exposure situations allows accounting for risk peaks.

The *consequence analysis* combines the hazard and exposure analysis yielding the expected damage or loss including all considered scenarios. In a first step, for each scenario, the expected loss is calculated individually for every object at risk. In a second step, the expected loss in this scenario is obtained by summing over the expected loss of the individual objects. The expected loss is calculated as a product of number and value of objects, their damage susceptibility (Uzielli et al. 2008), and their probability of presence in the endangered area.

In the last step, the societal risk and the risk to individuals (individual risk) are calculated. The *societal risk* covers the risk of a group of persons or objects in the hazard area under investigation. It represents the damage or the number of fatalities that are statistically expected due to the hazard impact in each scenario in 1 year. The societal risk in each of the scenarios is obtained as product of the expected damage in this scenario and its frequency. Summing over all scenarios yields the total societal risk per year, which can be either expressed in monetary units per year or as number of fatalities or injured persons per year.

The calculation of the *individual risk* for persons is only necessary for highly endangered objects. It expresses the probability for an individual person to die during a reference period (mostly 1 year) by a specific hazard scenario. Summing over all scenarios gives the total individual risk for one person. A more detailed description of these working steps is given in the guideline "risk concept for natural hazards" (Bründl 2009).

3.2 Risk Evaluation

"What may happen" is the guiding question for risk evaluation and it reflects that results of risk analyses have to be compared with acceptance criteria of the society. Acceptance of risk is a controversially discussed issue (Klinke and Renn 2002; Renn et al. 1992; Siegrist and Cvetkovich 2000; Sjöberg 2000; Slovic 2000) and it cannot be regarded without the context of costs and benefits that are related to the risk. Risk considered in isolation is never acceptable because no rational person would accept any risk at all except possibly in return for the benefits that come along with working, living in or visiting a region endangered by natural hazards (Kaplan and Garrick 1981; Vrijling et al. 1995).

Evaluation of *societal risk* is based on the economic concept of efficiency (Weck-Hannemann 2006) determined by economic evaluation methods such as the cost-effectiveness analysis (CEA) or the cost-benefit analysis (CBA). Thereby, it can be assessed if a certain annual risk reduction can be achieved by certain annual cost of mitigation measures, which is accepted by the society. This holds for tangible assets but also for persons, for which the value of statistical life (VSL) as a non-market value expressing the willingness of a society to pay for averting a fatality is taken as a valuation (see Hammitt 2000, for an overview). In western societies, the VSL is between 5 and 10 million Euro (Baranzini and Ferro Luzzi 2001; Leiter and Pruckner 2008; Rheinberger 2009).

Individual risk is evaluated by thresholds defining the maximum probability that an individual dies within a defined time period (usually 1 year) as consequence of a negative impact, i.e. for example a natural hazard event. Suggested by the Dutch National Environmental Plan (VROM 1988), this concept postulates that an acceptable risk must not add 1% or more to the existing probability of death of the group with the lowest risk within a society. Related to the Dutch society with a natural mortality of the group of 6–20 year old males of 10^{-4}/year, this results in a threshold of 10^{-6}/year (VROM 1988).

Based on this, later concepts suggested including a policy factor which reflects the degree of voluntariness a person gets involved in a risk (Vrijling et al. 1995). It was argued that persons taking risks with a high degree of voluntariness are willing to accept a higher individual risk as if they are obliged to bear a risk without any possibility of self-control.

Evaluation of individual risk in national policy as for example in Switzerland follows the basic idea proposed of Pate-Cornell (2002). She suggested three domains for dividing the individual risk: (1) the risk domain above the threshold of maximum acceptable individual risk ($10^{-3}/10^{-4}$/year), where reduction of individual risk is imperative, but should be achieved with minimal costs. This domain is therefore also called the cost-effectiveness domain, wherein the most cost-effective reduction method is applied to reach the acceptable threshold; (2) in the risk domain below the threshold of acceptable risk but above the *de minimis* threshold, risks should only be reduced if the costs of risk reduction are less than the benefits provided by risk reduction. Hence, the cost-benefit criterion of maximizing the net benefit of risk reduction holds; (3) below the *de minimis* threshold ($10^{-6}/10^{-7}$/year), risks should not be addressed by public authorities, but left to individual responsibility and prevention.

3.3 *Mitigating and Dealing with Risks*

Assessing the risks in an area under investigation indicates whether risks can be accepted by the society or not. If risks exceed the safety goals they have to be reduced until the required level of safety is reached. According to the principles ALARP ("as low as reasonably practical") and ALARA ("as low as reasonably achievable") all possible structural or non-structural measures, i.e. technical, biological or organizational measures and land-use planning should be evaluated and optimally combined to reach the intended level of safety, respecting general conditions such as economic commensurability, social acceptability and environmental compatibility.

With respect to the risk cycle (Fig. 1), structural measures are mainly applied in the field of prevention but are not exclusively limited to it, whereas non-structural measures may be equivalently used along the whole cycle, e.g. land-use planning in prevention, warning and rescue in intervention and loss compensation by insurance companies in recovery.

There are three main conceptual approaches that were developed to evaluate mitigation projects and policies. The common approach in environmental economics is cost-benefit analysis (CBA), which estimates societal willingness-to-pay for changes in the risk to human life and property in monetary terms. By contrast, a cost-effectiveness analysis (CEA) evaluates alternative mitigation projects by either maximizing risk reduction from a given budget or by minimizing costs to reach a default safety standard. For cases where monetizing of the utility of certain measures is not feasible, multi-criteria analysis (MCA) are used as a structured decision-making process that integrates multiple criteria in a formal way (Eisenführ and Weber 2003; Keeney and Raiffa 1993).

4 Sensitivity of Various Spatial Extent of Intensity Maps on the Result of Benefit-Cost-Analyses

Limited public budgets require a prioritization of mitigation measures by comprehensible criteria in most countries. In Switzerland, one of these criteria is the benefit-cost-ratio (BCR) of protection measures. Risk analyses are the basis for benefit-cost-analyses (BCA). Their results highly depend on input factors, like e.g. vulnerability of objects, which can vary due to different assumptions. The Federal Office for the Environment in Switzerland (FOEN) initiated the development of a tool with fixed parameters for the calculation of the BCR, allowing for a comparable evaluation and prioritization of mitigation measures. Based on earlier guidelines (Borter 1999; Borter and Bart 1999; Wilhelm 1999) and the risk concept as general framework, the online tool "EconoMe" (www.econome.admin.ch) was developed. EconoMe is a step-by-step procedure aiming at comparing the annual risk reduction of a mitigation measure or a combination of measures (regarded as annual benefit) with the associated annual costs resulting in a BCR. Projects with a BCR ≥ 1 are regarded to be economically efficient (Bründl et al. 2009). For mitigation projects with a capital expenditure of more than 1 million CHF (app. 900'000 EUR) the verification of economic efficiency with EconoMe is mandatory in Switzerland since 2008.

The calculated risk in EconoMe depends on two types of input parameters: first, intensity maps for different scenarios characterizing potential future events, and second, parameters describing the damage potential, like the value and the vulnerability of potentially affected objects and the probability of presence of people and objects. Both types of input parameters can have decisive influence on the BCR. Schaub and Bründl (2010) investigated the sensitivity of these input parameters on the calculated risk and the BCR of mitigation measures for two mitigation projects each for snow

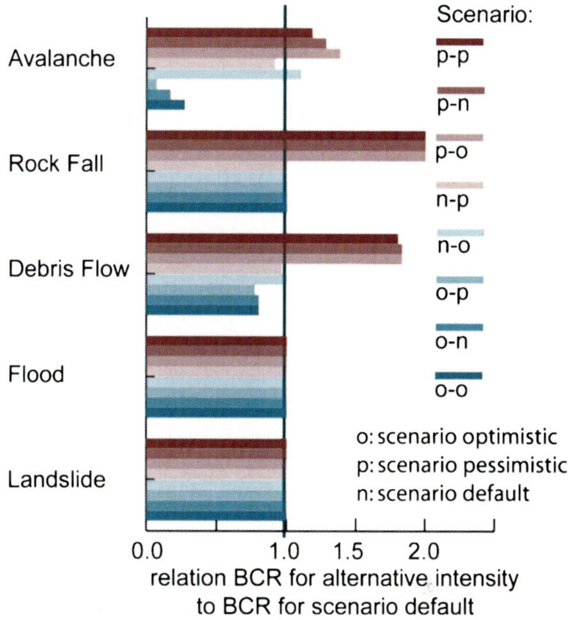

Fig. 4 Influence of varying intensity maps on the benefit-cost-ratio (BCR) of mitigation measures. *p* stands for "pessimistic" and *o* for an "optimistic" intensity maps. Both are thinkable alternatives of an intensity map suggested by the local hazard mapping expert. Pessimistic maps for rock fall and debris flow have the highest impact on the BCR in this study (Source: Schaub and Bründl 2010)

avalanches, debris flows, rock fall, flood and shallow landslides using the tool EconoMe. They checked the influence of the intensity on the BCR by calculating the risk with intensity maps as they were determined by the local hazard mapping expert (scenario default) compared to an optimistic (scenario optimistic) and a pessimistic version (scenario pessimistic) of the same intensity map. "Optimistic" means that the spatial extent was reduced to a possible minimum area, "pessimistic" means that the intensity zones were enlarged up the probable maximum area, both alternatives in agreement with the experiences of the local expert. The results indicate that variations of the spatial extent of the intensity maps have significant influence on the BCR. However, the influence differs on the process, which is to some extent an artifact of the chosen cases in this study (Fig. 4). Ranking the influence of the parameters intensity maps, vulnerability of objects, lethality of persons, spatial probability of the process regarding their influence on the BCR, the results show that varying intensity maps have the largest influence. This underlines that the quality of the intensity maps and therefore the quality of F-M analysis as basis for considered scenarios are of crucial importance for risk-based decision making.

5 Final Remarks

The way decision makers in European countries and worldwide deal with natural hazards has changed from a direct response to an event with the goal that the experienced disasters should never happen again, to strategies, which relate the potential consequences of future events to the cost of the corresponding mitigation measures. Risk serves as a construct to quantify undesirable consequences in future providing evidence for effectiveness and efficiency of mitigation measures. However, risk reduction may not be taken as the only criterion for decision-making. The success of a mitigation measure or a protection strategy in a long-term perspective highly depends on its acceptance by affected or involved people, its political practicability, its technical feasibility, its reliability and its maintenance, as well as other, e.g. ecological, indicators.

The most important basis for a risk assessment is an exact, comprehensible and state-of-the-art hazard analysis with an identification of scenarios representing the most probable courses of a hazard process. Sophisticated techniques for dating past events become increasingly important, since they can directly influence the economic efficiency of mitigation measures. We agree with Jakob (2012, *this volume*) that less accurate or even wrong hazard analysis can have drastic consequences on the safety of affected people and objects. Therefore, improvement of dating techniques methods significantly contributes to safety against natural hazards.

Acknowledgements The author gratefully acknowledges contributions by and/or discussions with Walter Ammann, Nicole Bischof, Sven Fuchs, Peter Gutwein, Bernhard Krummenacher, Hans Merz, Christoph Rheinberger, Jakob Rhyner, Hans Romang, Yvonne Schaub, Thomas Schneider, Christian Wilhelm, and Cornelia Winkler during the past years which all contributed in one way or another to the thoughts presented in this chapter.

References

Ammann WJ (2006) The risk concept. In: Dannenmann S, Ammann WJ, Vulliet L (eds) Risk 21 – coping with risks due to natural hazards in the twenty-first century. Taylor & Francis, London, pp 3–23

Baranzini A, Ferro Luzzi G (2001) The economic value of risks to life: evidence from the Swiss labour market. Swiss J Econ Stat 137(2):149–170

Borter P (1999) Risikoanalysen bei gravitativen Naturgefahren. Methode, Umwelt-Materialien 107/I. Bundesamt für Umwelt, Wald und Landschaft, BUWAL, Bern

Borter P, Bart R (1999) Risikoanalysen bei gravitativen Naturgefahren. Fallbeispiele und Daten, Umwelt-Materialien 107/II. Bundesamt für Umwelt, Wald und Landschaft, BUWAL, Bern

Bründl M (ed) (2009) Risikokonzept für Naturgefahren. Einzelprojekt A1.1. Leitfaden. Nationale Plattform Naturgefahren PLANAT, Bern

Bründl M, Romang H, Bischof N, Rheinberger CM (2009) The risk concept and its application in natural hazard risk management in Switzerland. Nat Hazards Earth Syst Sci 9(3):801–813

Carson WN (1991) Disaster management: a disaster manager's handbook. Asian Development Bank, Manila

Eisenführ F, Weber M (2003) Rationales Entscheiden, 4th edn. Springer, Berlin

Hammitt JK (2000) Valuing mortality risk: theory and practice. Environ Sci Technol 34(8): 1396–1400

Jakob M (2012) Events on fans and cones: recurrence interval and magnitude. In: Schneuwly-Bollschweiler M, Stoffel M, Rudolf-Miklau F (eds) Dating torrential processes on fans and cones – methods and their application for hazard and risk assessment, Advances in global change research. Springer, Dordrecht/Heidelberg/London/New York

Kaplan S, Garrick B (1981) On the quantitative definition of risk. Risk Anal 1(1):11–27
Keeney R, Raiffa H (1993) Decisions with multiple objectives. Preferences and value tradeoffs. Cambridge University Press, Cambridge
Kienholz H, Hübl J, Loipersberger A (2012) Documentation of torrential events. In: Schneuwly-Bollschweiler M, Stoffel M, Rudolf-Miklau F (eds) Dating torrential processes on fans and cones – methods and their application for hazard and risk assessment, Advances in global change research. Springer, Dordrecht/Heidelberg/London/New York
Klinke A, Renn O (2002) A new approach to risk evaluation and management: risk-based, precaution-based, and discourse-based strategies. Risk Anal 22(6):1071–1094
Leiter A, Pruckner G (2008) Dying in an avalanche: current risks and valuation. Technical report, Fakultät für Volkswirtschaft und Statistik, University of Innsbruck. http://econpapers.repec.org/paper/innwpaper/2008-11.htm
Pate-Cornell M (2002) Risk and uncertainty analysis in government safety decisions. Risk Anal 22(3):633–646
PLANAT (2004) Sicherheit vor Naturgefahren. Vision und Strategie, Bern
Rasmussen NC et al. (1975) Reactor safety study. An assessment of accident risks in U.S. Commercial Nuclear Power Plants. WASH-1400 (NUREG-75/014). Federal Government of the United States, U.S. Nuclear Regulatory Commission, Rockville, MD, USA
Renn O, Burns WJ, Kasperson JX, Kasperson RE, Slovic P (1992) The social amplification of risk – theoretical foundations and empirical applications. J Soc Issues 48(4):137–160
Rheinberger CM (2009) Preferences for mitigating natural hazards on alpine roads: a discrete choice approach. Diss ETH Zurich No. 18476, Zurich
Schaub Y, Bründl M (2010) Zur Sensitivität der Risikoberechnung und Massnahmenbewertung von Naturgefahren. Schweizerische Z Forstwes 161(2):27–35
Schneuwly-Bollschweiler M, Stoffel M, Rudolf-Miklau M (2012) Dating torrential processes on fans and cones – methods and their application for hazard and risk assessment, Advances in global change research. Springer, Dordrecht/Heidelberg/London/New York
Siegrist M, Cvetkovich G (2000) Perception of hazards: the role of social trust and knowledge. Risk Anal 20(5):713–719
Sjöberg L (2000) Factors in risk perception. Risk Anal 20(1):1–11
Slovic P (2000) The perception of risk. Earthscan, London
Uzielli M, Nadim F, Lacasse S, Kaynia AM (2008) A conceptual framework for quantitative estimation of physical vulnerability to landslides. Eng Geol 102(3–4):251–256
Vrijling JK, van Hengel W, Houben RJ (1995) A framework for risk evaluation. J Hazard Mater 43:245–261
VROM (1988) Dutch national environmental policy plan. Premises for risk management. Second chamber of the states general, session 1988–1989, 21 137, no. 5. The Hague
Weck-Hannemann H (2006) Efficiency of protection measures. In: Dannenmann S, Ammann WJ, Vulliet L (eds) Risk 21 – coping with risks due to natural hazards in the twenty-first century. Taylor & Francis, London, pp 147–154
Wilhelm C (1999) Kosten-Wirksamkeit von Lawinenschutz-Massnahmen an Verkehrsachsen, Vorgehen, Beispiele und Grundlagen der Projektevaluation. BUWAL, Bern

Hazard Mapping and Land-Use Planning – A Swiss Perspective

Thomas Egli and Maja Stucki

1 Introduction

1.1 Intention of Hazard Maps and Land-Use Planning

Floods and debris flows have the potential to cause fatalities, displacement of people and damage to the environment, to severely compromise economic development and to undermine economic activities. As floods and debris flows are natural phenomena they cannot be prevented. However, some human activities (such as the increase in human settlements and economic assets in floodplains and the reduction of the natural water retention by land use) and climate change contribute to an increase in the likelihood and adverse impacts of (flash) flood and debris-flow events. Damage caused by these processes may vary across countries and regions.

In order to provide an effective tool for information, as well as a valuable basis for priority setting and further technical, financial and political decisions regarding flood and flow risk management on fans and cones, the realization of maps illustrating the potential adverse consequences associated with (flash) flood and debris-flow scenarios is compulsory. These maps should also include information on potential sources of environmental pollution as a consequence of floods.

In a following step this information needs to be implemented in land-use planning. Land-use planning guarantees an appropriate utilization of endangered surfaces and pre-emptively contributes to risk reduction. Following the principles

T. Egli (✉)
Egli Engineering AG, Lerchenfeldstrasse 5, 9014 St. Gallen, Switzerland
e-mail: egli@naturgefahr.ch

M. Stucki
Egli Engineering AG, Schwarztorstrasse 87, 3007 Bern, Switzerland
e-mail: stucki@naturgefahr.ch

M. Schneuwly-Bollschweiler et al. (eds.), *Dating Torrential Processes on Fans and Cones*, Advances in Global Change Research 47, DOI 10.1007/978-94-007-4336-6_25,

of hazard avoidance, the realization of new building will be avoided in highly endangered areas whenever possible. In areas where constructions already exist risk will be reduced to an acceptable level using different measures.

1.2 Hazard Maps

Hazard maps need to cover all geographic areas susceptible to flooding under the following scenarios:

(a) floods with a low probability, or extreme event scenarios;
(b) floods with a medium probability (return period e.g., $\geq$100 years in Switzerland, $\geq$150 years in Austria)
(c) floods with a high probability, where appropriate.

For each of the scenarios the following elements need to be illustrated: extent; water depths or debris depths, as appropriate; and, where appropriate, flow velocity and direction of the relevant water flow. Further elements might be taken into account, as, for instance, the level of the energy line and erosion depths. The latter are considered relevant criteria for the delineation of hazard zones in Austria. The proper use and application of hazard maps into planning processes and awareness campaigns require the consideration of very basic information on the map such as a clear map title where clear reference shall be made to the content of the map. Such information should include indications on:

- Parameter: Extent, depth, flow velocity, past events
- Probability considered: defining more precisely what mean low, medium and high probability of occurrence are taken into account
- Location of the map as part of the catchment or country: provision of a small inset map
- Legend

1.3 Land-Use Planning

Land-use planning is an important component of risk management. It presents a preventive measure leading to the reduction of damage potential without changing the hazard potential. Land-use planning guarantees an appropriate use of land through the ordered sitting of a country, takes account of the hazard situation and helps to reduce existing risks. Recognition of hazard situations is a prerequisite for land-use planning; this can be achieved through the creation of hazard maps.

2 Creation of Hazard Maps

In Switzerland, several laws oblige the regional (cantonal) authorities to create hazard maps and to take them into account in regionally significant activities. The guidelines for the procedural steps in the creation of hazard maps have been defined by the national (federal) authorities. The procedure is separated into hazard recognition (*what can happen and where?*) and hazard evaluation (*how often can it occur and how intense will it be?*).

2.1 *Hazard Recognition*

Hazard recognition comprises above all a documentation of past activity in the area under investigation and should include all relevant processes and their spatial extent, damage, as well as indications on meteorology and hydrology. Data on past events will be summarized in an event cadastre which needs to be updated continuously. Information on past events will help e.g. the understanding of frequencies and possible devolution of events as well as for the definition of relevant scenarios.

A cartographic illustration of morphologic characteristics of the site under evaluation is illustrated in the *map of phenomena*. This map contains information on the field assessment such as critical torrent segments (avulsion) or silent witnesses of past activity, indications on channel capacity, overbank sedimentation etc. Dated past events represent a valuable additional information on fans and cones and help to complement fragmentary time series and chronical data, especially in areas which are (or have) not (been) used by humans in the past; they also allow a better classification of intensity and probability of recurrence or may increase the level of confidence when it comes to the determination of design events.

A hydro-engineering evaluation of all factors relevant for the occurrence of (flash) flood and debris-flow processes (i.e. hydrology, hydraulics, solid charge) forms a further element of the hazard recognition process. Special care needs to be addressed to hydrology. The hydro-engineering evaluation should also include an analysis of flaws within the water course such as narrow culverts. Knowledge on flaws is not only relevant for the recognition and evaluation of hazards but also key for the planning of countermeasures.

2.2 *Hazard Evaluation*

Based on the information from the hazard recognition *hazard indication maps* and *hazard maps* can be developed.

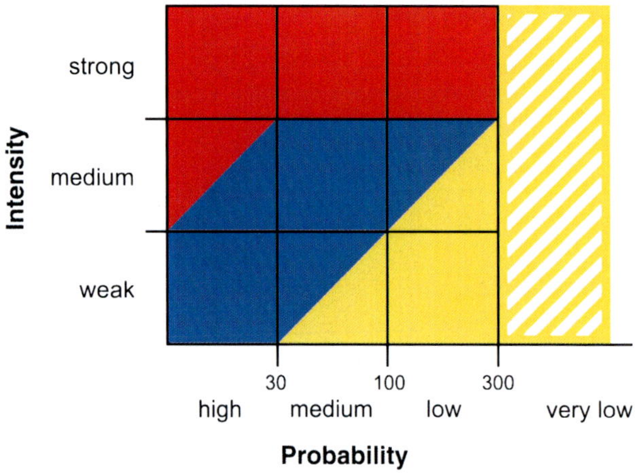

Fig. 1 Hazard levels indicating the result of intensity and probability

2.2.1 Hazard Indication Map

Hazard indication maps provide a rough overview about existing hazards; they are produced at a small scale but covers entire areas without rating existing hazards. The hazard indication map serves a rapid recognition of areas where conflicts exist and is basically used during the framework development planning process and for the examination of building applications outside the perimeter considered in more detailed hazard maps.

2.2.2 Hazard Maps

The hazard map serves the use zone planning and contains precise indications on the nature of hazard, its spatial extent and the degree of endangering. Analyses are not performed area-wide, but focus on clearly defined perimeters but in much greater detail. The level of endangering is obtained from the intensity and the probability of occurrence, as illustrated in Fig. 1. The different levels of hazards given in the figure indicate:

- red: considerable endangering, prohibition zone
- blue: average endangering, instruction zone
- yellow: small endangering, indication zone
- yellow-white hachure: residual endangering, indication zone
- white: no or negligible endangering based on current state of knowledge, no regulations

Hazard evaluations are based on scientific criteria and might, under certain circumstances, be complemented with model simulation runs. A technical report has to be prepared for each hazard map. It contains a general description of the

Table 1 Classification of intensities

Process	Flood	Debris flow
Strong intensity	$h > 2$ m or $v{*}h > 2$ m^2/s	$h > 1$ m and $v > 1$ m/s
Medium intensity	2 m $> h > 0.5$ m or 2 m^2/s $> v{*}h > 0.5$ m^2/s	$h < 1$ m or $v < 1$ m/s
Weak intensity	$h < 0.5$ m or $v{*}h < 0.5$ m^2/s	None

h water depth or thickness of debris-flow deposit, *v* flow velocity

methods, procedures and results of the hazard assessment. Hazard maps have to be updated periodically, especially in case that the hazard conditions have changed.

2.2.3 Intensity Maps

Intensity maps are another source of information and are produced within the process of hazard mapping. They show expected intensities for different probabilities (return periods of 30, 100 and 300 years) and are therefore of utmost value for the definition of object protection measures. The classification of intensities is based on the recommendations of the federal authorities following the criteria listed in Table 1.

3 Impact on Land-Use Planning

The background data that has been generated for the hazard maps will be taken into account in land-use planning. Endangered surfaces will be used appropriately so as to reduce risk.

The directoral plan (*Richtplan*) summarizes the goals of land-use planning and implements space-relevant duties; it serves as a spatial coordination and guidance tool. The communities are in charge of the elaboration of a zoning and a use zoning plan as well as construction regulations. A central duty of this task is the legally binding implementation of the hazard map. During the construction permit procedure, the cantonal and local authorities (i) will make sure that regulations are being followed and (ii) may also impose conditions such as object protection measures.

3.1 Directoral Plan

The directoral plan is a strategic and administrationally binding instrument of land-use planning serving both the coordination and prevention as well as the protection from natural hazards. The axiom of natural hazard prevention and the principles of natural hazard protection are stated in the directoral plan, such as the prioritization of land-use as compared to constructional measures. The directoral plan also contains the mandates for the elaboration of hazard (indication) maps, as well as timing and responsible for their realization. The directoral plan also gives the mandate for the implementation of hazard maps within use zoning plans to the respective communities and prescribes a periodic evaluation and actualization of the hazard assessment.

3.2 Use Zoning Plan

Use zoning plans exist at the community level and contain different zones foreseen for specific uses; the use is regulated for each allotment and binding for landowners, also with respect to the legal implementation of hazard maps. The different levels of endangering from the hazard map (red, blue, and yellow in the case of Switzerland) are therefore reproduced in the use zoning plan. For any kind of utilization of land in the red or blue zones (prohibition and instruction zones), prescriptions have to be defined for the construction and zoning regulations. In the yellow zone (indication zone) recommendations and prescriptions should be adapted to the local risks, with the goal of reducing existing and preventing new risks.

Construction zones cannot be chosen in the red hazard zone. Renovation of existing or the reconstruction of destroyed building are only possible exceptionally and based on prescriptions. In the blue hazard zone an examination for a creation of zones with constructional use can be possible provided that a superordinate interest is prevalent or that alternative locations are unavailable. In this case the construction permits will contain prescriptions. In the yellow and yellow-white hazard zones the realization of constructions with high damage potential should be avoided and recommendations are prepared for existing buildings.

The use zoning plan also contains regulations for allotments in endangered areas within a construction zone, recommendations on the handling of protection structures, special uses and for areas where hazard maps are not yet existing.

3.3 Construction Permits

During the construction permit procedure, authorities will, among others, verify the compliance of the project with the regulations on the protection from natural hazards, mainly by evaluating whether an endangering exists. Two scenarios are possible for allotments with existing hazards: Provided that risks can be reduced to an acceptable minimum with appropriate measures, necessary prescriptions will be decreed. On the other hand, if risks are too important a permit will not be emitted. The construction will also make sure that the prescriptions will be realized during construction. Possible prescriptions or conditions include:

- Limitations in use – for instance a prohibition of residential use in endangered parts of a building

- Object protection measures – see next chapter
- Construction standards – define requirements for new constructions with respect to natural hazards
- Technical protection structures – e.g., dams
- Recommendations – applicants are informed about optional protection measures

4 Protection Measures at the Very Local Level

Besides measures of land-use planning and protective measures at the source of the hazard, object protection represents another possibility to prevent people and their assets from natural hazards. Object protection measures should be applied in case endangering can neither be avoided nor reduced. Under such circumstances the building will be constructed in a way that hazards can cause only minimal damage to the construction. In the case of new buildings the planning of object protection measures requires little effort, but object protection measures can also be designed and implemented for existing buildings. The use of object protection measures is not applicable if endangering is considerable.

4.1 Assessment of Impacts

The design of object protection measures requires information on flow height and flow velocity of (flash) floods and debris flows. This information can be obtained from the intensity maps or the technical report, more detailed evaluations may however be required in most cases. Among the impacts of torrential activity are:

- Pressure from dynamic loading – depends on flow velocity, density and deflection angle
- Pressure from hydrodynamic and hydrostatic loading
- Backwater level at impact
- Vertical loading – in case that solid charge covers the building
- Pressure through friction – especially for deflection structures
- Impact forces of individual components – by large boulders in the front of debris flows

4.2 Different Measures

Measures can be separated into conceptual, reinforcement and shield measures. Conceptual measures can be found most frequently in new buildings, whereas reinforcement and shield measures are often realized *a posteriori* in existing buildings. Conceptual measures include:

- Terrain adjustment – e.g., through an elevated arrangement for small debris flows or a deepened arrangement for large debris flows
- Shape of mould of the construction – cuneiform or offset floor plan forms with respect to the main flow direction of debris flows are preferable, but tubes should be avoided on the outside facade
- Utilization concept of the interior space – plan rooms with short stopping time (corridor or plumbing unit) inside the facade exposed to hazards
- Position and height of openings – avoid windows (or keep them small) in the area of the exposed outside facade, openings should be reinforced, entrances should be avoided
- Utilization concept of the exterior space – perches and balconies should be planned opposite of the exposed exterior facade

The following reinforcement measures are useful:

- Reinforcement of exterior walls – dimensioning based on the impact and friction forces, a backfill can reduce the surface of the exposed exterior facade
- Protection of openings – dimensioning according to the impact forces, (deflector or impact plates)

Shield measures outside the construction include:

- Retention dam – measure against small debris flows, make sure that there is sufficient room for deposition
- Elevated arrangement– elevation of the terrain, only possible for new constructions
- Deflection walls/dams – to deflect the debris flows in a different direction
- Splitting wedge – debris masses will be divided and routed alongside the object

For the shield measures listed above, it is important to make sure that no additional endangering will result for neighbouring objects. For each object to be protected different measures, or combinations of several types of measures, are possible.

This contribution provided an overview on how land-use planning and management is addressed in Switzerland. Terminology and measures vary between the countries, as does the classification of hazards, relevant return periods and/or design events. In Austria, for instance, hazard terminology differs from that used in Switzerland, but the overall procedures remain comparable. Endangered areas are presented in red and yellow; blue indicates areas which should be reserved for the realization of countermeasures, brown surfaces are affected by processes other than torrential processes and snow avalanches (e.g. rockfall, landslide) and purple designates areas where a protective function can only be guaranteed if the soil or the slope will remain in its current state.

Design Criteria for Torrential Barriers

Florian Rudolf-Miklau and Jürgen Suda

1 Introduction: Safety Standards in Torrent Control Engineering

The term *torrent control* refers to the implementation of specific measures in the catchment or stream bed of a torrent. The purpose of torrent control is many fold – to stabilize the bed and adjacent slopes, to regulate discharge of floods, to dose runoff and solid transport, to filter large components (blocks, drift wood), to dissipate the energy of debris flow or to deviate (bypass) hazardous flow processes from objects or areas at risk (ONR 24800:2008). *Torrential barriers* (functional types, see Fig. 1) form part of the *structural protection measures* in torrent control that aim at reducing the risk to a tolerable level.

The design of these protection works must be in strict adherence to EUROCODE (compulsory in all CEN[1] member states). These standards require, in particular, the proof of *stability* and *usability* (including *durability*) of the protection works, as well as the determination of a *safety concept.* The safety concept is based on the probability and intensity of the impact processes, including the risks of consequences of a torrential event (Suda et al. 2009). This safety concept is part of new technical standards that are currently under development by an interdisciplinary working group (ON-AG 256.01) at the Austrian Standards Institute (ASI). The ONR

[1]European Committee for Standardization.

F. Rudolf-Miklau (✉)
Department IV/5 – Torrent and Avalanche Control, Federal Ministry for Agriculture, Forestry, Environment and Water Management, A-1030 Vienna, Austria
e-mail: florian.rudolf-miklau@lebensministerium.at

J. Suda
alp-infra, consulting + engineering GmbH, Lützowgasse 12/1, 1140 Wien, Austria
e-mail: juergen.suda@gmail.com

M. Schneuwly-Bollschweiler et al. (eds.), *Dating Torrential Processes on Fans and Cones*, Advances in Global Change Research 47, DOI 10.1007/978-94-007-4336-6_26,

Fig. 1 Types of torrential barriers (example): check dams (**a**; *above left*); dosing dam with double slot (**b**; *above right*); dosing dam with grill (**c**; *below left*); debris breaker (**d**; *below right*)

24800-series[2] focuses particularly on the regulation of the terminology and classification of torrent control measures, as well as on load models, design and construction, life cycle assessment and maintenance for torrential barriers.

Modern protection concepts in torrent control are scenario-oriented and try to optimize the function of protection works. The following functional types of torrential barriers can be distinguished:

- stabilizaton and consolidation
- retention
- dosing and filtering
- energy dissipation

A detailed classification of torrential barriers is given in Table 1.

[2]ONR 24800, Protection works for torrent control – Terms, definitions and classification (published 2008); ONR 24801, Protection works for torrent control – Actions on structures (in preparation); ONR 24802, Protection works for torrent control – Design of structures (published 2010); ONR 24803, Protection works for torrent control – Operation, monitoring, maintenance (published 2008).

Table 1 Classification of transverse protection structures according to their function and construction

	Stabilization Consolidation	Retention	Dosing	Filtering	Energy dissipation
Function	Check dam Sill	Retention dam	Dosing dam	Filtering dam	Debris breaker
	Bottom sill Ramp	Water … Debris …	Water … Debris …	Debris … Drift wood …	
Construction	Closed check dam	Simple check dam		Multiple check dam (cascade)	
	Conducting dams	Dam with unbroken crest		Dam with broken crest	
		Dam with small drain		Slit dam	Grid dam
		Dam with large openings		Single slit	Netting dam
				Multiple slit	Wire dam

Stresses on torrential barriers result from water (hydrostatic, dynamic), earth and debris-flow and woody debris impact pressures. In special cases effects from avalanches, falling rocks and earth quakes must also be considered. The relevant actions on torrential barriers result from the predominant displacement process in the torrent. Depending on flow density, water and sediment ratio, as well as flow behaviour, the four basic discharge processes can be distinguished as (Marco 2007; ONR 24800:2008) (Table 2):

- flood
- fluviatile sediment discharge
- debris flood and
- debris flow

The stresses within the building resulting from debris flows depend on the density, velocity, flow height and load. Since in nature these flow processes are affected by many factors, simplified models have been developed for calculation.

The torrential processes that act on torrential barriers can be defined as extreme events with a high intensity and small probability of occurrence. As a general rule, the more seldom an event occurs the stronger the impact forces (i.e. intensity) are. In torrent control engineering the desired level of safety is expressed as a *protection goal* (Rudolf-Miklau 2009), which corresponds to a so-called "*design event*" (DE; Rudolf-Miklau and Sereinig 2009). The *DE* depends on an assumed *return period* and represents the relevant scenario (Mazzorana et al. 2009) for the design and dimensioning of torrent control works, taking into account all process-related impacts (relevant impact combination) and the risk of functional or structural failure (Bergmeister et al. 2009).

In Austria and according to ÖNORM EN 1990 torrent control works generally have to be designed for a lifetime of 100 years. The dimensioning of torrent control structures integrates all possible impacts and influences that may occur during the construction and utilization phase. This requirement is directly related to the

Table 2 Physical properties of the basic torrential processes according to ONR 24800:2008

Discharge type	Flood types		Debris-flow types	
Discharge process	Flood	Fluvial sediment discharge	Debris flood	Debris flow
Type of process	Water	Low Sediment transport	High	Debris flow
Rheology	Newtonic		Approximately newtonic	Non newtonic
Vol. solid concentration	Order of parts per thousand	0–20%	20–40%	>40%
Max. grain size	mm–cm	–dm	–m	–m
Specific gravity	1,000 kg/m³	<1,300 kg/m³	1,300–1,700 kg/m³	>1,700 kg/m³
Relevant forces	Turbulence, shear stress		Buoyant forces, turbulences, shear stress, dispersive pressure	Buoyant forces, dispersive pressure, viscose and frictional forces
Solids dispersion in cross section	Solids transported near river bed and suspended solids distributed in cross section		Solids and suspended solids distributed in cross section	Solids distributed in cross section
Relevant damage components	Water and suspended solids	Water, suspended solids and bed load	Solids and water	Solids (less water)

stipulated proofs at limit state of stability and usability. In addition the design of structures has to take into account extraordinary (extreme) impacts, for example, extreme floods, debris flow, snow and rock avalanches, slope failure or earth quakes.

The integrity of torrent control works has to be ensured by the selection of appropriate building materials and by the adherence to the functional and structural design according to the "state-of-the-art", that is the currently accepted building standards. The design of protection works has to consider the risk of *structural* and *functional failure*, especially the failure of one or more structures in a function chain or bearing components. This can trigger a chain reaction, which, in the worst case, results in the failure of the whole protection system.

Torrent control works are regularly designed and constructed in an alluvial fan environment. The determination of a appropriate DE prerequisites a sufficient understanding of torrential processes as well as condensed data on historic events concerning their flow height, run out-distance, spread and impact forces. This article highlights the importance of these data from the point of view of torrent control engineering.

2 Design Event (DE) and Scenarios for Torrential Barriers

2.1 The Application of the Frequency-Intensity-Concept

The characteristic displacement processes of torrential events are defined by physical parameters such as the rheology (newtonian/non-newtonian), the volumetric concentration of solids, the density of the liquid–solid-mixture, the kinematic viscosity, the flow velocity or the relative discharge (ratio of total discharge to water discharge). Details on the physical properties of torrential processes are given in Table 2, additional information is given in e.g. Iverson (1997), Jakob and Hungr (2005) and Takahashi (2007).

A *torrential event* represents the entirety of these processes occurring in a temporal, areal and causal relationship and corresponds to a specific probability of recurrence and intensity. The underlying *concept of intensity and frequency* was originally established by Wolman and Miller (1960).

The *design event* (DE) is determined according to a defined *return period* (e.g. flood with return period of 100 years). *Intensity* in the colloquial use refers to strength or magnitude of a process or event.[3] Intensity of torrential events can be expressed by physical criteria such as discharge, solid transport, flow depth or area (mass) of deposited debris. In general the *frequency* represents the period

[3]In a physical context "intensity" here is not synonymous with "strength" but defined as power per unit area.

of recurrence between two events with comparable magnitude. Frequency is often expressed as a return period, which is equal to the reciprocal of the exceedance probability of extreme precipitation or discharge values (ONR 24800:2008). Frequency and intensity are functionally correlated.

For torrential processes the frequency-intensity-function shows an "emergent" behaviour (Schrott and Glade 2008) implying a limited predictability of discharge from extrapolations of hydrological data when a certain threshold value is exceeded.

The *event disposition* of a torrent catchment, defined as the entirety of all conditions essential for the emergence of hazardous processes (Hübl et al. 2006; Bergmeister et al. 2009; ONR 24800:2008), consists of the *basic disposition* comprising all factors immutable over a long range of time (e.g. geology, soils) and the *variable disposition*, which is the sum of all factors subject to a short-term or seasonal change (e.g. precipitation, land use; Hübl 2009). If the variable disposition of a torrent catchment is altered in the course of an event (e.g. the water storage capacity of soils is exceeded), the debris potential will increase erratically resulting in a transition of the predominant displacement process (e.g. solid transport is altered to debris flood) and a non-linear increase of discharge.

Hübl (2009) describes the emergent behaviour of torrential flows by a *multi-stage system status* (Fig. 2) as follows:

- *status I* comprises fluvial processes (floods, bedload transport)
- *status II* includes debris flows and debris floods
- *status III* represents excessive (extreme) events

It is doubtful that there is a linear relationship between the exceedance probability and the discharge in system status II and III according to the transient flow behaviour of debris flows. Furthermore, additional transitions of event disposition in the course of extreme events are likely.

The *disposition status model* proposed by Hübl (2009) has to be understood as a valuable, illustrative explanatory model rather than an exact correlation, as reliable data on extreme torrential events are rarely available. This deficiency concerning the physical understanding of torrential processes emphasizes the great importance of well documented historic events (Hübl et al. 2006).

2.2 Design Criteria for Fluvial Process (Status I)

According to ONR (24800:2008) the design of protection works in torrents governed by fluvial processes is measured by a *design flood*. Generally, a design flood (discharge in m^3/s) with a return period of 100 years represents the desired level of safety for flood and torrent control measures in European countries.[4] *Expected*

[4] In Austria a return period of 150 years is applied for torrent control engineering to account for uncertainties of hydrologic models and the transient behavior or torrential processes.

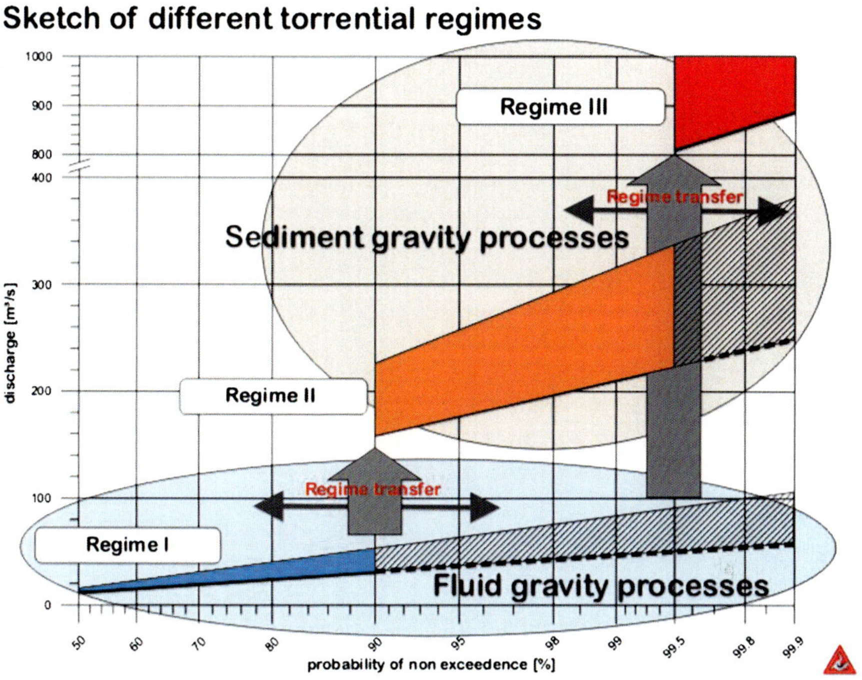

Fig. 2 System states (I–III) of a torrential catchment as a function of exceedance probability and intensity (discharge) showing the transition from fluvial processes to hyperconcentrated solid transport (debris flow, debris flood) like processes (Hübl 2009)

values for rainfall and flood events of a defined return period (including a corresponding confidence interval) can be derived from the hydraulic extreme value statistics (Rudolf-Miklau and Sereinig 2009).

Flood statistics are based on the assumption that the observation period is representative of the long-term runoff behaviour of the watershed. However extreme flood events are classified as "*statistical outliers*" *that* are not represented by the measured data collective (due to limited observation periods), but that nevertheless contribute valuable information on hydrological extremes (Merz 2007). Consequently the statistically deduced *design criterion* should be supported by additional information of temporal, spatial or causal reference (Blöschl and Merz 2008; Bergmeister et al. 2009). The *dating of historic flood events* from chronicles or traces in nature (flood marks, "silent witnesses") can provide precious additional information on return periods, levels of flooding and/or peak flood discharge. By dating historic events, extreme floods can be approximately related to a certain return period. A causal supplement of information is gained if observed floods are analyzed with respect to their emergence regarding the weather conditions, the behaviour of precipitation and the disposition of the catchment area (Merz 2007).

Flood statistics are normally only available for larger watercourses. The lack of flood statistics especially for ungauged torrent catchments is regularly substituted by *rainfall-runoff-models* (RRM). Nevertheless the application of RRMs requires to calibrate them individually for each torrent catchment by integrating natural conditions such as topography, geology, soils, vegetation or land use (Blöschl and Merz 2008). The temporal distribution of rainfall (centre of gravity of the distribution of precipitation at the beginning or end of the rainfall event) has a particularly decisive influence on the results of RRMs.

The *procedure of determination* of the design flood requires, as a first step, the specification of the expected value of discharge by means of flood statistics and additional hydrological methods. Based on this basic discharge, the design flood can be derived by considering solid transport, transient flow conditions and influences of stream morphology (Rudolf-Miklau and Sereinig 2009). The graphic presentation of discharge in the course of a flood event is given by a *flood hydrograph* (x-axis: time; y-axis: discharge). The flood volume can be calculated by integrating the hydrograph above the base flow.

Solid transport (bedload transport) can be included either by surcharges to the water discharge or by sediment transport models (e.g. Smart and Jäggi 1983; Rickenmann 1990; Palt 2001) that yield the *bedload transport rates* [kg/s*m or m^3/s*m]. The determination of the *discharge of solids* [m^3/s] is based on the assessment of *debris potential* [m^3] of a torrent catchment in the field. The standard method includes the investigation of each source of debris material regarding its disposition for erosion, the instantly displaceable mass of debris and the particle size distribution. The results are presented in a *profile of debris potentials* from which the *bedload discharge* can be deduced by considering the transport capacity of the relevant displacement process and the rate of erosion and deposition in the regarded cross section. The results can be displayed as a *debris balance* or a *bedload discharge profile* along the course of the torrent (Bergmeister et al. 2009; Moser 2010). Bedload discharge profiles allow the selection of the appropriate location for torrential barriers.

2.3 Design Criteria for Hyperconcentrated Solid Transport (Status II)

As *hyperconcentrated flows* (or debris flood) show a transient flow behaviour (Rickenmann 1995; Iverson 1997; Hübl 2009), a prediction of discharge based on hydraulic statistics is inaccurate. Consequently, flood events of status II can only be attributed to a certain return period with restrictions.

For the estimation of peak discharge of debris-flow regimes, empirical formulas (e.g. Takei 1980; Mizuyama et al. 1992; Rickenmann 1995) based on a simple correlation of discharge to bedload (solid transport) can be applied. According to Rickenmann (1999, 2001) the peak discharge Q [m^3/s] (as a function of the load of

Table 3 Process-related *event coefficient* (*EC*) for the rough estimation of a design flood based on water discharge according to Hübl (2009) and ONR 24802:2011

Displacement process	Event coefficient (EC)
Flood	1.00–1.05
(Fluvial) bedload transport	1.05–1.40
Debris flood	1.40–3.50
Debris flow	>3.50

the debris flow (D; m^3)) can be estimated for granular debris flows with Eq.1 and for muddy debris flows with Eq. 2:

$$Q = 0.135 \cdot D^{0.78} \quad (1)$$

$$Q = 0.0188 \cdot D^{0.79} \quad (2)$$

For the determination of *magnitude* (intensity) of debris flows, several physical (e.g. Flo2D: O'Brien 2003; Moser 2010) or empirical models (e.g. TopRunDF: Rickenmann and Scheidl 2010) developed in the past few years. They focus on predicting the run out distance or depositional area.

According to ONR (24802:2010) a rough estimation of the *design discharge* for debris floods can be obtained by multiplying the flood discharge with a dimensionless *event coefficient* (EC) relative for each process type (Table 3). The EC takes into account that in the course of hyperconcentrated displacement processes large masses of debris are transported within a short time. The characteristic and intensity of the debris displacement have a decisive influence on the shape of the flood hydrograph. Excessive displacement processes tend to generate hydrographs with extreme peak discharge but very short transit time, whereas debris flows are characterized by multiple consecutive, steep and short "hydrographs" comparable to roll waves (Hübl 2009).

There are two traditional simplified models to calculate *debris-flow impact forces* on barrier structures (Hübl 2006). The model according to Lichtenhahn (1973) is based on a triangular load distribution and a load increase factor (k_{deb}). The second approach is based on mechanical models and rheological properties of the displacement process. The impact on the structure in this case has a constant load distribution (rectangular load distribution). Alternative load models (distribution) are shown in (Hübl and Holzinger 2003; Hübl 2006; Suda et al. 2009). The relevant actions on buildings, their quantification and load models will be arranged in the ONR 24801:2012.

The peak impact load p_{deb} [kN/m²] can be defined as in (3). This formulation proposed by Hübl and Holzinger (2003) is based on laboratory experiments compared with field data. This empirical formulation considers the peak impact load (p_{deb}) [kN/m²], the velocity of the debris-flow front (v) [m/s] and the runoff height (h_{fl}) [m]. This model has a range of FROUDE-values from 1 to 15.ρ

$$p_{deb} = 4.5 \cdot \rho_{deb} \cdot v^{0.8} \cdot (g \cdot h_{fl})^{0.6} \cdot 10^{-3} \quad (3)$$

2.4 Design Criteria for Excessive Torrential Events (Status III)

Excessive torrential events (status III) generally exceed the design criteria for torrent control works. The impact of these extreme events can result in structural or functional failure of dams although only few cases of dam breakages are documented in the history of torrent control engineering. Excessive torrential events have to be considered for the determination of dam safety and the residual risk.

For large dams[5] European ICOLD[6] standards require that an extreme flood event of a return period of 5,000 (10,000) years can safely overflow the structure. In Austria this regulation applies for dams with a height exceeding 15 m or reservoirs with a storage capacity of more than 500,000 m³ (BMLFUW 2009). Torrent dams have to be dimensioned according to these standards if a longer lasting backwater can be expected. For "classical" torrential barriers, the appropriate design for extreme floods is achieved through constructive measures such as oversized or double-profiled discharge cross sections (ONR 24802:2010).

The hydraulic effects and consequences of a dam break can be estimated (calculated) with *flash flood models*. The failure of reservoirs upstream of steep torrential watercourses may trigger debris flows, which consequently have to be taken into account in a safety concept for flood retention basins.

3 Limit States and Impacts of Barriers

Impacts on torrent control works according to ÖNORM EN 1990 are categorized as *permanent*, *changeable* and *extraordinary* loads. According to ONR 24802:2010 extraordinary impacts are either caused by extreme events exceeding the design event or correlated to processes that are not covered by the systematical function of the structure (e.g. earth quakes). The proof of safety and function of barrier structures has to consider stability and usability, which primarily includes the retention (dosing) function, the hydraulic capacity and the ability of the structure to dissipate the energy of debris processes.

The *proof of the storage capacity* of the retention basin is fulfilled if the available deposition volume exceeds the bedload potential (including drift wood) of a DE plus a surcharge of 10%. Regular maintenance as well as the spilling or excavation of the retention basin has to be guaranteed (ONR 24802:2010).

The *proof of the hydraulic capacity* of a torrent control work based on the design discharge includes the following verifications:

- proof of the *capacity of the discharge section* (hydraulic capacity has to exceed the design flood including a freeboard)

---

[5] Height more than 15 m according to ICOLD definition.

[6] International Commission of Large Dams.

- proof of the *capacity of large openings* (slits)
- dimensioning of *spilling pool* (overflow)

There are no definite minimum standards for the freeboard at torrential barriers according to the excessive characteristic of extreme torrential processes; nevertheless, the relevant technical standards contain the recommendation to oversize discharge sections and openings (including rakes) in order to keep the barrier permeable as long as possible in the course of an event.

ONR 24802:2010 defines the *limit states* for the *proof of stability and usability* of the torrent control works and bearing structures. The failure types of torrential barriers are related to those of retaining structures. Both external and internal failure can occur in torrential barriers (Suda and Rudolf-Miklau 2010). The following *ultimate limit states* must be considered:

External stability:

- EQU: equilibrium limit state
- GEO: geotechnical limit state, failure or excessive deformation of the ground (sliding at the base, toppling, bearing resistance failure of the soil below the base, loss of overall stability)
- UPL: uplift limit state
- HYD: heave limit state

Internal stability:

- STR: structural failure, failure or excessive deformation of the structure or one of its elements (bending, shear and stability failure).

For assessments, torrential dams are usually treated as retaining structures. For ULS design the semi-probabilistic assessment concept is used. The models for assessment and the values of the partial safety factors for actions and resistances for GEO, UPL, EQU and HYD limit states are equal to ÖNORM B 1997-1-1.

The following additional impacts must also be considered in order to provide these proofs:

- own weight of structure
- soil pressure
- water pressure (design water level, water pressure in tension cracks)
- ground water pressure (water pressure at basis)
- dynamic water pressure of design event (flood, debris flow)
- dynamic water pressure exceeding the design event
- water pressure due to unplanned backwater
- traffic loads

Additional impacts such as slope failure, rock and snow avalanches, earth quakes or extreme floods may also have to be considered.

The *principal load model* for torrential barriers is shown in Fig. 3 (Suda and Rudolf-Miklau 2010). Details on the calculation of the specific loads and their distribution are given in Bergmeister et al. (2009), and detailed standards will be

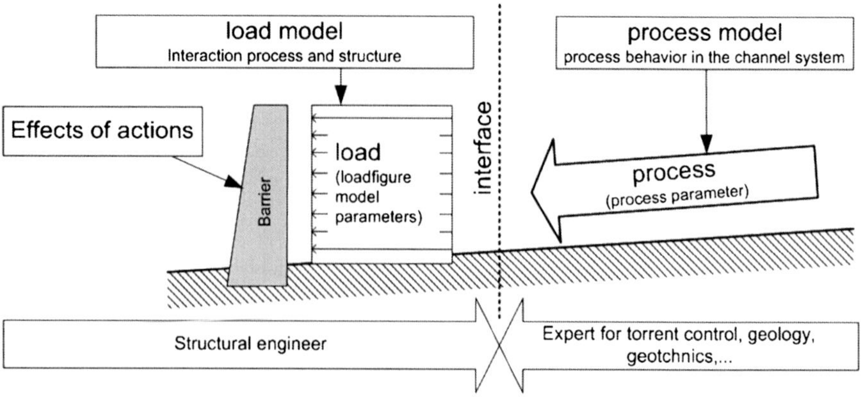

Fig. 3 Systematic depiction of the principal load model for torrential barriers (Suda and Rudolf-Miklau 2010)

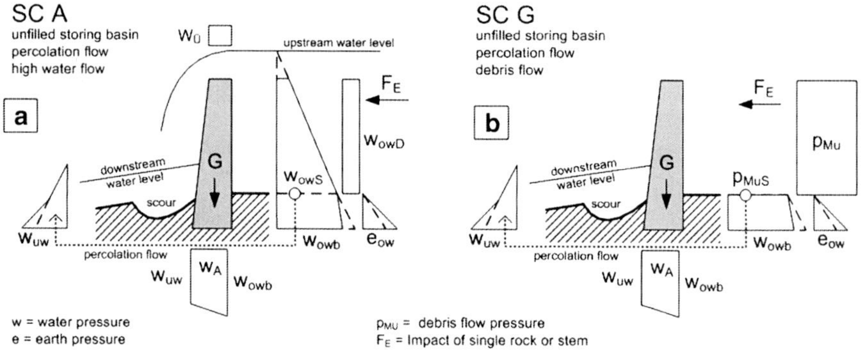

Fig. 4 Standardized stress combination (SC) for barrier structures at an unfilled storage basin for the load scenarios A (flood and percolating flow) and G (debris flow and percolating flow) according to ONR 24802:2011

included in ONR 24801:2011. According to ONR 24802:2010 the relevant impacts for the dimensioning of torrent control structures have to be combined according to the predominant displacement process (status I or II). These characteristic combinations of loads are specified as standardized stress combinations (SC; Fig. 4).

4 Conclusions

The design of torrent control works follows criteria that are based on the frequency and intensity of torrential events. The safety of these structures and the accuracy of dimensioning procedures are therefore highly dependent on the quality of process scenarios. Incorrect (inaccurate) assumptions concerning the scenarios may result in uneconomic (inefficient) design or in a partial or total failure of the structure or function of torrent control works, and can lead to negative consequences for the protected area.

For torrent processes the frequency-intensity-function shows emergent behaviour, which means that displacement processes can be erratically transformed (e.g. fluvial transport to debris flow) if the disposition of the catchment area changes in the course of a torrential event. For design procedures in torrent control engineering the knowledge of the predominant process regime (*status I*: floods and fluvial transport; *status II*: debris flood and debris flow; *status III*: extreme/excessive events) is of crucial importance for the selection of the appropriate hydraulic models, the accurate methods for the assessment of solid transport and/or the selection of the substantial stress combination.

As hydrological measurement data of flood and debris-flow events in torrent catchments are rarely available, statistically deduced *design criteria* should be supported by additional information of temporal, spatial or causal reference. The *dating of historic flood events* from chronicles or traces in nature on alluvial fans (flood marks, "silent witnesses", and geomorphic dating methods) can provide precious additional information on return period, predominant process, levels of flooding, peak flood discharge or debris volume. In the future, inclusion of such data will help in overcoming the central problem of torrent control engineering, namely the uncertainties in the determination of the return period (and related intensity) of design events (DE). Thus this article proofs the paramount importance of dating methods of historic torrential events for engineering purposes.

Acknowledgements The authors gratefully acknowledge the support of Emily Procter concerning her helpful "linguistic criticism" and Michelle Schneuwly-Bollschweiler, Barbara Kogelnig, Johannes Hübl, Markus Stoffel, and Christoph Skolaut for technical remarks.

References

Bergmeister K, Suda J, Hübl J, Rudolf-Miklau F (2009) Schutzbauwerke der Wildbachverbauung. Verlag Ernst und Sohn (Wiley VCH), Berlin

Blöschl G, Merz R (2008) Bestimmung von Bemessungshochwässern gegebener Jährlichkeit – Aspekte einer zeitgemäßen Strategie. Wasserwirtschaft 11/2008:12–18

BMLFUW (2009) Leitfaden zur Hochwassersicherheit von Stauanlagen. Bundesministerium für Land- und Forstwirtschaft, Umwelt und Wasserwirtschaft, Technische Universität Wien, Institut für Wasserbau und Ingenieurhydrologie

Hübl J (2006) Vorläufige Erkenntnisse aus 1:1 Murenversuchen: Prozessverständnis und Belastungsannahmen. In: FFIG, Reiser G (Hrsg) Geotechnik und Naturgefahren: Balanceakt zwischen Kostendruck und Notwendigkeit. Institut für Geotechnik, BOKU Wien, Geotechnik und Naturgefahren, Wien

Hübl J (2009) Hochwässer in Wildbacheinzugsgebieten. Wiener Mitteilungen Wasser Abwasser Gewässer 216:45–58

Hübl J, Holzinger G (2003) Entwicklung von Grundlagen zur Dimensionierung kronenoffener Bauwerke für die Geschiebebewirtschaftung in Wildbächen: Kleinmaßstäbliche Modellversuche zur Wirkung von Murbrechern, WLS Report 50 Band 3. Institut für Alpine Naturgefahren, Universität für Bodenkultur, Wien (unpublished)

Hübl J, Scheidl C, Schnetzer I, Kollarits S (2006) Standardized event documentation, a starting point for modeling and design of natural processes. Geophys Res Abstr 8 (SRef-ID: 1607-7962/gra/EGU06-A-09540)

Iverson RM (1997) The physics of debris flows. Rev Geophys 35: 245–296. American Geophysical Union

Jakob M, Hungr O (2005) Debris flow hazards and related phenomena. Springer, Berlin/Heidelberg/New York

Lichtenhahn C (1973): Die Berechnung von Sperren in Beton und Eisenbeton. In: Kolloquium über Wildbachsperren, Mitteilungen der forstlichen Bundesversuchsanstalt Wien, Heft 102:91–127

Marco F (2007) Torrential processes. In: Campus S, Forlati F, Barbero S, Bovo S (eds) Evaluation and prevention of natural risks. Taylor and Francis/Balkema, Leiden:169–199

Mazzorana B, Fuchs S, Hübl J (2009) Improving risk assessment by defining consistent and reliable system scenarios. Nat Hazards Earth Syst Sci 9:145–159

Merz R (2007) Hochwasserstatistik – Das Ausreisserproblem. Wiener Mitteilungen: Wasser, Abwasser, Gewässer. Institut für Wasserbau und Ingenieurhydrologie, Technische Universität Wien:181–194

Mizuyama T, Kobashi S, Ou G (1992) Prediction of debris flow peak discharge. In: Proceedings of the International Symposium Interpraevent, Bern, Switzerland, 4:99–108

Moser M (2010) Geschiebetransportberechnungen am Beispiel des Projektes Thomataler Bundschuhbach – Kombination von Geschiebefrachtdiagramm – Ingenieurmäßiger Ansatz (profilweise Berechnung) und 1d-numerische Simulation. Wildbach- und Lawinenverbau: Heft 165, 74. Jg., Salzburg

O'Brien JS (2003) FLO-2D users' manual version 2003.06, July 2003. FLO-2D Software Inc., Nutrioso, Arizona, 232 p

Önorm EN (1990) Eurocode – Grundlagen der Tragwerksplanung. Austrian Standards Institute, Vienna

ONR 24800 (2008) Schutzbauwerke der Wildbachverbauung – Begriffe und ihre Definition sowie Klassifizierung. Austrian Standards Institute, Vienna

ONR 24801 (2012) Schutzbauwerke der Wildbachverbauung – Statische und dynamische Einwirkungen. Austrian Standards Institute, Vienna (in process)

ONR 24802 (2011) Schutzbauwerke der Wildbachverbauung – Projektierung, Bemessung und konstruktive Durchbildung. Austrian Standards Institute, Vienna

ONR 24803 (2008) Schutzbauwerke der Wildbachverbauung – Betrieb, Überwachung und Instandhaltung. Austrian Standards Institute, Vienna

Palt SM (2001) Sedimenttransporte im Himalaya-Karakorum und ihre Bedeutung für Wasserkraftanlagen. Mitteilungen des Institutes für Wasserwirtschaft und Kulturtechnik der Universität Karlsruhe, vol 209

Rickenmann D (1990) Bedload transport capacity of slurry flows at steep slopes. Mitteilungen der Versuchsanstalt für Wasserbau, Hydrologie und Glaziologie der ETH Zürich, vol 103

Rickenmann D (1995) Beurteilung von Murgängen. Schweizer Ingenieur und Architekt 48: 1104–1108

Rickenmann D (1999) Empirical relationships for debris flows. Nat Hazard 19:47–77

Rickenmann D (2001) Estimation of debris-flow impact an flexible wire rope barriers. Swiss Federal Research Institute WSL, Bern

Rickenmann D, Scheidl C (2010) Modelle zur Abschätzung des Ablagerungsverhaltens von Murgängen. Wasser Energie Luft 102:17–26

Rudolf-Miklau F (2009) Naturgefahren-Management in Österreich. Verlag Lexis-Nexis Orac, Wien

Rudolf-Miklau F, Sereinig N (2009) Festlegung des Bemessungshochwassers: Prozessorientierte Harmonisierung für Flüsse und Wildbäche. ÖWAW 7–8:29–32

Schrott L, Glade T (2008) Frequenz und Magnitude natürlicher Prozesse. In: Flegentreff C, Glade T (eds) Naturrisiken und Sozialkatastrophen. Spektrum Akademischer Verlag/Springer, Berlin:134–150

Smart G, Jäggi M (1983) Sedimenttransport in steilen Gerinnen. Mitteilungen der Versuchs anstalt für Wasserbau, Hydrologie und Glaziologie der ETH Zürich, Nr. 103

Suda J, Rudolf-Miklau F (2010) Standards for design and maintenance of torrential barriers. In: Proceedings of the 8th ICOLD European Club Symposium, Innsbruck, Austria

Suda J, Strauss A, Rudolf-Miklau F, Hübl J (2009) Safety assessment of barrier structures. Struct Infrastruct Eng 5:311–324

Takahashi T (2007) Debris flow – mechanics, prediction and counter measures. Taylor & Francis/Baklema, London

Takei A (1980) Interdependence of sediment budget between individual torrents and a river system. In: Proceedings of the international symposium Interpraevent, Villach, Austria, 2:35–48

Wolman MG, Miller JP (1960) Magnitude and frequency of forces on geomorphic processes. J Geol 68(1):54–74

Forecasting, Early Warning and Event Management: Non-structural Protection Measures for Flash Floods and Debris Flows

Marco Borga

1 Introduction

In recent years, flood and flood-related disaster management has shifted from protection against flood to managing the risks of floods. In Europe, this shift is reflected in the Flood Risk Directive of October 2007 (2007/60/EC; FRD) (Quevauviller et al. 2009). The FRD requires EU Member States to undertake a preliminary assessment of flood risks and, for areas with a significant flood risk to prepare flood hazard and flood risk maps and flood risk management plans. In this framework, the chain of forecasting, warning and response systems is becoming an integral part of integrated risk management. Forecasting and warning are particularly interesting as risk management options for flash flood events and attendant geomorphic response like debris flows, which occupy a central position in this chapter.

Flash floods are generally triggered by convective precipitation. As such, these events are characterized by relatively small space and time scales. Rainfall duration is generally less than 24 h and size of the impacted catchments is less than 500 km^2 (Gaume et al. 2009; Marchi et al. 2010). Rough orography is often associated to flash floods, for two reasons. On one hand, orographic features often favour the anchoring of convective system, playing an important role in regulating of atmospheric moisture inflow to the storm and in controlling storm motion and evolution (Davolio et al. 2006). On the other hand, relief is also necessary for promoting flow concentration along drainage ways, which results in high unit discharges and relevant geomorphic effects in sloping watersheds (Marchi et al. 2010).

M. Borga (✉)
Department of Land and Agroforest Environments, University of Padova, AGRIPOLIS, via dell'Università, 16, Legnaro (PD) IT-35020, Italy
e-mail: marco.borga@unipd.it

M. Schneuwly-Bollschweiler et al. (eds.), *Dating Torrential Processes on Fans and Cones*, Advances in Global Change Research 47, DOI 10.1007/978-94-007-4336-6_27,

Debris flows are described as a flow of saturated non-plastic mineral and sometime organic debris in a steep channel that include some 50–70% solid grains by volume and can attain flow velocities in excess of 10 $m\,s^{-1}$ (Iverson 2009). Debris flows can have apparent viscosities that are five to six orders of magnitude greater than water, and fluid densities almost twice as great. As a consequence of the high fluid densities of debris flows, large rocks can be carried along and cause significant damage by impact (Wieczorek et al. 2000). Many debris flows in undisturbed landscapes begin as discrete landslides on steep (greater than about 15°) hillslopes as a result of a relatively rapid influx of large amounts of water (e.g. Iverson 1997). Large-scale sediment entrainment also has proven to be a process capable of generating debris flows. Material stored in channels can be mobilized by rainfall runoff or flash flood surges and act as the source sediment for debris flows.

Flash flood and debris flows share therefore a number of common hazard characteristics: (i) hazard sources are dispersed in space, (ii) event evolution is quick, with very limited time for warning. The above physical factors depending on exposure interact with the social, economical and environmental dimensions of the flash flood/debris-flows vulnerability. This is typically represented by dispersed urbanization, transportation axes, tourism structures, as well as urbanized areas downstream of small basins (particularly on alluvial fans and along the Mediterranean coast). Several important consequences arise in term of risk management strategies from the characterisation of hazard and vulnerability of flash floods and debris flows:

- Because the hazard sources are highly dispersed, the management of the flash- flood/debris-flows risk by means of structural measures aiming to reduce flood/sediment volumes and peaks is difficult (and often unfeasible in ecological or economic terms). Region-wide river training measures and land-use planning play an important role in flash flood/debris-flows risk management. Even in these cases, however, an often unquantified degree of residual risk will remain, which require 'acceptable' flood risk to be determined and mitigation solutions to be implemented.
- Flash flood/debris-flows forecasting, warning and emergency management (FWEM) are, by their nature, suitable to cope with the residual risk (NOAA-USGS 2005; Romang et al. 2008; Bründl et al. 2009). However, the local characteristics and the sudden nature of these events poses specific challenges to FWEM activities. The time available for communication is very limited and typically there is no time for learning as the event develops. Preparedness strategies play a central role in this context, adapting to the sudden nature of the event and to the large uncertainties affecting the forecast (Parker et al. 2007).
- Advancements in precipitation and flash flood/debris-flows forecasting are essential to improve emergency management. However, a focus on advances in forecasting alone will not be sufficient to reduce casualties and damages. A better understanding of the behaviour of people exposed to risk and of the organization of warning and rescue services is also essential. This understanding includes perceptions of risk and how such risk is communicated (Handmer 2001).

Against the backdrop of these developments it is the aim of this paper to review common elements of operational flash flood and debris-flows warning and emergency management in the context of available technologies, local scale requirements and local scale opportunities, providing indications on the prospects for improvement given likely future developments.

2 Forecasting and Warning Methodologies for Flash Floods and Debris Flows: The Value of the Local Information

Flash floods and debris-flow forecasting/warning are generally based on methods which aim to quickly assess localized flash flood/debris flow potential within a large area. These methods generally provide critical rainfall thresholds, often incorporating the influence of the initial soil moisture conditions. Threshold-based methodologies for flash floods and debris flows are summarized in the following sections.

2.1 Threshold-Based Methods for Flash Flood Warning

Threshold-based methods for flash flood warning are exemplified by the Flash Flood Guidance (FFG) system. FFG is the depth of rain of a given duration, taken as uniform in space and time on a certain basin, necessary to cause minor flooding at the outlet of the considered basin. This rainfall depth is computed by running in inverse mode a rainfall-runoff model, hence incorporating the soil moisture initial conditions into the assessment. The FFG is then compared to either real time-observed or forecasted rainfall of the same duration and on the same basin. If the nowcasted or forecasted rainfall depth is greater than the FFG, then flooding in the basin is considered likely. Georgakakos (2006) provided the theoretical basis of developing operational flash flood guidance systems by using analytical methods. Norbiato et al. (2008) provided an assessment of the method for regions located in Italy and France. Recently developed methodologies allow to specify the rainfall thresholds in terms of convenience of issuing an alert, through the minimization of expected costs, within the framework of a Bayesian approach (Martina et al. 2006). Figure 1 shows an example of rainfall thresholds i.e. accumulated volume of rain versus time of rainfall accumulation.

Direct comparison of quantitative precipitation forecasts with the rainfall thresholds (for specific rainfall durations) affords to issue flash flood warning, without the need to run the full chain of meteorological and hydrologic/hydraulic real-time forecasting models. These methods lead to an extremely simplified alert system to be used by non technical stakeholders and could also be used to supplement the traditional flood forecasting systems in case of system failures. These are important

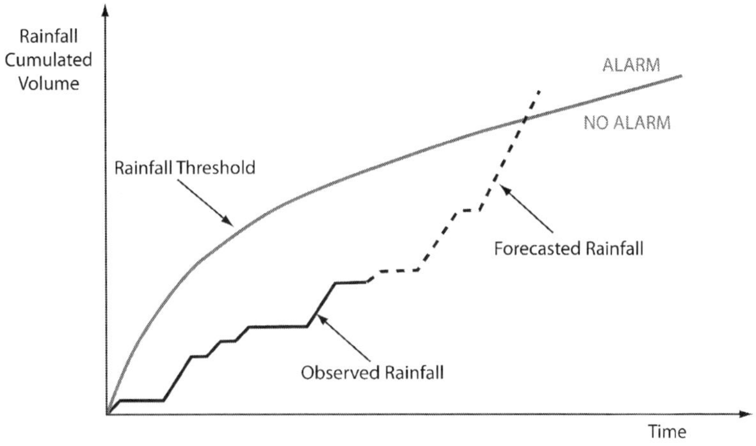

Fig. 1 Example of a flash flood rainfall threshold and its use

characteristics for use at local flood forecasting centres, and as such they increase the preparedness of communities exposed to flash flood risk. Furthermore, diagnostic methods based on the rainfall thresholds provides a useful concept which allows one to readily ingest local precipitation information and which promotes close collaboration between hydrologists and meteorologists by simplifying communication about the hydrological status of basins. Finally, they represent a potential benchmark for further development and inter-comparison purposes.

2.2 *Threshold-Based Methods for Debris-Flows Warning*

Rainfall thresholds, defined on empirical, statistical, or physical bases, identify the minimum or maximum condition of a particular triggering (or controlling) variable affecting slope failure (Guzzetti et al. 2008). In other words, any type of rainfall threshold separates rainfall events that trigger landslides from those which fail to trigger landslides (see http://rainfallthresholds.irpi.cnr.it/ for a comprehensive list of rainfall thresholds). Thresholds are especially useful for predicting landslide occurrence at the regional scale.

Thresholds commonly take the form of a power-law function, and because of variations in the rainfall conditions that trigger debris flows in different regions, such thresholds typically are region specific.

Most research in this area has followed Caine's (1980) approach by plotting rainfall duration against intensity for a large range of durations and plotting those events that have led to landslides or debris flows. Worldwide observations have identified the minimum and maximum rainfall over various periods of time critical for triggering debris flows. A minimum rainfall threshold defines the lowest amount of rainfall capable of initiating a landslide. A maximum threshold defines the amount of rainfall that has always triggered landslides historically. The probability of debris flow occurrence increases as rainfall progresses from the minimum threshold towards the maximum threshold. While this approach is a reasonable first step in identifying debris flow-producing storms, it shows a number of limitations as a warning tool, mostly because exceedance of the lower threshold envelope would result in too many warnings to be taken seriously by residents. This calls for more sophisticated approaches that integrate antecedent conditions or use statistical techniques to distinguish storm characteristics from storms that have produced landslides from those that have not (Chleborad et al. 2008; Cannon et al. 2008). If coupled with reliable rainfall forecasts, these methodologies can be embedded into real-time warning systems for the management of debris-flows risk (NOAA-USGS 2005).

An assumption inherent to most debris-flow forecasting and warning systems is unlimited sediment supply. However, the availability of sediments in different types of sediment storages (Schrott et al. 2002) has important consequences for any hazard analysis. Clearly, meteorological and climatic triggering factors and mechanics of debris flows are important considerations in many situations; however, if the sediment has been removed in the source areas, further hazard in the near future can be ranked as low. A different hazard ranking might be derived from physical or empirical models based on past events because they assume unlimited sediment availability.

2.3 The Value of the Local Information

An important aspect of threshold-based approaches is that these methods permits assimilation of local information, concerning both precipitation and streamflow or slope instability (Blöschl 2008). While in meteorological forecasts, computer based generation of uncertainty products is often considered essential in communicating uncertainty to the public (NRC 2006), for flooding/debris flows it is the assessment of the local situation that matters most. Parker and Handmer (1998) noted: "(...) those at risk may obtain much of their flood-related information from unofficial sources, such as personal networks and direct observation. Despite this, informal or unofficial systems are often afforded little official credence, even though empirical evidence indicates that formal flood warnings often fail. Exploration of the value of 'folk' or local, as opposed to specialist-technical, knowledge suggests that such knowledge satisfies a range of important needs which are likely to be unfulfilled by official warnings. The scope for personal networks to relay warnings and to contribute local knowledge towards system design appears to be large." The Swiss flood hazard information and warning system, known as IFKIS-Hydro (Romang et al. 2011), puts significant emphasis on local information, such as the water levels during events, and integrates it with more regional information on a web-based platform. The system involves also local observers, whose task is to provide an assessment of the local situation which feeds back into the system. Local observers have the added ability to report a wide variety of information (such as soil slips in the area) much of which cannot be captured by telemetered sensors.

3 Event Risk Management and Preparedness Methodologies

The consequences of flash floods/debris flows differ from plain floods because of the hazard characteristics. Short warning and reactions times especially expose people and require preparedness for immediate management response by organizations and people. Preparedness comprises actions taken to limit the impact of floods and debris flows. Preparedness structures responses to events well in advance by establishing mechanisms for rapid and orderly action which limits impacts (Faulkner and Ball 2007). Several may contribute to preparedness.

First, risk communication (i.e. public education, information and awareness) aims to place exposed and vulnerable populations in a position to respond safely, and possibly to reduce damage, when a flood/debris flows occurs.

Second, the public organizations are responsible for planning approval of construction (of building and infrastructure), promoting public awareness and knowledge, flood/debris flows forecasting and warning, emergency plan preparation and maintenance, emergency response including evacuation. They are closely attuned to the opportunities for reducing their own flash flood/debris flows impacts and those of the community which they serve. In particular it is important that opportunities are identified and taken to increase the resilience of communities, their buildings and their infrastructure to floods/debris flows.

Third, flood/debris flows forecasting and warning systems must be closely integrated with the identified needs and capabilities of members of the public and emergency service organizations, in order to maximize the opportunities for appropriate response and loss avoidance (Montz and Gruntfest 2002).

In order to reduce the impacts of flash floods/debris flows, preparedness measures need to be structured in accordance with the characteristics of the risks (e.g. compressed timescales; short to negligible warning lead times; immediate threat to life as well as property; requirement for refuges and safe places; high degree of public knowledge about appropriate responses; spontaneous and efficient organizational response requirement). Although valuable experience and knowledge has already been gained in flash flood locations, e.g. in Alpine locations (Bavarian State Ministry of the Environment 2006), about how best to set up preparedness, it is recognized that these events still present challenging lacks. Many flood preparedness methods and tools are designed for too loose timescales and less severe conditions (Drobot and Parker 2007). An additional challenge for professional agencies is to fully integrate 'risk' as formulated and predicted by the scientific community into the more uncertain setting of an often binary management decision (Drobot and Parker 2007).

An example of integration between preparedness-building strategies and decision making during extreme flood and debris-flows events is described by Romang et al. (2011) in the context of flood events occurred recently in Switzerland.

4 Conclusions

A review of non-structural risk management strategies for flash floods and debris flows has been provided in this contribution, with a focus on forecasting and warning methods and preparedness methodologies. Advancing methodologies for flash flood/debris flows forecasting and warning requires development of verification efforts methodologies. Verification is required to measure current performance and to establish a baseline to be used to quantify the effectiveness of future enhancements. Demonstrated improvements in flash flood/debris flows forecasting will lead to improved risk management. However, as flash floods/debris flows are locally rare events and are difficult to capture with existing conventional hydrometeorological networks, it is difficult to establish consistent verification of forecasting and warning accuracy with existing data archives. Efforts should be directed to (i) establish protocols for flash flood/debris flow observations, and (ii) develop archives of flash flood/debris flows across Europe. Post-event surveys have been shown to provide most of the required information. Furthermore, post-event survey may be extended to incorporate information on social response to warning and use of available information to mitigate risks. This will provide essential information to improve preparedness strategies, including the development of technologies which fill the current gap between forecasting and warning technologies and preparedness tools.

Acknowledgements The author gratefully acknowledges Günter Blöschl, Dominique Creutin, Giancarlo Dalla Fontana and Lorenzo Marchi for fruitful discussions and comments on a first draft of the paper.

References

Bavarian State Ministry of the Environment (2006) River basin agenda for Alpine space. Model and examples for sustainable development of the Alpine river basins. Bavarian State Ministry of the Environment, Public Health and Consumer Protection, Munich

Blöschl G (2008) Flood warning – on the value of local information. Int J River Basin Manag 6(1):41–50

Bründl M, Romang HE, Bischof N, Rheinberger CM (2009) The risk concept and its application in natural hazard risk management in Switzerland. Nat Hazard Earth Syst Sci 9(3):801–813

Caine N (1980) The rainfall intensity-duration control of shallow landslides and debris flows. Geogr Ann 62A:23–27

Cannon SH, Gartner JE, Wilson RC, Bowers JC, Laber JL (2008) Storm rainfall conditions for floods and debris flows for recently burned areas in southwestern Colorado and southern California. Geomorphology 96:250–269

Chleborad AF, Baum RL, Godt JL, Powers PS (2008) A prototype system for forecasting landslides in the Seattle, Washington area. Geol Soci Am Rev Eng 20:103–120
Davolio S, Buzzi A, Malguzzi P (2006) Orographic influence on deep convection: case study and sensitivity experiments. Meteorol Z 15:215–223
Drobot S, Parker DJ (2007) Advances and challenges in flash flood warnings. Environ Hazard 7:173–178
Faulkner H, Ball D (2007) Environmental hazards and risk communication: prologue. Environ Hazard 7:71–78
Gaume E, Bain V, Bernardara P, Newinger O, Barbuc M, Bateman A, Blaškovičová L, Blöschl G, Borga M, Dumitrescu A, Daliakopoulos I, Garcia J, Irimescu A, Kohnova S, Koutroulis A, Marchi L, Matreata S, Medina V, Preciso E, Sempere-Torres D, Stancalie G, Szolgay J, Tsanis I, Velasco D, Viglione A (2009) A collation of data on European flash floods. J Hydrol 367:70–78. doi:10.1016/j.jhydrol.2008.12.028
Georgakakos KP (2006) Analytical results for operational flash flood guidance. J Hydrol 317(1–2):81–103. doi:10.1016/j.jhydrol.2005.05.009
Guzzetti F, Peruccacci S, Rossi M, Stark CP (2008) The rainfall intensity–duration control of shallow landslides and debris flows: an update. Landslides 5(1):3–17
Handmer JW (2001) Improving flood warnings in Europe. Environ Hazard 3:19–28
Iverson RM (1997) The physics of debris flow. Rev Geophys 35:245–296
Iverson RM (2009) Elements of an improved model for debris flow motion. Invited contribution for Powders and Grains 2009. American Physical Society
Marchi L, Borga M, Preciso E, Gaume E (2010) Characterisation of selected extreme flash floods in Europe and implications for flood risk management. J Hydrol 394(1–2):118–133. doi:10.1016/j.jhydrol.2010.07.017
Martina MLV, Todini E, Libralon A (2006) A Bayesian decision approach to rainfall thresholds based flood warning. Hydrol Earth Syst Sci 10:413–426
Montz BE, Gruntfest E (2002) Flash flood mitigation: recommendations for research and applications. Environ Hazard 4:15–22
NOAA-USGS (2005) NOAA-USGS Debris flow warning system – final report. U.S. Geological Survey, Reston, Virginia. http://pubs.usgs.gov/circ/2005/1283/. Accessed on 21 Oct 2010
Norbiato D, Borga M, Degli Esposti S, Gaume E, Anquetin S (2008) Flash flood warning based on rainfall depth-duration thresholds and soil moisture conditions: an assessment for gauged and ungauged basins. J Hydrol 362(3–4):274–290. doi:10.1016/j.jhydrol.2008.08.023
NRC (2006) Completing the forecast: characterizing and communicating uncertainty for better decisions using weather and climate forecasts. The National Academies Press, Washington, DC, p 112
Parker DJ, Handmer JW (1998) The role of unofficial flood warning systems. J Conting Crisis Manag 6(1):45–60
Parker DJ, Tapsell S, McCarthy S (2007) Enhancing the human benefits of flood warning. Nat Hazard 43:397–414
Quevauviller P, Fragakis C, Balabanis P (2009) General features of the EU water policy and related scientific framework. In: Quevauviller P (ed) Water systems science and policy interfacing. RSC Publishing, Cambridge, pp 52–62
Romang HE, Wilhelm C, Fred V, Berod D, Hegg C, Rhyner J (2008) Emergency management tools for floods and debris flow events in smaller catchments. In: Interpraevent: Tagungsband Internationales Symposium Interpraevent, 2008, Dornbirn
Romang HE, Zappa M, Hilker N, Gerber M, Dufour F, Frede V, Berod D, Oplatka M, Rhyner J (2011) IFKIS-Hydro: an early warning and information system for floods and debris flows. Nat Hazard 56:509–527. doi:10.1007/s11069-010-9507-8
Schrott L, Niederheide A, Hankammer M, Hufschmidt G, Dikau R (2002) Sediment storage in a mountain catchment: geomorphic coupling and temporal variability (Reintal, Bavarian Alps, Germany). Z Geomorphol Suppl 127:175–196
Wieczorek GF, Morgan BA, Campbell RH (2000) Debris-flow hazards in the Blue Ridge of central Virginia. Environ Eng Geosci 6:3–23

Climate Research and Adaptation Strategies – Examples from the European Alps

Gernot R. Koboltschnig

1 Introduction

There is no doubt, the world's climate is changing. But, as soon as we are interested in the impacts at a special location and a certain time in future uncertainties grow tremendously. Concerning meteorological parameters affected by climate change IPCC already uses verbal phrases to describe uncertainties (IPCC 2007). The highest degree of certainty ("virtually certain"), and therefore the most reliable proof of climate change can be found for the temperature increase. The uncertainty for other climate components is much higher. Simple physically based assumptions seem to be able to describe further effects (e.g. warmer temperature – increased water vapour – increased precipitation – more floods). But the climate system is more complex. Concerning the impact on increasing Alpine natural hazards such as flood or debris-flow magnitudes and frequencies there is no clear and direct relation at the first glance. For the Danube River in Austria it was found out that there is an upward trend for floods. But, a deeper insight shows that "small floods" are responsible for this trend. It is more likely that – beside other reasons – river regulation measures for flood protection and hydropower use lead to a channelization of the river and a reduction of retention areas, which result in a reduced flow time (Blöschl and Merz 2008).

Nevertheless, it is more complicated to respond to the question, whether or not there is an increase of natural hazards on the regional level because commonly floods and debris flows in torrential catchments are triggered by convective rainfall events with high intensities while floods in rivers are triggered by larger scale precipitation. When comparing long-term trends of annual, seasonal or monthly

G.R. Koboltschnig (✉)
International Research Society INTERPRAEVENT, c/o Department for Water Management, Provincial Government of Carinthia, Flatschacher Strasse 70, 9020 Klagenfurt, Austria
e-mail: gernot.koboltschnig@ktn.gv.at

M. Schneuwly-Bollschweiler et al. (eds.), *Dating Torrential Processes on Fans and Cones*, Advances in Global Change Research 47, DOI 10.1007/978-94-007-4336-6_28,

precipitation data a correlation (at least for Austria) with short-term effects, such as heavy rainfall events or drought periods, can be seen (Böhm 2009). This correlation is not entirely clear in the moment but is the only possibility to use long term and reliable time series to get evidence for extreme events.

This chapter gives an overview of possible climate change impacts on natural hazard based on actual studies by Böhm (2009) with a focus on available long term observations, Blöschl and Merz (2008) and Blöschl et al. (2008) focusing on new ways for the estimation of design values for flood protection measures.

2 Climatic Aspects

When talking about climate change we normally think about the long term temperature increase, which globally was nearly 1°C within the past 100 years. Due to several circumstances we can assume that the global warming is not only a natural effect. Primarily emissions of CO_2, CH_4 and N_2O are responsible for the greenhouse effect (IPCC 2007). There are regional deviations from this global +1°C trend. Especially in the greater Alpine region the warming is about double (Auer et al. 2007). The most significant fact is that warmer air has the ability to absorb more water vapour than colder air (at sea level 5 g/m³ at 0°C, 10 g/m³ at 10°C and 30 g/m³ at 30°C). Therefore more precipitation can be expected. But at this point an important uncertainty has to be taken into account: Where will this additional water be transported to and where will it go down as precipitation?

Due to long term meteorological observations in the Alps we already know about huge regional differences of precipitation. A long-term, homogenized database of observed historical meteorological data can help to understand possible future changes. Within the HISTALP project (Auer et al. 2007), which was supported and elaborated by 13 states in and around the Alps, time series of the main meteorological variables (monthly data of temperature, pressure, precipitation, sunshine and cloudiness) were homogenized and analysed. Monthly precipitation time series go back to 1800. Due to this long term observations and the regional differences which came out it was possible to subdivide the greater Alpine region into four major sub-regions (see Fig. 1). Austria is part of all of the four sub-regions.

Relative changes of monthly precipitation for the four sub-regions can be seen in Fig. 2. For the long term trend detection the fat lines are of special interest. From 1860 to 1980 there was a reverse development of precipitation trends in the Alps. There was an increase in NW but a decrease in SE. All in all a decrease in precipitation can be seen in the two southern sub-regions. But, the bandwidth of extreme up- and downward amplitudes did not change so far in these 200 years of instrumental observations.

Studying the impact of a possible changing precipitation on the extremes of precipitation is of highest interest. A possibility to show that is to analyse the variability of precipitation. Table 1 displays the relative change of the variability of air temperature and precipitation. The two southern sub-regions have been

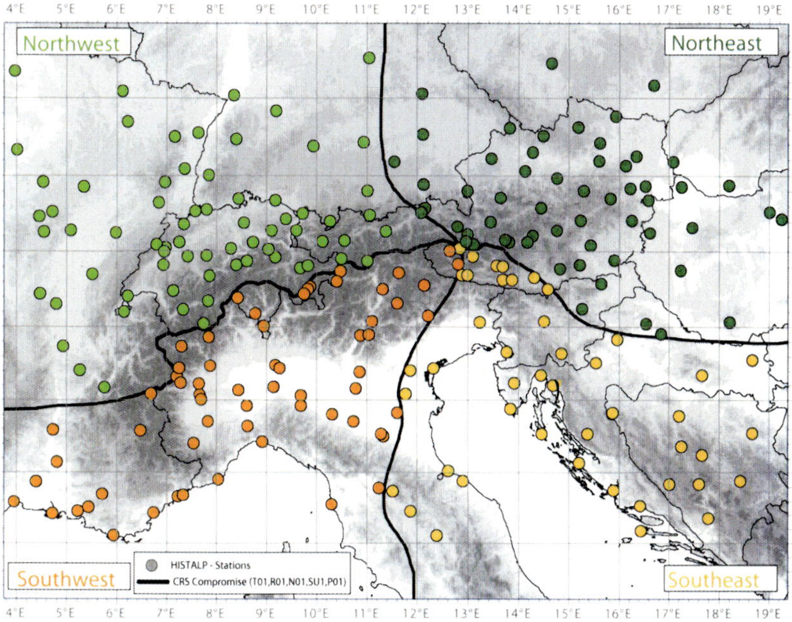

Fig. 1 The greater Alpine region (GAR) in south-central Europe with its five principal horizontal and vertical "coarse resolution subregions" (CRSs) and the network of HISTALP-stations (*dots* with different colours for regions) of long climate time series. The network consists of low and high elevation stations (figure after Böhm 2009 with the permission of the author)

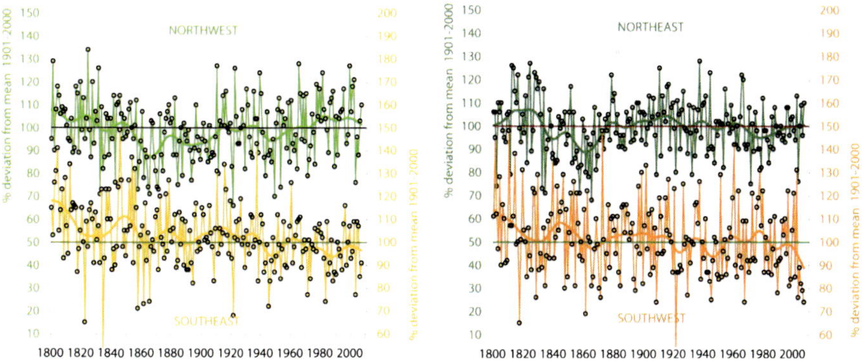

Fig. 2 Regional annual precipitation series 1800–2007 in the greater alpine region. *Left chart*: CRS-Northwest (*green*) and CRS-Southeast (*yellow*), *right chart*: CRS-Northeast (*green*) and CRS-Southwest (*orange*). Single years and 30-years smoothed (Gaussian low pass), anomalies in percent of twentieth century means (figure after Böhm 2009 with the permission of the author)

combined to one region to have enough long term time series available. Concerning air temperature it can be clearly seen that the climate in the twentieth century is of a more settled type during all seasons. Concerning precipitation it is not so clear but generally the variability is decreasing, especially in spring. Only in the NW, a

Table 1 Relative changes (in respect to the nineteenth century mean) of annual and seasonal climate variability (90% interquantile range of the detrended values) from nineteenth to twentieth century in three subregions of the GAR in respect to the two climate elements temperature and precipitation

	NE SPR.	NW SPR.	S SPR.	NE SUM.	NW SUM.	S SUM.	NE AUT.	NW AUT.	S AUT.	NE WIN.	NW WIN.	S WIN.	NE YEAR	NW YEAR	S YEAR	
TEMP.	−16.0	−14.9	−13.2	−20.0	−7.6	−2.3	−3.2	−9.7	+0.3	−7.4	−12.0	−6.2	−28.9	−23.8	−11.5	%
PREC.	−6.1	−3.0	−9.0	+1.7	+8.9	−2.4	−0.3	+14.1	−5.8	+8.8	+13.5	−2.6	−4.6	+5.3	−3.7	%

Table after Böhm (2009) used with the permission of the author region with increasing precipitation (see Fig. 2) the variability is increasing as well. Historical long term trends even for seasonal characteristics are comparable to what we expect in future of the next 100 years (Böhm 2009).

All these analyses have been done on historical long term series of monthly data. In the moment Austrian climatologists have started to analyse daily precipitation series. Böhm (2009) states that it already can be shown that there is a close correlation between the long term trend of yearly and monthly series and the trend of extreme events such as heavy rainfall or dry spells. This means that in the western part of Austria with increasing precipitation more heavy rainfall events are possible but together with less dry spells. For the south-eastern part of Austria the opposite is possible. Nevertheless we have to be careful because the extremes of the extremes, which can lead to large scale flooding are by definition very seldom events, and can statistically not be analysed using these "short" time series. Furthermore we know that events of 50 and 500 years return period are very close concerning their absolute values.

3 Hydrological Aspects

A typical method to simulate possible future floods and trends for runoff is to model runoff based on climatic projections. Based on the change of climatic variables precipitation and temperature, floods will appear more often or not. But, do we take into account that floods use to appear in a higher temporal density during flood decades? Known as the Hurst effect (Hurst 1951 cited after Blöschl and Montanari 2010) there are different periods with a lower and higher numbers of flood events in well and long term observed catchments. If we are interested in future changes of design values, one has to take into account uncertainties of current design values due to short observation periods. To exceed these uncertainties the magnitude of a possible climate change impact must be of an implausible value. Therefore Blöschl and Montanari (2010) make hydrologists aware to rely on their knowledge on processes instead of blindly trusting model results. Especially trends in observed runoff time series lead us to extrapolate these trends into future. Considering the Hurst effect and the possibility of flood decades the question is if it is feasible to do so?

One of the main problems for the planning of flood protection and torrent control measures is the uncertainty of the design flood definition. The usual system based on calculated discharge values, which are the result of statistical analyses for the given return period has the main problem that values will always change the more observations are available. Although a new initiative of reconsidering the methods for the statistical determination of design values was started in Austria (Sereing 2009; Mayer et al. 2009) the last innovation has been to additionally serve flood protection planners with an uncertainty bandwidth for design values. This is an appropriate and correct way for hydrologists but does not solve the problem for planning engineers. Very often the upper limit of the design value bandwidth is taken as "the truth". Merz and Blöschl (2008) describe a new approach and call

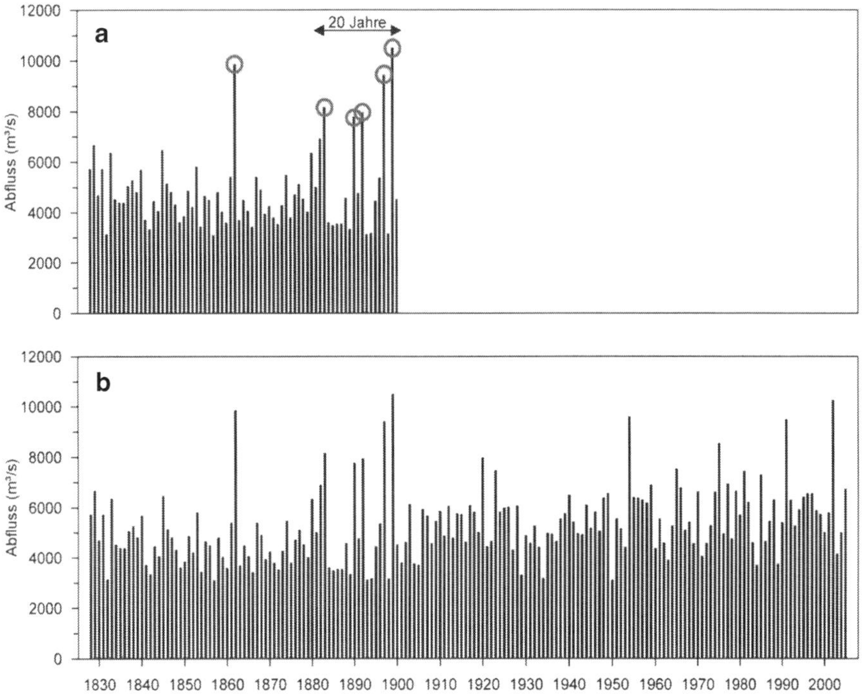

Fig. 3 (**a**) Annual floods of river Danube in Vienna for 73 years, five of the six highest floods appeared in the past 20 years. (**b**) Entire time series from 1828 to 2005 (figure after Blöschl and Merz 2008 with the permission of the authors)

it the "extreme value hydrology" instead of extreme value statistics. Hence, the information basis is extended and the expert knowledge can be implemented based on objective documents. Climate change in this consideration is another uncertainty. In most cases the climate signal is much smaller in comparison to uncertainties coming from extrapolations used in statistical analyses of events with higher return periods (e.g. 100 year or more). An example, which can occur using limited time series, is shown in Fig. 3. Figure 3a is a time series of 73 years, which is normally a very long period of hydrological observations and could be perfectly used to calculate statistical parameters. If a trend analysis is performed we could see that five out of the six highest annual floods appear in the end of the observation period. An extrapolation would predict even higher floods in future. The time series in Fig. 3a starts in 1828 and ends in 1900. Figure 3b shows the entire time series of observations from 1828 to 2005 of the gauging station at river Danube in Vienna. There is no more trend for higher floods detectable. But there is a higher frequency of small flood events observable. Even this trend is most probably not caused by climate change. It is more likely being a consequence of technical river regulations measures, which cut natural flood plains and decreased the overbank flow of smaller floods (Blöschl and Merz 2008; Böhm 2008). All in all the flow time of river Danube has dramatically been increased by technical measures.

A possibility to differentiate between trends coming from land cover changes and climate change was proposed by Blöschl et al. (2007). They found out that land cover changes affect only smaller areas and climate change affects smaller and larger areas. Hence, trends have to be compared for areas of different scale. While on the one hand changes in the flood regime induced by technical measures are mostly reproducible regime changes induced by climate change must not be reproducible. Therefore Blöschl and Merz (2008) recommend to explicitly identifying uncertainties instead of trying to make forecasts for these changes.

Land cover change can be influenced by climate change too. Glacierized areas will show different behaviour due the continuous retreat of glaciers. Actual studies in Austria have shown that glacier melt has rather a balancing influence during low flow conditions than on floods (Koboltschnig et al. 2008). It is more likely that the timing of snow accumulation and snowmelt will change in future.

As the uncertainties of climate change concerning floods are too big to make serious forecasts a good possibility to be prepared for possible future changes is to simulate scenarios, which lie in a manageable bandwidth. Within the "Floodrisk II" project (Habersack et al. 2009), which has been elaborated after the heavy 2002, 2005 and 2006 floods in Austria, Blöschl et al. (2008) did scenario analyses for two catchment areas in the north and the south of the Alps to see a possible upward shifting of floods due to climate change. They concluded that based on a plausible climate change input scenario floods could increase but uncertainties in the scenario definition as well as uncertainties in the modelling have to be taken into account. Based on the analysis of available flood discharge observations information on the significance for the determination of design values were derived and approximations for the uncertainties could be given. It has shown that there is always a possibility for flood decades.

4 Climate Change Impacts on Natural Hazards and Adaptation Measures

Due to the above mentioned climatic background we know that meteorological parameters (air temperature, precipitation) do not show the same trend for observed time series and we have seen a different degree of certainty concerning future model predictions. Regarding various types of natural hazards we should be aware that there is a long chain of dependencies and uncertainties from a changing climate to the impact on natural hazards. The central question is: how and where can climate change have an impact on natural hazards? For Southern Germany coupled climate and hydrological models have predicted an upward trend for precipitation and flooding (KLIWA 2006). As a consequence and adaptation measure design values for flood protection measures have now to be increased generally by 15% in Bavaria and similar in Baden-Württemberg (KLIWA 2006). Based on a large Austrian study (BMLFUW 2010) where no clear and direct impact on flooding due to climate change has been detected the decision was taken not to do any general

Fig. 4 Completes adaptation works at Hoher Sonnblick observatory (3,106 m a.s.l.) to reduce problems of permafrost degradation (picture by Schöner W.)

adaptation for design values in Austria. In the beginning of the climate change discussion increasing trends of damages caused by natural hazards, e.g. shown by MunichRe (2010), were correlated with increasing air temperature. Nowadays we know that this was primarily caused by an increasing damage potential. Furthermore in the USA an increase of Hurricane activities and damages seemed to occur especially in the past decade. By applying two normalization methods for damage records from 1900 to 2005, which on the one hand provide an estimate of the damage that would occur if storms from the past made landfall under another year's societal conditions and on the other hand use changes in inflation and wealth at the national level and changes in population and housing units at the coastal county level, Pielke et al. (2008) showed that there is no remaining absolute trend in absolute Hurricane damage. A similar result was shown investigating the European storminess (Matulla et al. 2008). There is no increasing trend for the storminess. Concerning water related natural hazards such detailed investigations need to be performed in future.

Hence, there is a need to study the trend and the degree of certainty of each climatic parameter, which can have an influence or the ability to change the behaviour of natural hazards separately, before implying a general trend of natural hazards to be increased in future. We also need to know if an impact is significant compared to the uncertainties of past observations and model simulations.

A sub project of the AdaptAlp project (Adaptation to Climate Change in the Alpine Space, www.adaptalp.org) had the goal to find adaptation strategies against natural hazards in the European Alps, which already had to be applied in the past due to the impact of climate change (Schöner and Binder 2010). The main requirement for these case studies to be analysed was that there must have been an observed climate change impact. As the degree of certainty is the highest for the air temperature change and therefore the hardest fact, Schöner and Binder (2010) searched for examples where the air temperature increase was the cause for an increasing natural hazards endangerment. All examples of implemented adaptations to climate change were found in high alpine areas. The description goes from naturally dammed glacier lakes, which had to be drained to reduce the risk of spontaneous outburst floods (e.g. thermo karstic lakes in Wallis/Switzerland or the glacier Rochemelon in France) to problems of permafrost degradation where the underground had to be stabilized using concrete injections and retaining walls (e.g. Hoher Sonnblick observatory in Austria, see green ovals in Fig. 4). Several examples can be found in alpine terrain beneath permafrost degradation zones, where hiking tracks had to be displaced to reduce the endangerment of rock fall (Behm et al. 2006).

5 Conclusions

- It is most likely that the knowledge on the impact of climate change is quite consolidated for the parameter air temperature but not that good for precipitation.
- All meteorological parameters can show different regional behaviour. Due to long term historical time series analyses for the Alps future regional differences can be expected.
- It seems to be possible that historical short-term extreme values of precipitation (e.g. daily values) are represented in monthly precipitation observations. This method, which was applied for historical data could be used for simulations of future scenarios. An extrapolation on the "extremes of the extremes" will not be possible.
- To be able to make hydrological conclusions based on climatic simulations all uncertainties, which appear in the methodological approach, have to be taken into account. Therefore the extrapolation of trends for future floods has to be handled with care.
- Uncertainties for the determination of hydrological design values, starting with instrumental, mathematical and statistical uncertainties, have to be explicitly identified instead of trying to make forecasts for climate change impacts on hydrological design values.
- Uncertainties, which grow during the process of design value determination, seem to exceed the expected increase due to the impact of climate change.
- The adaptation to climate change has already taken place, as hydrological design values have always to be reconsidered as soon as the knowledge on observed flood data grows.

- As the most significant climate change in the European Alps in the past and in future regards an increase of air temperature, impacts on and an increase of natural hazards are most likely to be found for temperature sensitive processes.
- To increase the preparedness for the unexpected it is useful to work with "what if" scenarios covering the range given from climate models. Results should be communicated as scenarios and not predictions.
- Especially under "climate change-research-conditions" the knowledge on dominant processes being able to trigger natural hazards – including realistic (hard fact) climate change impacts on them – is more and more necessary and needed instead of blindly relying on any model outputs.

Acknowledgments The author gratefully acknowledges Reinhard Böhm (Central Agency for Meteorology and Geodynamics) and Günther Blöschl (Vienna Technical University) to give the permission for the use of figures and tables from their original papers

References

Auer I, Böhm R, Jurkovic A, Lipa W, Orlik A, Potzmann R, Schöner W, Ungersböck M, Matulla C, Briffa K, Jones PD, Efthymiadis D, Brunetti M, Nanni T, Maugeri M, Mercalli L, Mestre O, Moisselin J-M, Begert M, Müller-Westermeier G, Kveton V, Bochnicek O, Stastny P, Lapin M, Szalai S, Szentimrey T, Cegnar T, Dolinar M, Gajic-Capka M, Zaninovic K, Majstorovic Z, Nieplova E (2007) HISTALP – Historical instrumental climatological surface time series of the greater Alpine region 1760–2003. Int J Climatol 27:17–46

Behm M, Raffeiner G, Schöner W (2006) Auswirkungen der Klima- und Geltscheränderung auf den Alpinismus. Umweltdachverband, Wien, 96p

Blöschl G, Merz R (2008) Bestimmung von Bemessungshochwässern gegebener Jährlichkeit – Aspekte einer zeitgemäßen Strategie (Estimating design floods of a given return period – facets of a contemporary strategy. Wasserwirtsch 98(11):2–11

Blöschl G, Montanari A (2010) Climate change impacts—throwing the dice? Hydrol Processes 24:374–381

Blöschl G, Ardoin-Bardin S, Bonell M, Dorninger M, Goodrich D, Gutknecht D, Matamoros D, Merz B, Shand P, Szolgay J (2007) At what scales do climate variability and land cover change impact on flooding and low flows? Hydrol Processes 21:1241–1247

Blöschl G, Viglione A, Heindl H (2008) Vertiefung und Vernetzug zukunftsweisender Umsetzungsstrategien zum integrierten Hochwassermanagement. Floodrisk II, Workpackage Meteorologie/Hydrologie, TP6.2 – Dynamik von Hochwasserbemessungsgrößen und Konsequenzen – Klimawandel. Bundesministerium für Land- und Forstwirtschaft, Umwelt und Wasserwirtschaft, Austria

BMLFUW (2010) Anpassungsstrategien an den Klimawandel für Österreichs Wasserwirtschaft - Kurzfassung. Edited by Schöner W, Böhm R, Haslinger K (Central Institute of Meteorology and Geodynamics – ZAMG) and Blöschl G, Merz R, Blaschke AP, Viglione A, Parajka J, Kroiß H, Kreuzinger N (Vienna Technical University). Austrian Federal Ministry for Agriculture, Forestry, Environment and Water Management

Böhm R (2008) Heiße Luft, Reizwort Klimawandel – Fakten-Ängste-Geschäfte. VABENE, Wien-Klosterneubeurg, 261 pp

Böhm R (2009) Geändertes Umfeld durch Klimawandel? (Modified environment due to climate change?). Wildbach- und Lawinenverbau 163:35–50

Habersack H, Bürgel J, Kanonier A (2009) FloodRisk II, Vertiefung und Vernetzung zukunftsweisender Umsetzungsstrategien zum integrierten Hochwassermanagement, Synthesebericht. Austrian Federal Ministry for Agriculture, Forestry, Environment and Water Management. Available online: http://www.forstnet.at/filemanager/download/47799/. Accessed 18 June 2010

Hurst HE (1951) Long-term storage capacity of reservoirs. Trans Am Soc Civil Eng 116:770–808

IPCC (2007) Climate change 2007 – the physical science basis. Working group I contribution to the 4th assessment report of the intergovernmental panel on climate change. Cambridge University Press, 996 p

KLIWA (2006) Our climate is changing. Consequences–Extend–Strategies. Brochure published by the State Environment, Measurement and Nature Conservation Agency, Baden-Württemberg and the Bavarian State Environment Agency (LfU). Available online: http://www.kliwa.de/download/KLIWA_en.pdf. Accessed 18 June 2010

Koboltschnig GR, Schoener W, Zappa M, Kroisleitner C, Holzmann H (2008) Runoff modelling of the glacierized alpine Upper Salzach basin (Austria): multi-criteria result validation. Hydrol Processes 22:3950–3964

Matulla C, Schöner W, Alexandersson H, von Storch H, Wang XL (2008) European storminess: late nineteenth century to present. Clim Dyn 31:125–130. doi:10.1007/s00382-007-0333-y

Mayer B, Rudolf-Miklau F, Korck J (2009) Tagungsbericht: Expertenworkshop Bemessungsereignis und Klimawandel (Expert hearing design events and impact of climate change). Hydrol Wasserbewirtsch 53(6):397–398

Merz R, Blöschl G (2008) Informationserweiterung zur Bestimmung von Hochwasserwahrscheinlichkeiten (Extended information basis for the determination of flood probabilities). Hydrol Wasserbewirtsch 52(6):300–309

Pielke RA, Gratz J, Landsea CW, Collins D, Saunders MA, Musulin R (2008) Normalized hurricane damage in the United States: 1900–2005. Nat Hazards Rev 9(1):29–42. doi:10.1061/(ASCE)1527-6988(2008)9:1(29)

Schöner W, Binder D (2010) Examples for adaptation measures of the risk management for natural hazards to Climate Change in the Alps. Interim report, ZAMG – Central Agency for Meteorology and Geodynamics, Vienna, 28p

Sereing N (2009) Definition of design events in different process types. Report of the expert hearing in Vienna April 24th Vienna within the WP4 of the AdaptAlp project. Available online: http://www.adaptalp.org ->Project ->WP4-Water Regime -.WP4 Documents

Dating Methods Overview: Checklist for Practitioners

Michelle Schneuwly-Bollschweiler, Markus Stoffel, and Florian Rudolf-Miklau

Dating methods and their application are of interest to scientists who want to gain deeper knowledge about various aspects of past hydrogeomorphic activity such as changes in the frequency of events, aggradation and erosion phases or the impacts of global change on process activity. Similarly, past events are interesting for rheologic studies, forces involved or model improvements. These aspects may be highly interesting for researchers; the results of such studies remain, however, often restricted to the scientific community.

Practitioners on the other hand, need data on past events for hazard and risk assessments or for planning of mitigation measures. Therefore, reliable knowledge about the frequency, magnitude, breakout locations and reach of damaging events in the past is essential. Accurate and precise information about statistical and physical properties for processes causing extreme and damaging events are needed but are frequently missing due to their episodic occurrence. Fragmentary information results in large uncertainties which lie behind each of the relevant parameters.

The dating methods described earlier in this book (Schneuwly-Bollschweiler et al. 2012) are appropriate to provide at least part of the data required by practitioners. To give an overview on the type of data that can be obtained through the different methods and to display and compare the advantages and limitations

M. Schneuwly-Bollschweiler (✉) • M. Stoffel
Laboratory of Dendrogeomorphology (dendrolab.ch), Institute of Geological Sciences, University of Berne, CH-3012 Berne, Switzerland

Chair for Climate and Climatic Impacts, Environmental Sciences, University of Geneva, CH-1227 Carouge-Geneva, Switzerland
e-mail: michelle.schneuwly@dendrolab.ch; markus.stoffel@dendrolab.ch

F. Rudolf-Miklau
Department IV/5 – Torrent and Avalanche Control, Federal Ministry for Agriculture, Forestry, Environment and Water Management, A-1030 Vienna, Austria
e-mail: florian.rudolf-miklau@lebensministerium.at

M. Schneuwly-Bollschweiler et al. (eds.), *Dating Torrential Processes on Fans and Cones*, Advances in Global Change Research 47, DOI 10.1007/978-94-007-4336-6_29,

of each method in a structured way, an overview ("checklist for practitioners") is presented in the following.

The checklist is composed of ten major criteria explained in the following:

1. *Time period*: for which time scale can the respective method provide useful data?
2. *Dating of events*:
   (a) Chronometric: chronometric dating methods allow determination of an approximate computed age. Earlier, the term "absolute" dating was used, however, this term is not used anymore as it implies an unwarranted certainty and precision.
   (b) Stratigraphic: stratigraphic (old: "relative") dating methods determine the relative order of past events, without necessarily determining their absolute age.
3. *Temporal resolution*: this category describes how much temporal detail the method is able to provide. Can we get information on the day of the event or just on the century or even the millennium?
4. *Data type*: what information on torrential processes can be obtained by the dating method? Four different answers were possible: yes (=almost always), possible (=the method may provide data under certain circumstances), rarely (=it is not normally possible to obtain data), no (=never).
   (a) Frequency: how often did hydrogeomorphic events occur in the past on a cone or fan?
   (b) Deposition thickness: is the method able to determine the importance of deposited material?
   (c) Process energy: does the method allow the reconstruction of energies involved in individual events?
   (d) Spread/reach: does the method allow the spatial reconstruction on where and how far the material was flowing? Can breakout locations be reconstructed?
   (e) Process type: does the method allow the precise determination of the type of process (e.g. debris flow, debris flood, sheet flow, . . .) of past events?
5. *Advantages/major benefits*: what is the most important advantage or benefit of a method as compared to other dating methods?
6. *Limitations/prerequisites*: which conditions have to be met for a method to provide valuable data for practitioners? What is the main limitation?
7. *Practical use*:
   (a) Hazard zoning: does the method provide data that is of use for hazard zoning and hazard maps?
   (b) Planning of mitigation measures: does the method provide data that is of use for the planning of mitigation measures?
   (c) Disaster management: does the method provide data that is of use for the catastrophe management?
8. *Example of application*: this category provides a reference to a study where a method has been applied. The example should stem from an "average" fan/cone and be reflective of a "standard" case. The size, name and reference for the fan/cone are provided.

9. *costs*: costs refer to mean costs necessary for an investigation of a "standard" fan/cone. Costs are provided in three categories (<10,000 €, 10,000–30,000 € or >30,000 €) as costs will vary in different countries. The figures should be seen as relative information for the comparison of different approaches rather than as absolute values of real costs.
10. *Duration*: how long (in person months) does it take to obtain data for the relevant "standard" fan or cone including preparation, fieldwork, analyses, interpretation and reporting?

As can be seen in the Table 1 and in Fig. 1, the dating methods can be divided into two main groups as far as the temporal scale is concerned. On one hand, there are methods providing data on events of the recent past and up to some centuries (silent witnesses, historical sources, digital images, dendrogeomorphology, vegetation analysis, lichenometry, lake sediments/varves). Methods of the second group (cosmogenic nuclides, radiocarbon dating, luminescence dating) are able to provide data on much longer timescales (up to 1 million years), in contrast, they fail to shed light on the recent activity in torrents and on fans and cones.

Data on very recent events is most often easily available through reports or real-time observation. For the past few years to sometimes decades, archival records, silent witnesses or digital images may provide detailed information on events. The assessment of such data is normally rather cost-effective and quick to perform. Therefore, these methods normally represent the first step in the assessment of past events.

To have an idea of recurrence intervals for more than just a few years and for events that have occurred during the past decades to centuries, additional information needs to be reconstructed with other approaches. Here, dendrogeomorphology, vegetation analyses, lichenometry or lake sediments/varves represent the appropriate tools to gain more in-depth data on past activity. These methods are more time consuming as compared to historical sources and are therefore more expensive, especially in terms of personnel costs. They are, however, also able to provide additional data as compared to historical sources, for instance.

Cosmogenic nuclide, radiocarbon dating or luminescence dating are helpful to illustrate long-term changes in the activity on the cone or fan or on the alternation of deposition and erosion phases in time and space. Similarly, long-term climate change and its effect on activity are better assessed if the method provides data over centuries to millions of years. These methods, however, require high time investment and high material costs. They are therefore normally less often applied for the assessment of hazards and risk on fans and cones and do not currently represent standard engineering methods of event documentation.

In conclusion, we recommend a combination of several complementary methods in a three-step approach for the identification of past hydrogeomorphic events on fans and cones:

1. Historical sources: all data available from archives should be considered.
2. Silent witness and digital images complement the data from archives

Table 1 Checklist for practitioners

Method		Silent witnesses	Historical sources	Digital images	Dendrogeomorphology	Vegetation analyses	Lichenometry	Lake sediments/ varves	Cosmogenic nuclides	Radiocarbon dating	Luminescence dating
Time period covered		$10–10^3$ years	$1–1,5\times10^2$ years	10–50 years	$1–10^3$ years	$10^1–10^3$ years	$10–5\times10^2$ years	$1–10^4$ years	$10^2–10^6$ years	$10^2–4\times10^4$ years	$10^2–10^5$ years
Dating of events	Chronometric		x		x			x	x	x	x
	Stratigraphic	x		x		x	x	x			
Temporal resolution		Yearly – millennial	Daily – yearly	Single point in time unless multiple surveys are conducted	Monthly – yearly	Decadal	Sub decadal	Seasonal-yearly	Decadal - multi millennial	Decadal – centennial	Decadal – millennial
Data type	Frequency	Possible	Possible	Possible	Yes	Rarely	Possible	Yes	Possible	Possible	Possible
	Deposition thickness	Possible	Rarely	Possible	Possible	Rarely	Rarely	Rarely	No	Rarely	Rarely
	Process energy	Rarely	Rarely	Possible	No	Rarely	Rarely	Possible	No	Rarely	Rarely
	Spread/reach	Possible	Possible	Yes	Yes	Possible	Possible	Possible	Possible	Possible	Possible
	Process type	Yes	Possible	Yes	No	Possible	No	No	No	No	No
Advantages/ major benefits		Various kinds of data can be analyzed, spatial pattern of process types	Often detailed data for localized and/or regional extreme events, overall evaluation of damages and affected areas	Broad spatial extents, high resolution, ability to capture processes in inaccessible terrain, possible to conduct multiple surveys to record fan/cone evolution over time.	High temporal and spatial resolution	Information on channel evolution and migration	Covers high-elevation, unvegetated areas	High temporal resolution and continuous reconstruction of events possible	Rock surfaces directly related to the event are dated. Very long time period covered	Long time period covered	Long time period covered

(continued)

Table 1 (continued)

Method		Silent witnesses	Historical sources	Digital images	Dendrogeomorphology	Vegetation analyses	Lichenometry	Lake sediments/ varves	Cosmogenic nuclides	Radiocarbon dating	Luminescence dating
Limitations/prerequisites		Traces of past events should not be reworked	Assessment of the land conditions before the records, data are often fragmentary, sometimes contradictory information	Availability of existing data; Data quality dependent upon sensor-type, presence of vegetation and post-processing methods.	Availability of old-growth trees with impacted by process in question, no other geomorphic impacts	Vegetation has to be undisturbed from other influences	Dated deposits are necessary for the calibration	Outlet of the torrent under investigation into a lake	Availability of suitable rock surfaces	Availability of suitable organic material	Availability of quartz or feldspar. Pre-depositional bleaching
Practical use	Hazard zoning	Yes	Possible	Yes	Yes	Yes	Yes	No	No	Possible	No
	Planning of mitigation measures	Possible	Possible	Yes	Yes	Possible	Possible	No	No	Possible	No
	Disaster management	Possible	Possible	Yes	Possible	No	No	No	No	No	No
Example of application	Cone size		284 ha	1–1,000 Ha	20 ha	10 ha			1,000 ha	10–1,000 ha	–
	Cone name		Chieppena torrent (IT)	Stevens Gulch (Colorado USA), Death Valley (California USA)	Birchbach (Valais Alps, Switzerland)	Malga Caldea (Valle dell'Avio, Italy)	La Selle catchment (Massif des Ecrins, France)	Catchment of Lake Thun (Switzerland)	Owens Valley (California, USA)	Several fans in NW UK, Cheekye fan (British Columbia, Canada)	Pisco Valley (Peru)

(continued)

Table 1 (continued)

Method			Silent witnesses	Historical sources	Digital images	Dendrogeom-orphology	Vegetation analyses	Lichenometry	Lake sediments/ varves	Cosmogenic nuclides	Radiocarbon dating	Luminescence dating
	Reference		Standard procedure for hazard assessment in Austria	Marchi and Cavalli (2007)	Staley and Wasklewicz (2012), this volume; Staley et al. (2006)	Bollschweiler and Stoffel (2010)	Baroni et al. (2007)	Jomelli (2012), this volume	Wirth et al. (2011)	Ivy-Ochs et al. (2012), this volume	Chiverell and Jakob (2012), this volume	Schlunegger et al. (2012), this volume
Costs	Personnel costs	<10,000 €	×	×	×							
		10,000–30,000 €				×	×	×				
		> 30,000 €			×				×	×	×	×
	Material/ other costs	<10,000 €	×	×		×	×	×				
		10,000–30,000 €			×				×	×		×
		>30,000 €			×						x	
Duration (man months)			2	2	6	4	4	4	6	6	6	5

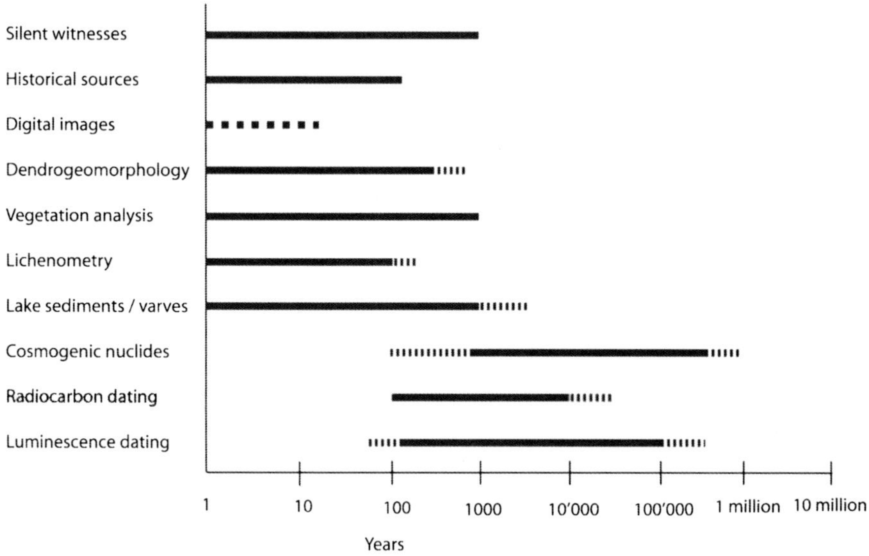

Fig. 1 Timescales covered by the different dating methods described in this book

3. Reconstruction of past activity using methods such as dendrogeomorphology, vegetation analyses, lichenometry or lake sediments/varves for the definition of recurrence intervals, spatial spread and changes in runoff.

References

Baroni C, Armiraglio S, Gentili R, Carton A (2007) Landform-vegetation units for investigating the dynamics and geomorphologic evolution of alpine composite debris cones (Valle dell'Avio, Adamello Group, Italy). Geomorphology 84:59–79

Bollschweiler M, Stoffel M (2010) Variations in debris-flow occurrence in an Alpine catchment – reconstruction and implications for the future. Global Planet Change 73(3–4):186–192

Chiverell R, Jakob M (2012) Radiocarbon dating: alluvial fan/debris cone evolution and hazards. In: Schneuwly-Bollschweiler M, Stoffel M, Rudolf-Miklau M (eds) Dating torrential processes on fans and cones – methods and their application for hazard and risk assessment, Advances in global change research. Springer, Dordrecht/Heidelberg/London/New York

Ivy-Ochs S, Dühnforth M, Densmore AL, Aflimov V (2012) Dating fan deposits with cosmogenic nuclides. In: Schneuwly-Bollschweiler M, Stoffel M, Rudolf-Miklau M (eds) Dating torrential processes on fans and cones – methods and their application for hazard and risk assessment, Advances in global change research. Springer, Dordrecht/Heidelberg/London/New York

Jomelli V (2012) Lichenometric dating of debris avalanche deposits with an example from the French Alps. In: Schneuwly-Bollschweiler M, Stoffel M, Rudolf-Miklau M (eds) Dating torrential processes on fans and cones – methods and their application for hazard and risk assessment, Advances in global change research. Springer, Dordrecht/Heidelberg/London/New York

Marchi L, Cavalli M (2007) Procedures for the documentation of historical debris flows: application to the Chieppena Torrent (Italian alps). Environ Manage 40(3):493–503

Schlunegger F, Norton K, Steffen D (2012) Timing the interface between mass wasting and fluvial processes with OSL. In: Schneuwly-Bollschweiler M, Stoffel M, Rudolf-Miklau M (eds) Dating torrential processes on fans and cones – methods and their application for hazard and risk assessment, Advances in global change research. Springer, Dordrecht/Heidelberg/London/New York

Schneuwly-Bollschweiler M, Stoffel M, Rudolf-Miklau M (2012) Dating torrential processes on fans and cones – methods and their application for hazard and risk assessment, Advances in global change research. Springer, Dordrecht/Heidelberg/London/New York

Staley DM, Wasklewicz TA (2012) Deciphering the geomorphic processes influencing alpine debris fans using airborne laser swath mapping data: an example from the Colorado Front Range. In: Schneuwly-Bollschweiler M, Stoffel M, Rudolf-Miklau M (eds) Dating torrential processes on fans and cones – methods and their application for hazard and risk assessment, Advances in global change research. Springer, Dordrecht/Heidelberg/London/New York

Staley DM, Wasklewicz TA, Blaszczynski JS (2006) Surficial patterns of debris flow deposition on alluvial fans in Death Valley, CA using airborne laser swath mapping data. Geomorphology 74(1–4):152–163

Wirth S, Girardclos S, Rellstab C, Anselmetti FS (2011) The sedimentary response to a pioneer geo-engineering project: tracking the Kander River deviation in the sediments of Lake Thun (Switzerland). Sedimentology. doi:10.1111/j.1365-3091.2011.01237.x

Summary and Outlook

Michelle Schneuwly-Bollschweiler, Markus Stoffel, and Florian Rudolf-Miklau

Torrential activity on alluvial fans and debris cones represents a major hazard in many regions where human settlements and infrastructure expand into potentially endangered areas. For the assessment of hazard and risk, detailed knowledge about past torrential activity and reliable predictions of possible future evolutions are of crucial importance. Data can be obtained from various data sources as has been shown in the previous chapters of this book (Schneuwly-Bollschweiler et al. 2012).

First of all, field data and monitoring of recent events provides good knowledge about process dynamics, spatial behaviour, and magnitude of contemporary events. Archival records may provide similar, even though normally less detailed, data on events of the recent past. Field evidence of past events (silent witnesses), such as mud traces or sediment deposits in a channel, may complement archival data. Digital images have the potential to show the evolution of a catchment at single points in the past and therefore provide information on past hydrogeomorphic activity. The analysis of trees and tree rings on forested fans and cones (dendrogeomorphology) helps gaining more information on frequencies and spatial dynamics of earlier/older events. Similarly, the analysis of lichens (lichenometry) on rock surfaces allows dating of older deposits and therefore the assessment of past activity. If material of hydrogeomorphic activity is deposited in a lake, the analysis of lake sediments (varves) offers valuable clues on the past occurrence of events as well.

M. Schneuwly-Bollschweiler (✉) • M. Stoffel
Laboratory of Dendrogeomorphology (dendrolab.ch), Institute of Geological Sciences, University of Berne, CH-3012 Berne, Switzerland

Chair for Climate and Climatic Impacts, Environmental Sciences, University of Geneva, CH-1227 Carouge-Geneva, Switzerland
e-mail: michelle.schneuwly@dendrolab.ch; markus.stoffel@dendrolab.ch

F. Rudolf-Miklau
Department IV/5 – Torrent and Avalanche Control, Federal Ministry for Agriculture, Forestry, Environment and Water Management, A-1030 Vienna, Austria
e-mail: florian.rudolf-miklau@lebensministerium.at

M. Schneuwly-Bollschweiler et al. (eds.), *Dating Torrential Processes on Fans and Cones*, Advances in Global Change Research 47, DOI 10.1007/978-94-007-4336-6_30,

Radiocarbon dating will probably not yield details on the recent past; however, the method allows the dating of organic material found in the fan or cone and can therefore help to improve the understanding of older activity. Cosmogenic nuclides and luminescence dating also cover huge time spans and may shed light onto long-term changes in torrential activity.

In addition to the dating methods illustrated in this book, there are several other dating methods that can, under certain circumstances, be applied on fans and cones, such as Thorium-Uranium disequilibrium dating (e.g. Schwarcz 1989; Sharp et al. 2003), varnish microlamination dating (e.g. Liu and Broecker 2007, 2008) or pedogenesis (e.g. Birkeland 1978, 1990; Walker et al. 1996). These methods will only provide dates of a surface or deposit and do neither include the dating of subsequent floods over that surface nor will they yield direct information on hydrogeomorphic activity. As the ageing of a landform *per se* will not necessarily improve hazard assessment, we do not present these techniques in greater detail.

It is obvious that all dating methods have their advantages and their limitations and that their application always depends on site specific conditions and on the type of information sought. To obtain the best possible data with highest accuracy for a better estimate of hazards and risks on a fan or cone, the combination of complimentary methods often represents an efficient and tailored solution. In a first step, data from recent events, available in archives, silent witnesses or digital images, should be analyzed as they normally provide the best current-day information and are relatively easy to obtain. Data on very recent events is often readily available from reports or direct observation. These methods have also the advantage of being comparably cheap as neither personnel nor material costs are high.

This kind of data, however, often represents a "snapshot" of a single event and therefore is of limited temporal reach. If hazard assessment is based on a single or a very limited number of events, the risk of misjudgement is high. Similarly, it appears difficult to determine a design event or protection goal based on only few events known from historical sources. In order to gain a realistic order of recurrence intervals, changes in runoff or the spatial spread for events that have occurred during the past decades to centuries, other methods need to be applied. Here, dendrogeomorphology, vegetation analyses, lichenometry or lake sediments (varves) represent appropriate techniques to gain more in-depth data on past activity. The choice of the method is strongly dependent on the site specific setting.

Cosmogenic nuclides, radiocarbon dating or luminescence dating are, in contrast, able to answer questions on long-term changes of activity on the cone or fan or on the alternation of deposition and erosion phases in time and space. Similarly, long-term climate change and its effect on activity are better assessed if the method provides data over centuries to millions of years. These types of information are probably and in most cases of rather limited importance for practitioners in natural hazard management. In addition, the application of these approaches involves high personnel and material costs as compared to the cost-effective methods referred to before.

It is also important to notice that current frequencies and magnitudes will not necessarily remain the same in the future. We should be aware that there is a long chain of dependencies and uncertainties from a changing climate to the impact on natural hazards (Koboltschnig 2012, *this volume*). Consequently the extrapolation of possible future trends from past data using extreme value statistics should take account of uncertainties and should therefore be supported by as much as possible additional information (such as the dating methods proposed in this book).

Hydrogeomorphic events are normally triggered by intense or prolonged precipitation events. Changes in the precipitation regime (e.g. more extreme precipitation) are likely to affect the release of hydrogeomorphic processes. In addition, the occurrence and also size of events depends on catchment properties and changes in the catchment and release zone of hydrogeomorphic events will possibly affect the release of events. The projected temperature increase can affect permafrost bodies and therefore more loose material can be made available for the entrainment of debris flows and other hydrogeomorphic events. On the other hand, temperature increase may also help vegetation to grow on previously bare surfaces and therefore stabilize loose material. The answer of catchments to climate change is therefore complex and not univocal.

Not only climate change but also alteration of land use as well as socio-economic trends (e.g. urbanization, rural exodus, demographic change, sealing of soils) may affect catchment areas and therefore alter the probability of hydrogeomorphic events to be released. Areal development and land use planning on the cone or fan surface have, in addition, an enormous influence on the hazard potential of a torrent. Increased land demands and an extension of settlements closer to torrents increases the hazards and risks related to torrential activity. Foresighted and proactive land-use planning therefore plays an important role in future risk reduction.

Apart from the changes in the catchment area and on the fan or cone surface, climatic changes play an important role in the future evolution of hydrogeomorphic events. As described by Koboltschnig (2012, *this volume*), many uncertainties remain in the prediction of future climate parameters. The influence of a future climate on the occurrence, magnitude and behaviour of future hydrogeomorphic events can only be predicted once we understand how past events have been triggered at the local scale. Therefore, the importance of a detailed, long and dense database for hazard assessment on fans and cones cannot be stressed enough.

Past data alone does not allow the definition of hazards and risks. Modelling of hydrogeomorphic events under current conditions of the channel and the fan or cone or for future scenarios may provide additional data on breakout locations, the spread of events or runout distances (e.g. Rickenmann and Scheidl 2012, *this volume*). Models should, however, be calibrated on real event data at the site in question and the reconstruction of past events can help this calibration (Ballesteros et al. 2011a, b).

To improve knowledge about torrential activity for the next generations, it is important to document current-day events systematically and in detail. As described by Marchi and Tecca (2012, *this volume*), many torrents are currently being monitored so as to improve our understanding of processes in general and at specific sites in particular. In addition, a new generation of historical documentation is in progress (see e.g. Kienholz et al. 2012, *this volume*) thanks to the intensification of post-event surveys, and to technologic-scientific progress in data acquisition, processing and archiving.

For the purposes of natural hazard engineering, condensed datasets on torrential events (frequency, magnitude, spread) are essential to meet the quality/safety requirements of hazard maps (Egli 2012, *this volume*) and protection works (Rudolf-Miklau and Suda 2012, *this volume*). The reliability of predictions and warning systems is the prerequisite of gain trust of people in these non-structural protection concepts (Borga 2012, *this volume*). Accuracy in safety concepts, based on reliable data, provides legal and economic confidence.

References

Ballesteros JA, Bodoque JM, Díez A, Sánchez M, Stoffel M (2011a) Calibration of floodplain roughness and estimation of palaeoflood discharge based on tree-ring evidence and hydraulic modeling. J Hydrol 403:103–115

Ballesteros JA, Eguibar M, Bodoque JM, Díez A, Stoffel M, Gutíerrez I (2011b) Estimating flash flood discharge in an ungauged mountain catchment with 2D hydraulic models and dendrogeomorphic paleostage indicators. Hydrol Processes 25:970–979

Birkeland PW (1978) Soil development as an indication of relative age of quaternary deposits, Baffin Island, Nwt, Canada. Arctic Alpine Res 10(4):733–747

Birkeland PW (1990) Soil-geomorphic research – a selective overview. Geomorphology 3(3/4):207–224

Borga M (2012) Forecasting, early warning and event management: non-structural protection measures for flash floods and debris flows. In: Schneuwly-Bollschweiler M, Stoffel M, Rudolf-Miklau M (eds) Dating torrential processes on fans and cones – methods and their application for hazard and risk assessment, Advances in global change research. Springer, Dordrecht/Heidelberg/London/New York

Egli (2012) Hazard mapping and land-use planning (management) – a Swiss perspective. In: Schneuwly-Bollschweiler M, Stoffel M, Rudolf-Miklau M (eds) Dating torrential processes on fans and cones – methods and their application for hazard and risk assessment, Advances in global change research. Springer, Dordrecht/Heidelberg/London/New York

Kienholz H, Huebl J, Loipersberger A (2012) Documentation of events. In: Schneuwly-Bollschweiler M, Stoffel M, Rudolf-Miklau M (eds) Dating torrential processes on fans and cones – methods and their application for hazard and risk assessment, Advances in global change research. Springer, Dordrecht/Heidelberg/London/New York

Koboltschnig GR (2012) Climate research and adaptation strategies – examples from the European Alps. In: Schneuwly-Bollschweiler M, Stoffel M, Rudolf-Miklau M (eds) Dating torrential processes on fans and cones – methods and their application for hazard and risk assessment, Advances in global change research. Springer, Dordrecht/Heidelberg/London/New York

Liu T, Broecker WS (2007) Holocene rock varnish microstratigraphy and its chronometric application in the drylands of western USA. Geomorphology 84(1–2):1–21

Liu T, Broecker WS (2008) Rock varnish microlamination dating of late Quaternary geomorphic features in the drylands of western USA. Geomorphology 93(3–4):501–523

Marchi L, Tecca PR (2012) Debris-flow monitoring in Italy. In: Schneuwly-Bollschweiler M, Stoffel M, Rudolf-Miklau M (eds) Dating torrential processes on fans and cones – methods and their application for hazard and risk assessment, Advances in global change research. Springer, Dordrecht/Heidelberg/London/New York

Rickenmann D, Scheidl C (2012) Debris-flow runout and deposition on the fan. In: Schneuwly-Bollschweiler M, Stoffel M, Rudolf-Miklau M (eds) Dating torrential processes on fans and cones – methods and their application for hazard and risk assessment, Advances in global change research. Springer, Dordrecht/Heidelberg/London/New York

Rudolf-Miklau F, Suda J (2012) Design criteria for torrential barriers. In: Schneuwly-Bollschweiler M, Stoffel M, Rudolf-Miklau M (eds) Dating torrential processes on fans and cones – methods and their application for hazard and risk assessment, Advances in global change research. Springer, Dordrecht/Heidelberg/London/New York

Schneuwly-Bollschweiler M, Stoffel M, Rudolf-Miklau F (2012) Dating torrential processes on fans and cones – methods and their application for hazard and risk assessment. Springer, Dordrecht/Heidelberg/London/New York

Schwarcz HP (1989) Uranium series dating of Quaternary deposits. Quat Int 1:7–17

Sharp WD, Ludwig KR, Chadwick OA, Amundson R, Glaser LL (2003) Dating fluvial terraces by Th-230/U on pedogenic carbonate, Wind River Basin, Wyoming. Quat Res 59(2):139–150

Walker MD, Everett KR, Walker DA, Birkeland PW (1996) Soil development as an indicator of relative Pingo Age, Northern Alaska, USA. Arctic Alpine Res 28(3):352–362

Printed by Publishers' Graphics LLC
MO20120918-2013-411